P9-DUZ-934

ALTERNATE REALITIES

ALTERNATE REALITIES

MATHEMATICAL MODELS OF NATURE AND MAN

John L. Casti
Institute for Econometrics, Operations Research, and System Theory
Technical University of Vienna
Vienna, Austria

A Wiley-Interscience Publication
JOHN WILEY & SONS
New York • Chichester • Brisbane • Toronto • Singapore

To the scientific ideals of the
INTERNATIONAL INSTITUTE FOR APPLIED SYSTEMS ANALYSIS
May aspirations and achievements someday coincide

Library of Congress Cataloging in Publication Data:

Casti, J.L.
Alternate realities: mathematical models of nature and man/
John L. Casti.

p. cm.
"A Wiley-Interscience publication."
Bibliography: p.
Includes index.
ISBN 0-471-61842-X
1. Mathematical modeling. 2. System analysis I. Title.
QA401.C358 1988
001.4′34—dc19 88-22337
CIP

Printed in the United States of America

10 9 8 7 6 5 4 3 2 1

PREFACE

In experimental science, the principal focus is upon developing new instruments and techniques for measuring various aspects of natural phenomena, with all other considerations subservient to this primary task. So it is also in theoretical science, where the overarching problem is how to capture the processes of Nature and man in formal mathematical representations, with all matters of technique and choice of methods being of secondary consideration. This book addresses the interface between these complementary aspects of scientific practice.

If the natural role of the experimenter is to generate new observables by which we know the processes of Nature, and the natural role of the mathematician is to generate new formal structures by which we can represent these processes, then the system scientist finds his niche by serving as a broker between the two. In this volume my goal is to show (more by example than by precept) how to bridge the gap between experimental science, on the one hand, and the world of mathematics on the other. The objective is to provide both the student and the scientific practitioner with an attitude, or philosophy, as well as a set of tools, with which to probe the workings of both Nature and man.

I hasten to point out that this is *not* a book on systems analysis, only a volume on the theory and practice of mathematical modeling. As a result, the focus is on how to go from observing the behavior of natural and human systems to formal mathematical models of such behavior, and it will not be our concern here to enter into the pedestrian (although often very useful) issues surrounding collection of data, design of information systems, consultation with the client, selection of equipment and the myriad other nonscientific activities involved in what is sometimes termed "systems analysis." Thus the principal aim of this book is to bring modern mathematics into the service of applied system modeling, with primary emphasis upon the word "modern."

Should you have the misfortune to pick up a typical current textbook purporting to address the arcane arts of mathematical modeling, the chances are overwhelmingly high that the author will transport you back into the 1950s with an account of how to model an oscillating pendulum, freeway traffic, or dogfood mixing using the static, equilibrium-centered, linear techniques of mathematical programming, regression analysis or, perhaps, elementary functional analysis. My feeling is that the time is long overdue to bring the mathematics of the 1980s into contact with the students of the 1980s and offer courses on modeling that stress dynamics rather than statics, nonlinearity rather than linearity, and possibility rather than optimality.

The concepts noted above constitute the "fingerprints" of what has been termed elsewhere the *system-determined* sciences. In the language of Aristotelian causes, the system-determined sciences emphasize the manipulation of *formal* cause in contrast to classical physics and engineering, which concern themselves primarily with *material* and *efficient* cause. In more familiar terms, our focus is upon issues of information as opposed to a concern with matter or energy *per se.* As a result, our motivation and examples come mainly from the social, behavioral and life sciences rather than the physical sciences, an emphasis that is, in my opinion, long overdue in textbook accounts of mathematical modeling.

In 1985 and 1986 I was invited to put these radical notions to the test and gave just such a course of lectures, both at the University of Vienna and at the Technical University of Vienna. The focus of these lectures was upon system-theoretic concepts such as complexity, self-organization, adaptation, bifurcation, resilience, surprise and uncertainty, coupled with the kinds of mathematical structures needed to pin down these notions within the confines of a formal mathematical system. This book is an outgrowth of those lectures.

While it's not normal practice in a book's Preface to already start anticipating reviewer opinion and commentary, if referees' reports on the original manuscript are taken as reliable indicators, this is no normal book. Moreover, taking past experience as a guide, when one starts tip-toeing through the no-man's land between mathematics and the real world, which any book on mathematical modeling necessarily must, it's a narrow path indeed that must be tread to avoid landmines and sniper fire from both sides of the divide. So to set this volume in perspective, let me first say what it's not. It is not a treatise about the *mathematics* of modeling, nor is it an account of *case studies* in the real-world application of mathematical concepts. Neither is the book intended as a compendium or an encyclopedic survey of the entire field of theoretical system thinking. In this sense I have been remarkably self-indulgent, choosing to discuss those areas in which I have an interest and ignoring those in which I don't. With all these negatives and disclaimers, what then **is** this book? Perhaps the best way to describe what it is (in my view, at least) is to term it a book on the **theory of models**. At this point you might well ask: "What in heaven's name could this possibly mean?"

The genesis of the idea from which this book took form was my desire to put together a course of lectures on modeling for mathematically trained students in disciplines like economics, biology, computer science and so forth—but not mathematics. The underlying goal was to expose these students to a spectrum of *concepts* and *ideas* from across the landscape of modern applied mathematics, and to indicate some of the ways in which these notions had been used to represent toy versions of real-world situations. Thus the emphasis was on the ideas underlying cellular automata, chaos, catastrophe theory and the like, as well as on the *kinds* of situations where they had been used, but not on the fine-grained technical details. But since the students all had a good grounding in basic mathematics, I saw no reason to present these ideas in the language of popular science using no mathematics at all. On the other hand, the technical finery and rigor demanded by dues-paying members of the mathematical community was also inappropriate, since the students were really interested in what kinds of questions these concepts might illuminate, not in delicate webs of definitions or intracacies of proofs. The book before you represents my compromise between the Scylla of mathematical precision and the Charybdis of messy real-world detail.

No doubt there will be readers who will take umbrage at what they see as my cavalier disregard for the delicacies of their art, and perhaps rightly so. They have my apologies along with the recognition that even in a 500-page volume such as this, doing justice to the way things "really are" in so many areas is just plain impossible. My hope is that even St. Simon the Stylite will still find something of interest here, a point or concept that piques his curiosity, at least partially compensating for the book's obvious faults elsewhere. On balance, I think of the book as more of an Impressionist painting than a photographic account of the philosophy and ideas of modeling, and I'll consider my time in writing it well spent if the reader ends up saying to himself, "yes, there really is a legitimate intellectual activity here called 'modeling,' and it's not mathematics and it's not applications." Elicitation of this very sentence is the real goal of this book.

Since the Table of Contents speaks for itself, I will content myself at this point only with the admonition to the reader that each chapter's Discussion Questions and Problems and Exercises are an integral part of the book. The

Discussion Questions amplify many of the points only touched upon in the text and introduce topics for which there was no room for a more complete account elsewhere. The Problems and Exercises provide challenges ranging from simple drill to state-of-the-art research and are deliberately worded so that the reader will know what the answer should be—even if he can't see how to get it! In any case, my message in this volume will be only half-transmitted if just the chapter text is read. Although the mathematical *ideas* used in the book are quite contemporary, the actual *technique* involved in using these ideas is, for the most part, rather modest. As a result, most of the sections of each chapter should be accessible to that proverbial seeker of knowledge, "the well-trained undergraduate," with a **working** knowledge of ordinary differential equations, linear algebra, matrix theory and, perhaps, abstract algebra. Note, however, the emphasis on the term "working." To really understand the ideas presented here, it will not be sufficient just to have sat through a course in the above topics; the student will have to actually recall the material of the course and know how to apply it. With this caveat, the book should serve as a text-reference for a course in modern applied mathematical modeling for upper-division undergraduates, not to mention their presumably more mathematically sophisticated colleagues in graduate school.

It's impossible to put together a book such as this without the generous help, encouragement, and cooperation of numerous friends and colleagues. At the head of this list come Profs. Karl Sigmund of the Mathematics Institute of the U. of Vienna and Manfred Deistler, my colleague at the Institute for Econometrics, Operations Research, and System Theory of the Technical U. of Vienna. Each offered me not only a forum in which to present these slices of modern system theory, but also a group of talented students who in some cases were willing to serve as enthusiastic "guinea pigs" for the particular brand of system-theoretic medicine that I was serving up. Also occupying a prominent position on this roll call of honor is Steven Dunbar, who courageously took on the task of keeping me technically honest with a thorough review of much of the mathematical finery in the book. Naturally, and as always, I accept final responsibility for the book's inevitable *faux pas.* But Steve's hard work has insured me a far lower embarrassment level than would otherwise have been the case.

Others who contributed by providing examples, software consultations, stimulating discussions or just plain friendship include Hugh Miser, Paul Makin, Hans Troger, Robert Rosen, René Thom, Nebojsa Nakicenovic, Lucien Duckstein, Mel Shakun, George Klir, George Leitmann, Alain Bensoussan, Myron Allen, Don Saari, and Clifford Marshall. In addition, the enthusiasm and encouragement of Maria Taylor, my editor at Wiley, has been a constant source of support. Last, but far from least, kudos to my wife Vivien, who tolerated my spending more time with the vagaries of computers and system modeling than with her for much longer than any companion should. Finally, in a work of this kind it's impossible for me not to cite the role of the International Institute for Applied Systems Analysis (IIASA), where I spent far more years than was good for me in developing and promoting much of the philosophy and some of the technique appearing in this *"magnum opus."* In particular I must acknowledge the various managements of the Institute, whose understanding and appreciation of the need for theoretical work and conceptual foundations served as a never-ending source of motivation for the preparation of this volume. Without remarkable views such as theirs on the value of system thinking as an intellectual undertaking, books like this would never have to be written.

JLC
Vienna, Austria
October 1988

CREDITS

Grateful acknowledgment is made to the following sources for permission to reproduce copyrighted material. Every effort has been made to locate the copyright holders for material reproduced here from other sources. Omissions brought to our attention will be corrected in future editions.

American Physical Society for Figures 2.4–2.8 which originally appeared in Wolfram, S., "Statistical Mechanics of Cellular Automata," *Reviews of Modern Physics,* 55 (1983), 601–644, and for Figure 5.11 from Eckmann, J. P., "Roads to Turbulence in Dissipative Dynamical Systems," *Reviews of Modern Physics,* 53 (1981), 643–654.

Corgi Books for Figure 2.9 which appeared in Gribbin, J., *In Search of the Double Helix,* 1985.

Alfred Knopf, Inc. for Figures 2.10–2.12 from Eigen, M. and R. Winkler, *The Laws of the Game,* 1981.

William Morrow and Co. for Figure 2.13 which appeared originally in Poundstone, W., *The Recursive Universe,* 1985.

Elsevier Publishing Co. for Figure 2.14 which appeared originally in Young, D., "A Local Activator-Inhibitor Model of Vertebrate Skin Patterns," *Mathematical Biosciences,* 72 (1984), 51–58, and for Figures 5.9–5.10 taken from Swinney, H., "Observations of Order and Chaos in Nonlinear Systems," *Physica D,* 7D (1983), 3–15.

Heinemann Publishing, Ltd. for Figure 8.9 from Atkin, R., *Mathematical Structure in Human Affairs,* 1974.

MIT Press for Figures 5.1–5.2 found originally in Arnold, V. I., *Ordinary Differential Equations,* 1973.

Springer Verlag, Inc. for Figure 5.3 taken from Arnold, V. I., *Mathematical Methods in Classical Mechanics,* 1978, and for Figure 5.5 found in Lichtenberg, A., and M. Lieberman, *Regular and Stochastic Motion,* 1983.

Mathematical Association of America for Figure 5.6 from Oster, G., "The Dynamics of Nonlinear Models with Age Structure," in *Studies in Mathematical Biology, Part II,* S. Levin, ed., 1978.

Academic Press, Inc. for Figure 6.1 from Vincent, T., and J. Brown, "Stability in an Evolutionary Game," *Theoretical Population Biology,* 26 (1984), 408–427.

Professors E. C. Zeeman for the figure in Discussion Question 15 of Chapter Four and J. H. Johnson for Figure 8.8.

CONTENTS

Chapter One THE WAYS OF MODELMAKING: NATURAL SYSTEMS AND FORMAL MATHEMATICAL REPRESENTATIONS

Chapter Two PATTERN AND THE EMERGENCE OF LIVING FORMS: CELLULAR AUTOMATA AND DISCRETE DYNAMICS

Chapter Nine HOW DO WE KNOW?: MYTHS, MODELS AND PARADIGMS IN THE CREATION OF BELIEFS

CHAPTER ONE

The Ways of Modelmaking: Natural Systems and Formal Mathematical Representations

1. *A Theory of Models*

What do you think of when you hear the word "model"? An elegant mannequin from the pages of *Vogue* perhaps? Or maybe a miniature version of that super-exotic Ferrari or Lamborghini that you've been drooling over? To a system scientist or applied mathematician, quite a different picture comes to mind. For such practitioners of the "black arts," a *model* means an encapsulation of some slice of the real world within the confines of the relationships constituting a formal mathematical system. Thus, a model is a mathematical representation of the modeler's reality, a way of capturing some aspects of a given reality within the framework of a mathematical apparatus that provides us with a means for exploring the properties of that reality mirrored in the model.

This book is about the ways and means of constructing "good" models of reality, the properties of such models, the means for encoding specific realities into definite formal systems and the procedures for interpreting the properties of the formal system in terms of the given real world situation. In short, we're interested in the ways of modelmaking. Before embarking upon a more detailed account of what we mean by the terms *model, encoding, formal system,* and so forth, it's necessary to examine some basic epistemological and operational issues lying at the heart of what we shall term the *theory of models.*

According to the great nineteenth-century British physicist Maxwell, "the success of any physical investigation depends upon the judicious selection of what is to be observed as of primary importance." This view suggests the notion that what constitutes one's reality depends upon one's capacity for observation. Here we adopt the position that since natural phenomena impinge upon our consciousness only through instruments of observation, then, to paraphrase Maxwell, "the success of any modeling venture depends upon a judicious selection of observables and means for encapsulating these observables within the framework of convenient formal mathematical systems."

As noted by Rosen, in dealing with the idea of a natural system, we must necessarily touch on some basic philosophical questions of both an ontological and epistemological character. This is unavoidable in any case and must be addressed at the outset of a work such as this, because our tacit assumptions in these areas determine the character of our science. It's true that many scientists find an explicit consideration of such matters irritating, just as many working mathematicians dislike discussions of the foundations of mathematics. Nevertheless, it's well to recall the remark of David Hawkins: "Philosophy may be ignored but not escaped; and those who most ignore escape least."

Our viewpoint is that *the study of natural systems begins and ends with the specification of observables belonging to such a system, and a characterization of the manner in which they are linked.* Purely theoretical issues may be pursued in the process of investigating a system, but ultimately contact with reality occurs through the observables. During the course of this book, it will be argued that the concept of a model of a natural system N is a generalization of the concept of a subsystem of N, and that the essential feature of the modeling relation is the exploration of the idea that there is a set of circumstances under which the model describes the original system to a prescribed degree of accuracy. In other words, a particular facet of system behavior remains *invariant* under the replacement of the original system by a proper subsystem.

At this point it is well to consider why one constructs models of natural phenomena in the first place. Basically, the point of making models is to be able to bring a measure of order to our experiences and observations, as well as to make specific predictions about certain aspects of our experienced world. The central question surrounding the issue of model credibility is to ask to what extent "good" predictions can be made if the best the model can do is to capture a subsystem of N. The answer is wrapped up in the way in which the natural system is characterized by observables, the procedure by which observables are selected to form the subsystem, and the manner in which the subsystem is encoded into a formal mathematical system F which *represents,* or "models," the process of interest. These notions will be made more explicit later, but before doing so, let us consider a familiar example that illustrates many of these points.

We consider an enclosed homogeneous gas for which the observables are taken to be the volume V occupied by the gas, the pressure P and the temperature T. By this choice, the abstract states of the system, i.e., the actual *physical states* comprising the volume, pressure and temperature, are encoded into a three-dimensional euclidean space in the familar manner. At equilibrium, these three observables are not independent but are linked by the relation (equation of state) $PV = T$, the *ideal gas law.* Of course, we know that such a selection of observables represents an abstraction in the sense that many other observables have been omitted that, in principle, influence the gas (external radiation, properties of the container, etc.). Experience has shown, however, that the subsystem consisting of P, V and

T, together with its encoding into the region of R^3 defined by the ideal gas law, enables us to make very accurate predictions about the *macroscopic* behavior of the system. Should we desire to make predictions about the gas at the molecular level, it would be necessary to choose an alternate set of observables. In the *microscopic* context, the positions and momenta of the $\bar{N} \cong 10^{24}$ molecules composing a mole of the gas would be a natural choice, and these observables would be encoded into the euclidean space $R^{6\bar{N}}$. Linkages between the observables in this case are specified by various conservation laws operating at the microlevel.

Our primary objective in this chapter is to provide a framework within which we can speak of fundamental issues underlying any theory of modeling. Among matters of greatest concern, we find:

- What is a model?
- What features characterize "good" models?
- How can we represent a natural process N in a formal system F?
- What is the relationship between N and F?
- When does the similarity of two natural systems N_1 and N_2 imply that their models F_1 and F_2 are similar?
- How can we compare two models of the same natural process N?
- Under what circumstances can we consider a linkage between observables as constituting a "law" of Nature?
- What procedures can we invoke to identify key observables and thereby simplify a model?
- How does a given system relate to its subsystems?
- When can two systems that behave similarly be considered as models of each other?

Such a list (and its almost infinite extension) represents issues in the philosophy of science and, in particular, the theory of models. No uniform and complete answers to these issues can ever be expected; the best we can hope for is to provide a basis for considering these matters under circumstances appropriate to a given setting.

In what follows, we sketch a formalism suitable for studying the foregoing questions and indicate by familar examples some of the advantages to be gained by looking at system modeling questions in such generality. In essence, the argument is that the detailed study of a given natural system N cannot be suitably interpreted and understood within the level of the phenomenon N itself. A more general metalevel and a metalanguage provided by a theory of models is required.

2. *States, Observables and Natural Systems*

Consider a particular subset S of the observable world, and assume that S can exist in a set of physically distinct states $\Omega = \{\omega_1, \omega_2, \ldots\}$. Note that an observer probing the behavior of S may or may not be able to determine whether S is in state ω_i or ω_j, $j \neq i$. It all depends upon the resolution of the measuring apparatus (observables) at his disposal. Also note that the set Ω may be finite or infinite (uncountable, even). We call Ω the set of *abstract states* of S. Here it's important to emphasize that what counts as a "physically distinct state" is not an intrinsic property of the system, but depends crucially upon the observer and the ways he has of probing the system and distinguishing one state from another. Since this notion is crucial for all system modeling, let's look at some examples to firmly fix the idea.

Examples

1) If S is the on-off switch for the table lamp on my desk, then the set of abstract states of S might be

$$\Omega = \{\omega_1 = \text{OFF}, \omega_2 = \text{ON}\}.$$

Here Ω is a finite set.

2) Let S be the equilateral triangle $\triangle$ with vertices denoted $[a, b, c]$ moving counterclockwise from the top, and assume that S can be distinguished in three forms corresponding to rotations by $0, 2\pi/3$ and $4\pi/3$ radians. Then one possibility for the set of abstract states Ω of the triangle is

$$\Omega_1 = \{\omega_1 = [a, b, c],\ \omega_2 = [c, a, b],\ \omega_3 = [b, c, a]\},$$

corresponding to an ordered labeling of the vertices of S. But an equally valid set of abstract states is

$$\Omega_2 = \left\{\omega_1 = 0,\ \omega_2 = \frac{2\pi}{3},\ \omega_3 = \frac{4\pi}{3}\right\},$$

or even

$$\Omega_3 = \{\omega_1 = 0,\ \omega_2 = 1,\ \omega_3 = 2\},$$

where each state $\omega_i \in \Omega$ is just a *label* for a *physical* state of S.

This example illustrates several vital points about abstract state-spaces:

- A given system S usually has many sets of abstract states Ω.
- The set of abstract states for S need not be and, in general, is not a set of numbers.
- All sets of abstract states are equivalent insofar as they characterize the distinct physical states of S; for Nature there is no preferred space of states, although for modelers some sets Ω may be more *convenient* to use than others.

Now let's turn to the notion of an observable. Assume that we are given the system S, together with a set Ω comprising the abstract states of S. Then a rule f associating a real number with each $\omega \in \Omega$ is called an *observable* of S. More formally, an observable is a map $f: \Omega \to R$.

Examples

1) Let S be a ball constrained to roll along an inclined plane of unit length, and let $\Omega = \{\omega_1, \omega_2, \omega_3, \omega_4\}$, where ω_1=ball at the top of the plane, ω_2=ball one-third of the way down the plane, ω_3= ball two-thirds of the way down the plane, ω_4= ball at the bottom of the plane. Define the observable $f: \Omega \to R$ by the rule

$$f(\omega) = \text{distance of the ball from the top of the plane.}$$

Then

$$f(\omega_1) = 0, \quad f(\omega_2) = \tfrac{1}{3}, \quad f(\omega_3) = \tfrac{2}{3}, \quad f(\omega_4) = 1.$$

2) Using the same system S and state-space Ω as above, let the observable $g: \Omega \to R$ be defined by the rule

$$g(\omega) = \text{distance of the ball from the middle of the plane.}$$

Then

$$g(\omega_1) = \tfrac{1}{2}, \quad g(\omega_2) = \tfrac{1}{6}, \quad g(\omega_3) = \tfrac{1}{6}, \quad g(\omega_4) = \tfrac{1}{2}.$$

In this case it can be seen that the states ω_1 and ω_4, as well as ω_2 and ω_3, are indistinguishable using the observable g, while the observable f distinguishes (separates) all states. Consequently, we can "see" more of the system S using the observable f than by using g. This is a crucial point that we will return to often. Note also the *linkage* relationship between f and g given explicitly by the rule

$$f(\omega) = \begin{cases} \frac{1}{2} - g(\omega), & \omega = \omega_1, \omega_2, \\ \frac{1}{2} + g(\omega), & \omega = \omega_3, \omega_4. \end{cases} \tag{$*$}$$

Linkage generalizes our usual notions of one observable being a *function* of another. Here we see that although the observable g alone contains less information about the system than f, this lack of information can be compensated for if we know the linkage relation above. Intuitively speaking,

$$g + (*) = f.$$

Generally speaking, in order to "see" the complete system S, we would need an infinite number of observables $f_\alpha: \Omega \to R$, where α ranges over some possibly uncountable index set. Thus, the complete system S is described by Ω and the entire set of observables $\mathcal{F} = \{f_\alpha\}$. But for practical modeling purposes, it's inconvenient to work with such a large set of observables, so we boldly just throw most of them away and focus our attention on a proper subset A of $\mathcal{F}$. We call A an *abstraction* of S, since the view we have of S using the observables A is necessarily a partial view formed by abstracting, i.e., throwing away all the information contained in the observables $\mathcal{F} - A$. It's an amusing aside to wonder why self-styled practically-oriented people reserve their greatest scorn for "useless abstractions," when the very essence of an abstraction is to reduce the description of a system to a simpler, and presumably more tractable form. Thus, in many ways there is nothing more useful and practical than a good abstraction. This calls to mind Hilbert's profound observation that "there is nothing more practical than a good theory." Much of our subsequent development is focused upon tricks, techniques and subterfuges aimed at finding good abstractions.

We are now in a position to put forth our notion of a *natural system* N. For us, N consists of an abstract state-space Ω, together with a finite set of observables $f_i \colon \Omega \to R$, $i = 1, 2, \ldots, n$. Symbolically,

$$N = \{\Omega, f_1, f_2, \ldots, f_n\}.$$

We employ the term "natural" to distinguish N from the idea of a "formal" mathematical system to be discussed below. It should not be interpreted to mean restriction of N to the class of systems studied in the "natural" sciences (chemistry, physics, astronomy, etc.); it includes social, behavioral and living systems as well. In all that follows, we shall use the term *natural system* in this extended sense.

3. *Equations of State*

The observables $\{f_1, f_2, \ldots, f_n\}$ provide the percepts by which we see a natural system N, the raw data, so to speak. But there is more to the "system-ness" of N than just the separate observables by which we see it, just as a book is more than the individual words that comprise it. The essential system nature of N is contained in the relationships linking the observables $\{f_1, f_2, \ldots, f_n\}$. We term such a set of relationships the *equation of state,* or the *description,* for N. Formally, the equations of state can be written as

$$\Phi_i(f_1, f_2, \ldots, f_n) = 0, \qquad i = 1, 2, \ldots, m,$$

where the $\Phi_i(\cdot)$ are mathematical relationships expressing the dependency relations among the observables. We can write this more compactly as

$$\Phi(f) = 0.$$

Example: The Ideal Gas Law

Let N consist of one mole of an ideal gas contained in a closed vessel. Take Ω to be the positions and velocities of the 10^{24} or so molecules making up the gas, and define the three observables

$$P(\omega) = \text{pressure of the gas when in state } \omega,$$
$$V(\omega) = \text{volume of the gas when in state } \omega,$$
$$T(\omega) = \text{temperature of the gas when in state } \omega.$$

Then the ideal gas law asserts the single equation of state

$$\Phi(P, V, T) = 0,$$

where

$$\Phi(x, y, z) = xy - z.$$

The foregoing example, simple as it is, serves to illustrate another deep epistemological issue in the theory of models, namely, the distinction between *causality* and *determinism.* An equation of state, $\Phi(f) = 0$, necessarily establishes a *deterministic* relationship between the observables, but it contains no information whatsoever about any possible *causal* implications among the elements of the set $\{f_i\}$. Thus, the ideal gas law asserts that once we know any two of the three observables P, V and T, the remaining observable is *determined* by the other two, but not *caused* by them.

The difficulty in assigning a causal ordering to the set of observables is one of the principal difficulties in economic modeling, for instance, where we often have available theoretical and/or empirical equations of state, but little knowledge of how to separate observables into so-called exogenous and endogenous subsets. This same obstacle stands in the way of making effective use of system-theoretic techniques in many other areas of the social and biological sciences, and deserves far more formal attention than it has received thus far.

Now imagine that there are r observables that remain constant for every state $\omega \in \Omega$. For simplicity of exposition, assume these are the first r observables. Such quantities are usually termed *parameters* and arise as a matter of course in virtually all natural systems. Typical examples are the stiffness constant of a spring, the prime interest rate in a national economy or the gravitational constant in orbital mechanics. If we let $f_i(\omega) = \alpha_i$, $\omega \in \Omega$, α_i a real number, $i = 1, 2, \ldots, r$, we can write the equation of state as

$$\Phi_{\alpha_1, \alpha_2, \ldots, \alpha_r}(f_{r+1}, f_{r+2}, \ldots, f_n) = 0, \qquad (\dagger)$$

indicating explicitly the dependence of the description upon the parameter values $\alpha_1, \ldots, \alpha_r$. In other words, we have an *r-parameter family* of descriptions, and for each set of values of the set $\{\alpha_i\}$, $(\dagger)$ describes a *different* system.

We now introduce the additional assumption that the last m observables $f_{n-m+1}, \ldots, f_n$ are functions of the remaining observables $f_{r+1}, \ldots, f_{n-m}$, i.e., we can find relations $y_i(\cdot)$, $i = 1, 2, \ldots, m$, such that

$$\begin{aligned} f_{n-m+1}(\omega) &= y_1(f_{r+1}(\omega), \ldots, f_{n-m}(\omega)), \\ &\vdots \\ f_n(\omega) &= y_m(f_{r+1}(\omega), \ldots, f_{n-m}(\omega)). \end{aligned}$$

If we introduce the notation

$$\begin{aligned} \alpha &\doteq (\alpha_1, \alpha_2, \ldots, \alpha_r), \\ u &\doteq (f_{r+1}, f_{r+2}, \ldots, f_{n-m}), \\ y &\doteq (f_{n-m+1}, f_{n-m+2}, \ldots, f_n), \end{aligned}$$

the equations of state (†) become

$$\Phi_\alpha(u) = y. \tag{‡}$$

In (‡) it's natural to think of the observables u as representing the *inputs* to the system, with the observables y being the resulting *outputs*. The vector α is, as before, the set of parameters. Note also the assumption that we can solve for the observables $f_{n-m+1}, \ldots, f_n$ in terms of the observables $f_{r+1}, \ldots, f_{n-m}$ introduces a notion of *causality* into our previous acausal relationship (†). We can now think of the inputs as somehow "causing" the outputs. The problem, of course, is that there may be many ways of solving for some of the observables in terms of the others, so it's usually possible to have many distinct separations of the observables into inputs and outputs and, correspondingly, many different causal relationships. As already noted, in some parts of the natural sciences (e.g., classical physics), there are natural and useful conventions established for making this separation of observables; in other areas like the social and behavioral sciences, there is no clear-cut procedure or body of past evidence upon which to base such a classification of observables, and one ends up with many distinct, often contradictory, theories depending upon the choice that's made. For instance, is unemployment caused by inflation or is it the other way round, or perhaps neither. No one really seems to know, yet far-reaching economic and social policy is made on the basis of one assumption or the other. We shall return to this crucial issue in various guises in almost every chapter of this book. But for now, let's shift attention back to the parameterized equation of state (‡) and give an *interpretation* of it that will be of considerable intuitive use in later discussions.

If we adopt a biological view of (‡), it's natural to think of the parameter vector $\alpha = (\alpha_1, \alpha_2, \ldots, \alpha_r)$ as representing the "blueprint" or "program" for the system N. Biologically speaking, α represents the *genetic* make-up, or the *genome*, of N. Similarly, the inputs u correspond to the *environment* in which N operates and with this view of u, we can only conclude that the output y corresponds to the "form" of N that emerges from the interaction of the genome and the environment. In more biological terms, y is the *phenotype* of N.

The foregoing development shows explicitly how we can start from the fundamental description of N given by the equations of state (†), and by making natural assumptions about the dependencies of observables upon the states of Ω and upon each other, arrive at a standard parameterized family of input/output descriptions of N that can be given several types of interpretations. We now explore the implications of this set-up for determination of when two descriptions are the "same," and for the associated questions of bifurcation, complexity and error.

4. *Equivalent Descriptions and the Fundamental Question of Modeling*

The family of descriptions (‡) leads immediately to what we term *The Fundamental Question of System Modeling:* "How can we tell if two descriptions of N contain the same information?" Or: "When are two descriptions of N equivalent?" In the pages that follow, we shall justify the lofty plane to which this question has been elevated by showing its central role in the analysis of *all* issues pertaining to matters of system complexity, bifurcation, error, self-organization, adaptation, and so on. For now, let's focus upon means for formulating the Fundamental Question in more tractable mathematical terms.

To motivate the intuitive concept of equivalent descriptions, consider the descriptions of an ellipse in the $x - y$ plane. From elementary analytic geometry, we know that the expression

$$\frac{x^2}{a^2} + \frac{y^2}{b^2} = 1, \qquad a, b \text{ real},$$

describes a family of ellipses having semi-axes of lengths a and b [Fig. 1.1(a)]. In Fig. 1.1(b), we show the same ellipse rotated through an angle θ. On the one hand, it's clear that the two figures display different closed curves E and $\hat{E}$ in the plane; on the other hand, E and $\hat{E}$ are related to each other through the simple act of rotating the coordinate axes through an angle θ, i.e., they can be made congruent to each other by rotating the axes by which we *describe* the curve. It is in this sense that the two descriptions

$$\frac{x^2}{a^2} + \frac{y^2}{b^2} = 1,$$

and

$$\frac{\hat{x}^2}{a^2} + \frac{\hat{y}^2}{b^2} = 1,$$

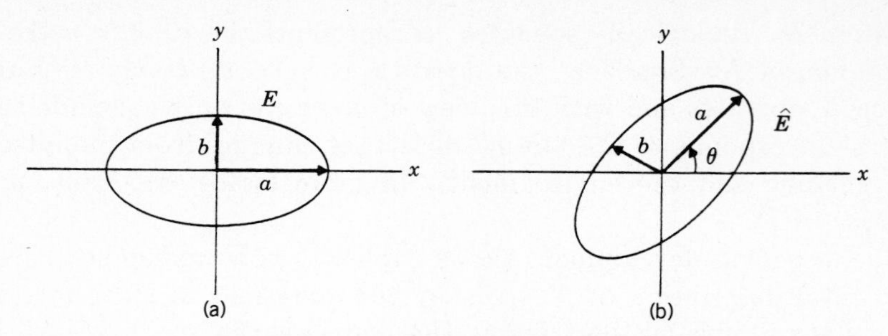

Figure 1.1 A Family of Ellipses

are *equivalent* with $\hat{x} = x\cos\theta - y\sin\theta$, $\hat{y} = x\sin\theta + y\cos\theta$. These two descriptions contain exactly the same information about the ellipse. The curves E and $\hat{E}$ are for all practical purposes identical; they differ only in the way we choose to describe them.

Note in the foregoing example that the numbers a and b are *intrinsic* properties of the curve; they remain the same in both descriptions. Such *invariants* represent properties of the system that are independent of the choice of coordinates used to describe the ellipse. It is the invariants that represent the true system-theoretic properties of the curve; all other coordinate-dependent properties are artifacts of the description, not properties of the system. We shall emphasize this point as we go along during the course of developing alternate descriptions later in the book.

In the preceding development, we have regarded the parameters a and b as being fixed. Assume now that we make the change $a \to \hat{a}$, $b \to \hat{b}$, i.e., we pass from a system (ellipse) with semi-axes a and b to one with semi-axes $\hat{a}$ and $\hat{b}$. It's clear geometrically that if we start with the ellipse having semi-axes $\hat{a}$ and $\hat{b}$, we can make a coordinate transformation $x \to \hat{x}$, $y \to \hat{y}$ so that in the $(\hat{x}, \hat{y})$-plane, the ellipse now has semi-axes equal to a and b. In fact, from the formula

$$\frac{x^2}{\hat{a}^2} + \frac{y^2}{\hat{b}^2} = 1,$$

it follows that the required coordinate change is

$$\hat{x} = \left(\frac{a}{\hat{a}}\right) x, \qquad \hat{y} = \left(\frac{b}{\hat{b}}\right) y.$$

What has been accomplished by the change to the $(\hat{x}, \hat{y})$ coordinates is to neutralize the effect of the change of parameters $(a, b) \to (\hat{a}, \hat{b})$. When such a change of variables can be made, we call the two ellipses *equivalent.* Note that the above arguments break down whenever any of the parameters $a, b, \hat{a}, \hat{b}$ are zero, i.e., when the family of ellipses degenerates into a different conic section. In these cases, such distinguished values of the parameters are called *bifurcation points.*

Generalizing the above set-up to our description of N given by (‡), we consider two descriptions $\Phi_\alpha, \Phi_{\hat{\alpha}}$ corresponding to the "genomes" α and $\hat{\alpha}$, i.e., we have the descriptions

$$\Phi_\alpha : U \to Y, \qquad \Phi_{\hat{\alpha}} : U \to Y,$$

where U and Y are the input and output spaces, respectively. We will consider these descriptions to be *equivalent* if it's possible to find coordinate changes (usually nonlinear) in U and/or Y such that the description Φ_α is transformed into $\Phi_{\hat{\alpha}}$. Diagrammatically, we seek bijections (one-to-one and onto) $g_{\alpha,\hat{\alpha}} : U \to U$ and $h_{\alpha,\hat{\alpha}} : Y \to Y$ such that the following diagram commutes

$$\begin{array}{ccc} U & \xrightarrow{\Phi_\alpha} & Y \\ {\scriptstyle g_{\alpha,\hat{\alpha}}}\downarrow & & \downarrow{\scriptstyle h_{\alpha,\hat{\alpha}}} \\ U & \xrightarrow[\Phi_{\hat{\alpha}}]{} & Y \end{array}$$

i.e., can be traversed in either path from U to Y starting in the upper left-hand corner. What we are saying here is that the change from the system described by Φ_α to the system described by $\Phi_{\hat{\alpha}}$ can be "undone" or "neutralized" by a corresponding change of coordinates in U and/or Y. Biologically, this would say that two organisms are equivalent if a genetic change could be reversed by means of a suitable change of environment and/or phenotype. Since the idea of a commutative diagram plays an important role at various points in our narrative, let's pause for a moment to examine the underlying notion in a bit more detail.

Suppose we are given sets A, B, C and D, together with well-defined maps f, g, α and β acting in the following manner:

$$\begin{aligned} f : A \to B, \quad g : C \to D, \\ \alpha : A \to C, \quad \beta : B \to D. \end{aligned}$$

Pictorially, it's convenient to represent this situation with the diagram

$$\begin{array}{ccc} A & \xrightarrow{f} & B \\ {\scriptstyle \alpha}\downarrow & & \downarrow{\scriptstyle \beta} \\ C & \xrightarrow[g]{} & D \end{array}$$

We say the above diagram is *commutative,* or *commutes,* if the various maps satisfy the following relation: $g \circ \alpha = \beta \circ f$. In other words, the diagram is commutative if we can start at any corner (i.e., with any set) and traverse the diagram by following the arrows. The general idea can be extended to cover diagrams involving an arbitrary number of sets and maps, and forms the basis for the mathematical field called *category theory.* Readers interested in knowing more about this kind of abstract "diagram-chasing" are urged to consult the chapter Notes and References, as well as the treatment given in Discussion Question #2.

Example: d'Arcy Thompson's Theory of Biological Transformations

In the early twentieth century, d'Arcy Thompson proposed a theory of biological structure whose basic idea was to regard a biological organism as a geometrical object. Each phenotype was associated with a point in a topological space Y, and Thompson considered two organisms to be "close" in this space if there was a *homeomorphism* (continuous coordinate change) in Y that mapped them one to the other. Put more loosely, "phenotypes of closely related organisms can be deformed continuously into each other."

To make contact with our earlier ideas, let two organisms be "close" if their genomes are close. Assuming we have a measure on the space of genomes, we can rephrase Thompson's requirement in more specific terms using our diagram for equivalence. Consider the two organisms described by the equations of state

$$\Phi_\alpha : U \to Y,$$
$$\Phi_{\hat{\alpha}} : U \to Y,$$

where U is the space of environments and Y is the topological space of phenotypes. The genomes α and $\hat{\alpha}$ are assumed to be close, i.e., $\|\alpha - \hat{\alpha}\| < \epsilon$, where $\epsilon > 0$ is some prescribed distance representing genomic "closeness." Then we say the phenotypes Φ_α and $\Phi_{\hat{\alpha}}$ are "close" *in a fixed environment* U, if there exists a homeomorphism $h_{\alpha,\hat{\alpha}} : Y \to Y$ such that the diagram

$$\begin{array}{ccc} U & \xrightarrow{\Phi_\alpha} & Y \\ {\scriptstyle id_U}\downarrow & & \downarrow{\scriptstyle h_{\alpha,\hat{\alpha}}} \\ U & \xrightarrow[\Phi_{\hat{\alpha}}]{} & Y \end{array}$$

commutes, i.e., the small genetic change $\alpha \to \hat{\alpha}$ can be offset by a continuous phenotypic deformation. Here the condition that the environment remains fixed requires that we take our earlier transformation $g_{\alpha,\hat{\alpha}}$ = identity on $U \doteq id_U$.

As a concrete illustration of this theory, consider the skull phenotypes displayed in Fig. 1.2. Here Fig. 1.2(a) represents a chimpanzee's skull, and

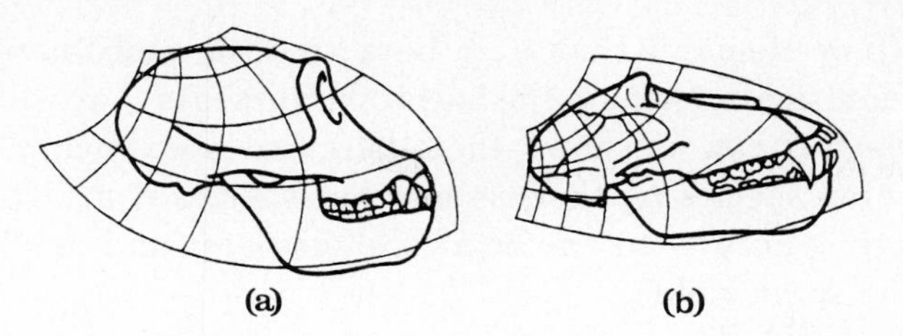

Figure 1.2 The Skulls of a Chimpanzee and a Baboon

Fig. 1.2(b) is the skull of a baboon. From the superimposed curvilinear coordinate frames, it's clear that the baboon skull is obtained from the chimpanzee's simply by a nonlinear change of variable, i.e., a transformation of the type $h_{\alpha,\hat{\alpha}}$ discussed above. Many other examples of this sort are found in the classic work of d'Arcy Thompson cited in the Notes and References.

Example: Forestry Yields

An empirical relation often used in forestry models for expressing timber yield as a function of tree diameter and height is

$$V = \epsilon + \beta D^2 H,$$

where

$$\begin{aligned} V &= \text{total tree volume exclusive of bark (in m}^3\text{)}, \\ D &= \text{tree diameter at breast height (in cm)}, \\ H &= \text{tree height from breast height (in m)}, \\ \epsilon, \beta &= \text{parameters characterizing the specific tree type.} \end{aligned}$$

The yield relationship assumes our standard form if we set $U = (D, H)$, $Y = V$, $\alpha = (\epsilon, \beta)$ and use the description

$$\begin{aligned} \Phi_\alpha &: U \to Y, \\ (D, H) &\mapsto \epsilon + \beta D^2 H. \end{aligned}$$

A change of parameter $\alpha \to \hat{\alpha}$ in this situation represents a change from one tree type to another, and it's of interest to ask whether such a change in tree type can be offset by a change in volume and/or diameter to keep the timber yield constant. Translating this question into a diagram, we ask if there is a map $g_{\alpha,\hat{\alpha}}: U \to U$, such that the diagram

$$\begin{array}{ccc} U & \xrightarrow{\Phi_\alpha} & Y \\ {\scriptstyle g_{\alpha,\hat{\alpha}}}\downarrow & & \downarrow{\scriptstyle id_Y} \\ U & \xrightarrow[\Phi_{\hat{\alpha}}]{} & Y \end{array}$$

commutes. If we demand that $g_{\alpha,\hat{\alpha}}$ be a smooth coordinate change, i.e., infinitely differentiable, then on the basis of arguments that will be developed in Chapter Four, it can be shown that there exists no such g. We conclude that for any tree species α, there is another species $\hat{\alpha}$ arbitrarily close by, such that there is no way to "deform" the diameter and height observables to preserve the same yield.

5. *Bifurcations and Catastrophes*

In the preceding discussions, we have seen that there can exist values of the system parameters α such that for all $\hat{\alpha}$ nearby to α, there exist *no* coordinate changes

$$g_{\alpha,\hat{\alpha}}: U \to Y, \qquad h_{\alpha,\hat{\alpha}}: Y \to Y,$$

making the diagram

$$\begin{array}{ccc} U & \xrightarrow{\Phi_\alpha} & Y \\ {\scriptstyle g_{\alpha,\hat{\alpha}}}\downarrow & & \downarrow{\scriptstyle h_{\alpha,\hat{\alpha}}} \\ U & \xrightarrow[\Phi_{\hat{\alpha}}]{} & Y \end{array}$$

commute. We call such values of α *bifurcation points*; all other values are called *regular* or *stable points.*

It's important here to note that whether or not a given point α is a bifurcation point depends upon the space of maps in which we seek the coordinate changes g and h. For instance, α may be a bifurcation point if we seek g and h in the space of *smooth* coordinate changes or *diffeomorphisms,* but may well be a regular point if we look for g and h in the much larger space of continuous coordinate changes. The particular space we choose is usually dictated by the analytic nature of the description Φ_α, as well as by the structure naturally imposed upon the spaces U and Y by the type of questions being considered. Illustrations of various cases will appear throughout the book, although we will usually seek g and h as diffeomorphisms unless stated otherwise.

If we assume that the possible parameter values α belong to some topological space A so that it makes sense to speak of points of A being "close," then we see that a point $\alpha \in A$ is a bifurcation point if Φ_α is *inequivalent* to $\Phi_{\hat{\alpha}}$ for all $\hat{\alpha}$ "near" to α, i.e., for all $\hat{\alpha}$ contained in a neighborhood of α. On the other hand, a point α is a regular point if Φ_α is equivalent to $\Phi_{\hat{\alpha}}$, for all $\hat{\alpha}$ in a neighborhood of α. Under reasonably weak conditions on the set A and the maps Φ_α, g and h, it can be shown that the bifurcation points form a "small" part of the set A (technically, a nowhere dense subset of A). Therefore, almost all points $\alpha \in A$ are regular points, and the qualitatively different behaviors that emerge only at bifurcation points are rare events as we wend our way through the space A.

Example: The Cusp Catastrophe

Here we let

$$\begin{aligned} U &= \{\text{all real cubic polynomials in a single indeterminate } z\}, \\ &= \{z^3 + az + b\colon a, b \text{ real}\}, \\ Y &= \{1, 2, 3\}, \\ A &= \{(a, b)\colon a, b \text{ real}\}, \end{aligned}$$

together with the map

$$\Phi_{(a,b)}\colon U \to Y,$$
$$z^3 + az + b \mapsto \text{the number of real roots of the cubic polynomial.}$$

Results from elementary algebra show that the parameter space A is partitioned by the root structure as shown in Fig. 1.3.

For this example, the bifurcation set consists of the points on the curve $4a^3 + 27b^2 = 0$. The space A is partitioned into three disjoint regions marked "1", "2" and "3" in Fig. 1.3, corresponding to the number of real roots of $z^3 + az + b$ in the particular region of A. In this situation, two cubics are equivalent if they have the same number of real roots. Since the set Y is discrete, two points $y_1, y_2 \in Y$ are "close" only if $y_1 = y_2$. Thus, we can have two cubics being close in the sense of their coefficients (a, b) being close in A, but they may be inequivalent in terms of their root structure. This can happen, however, only if one of the cubics corresponds to a bifurcation point and the other corresponds to a regular point, i.e., the coefficients (a^*, b^*) and (a, b) lie in differently numbered regions of the space A.

For reasons that will be made clear in Chapter Four, the set-up just described serves as the prototype for the development of Thom's theory of *elementary catastrophes.* Our case of cubic polynomials corresponds to the most well-known elementary catastrophe, the *cusp.* In Thom's theory, the set of bifurcation points is usually termed the *catastrophe set* or catastrophe manifold. The other elementary catastrophes (fold, butterfly, etc.) correspond to polynomials of different degrees, and their catastrophe sets are

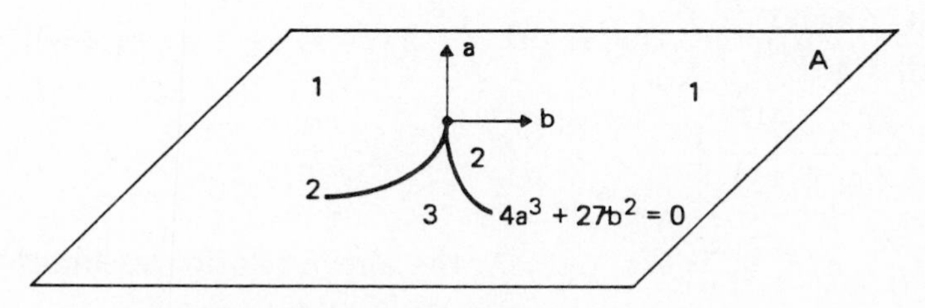

Figure 1.3 Cubic Polynomial Root Structure

determined in a fashion similar to what we have done above. Many more details on this example as well as this entire circle of ideas are given in Chapter Four.

Example: Stellar Structure

We consider a simplified version of the formation of stars. Let's focus upon the ellipsoidal states of relative equilibrium of a rotating liquid mass. Assume the liquid has uniform density ρ, with semi-axes a, b and c in the x, y and z directions, and let the mass be rotating about the z-axis with angular velocity W (think of an American football placed on the ground and then given a spin). The mass

$$M = \frac{4}{3}\pi abc\rho,$$

and the angular momentum

$$L = \frac{1}{5}M(a^2 + b^2)W \doteq IW,$$

are conserved, with the effective potential energy being given by

$$V = \frac{1}{2}\int \rho g \, dv + \frac{L^2}{2I},$$

where g is the gravitational potential.

The equilibrium configurations of this rotating mass are given by the stationary values of V. It's convenient to introduce the new variables

$$x_1 = \left(\frac{a}{c}\right)^2, \quad x_2 = \left(\frac{b}{c}\right)^2, \quad \lambda = 25\left(\frac{4\pi\rho}{3M}\right)^{1/3} \frac{L^2}{3GM^3},$$

where G is the gravitational constant. In these variables, it can be shown that the energy function takes the form

$$V = \frac{3}{10}\left(\frac{4\pi\rho}{3M}\right)^{1/3}\left(-(x_1x_2)^{1/6}\int_0^\infty [(1+s)(x_1+s)(x_2+s)]^{-1/2}\,ds + \frac{\lambda(x_1x_2)^{1/3}}{(x_1+x_2)}\right).$$

If we let $\alpha_1 = a/c$, $\alpha_2 = b/c$, $\alpha_3 = \lambda$, the above relation assumes the form

$$\Phi_\alpha : U \to Y,$$

where $U = R^3, Y = R, A = R^3$. There are two sets of solutions to the equilibrium equations

$$\frac{\partial V}{\partial x_1} = \frac{\partial V}{\partial x_2} = 0,$$

one given by Maclaurin spheroids with $a/c = b/c$, and the other by Jacobi ellipsoids with $a \neq b$. The Jacobi ellipsoids *bifurcate* from the Maclaurin spheroids at the point $\alpha_1^* = \alpha_2^* = 1.716$, $\lambda^* = 0.769$. It can be shown that the Jacobi ellipsoids are stable everywhere they exist, whereas the Maclaurin spheroids are stable for all $\lambda < 0.769$.

6. *Complexity*

The notion of one description of a system bifurcating from another also provides the key to begin unlocking one of the most important, and at the same time perplexing, problems of system theory: the problem of system complexity. What does it mean when we say a collection of atoms or a national economy or a vertebrate eye constitutes a "complex" system? Intuitively, it seems to mean that the system contains many interacting subsystems and/or we cannot offer an explanation for the system in terms of its observable subsystems. All of these intuitive notions, as well as many others, have been offered up as the basis for a theory of system complexity, and all have failed in the sense that there exist important natural systems that by any intuitive interpretation would be termed "complex," but that fall outside the scope of the proposed theory. So why does it seem so difficult to capture formally what appears to be such an easily grasped intuitive concept?

Our view is that the basis of the complexity difficulty lies in the implicit assumption that complexity is somehow an inherent or intrinsic property of the system itself, i.e., the complexity of a system is independent of any other system that the original system may be in interaction with. In particular, we question the claim that the system complexity is independent of any observer/controller that may be influencing the system. If we reject this unwarranted hypothesis, we arrive at the relativistic view of system complexity depicted in Fig. 1.4.

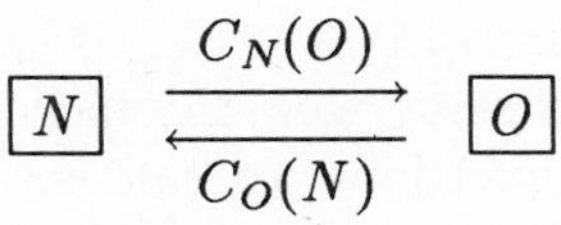

$$\boxed{N} \underset{C_O(N)}{\overset{C_N(O)}{\rightleftarrows}} \boxed{O}$$

Figure 1.4 Relative Complexity of System and Observer

Here we have the natural system N being observed by the system O. The two arrows labeled $C_N(O)$ and $C_O(N)$ represent the complexity of O as seen by N, and the complexity of N as seen by O, respectively, using whatever measure of complexity one chooses. The important point here is that the complexity of a given system is *always* determined by some other system with which the given system interacts. Only in extremely special cases where one of these reciprocal interactions is so much weaker than the other that it can be ignored (as in much of classical physics), can we justify the traditional approaches to complexity as an intrinsic system property. But such situations are almost exclusively found in physics and engineering, which tends to explain why most approaches to the problem emphasize classical notions like entropy and information. As soon as we pass from physics to biology and on to the social and behavioral sciences, we see the obvious flaws in the traditional view and it's the reciprocal relation *between* the quantities $C_O(N)$ and $C_N(O)$ that is of real interest, not one or the other in isolation.

But how should we measure $C_O(N)$ and $C_N(O)$? If we want to explore the relationship between them, we must first find some common basis for their expression. It's at this point that our earlier ideas of system equivalence come into play. Let's define the quantity $C_O(N)$ as

$$C_O(N) \doteq \# \text{ of inequivalent descriptions of } N \text{ formed by an observer } O.$$

The quantity $C_N(O)$ is defined analogously. Now let's make this definition a bit more explicit. From O's perspective, we have the family of descriptions of N given as $\Phi_\alpha : U \to Y$, $\alpha \in A$. If O can "see" the entire set A, then $C_O(N)$ is just equal to the number of equivalence classes into which A is split by the induced relation $\sim$ given by

$$\alpha \sim \alpha' \quad \text{if and only if} \quad \Phi_\alpha \sim \Phi_{\alpha'}.$$

Formally,

$$C_O(N) = \text{card } (A/_\sim).$$

However, if O can only see a proper subset $\hat{A}$ of A, then

$$C_O(N) = \text{card } (\hat{A}/_\sim) \leq \text{card } (A/_\sim),$$

since $\hat{A} \subset A$. Thus, the complexity of N as seen by O is, in general, less when O can form fewer descriptions of N. Of course, this conclusion is in complete accord with our feeling that the more we can "see" ("know") of a system, the finer discriminations we can make about its structure and behavior, and consequently, the more complex it will appear to us.

At first glance, it may appear that the complexity of many systems will be infinite, since a great many situations don't lead to a finite classification of the points of A under the relation $\sim$. But such situations are, for the most part, exactly like that faced in matrix theory when we classify the real $n \times n$ matrices according to similarity. There is an uncountable infinity of classes, with each class determined by the characteristic values (and their multiplicities) of the individual matrices in $R^{n \times n}$. So even though there are an infinite number of inequivalent matrices, we can label each class by a finite set of numbers (the characteristic values and their multiplicities) and operate to advantage in the set $R^{n \times n}/_{\sim}$ to determine system invariants, canonical forms and the like.

Example: The Cubic Polynomials

In the last section we considered the family of descriptions (e.g., relative to ourselves as observers) of cubic polynomials

$$\begin{aligned} \Phi_\alpha : U \to Y, &\qquad \alpha \in A, \\ z^3 + az + b &\mapsto \text{number of real roots,} \end{aligned}$$

where $A = R^2$. In Fig. 1.3, we saw the set A partitioned into equivalence classes by the cusp curve $4a^3 + 27b^2 = 0$. Each class is labeled by one of the integers "1," "2" or "3." Thus, if the natural system N is taken to be the set of cubic polynomials, our definition yields the complexity of N as $C_O(N) = 3$.

7. *Error and Surprise*

Another concept that is often associated with complexity is to say that a complex system "makes errors" or displays "surprising" or "unexpected" behavior. All of these notions ultimately derive from one description bifurcating from another. Let's illustrate this claim with an important situation in numerical analysis.

Example: Computer Roundoff Error

Consider the situation in which we have the system state-space $\Omega = R$, the real numbers, and the observables $f = (f_1, f_2, \ldots, f_n)$ are defined as

$$\begin{aligned} f_i : R \to R, &\qquad i = 1, 2, \ldots, n, \\ r &\mapsto i\text{th coefficient in the decimal expansion of } r. \end{aligned}$$

Then clearly $r_1, r_2 \in R$ are equivalent with respect to the observables f whenever r_1 and r_2 agree in the first n terms of their decimal expansions. Choose numbers r_1^*, r_2^* such that

$$r_1 \sim_f r_1^*, \quad r_2 \sim_f r_2^*.$$

Now let the 1-system interact with the 2-system through multiplication; that is, we form the products $r_3 = (r_1 r_2)$ and $r_3^* = (r_1^* r_2^*)$ and find that, in general, $r_3 \not\sim_f r_3^*$. In other words, the equivalence classes under f are split by the interaction, i.e., by the dynamics. The interaction generates a bifurcation of the f-classes, a bifurcation we usually term *roundoff error* in the above context. It's instructive to examine the source of this so-called error.

To see the way the error is introduced in the above situation, let's consider a numerical example. Suppose

$$r_1 = 123,\ r_1^* = 124,\ r_2 = 234,\ r_2^* = 235,$$

and we use $f = (f_1, f_2)$; that is, the equivalence relation generated by f is such that two numbers are equivalent if they agree in their first two places. Here we have

$$r_1 r_2 = 28,782 \not\sim_f r_1^* r_2^* = 29,140,$$

which disagrees with our expectation based upon f-equivalence. Our surprise at finding $r_1 r_2 \not\sim_f r_1^* r_2^*$ occurs because the set of chosen observables $f = (f_1, f_2)$ is too limited, thereby causing an unrealistic expectation concerning the interaction between the 1- and 2-systems. If we had expanded our set of observables to $\hat{f} = (f_1, f_2, f_3)$, then no such discrepancy would have occurred since under this set of observables r_1 and r_1^* are not equivalent. Thus, the entire source of our observed error is due purely to the incompleteness in the description of the system.

The preceding arguments are entirely general: error (or surprise) always involves a discrepancy between the objects (systems) *open* to interaction and the abstractions (models, descriptions) *closed* to those same interactions. In principle, the remedy is equally clear: just supplement the description by adding more observables to account for the unmodeled interactions. In this sense, error and surprise are indistinguishable from bifurcations. A particular description is inadequate to account for uncontrollable variability in equivalent states, and we need a new description to remove the error.

It's interesting to note that since bifurcation and error/surprise are identical concepts, and complexity arises as a result of potential for bifurcation, we must conclude that complexity implies surprise and error; that is, to say a system displays counterintuitive behavior is the same as saying that the system has the capacity for making errors, although the error is not *intrinsic* to an isolated system but occurs when the system interacts with another.

The concepts of bifurcation, complexity, similarity, error and surprise have all been introduced using the bare-bones minimum of mathematical machinery—essentially just the idea of a set and a mapping between sets. This is indeed a primitive level of mathematical structure, and we can certainly expect far more detailed and precise results if we make use of more elaborate and sophisticated mathematical representations. To this end, we consider the idea of a *formal* mathematical system.

8. *Formal Systems*

In simplest possible terms, a formal mathematical system F is a collection of abstract symbols, together with a rule (grammar) expressing how symbols can be combined in order to create new symbols and symbol strings. In addition, we also include as part of the definition a specification of a set of symbol strings that are assumed to be labeled TRUE without proof (the axioms), as well as a set of rules of logical inference enabling us to generate new TRUE statements (theorems) from earlier ones.

Example: Formal Addition

Let the symbols of the system F be all finite strings of dashes, e.g., $- - --$ and $--$ would be symbols of this system. Let the single rule of combination be given by

$$\underbrace{- - \cdots -}_{p \text{ times}} \oplus \underbrace{- - \cdots -}_{r \text{ times}} = \underbrace{- - \cdots -}_{p+r \text{ times}},$$

$p, r \geq 0$. Further, assume the single axiom to be $- \oplus 0 = -$.

This example serves as a formal *model* for the addition of nonnegative integers if we interpret the strings of dashes to be integers, i.e.,

$$\underbrace{- - \cdots -}_{r \text{ times}} \doteq (r) = \text{the integer } r,$$

with the rule of combination interpreted as the usual rule for addition, viz. $(p) + (r) = (p + r)$. But this is just an *interpretation* of the symbols, and the formal system could also serve to model other situations as, for example, the addition of apples or the mixing of cars on a highway. It's important to keep in mind the fact that a formal system doesn't really represent *anything* until its symbols, rules and axioms are interpreted; until that time F is literally nothing but a collection of marks on a piece of paper and a prescription for creating new marks from old.

The big advantage of a formal system is that it can often be interpreted in many different ways, thereby serving as a model for a variety of situations. In addition, a formal system contains an automatic mechanism for creating new theorems from old: the rules of logical inference. Furthermore, the need to interpret the symbols unambiguously and the need to specify precisely all rules of grammar and inference force us to be very specific and exact about our assumptions when using F to represent a natural system N.

Example: Computer Programming Languages

Consider a typical high-level computer programming language like Fortran. Here the symbols (words) of the formal system are the individual statements that can be written in the language, such as GO TO x, WRITE y,

STOP, etc. The number of such words is usually restricted to a very small vocabulary of a couple dozen words or so. Similarly, the grammar of such a language is also very restrictive, admitting such combinations of words as IF(x.LT.y).AND.IF(z.GT.r)GO TO g, but not statements (sentences) such as WRITE-READ. The individual words of the language constitute the axioms, while the theorems correspond to all correctly or well-formed programs.

Incidentally, the above example shows clearly the enormous difference in subtlety and quality between a computer programming language and a natural human language, displaying the dubious nature of proposals often seen in universities nowadays to substitute knowledge of a computer language for a foreign language in Ph.D. language competency requirements. Even the most elaborate and sophisticated computer languages pale by comparison with natural languages in their expressive power, not to mention the enormous cultural content carried along almost for free with a natural language. For reasons such as these, many see such proposals to replace knowledge of a natural language by knowledge of a computer language as at best a shallow, anti-intellectual joke, and the theory of formal systems provides a basis for construction of rational arguments against their adoption. But to develop this line of argument would take us too far afield, so let's get back to system theory.

9. *Modeling Relations*

Let's now consider the notion of a formal system and its role in system modeling. Loosely speaking, we begin with a collection of symbols together with a finite set of rules for assembling such symbols into strings of finite length. A set of axioms having been given, the axioms together with the rules of grammar and logical inference constitute our formal system F. Usually in modeling a system N, we take familiar mathematical objects such as groups, topological spaces, graphs and differential equations as the formal systems of interest. The important point to keep in mind regarding formal systems is that they are entirely constructions of the human mind. Unlike the situation with natural systems defined by observables, formal systems are defined solely in terms of symbols and rules for their manipulation (symbolic logic).

Formal systems provide the basis for making *predictions* about N, hence, their interest. If we can establish a "faithful" correspondence between the observables and linkages of N and the elements of the formal system F, then we can use the rules of inference in F to derive, in turn, new theorems interpretable as relations between the observables of N. Under such conditions we call F a *formal description* of N. Missing from this program is a specification of how to construct a faithful mapping, or *encoding,* of the system N into a particular formal system F.

The final step in the process of translating a specific natural system N into a particular formal system F is an *encoding map* $\mathcal{E}: N \rightarrow F$. In rough terms, $\mathcal{E}$ provides a "dictionary" to associate the observables of N with the objects of F. The most familiar example of such an encoding is to take F to be a system of ordinary differential equations, in which case $\mathcal{E}$ usually associates the observables of N with the dependent variables (coordinate functions) of the system of equations.

Since the encoding operation $\mathcal{E}$ provides the link between the real world of N and the mathematical world of F, it's essential that we be able to *decode* the theorems (predictions) of F if we hope to interpret those theorems in terms of the behavior of N. Basically, what we must try to ensure is that our encoding is consistent in the sense that the theorems of F become predictions about N that may be verified when appropriately decoded back into relations in N. This modeling relation is schematically depicted in Fig. 1.5.

In Fig. 1.5 we have explicitly indicated the separation between the real world of N and the mathematical world of F by the vertical dotted line. This diagram also makes explicit the role of the system scientist: he is neither a natural scientist living on the left side of the diagram nor a mathematician living on the right. Rather, the system scientist is the keeper of the abstract encoding/decoding operations $\mathcal{E}$ and $\mathcal{D}$. In short, he keeps the dictionaries whereby we translate back and forth between the real world of N and the mathematical world of F. Of course, in order to carry out this function, the system scientist has to be knowledgeable about both N and F, but not necessarily an expert in either. His expertise is centered upon making adroit choices of $\mathcal{E}$ and $\mathcal{D}$ so that in the transition from N to F and back again, as little information is lost as possible. To illustrate the steps in the modeling process, let's consider two familiar cases.

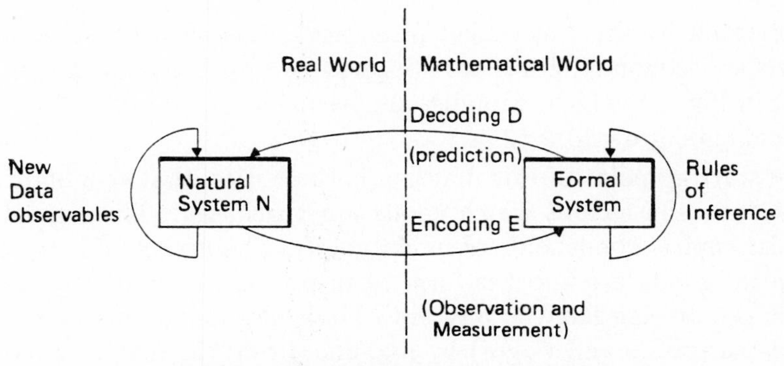

Figure 1.5 The Modeling Relation

Example: Classical Mechanics

The most familiar setting in which to expose the general ideas outlined above is for the case of classical Newtonian mechanics, where we consider systems of material particles.

Our first order of business is to specify the fundamental observables that are to be encoded and to define the corresponding state-space. Newton proposed that these fundamental observables could be taken to be (1) the displacement of the particles from some convenient reference point, and (2) the velocities of these particles. Thus, given a system of M particles, each abstract state of the system is represented by a set of $6M$ numbers. This gives an encoding of the abstract states into the points of R^{6M}—the *phase-* or *state-space* of an unconstrained classical M-particle mechanical system.

The next step in the Newtonian encoding is to postulate that every other observable of a particle system can be represented by a real-valued function on this phase-space. Consequently, every other observable is totally linked to these state variables on a set Ω of abstract states. This postulate removes the necessity of ever referring back to the set Ω of abstract states. We can now confine ourselves entirely to the state-space into which Ω is encoded by state variables.

Since the Newtonian framework assumes that *every* observable pertaining to a particle system is represented by a function of the state variables, it must be the case that the *rates* at which state variables themselves change are also functions of the state alone, and if such rates are themselves observables, then there must be equations of state expressing the manner in which the rates of change of the state variables depend on the values of the state variables. It is this fact that leads directly to the entire dynamical apparatus of Newtonian physics.

Example: Keynesian Dynamics and Equivalent Economies

Motivated by the Newtonian framework, let's now consider its application to an economic situation. Suppose we have the economic system depicted in Fig. 1.6 which embodies the basic ideas underlying Keynes' theory of economic growth.

The starting point of this diagram is the notion that economic activity depends upon the rate at which goods are purchased. These goods are of two kinds: capital goods and consumer goods. The next step is to recognize that capital goods are another form of investment. Finally, we note that money is the driving force to buy both kinds of goods. This money can be obtained (as profits and wages) by manufacturing the goods. However, *all* profits and wages do not automatically go back as investment; some money is saved. The system is prevented from grinding to a halt by new investment.

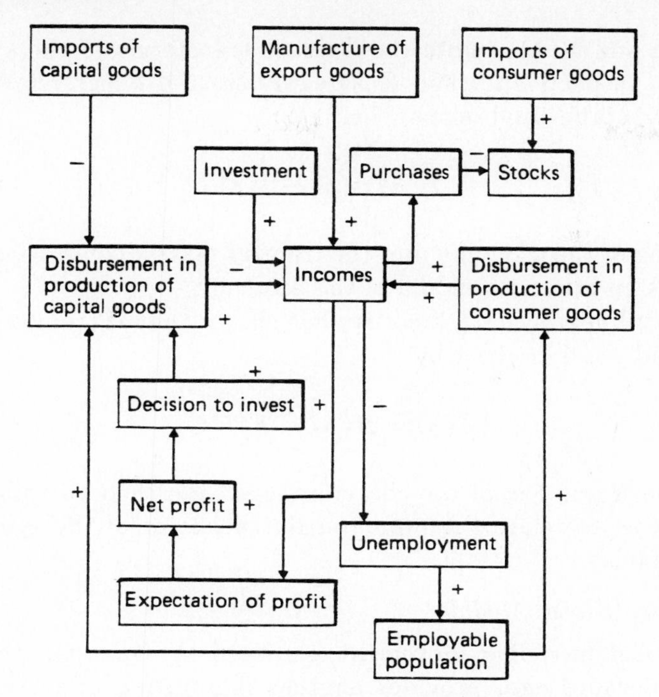

Figure 1.6 A Keynesian Economy

Thus, the levels of economic activity and employment depend upon the rate of investment.

If we assume that the signed arcs in the diagram indicate a linear relationship between the origin and destination of the arc, and if we regard investment as the driving force, with measurable outputs being imports of consumer and capital goods as well as manufacture of export goods, a dynamic Keynesian economy can be analytically represented by the linear system

$$\begin{aligned} \dot{x} &= Fx + Gu, \\ y &= Hx, \end{aligned} \qquad (F_1)$$

where $u(t)$ and $y(t)$ are the input and outputs, respectively, $x(t)$ is a vector representing the other components of the economy, and F, G and H are constant matrices whose entries are determined by the flows along the system arcs. Specification of the initial state of the economy $x(0)$, as well as the pattern of investment $u(t)$, then determines the future course of economic activity, including the observed outputs $y(t)$. Thus, the linear system F_1 represents an *encoding* of the economy into the formal mathematical system represented by the differential and algebraic relations F_1.

An alternate encoding is to associate the economy with a *transfer function* matrix $Z(\lambda)$, a rational matrix in the complex variable λ. Formally, if we let $U(\lambda)$ and $Y(\lambda)$ denote the Laplace transforms of the system input and output, respectively, then $Z(\lambda)$ expresses the linear relation between the economy's input and output, i.e.,

$$Y(\lambda) = Z(\lambda)U(\lambda). \qquad (F_2)$$

The encoding of the economy into the transfer matrix comprising the formal system F_2 is an alternate model of the economy.

Work by linear system theorists has shown that there is a relation between F_1 and F_2. It's given by

$$Z(\lambda) = H(\lambda I - F)^{-1}G.$$

This relation forms one of the cornerstones of the theory of linear systems and enables us to relate the input/output behavior of the economy to its internal behavior.

Example: Global Models

The global modeling movement, initiated by Forrester and Meadows about two decades ago, provides another illustration of the principles of this chapter. All global modeling attempts begin by abstracting a certain finite collection of observables from the essentially infinite number of observables defining the *world problematique.* For his original model, Forrester chose five observables: Population, Natural Resources, Capital, Pollution and Fraction of Capital Devoted to Agriculture. Various linkages were then postulated among these observables, leading to a formal system comprised of a collection of finite-difference equations. These equations contained various parameters representing elements of the system such as birth and death rates, pollution generation rates, and land yields. It's important to recognize that these parameters were not part of the basic observables (state variables) of the model. As a result, global modelers could create *families* of models. This simple observation provides the key to understanding one of the most severe criticisms aimed at such modeling efforts, namely, their almost pathological sensitivity to small changes in the defining parameters. We examine this issue from the vantage point of our general modeling framework.

Denote the abstract state-space of the global model by Ω, and let A be its parameter space. Generally, A is some subset of a finite-dimensional space like R^n. Then the class of models is represented by the Cartesian product $\Omega \times A$. (Such a class of models is a special case of what in mathematics is termed a *fiber bundle,* with base A and fiber space Ω). Pictorially, we can envision the situation as in Fig. 1.7. For each fixed $\alpha^* \in A$, the linkages between the observables on Ω are used (via the encoding into difference equations) to obtain a time-history of the observables. In other words, we

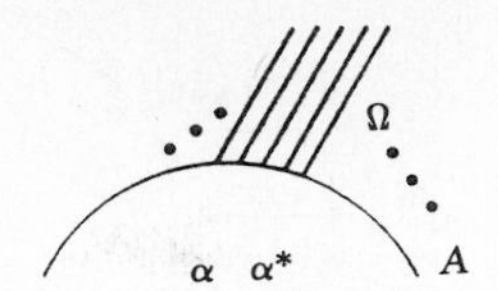

Figure 1.7 A Fiber Bundle of Global Models

obtain a fibering of the state-space Ω into trajectories determined by solutions of the equations of motion associated with the corresponding value of α. It then becomes an interesting question to relate the fiberings corresponding to different values $\alpha, \alpha^* \in A$. This leads to a special case of the overall notion of *structural stability.*

To each $\alpha \in A$ we associate a one-parameter family of transformations

$$T_t^\alpha : \Omega \to \Omega,$$

representing the dynamics of the system in the state-space Ω. If $\alpha^* \in A$ is another point, then it will also be associated with a corresponding one-parameter family

$$T_t^{\alpha^*} : \Omega \to \Omega.$$

Roughly speaking, if the effect of replacing α by α^* can be annihilated by coordinate transformations in Ω, then, by definition, the dynamic imposed by the difference equations represented by T_t is *structurally stable* with respect to the perturbation $\alpha \to \alpha^*$. Another way of looking at this question is to ask if we can find mappings g_{α,α^*} and h_{α,α^*} such that the following diagram commutes

$$\begin{array}{ccc} \Omega & \xrightarrow{T_t^\alpha} & \Omega \\ {\scriptstyle g_{\alpha,\alpha^*}}\downarrow & & \downarrow{\scriptstyle h_{\alpha,\alpha^*}} \\ \Omega & \xrightarrow[T_t^{\alpha^*}]{} & \Omega \end{array}$$

Since the situation described here is so important, let's recast it into a diagram analogous to that of Fig. 1.7. Figure 1.8 allows us to enlarge somewhat on our previous remark that $\Omega \times A$ itself possesses the structure of a fiber bundle. On each fiber Ω, there is imposed a group of transformations, namely, the corresponding dynamics T_t^α. The notion of structural stability allows us to consider mappings of the *entire* space $\Omega \times A$ that (1) leave A fixed, and (2) preserve dynamics on the fibers. Actually, it's also possible to consider a broader class of mappings that do not leave A fixed. Such mappings are intimately connected with what physicists call *renormalization,* but we shall not go into this detail here. However, it should be noted that a great deal of the *technical* criticism received by the Forrester-Meadows-Mesarovic

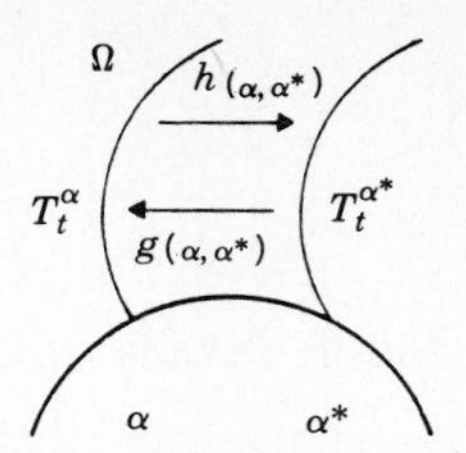

Figure 1.8 Fibering of the State-Space

exercises could have been avoided by paying more thorough attention to the structural stability properties of their models.

In closing, let us again emphasize that the structural stability of any model is contingent upon the choice of both Ω and A, as well as upon the dynamics T_t. Thus, the entire issue depends upon the abstractions that are made and the encodings that are chosen. No natural process is *absolutely* stable or unstable, and its stability depends initially upon nonmathematical considerations. To this extent, the matter of stability is a premethodological question. To illustrate this point at a nontechnical level, some of the most damaging critiques of the global models are that the modelers omitted variables (observables) that were centrally important for understanding the behavior of the observables that were actually employed. Thus, in the Forrester model there is no accommodation made for energy supply and consumption, an omission that dramatically affects agriculture, population and pollution. Here we see how the *abstraction* operation can go wrong. The modelers threw away observables vitally important to the goals of the model, and as a result no amount of technical virtuosity using the observables retained allows any sort of meaningful conclusions to be obtained. In short, the models were *invalidated* by virtue of faulty abstractions.

Discussion Questions

1. Consider the situation depicted below in which two natural systems, N_1 and N_2, code into the same formal system F. The specific features of this situation depend upon the degree of overlap in F of those propositions $\mathcal{E}_1(N_1)$ and $\mathcal{E}_2(N_2)$ representing qualities of N_1 and N_2.

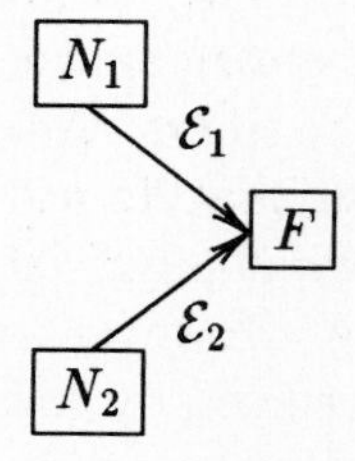

a) What does it mean if we have $\mathcal{E}_1(N_1) = \mathcal{E}_2(N_2)$? In this case, N_1 and N_2 share a common model. This situation is the basis for the idea of system *analogy.*

b) If we define the relation $N_1 \sim N_2$ if and only if N_1 and N_2 are analogous, show that $\sim$ determines an *equivalence* relation on the set of natural systems, i.e., $\sim$ is reflexive, symmetric and transitive.

c) What does the relation of system analogy have to do with schemes for analog or digital simulation?

d) Consider the cases $\mathcal{E}_1(N_1) \subset \mathcal{E}_2(N_2)$ and $\mathcal{E}_1(N_1) \cap \mathcal{E}_2(N_2) \neq \emptyset$, and give interpretations in terms of subsystem analogies.

2. Suppose we have the modeling relation

$$\boxed{N_1} \xrightarrow{\mathcal{E}_1} \boxed{F_1}$$

$$\boxed{N_2} \xrightarrow{\mathcal{E}_2} \boxed{F_2}$$

in which two systems encode into different formal systems.

a) Consider the case in which $\mathcal{E}_1(N_1)$ and $\mathcal{E}_2(N_2)$ are *isomorphic,* i.e., there exists a one-to-one, onto, structure-preserving mapping between the sets of encoded propositions $\mathcal{E}_1(N_1)$ and $\mathcal{E}_2(N_2)$. Here a mapping is considered to be structure-preserving if it can be extended in a unique way to all inferences in F_1, F_2 that can be obtained from $\mathcal{E}_1(N_1)$ and $\mathcal{E}_2(N_2)$, respectively. What interpretation can you give to the relationship between N_1 and N_2 in this case? Does the relationship that exists between N_1 and N_2 depend upon the isomorphism between $\mathcal{E}_1(N_1)$ and $\mathcal{E}_2(N_2)$? How?

b) Suppose $\mathcal{E}_1(N_1)$ and $\mathcal{E}_2(N_2)$ satisfy a weaker relation than isomorphism. For example, they may share a property like "stable" or "finite" or "finite-dimensional." In such cases, we have the basis for *metaphorically* relating N_1 and N_2 through their models sharing this common property.

In mathematics, the set of all objects possessing a given property forms what is termed a *category.* Thus we have the category "GROUPS," or the category "LINEAR SYSTEMS," and so forth. Typically, along with such objects come appropriate families of structure-preserving maps, providing us with the concept of morphism between objects of the category. For instance, the set of group homomorphisms in the category GROUPS. Show that the idea of a system metaphor defines a categorical relation upon the set of all natural systems.

3. Assume we have a single system N that encodes into two different formal systems, i.e.,

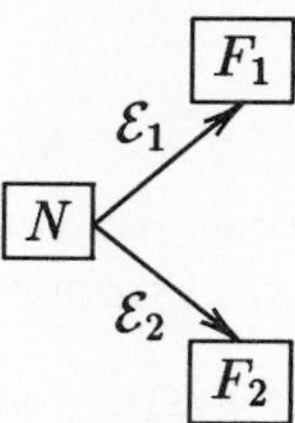

Here we will consider to what extent uncoded linkages of observables in N manifest themselves in the form of mathematical relations in F_1 and F_2.

a) *Case I:* $F_1 \to F_2$. In this situation, every proposition in F_1 can be associated with a unique proposition of F_2, but not necessarily conversely. Intuitively we would say that F_2 provides a more comprehensive description of N than does F_1; in fact, the encoding $\mathcal{E}_1(N)$ in F_1 can be *reduced* to the encoding $\mathcal{E}_2(N)$ in F_2.

What does the above situation have to do with the scientific principle of *reductionism,* which asserts that there is a universal way of encoding every natural system into a formal system, and that every other encoding is somehow reducible to the universal encoding? What conditions would have to be demonstrated in order to establish reductionism as a valid scientific principle? If such a universal encoding were displayed, do you think it would serve any *practical* purpose? Why? (*Remark:* In this connection, think of the practical importance of the Turing machine which serves as a universal computer.)

b) *Case II:* $F_1 \cap F_2 = \{\emptyset\}$. In this event, there are no nontrivial relations between F_1 and F_2, and the encodings $\mathcal{E}_1(N)$ and $\mathcal{E}_2(N)$ cannot, in principle, represent any common linkages between observables in N. What does this case have to do with the problem of "complementarity" in quantum mechanics, which asserts that an object can display wave-like behavior or particle-like behavior, depending upon how we look at it, but not both simultaneously? Discuss also the relationship of this case to the concept of system complexity.

4. The situation described in the preceding question can also serve as a point of departure for discussing problems of bifurcation. Suppose that a relation can be established between a *subset* of F_1 and a *subset* of F_2. Here certain qualities of N encoded into $\mathcal{E}_1(N)$ are related to corresponding qualities encoded into $\mathcal{E}_2(N)$. Thus, this mapping establishes a logical relation between the encodings, but one that does not hold universally. A *partial* reduction is possible, and outside the domain in which this partial relation holds, F_1 and F_2 are logically independent and the linkage between the qualities of N that are encoded is broken. Discuss this situation in relation to the process of bifurcation developed in the text.

5. Imagine we have a system N characterized by the values of the variables (observables) $x = (x_1, x_2, \ldots, x_n)$ and a particular set of parameters $\alpha = (\alpha_1, \ldots, \alpha_p)$. Suppose that the model describing the relationship between these quantities is given by the smooth function

$$f \colon R^n \times R^p \to R,$$

i.e., the system's equation of state is

$$f_\alpha(x) = \gamma, \qquad \alpha \in R^p,\ x \in R^n,\ \gamma \in R.$$

Catastrophe theory addresses the question: What does f look like in the neighborhood of a critical point, i.e., a point x for which $\text{grad}_x f_\alpha = 0$? We have already seen that $f_\alpha \sim f_{\alpha^*}$ if we can find smooth coordinate changes g_{α,α^*} and h_{α,α^*} such that the diagram

$$\begin{array}{ccc} X & \xrightarrow{f_\alpha} & R \\ {\scriptstyle g_{\alpha,\alpha^*}}\big\downarrow & & \big\downarrow{\scriptstyle h_{\alpha,\alpha^*}} \\ X & \xrightarrow[f_{\alpha^*}]{} & R \end{array}$$

commutes. Those α for which there is no such change of variables g, h for all α^* in a local neighborhood are the bifurcation points.

a) Relate the above catastrophe theory set-up to the general situation described in the previous question.

b) Suppose you wanted to study how the quality of life changed in an urban area as a function of factors like climate, budgetary levels, property tax rates, zoning regulations and recreational facilities. How would you define the urban "genotype," "environment" and "phenotype" in order to characterize such a situation in a catastrophe theory framework?

6. We have seen that the equation of state $\Phi(f_1, f_2, \ldots, f_n) = 0$ establishes a mathematical dependency among the observables of a system N. Given such a relation, what additional criteria would you impose to distinguish the relation as a *law of Nature,* as opposed to just an *empirical* relationship? (*Hint:* Consider the fact that, in general, when we change the encoding of N into the formal system F, we also change the linkages between the observables of N.)

7. Some of the characteristics used to describe "complex" systems are that they exhibit counterintuitive behavior, have many interactions, possess many feedback-feedforward loops, involve a high degree of decentralized decision making and are indecomposable. Give examples of natural systems displaying such features. Do these examples agree with your feeling as to what constitutes something being "complex"? How can you relate these properties to the concept of complexity as a "relative" notion that depends upon interactions with other systems?

8. The main point of making models in F representing a natural process N is to be able to make predictions about the behavior of N by logical deduction in F. To what extent can this program be effectively carried out if the best that F can do is capture a *subsystem* of N? Discuss this problem in connection with determining the *credibility* of a model.

9. There are two main strategies that can be employed for theory development: *reductionism* and *simulation.* In reductionism, we try to explain the behavior of N in terms of *structural* subunits. But in simulation we attempt to understand a limited family of behaviors as manifested simutaneously by a family of *structurally diverse* systems, i.e., the units of analysis are behaviors, not material units. Discuss these two dual modes of analysis, and give examples of situations in which one or the other appears to be the strategy of choice.

10. What would you consider to be good candidates for the abstract state-space Ω of a national economy? A human brain? A natural language? An automobile? Construct at least two different state-spaces for these natural systems, and examine the relationship, if any, between your state-spaces.

11. In the process of abstraction, we discard observables of the original system S, thereby obtaining a new system N that is *closed* to interactions with the outside world that S is open to, i.e., S can interact through the observables thrown away in forming N. In this sense, N is a *closed* approximation to S.

a) Discuss the proposition that "surprise" is a result of the discrepancy between the behavior of a system open to interactions with the outside and the behavior of a similar system closed to those same interactions.

b) Consider the degree to which the system N formed from S by abstraction can be used to predict the occurrence of "surprising" behavior in S. Does this idea involve a contradiction in terms?

Problems and Exercises

1. Suppose we have the two commutative diagrams

$$\begin{array}{ccc} A & \xrightarrow{f} & B \\ \varphi\downarrow & & \downarrow\psi \\ A & \xrightarrow[f']{} & B \end{array} \qquad \text{and} \qquad \begin{array}{ccc} B & \xrightarrow{g} & C \\ \varphi'\downarrow & & \downarrow\psi' \\ B & \xrightarrow[g']{} & C \end{array}$$

Thus, f is equivalent to f' and g is equivalent to g'. Show that the composite map $g \circ f: A \to C$ is not, in general, equivalent to the map $g' \circ f': A \to C$.

Thus, a model of individual subsystems doesn't necessarily allow us to model the composite system.

Show also that even if $g \circ f$ is equivalent to $g' \circ f'$, we will usually not have φ and ψ' as maps of $A \to A$ and $C \to C$, respectively. (This exercise bears upon the issue of reductionism vs. holism, since if we did have $g \circ f$ equivalent to $g' \circ f'$, then it would always be possible to reconstruct the behavior of the entire system by piecing together the behaviors of its component subsystems.)

2. Given a smooth equation of state $\Phi(f_1, f_2, \ldots, f_n)\colon R^n \to R$, show that there exists an integer p, $0 < p < n$, such that Φ can be rewritten as a *stable* p-parameter family, but not for any integer $k > p$, i.e., we can always stabilize any equation of state (Buckingham's Theorem). (*Recall:* A map $\Phi_\alpha\colon R^m \to R$ is called stable if, for any $\hat{\alpha}$ in a neighborhood of α, we can find smooth maps $g_{\alpha,\hat{\alpha}}$, $h_{\alpha,\hat{\alpha}}$ such that the following diagram

$$\begin{array}{ccc} R^m & \xrightarrow{\Phi_\alpha} & R \\ {\scriptstyle g_{\alpha,\hat{\alpha}}}\downarrow & & \downarrow{\scriptstyle h_{\alpha,\hat{\alpha}}} \\ R^m & \xrightarrow[\Phi_{\hat{\alpha}}]{} & R \end{array}$$

commutes.)

3. Let

$$\begin{aligned} \Omega &= \{\text{all cubic curves in one variable } z\}, \\ &= \{z^3 + az + b\colon a, b \text{ real}\}. \end{aligned}$$

Consider two descriptions of Ω given by

$$\begin{aligned} f_1 : \Omega &\to S = R^2, \\ \omega &\mapsto (a, b), \end{aligned}$$

and

$$\begin{aligned} f_2 : \Omega &\to \{1, 2, 3\} = \hat{S}, \\ \omega &\mapsto \text{the number of real roots of } \omega. \end{aligned}$$

Call a point $\alpha \in S$ (or $\hat{S}$) *generic* if there exists an open neighborhood U of α such that all $\hat{\alpha} \in U$ have the same root structure as α; otherwise, α is a *bifurcation point.*

Show that the generic points under the description f_1 are $\{(a, b)\colon 4a^3 + 27b^2 \neq 0\}$, while the generic points under f_2 are $\{\emptyset\}$, the empty set.

4. Consider two observables $f: \Omega \to S$ and $g: \Omega \to \hat{S}$. We say that f *bifurcates* from g on those points in S that are generic relative to f, but bifurcation points relative to g. Similarly, g bifurcates from f on those points in $\hat{S}$ that are generic relative to g, but not generic relative to f.

a) Prove that f is equivalent to g if and only if the bifurcation sets in S and $\hat{S}$ relative to the descriptions (g, f) and (f, g), respectively, are empty, i.e., f and g differ only on bifurcation sets.

b) If f and g are not equivalent, then we say one description *improves* upon the other. Show that f improves upon g if and only if all $\alpha \in S$ are generic relative to g, while the bifurcation set in $\hat{S}$ relative to f is nonempty.

c) Consider the two descriptions f_1 and f_2 of cubic curves given in Problem 3. Define the product description

$$\begin{aligned} f_1 \times f_2 : \Omega \to S \times \hat{S}, \\ \omega \mapsto (f_1(\omega), f_2(\omega)). \end{aligned}$$

Show that $f_1 \times f_2$ improves upon either f_1 or f_2 taken separately.

5. Let Ω be a set of states, and let $f, g: \Omega \to R$ be two observables. Define equivalence relations R_f and R_g on Ω by the rule:

$$\omega \, R_f \, \omega' \text{ if and only if } f(\omega) = f(\omega'),$$

with R_g defined similarly. Call the equivalence classes $[\omega]_f$ and $[\omega]_g$, respectively. Consider the set Z of R_g classes that intersect $[\omega]_f$. We say that:

- g is *totally linked* to f at $[\omega]_f$ if Z is a single g-class.
- g is *partially linked* to f at $[\omega]_f$ if Z is more than one g-class but not all of Ω/R_g.
- g is *unlinked* to f at $[\omega]_f$ if $Z = \Omega/R_g$.

a) Show that g is totally linked to f if and only if R_f refines R_g. (*Recall:* A relation R' *refines* a relation R if $\omega \, R' \, \bar{\omega}$ implies $\omega \, R \, \bar{\omega}$.)

b) Prove that f and g are totally unlinked if and only if every R_f-class intersects every R_g-class, and conversely.

(*Note:* The idea of linkage is crucial when we want to speak of prediction. If g is linked to f at a state $[\omega]_f$, then we obtain information about $g(\omega)$ by knowing $f(\omega)$. In fact, if g is totally linked to f at $[\omega]_f$, then we can say that there exists a map $h: R \to R$ such that $g(\omega) = h(f(\omega))$, i.e., g is a *function* of f. So in this case we can obtain $g(\omega)$ solely by computation; we need not measure g directly. In this sense, computation and measurement are identical processes, both of which depend critically on the existence of linkage relations between observables.)

6. A *dynamic* on a state-space Ω is a one-parameter family of transformations $T_t: \Omega \to \Omega, t \in R$. Given an observable f on Ω, we say that f is *compatible* with T_t if two states $\omega, \bar{\omega}$ that are equivalent under f, i.e., $f(\omega) = f(\bar{\omega})$, are such that $T_t(\omega) = T_t(\bar{\omega})$ for all $t \in R$. Thus, f and T_t are compatible if T_t preserves the equivalence classes of R_f. In other words, we cannot use the dynamics to distinguish between states that cannot be distinguished by f.

Now let g be another observable on Ω that is also compatible with the same dynamic T_t. In order for the two observers to conclude that they are actually observing the same system, there must be a mapping

$$\varphi: \Omega/R_f \to \Omega/R_g$$

such that the diagram

$$\begin{array}{ccc} \Omega/R_f & \xrightarrow{T_t^{(f)}} & \Omega/R_f \\ \varphi\downarrow & & \downarrow\varphi \\ \Omega/R_g & \xrightarrow[T_t^{(g)}]{} & \Omega/R_g \end{array}$$

commutes for all $t \in R$.

a) Show that the only candidate for such a map φ is the map that makes the following diagram commute:

$$\begin{array}{ccc} & \Omega & \\ \pi_f \swarrow & & \searrow \pi_g \\ \Omega/R_f & \xrightarrow[\varphi]{} & \Omega/R_g \end{array}$$

where

$$\pi_f: \Omega \to \Omega/R_f,$$
$$\omega \mapsto [\omega]_f,$$

and π_g is defined similarly, i.e., π_f and π_g are the natural projections of Ω onto its equivalence classes under f and g, respectively.

b) Show that such a map φ exists if and only if R_f refines R_g, or vice versa, i.e., each observer must acquire (or have) enough observables of the right type so that each of them sees the same relation R which simultaneously refines both R_f and R_g. Otherwise, f and g are *inequivalent* observers, and their descriptions of the system must necessarily bifurcate with respect to each other.

7. Consider the one-parameter family of *linear* maps $\Phi_\alpha \colon R^n \to R^m$, $\alpha \in R$. We can represent Φ_α by an $m \times n$ real matrix A_α, depending upon the single parameter α. Show that the bifurcation points of the diagram

$$\begin{array}{ccc} R^n & \xrightarrow{A_\alpha} & R^m \\ g \downarrow & & \downarrow h \\ R^n & \xrightarrow[A_{\hat{\alpha}}]{} & R^m \end{array}$$

are exactly those α for which rank $A_\alpha \neq$ rank $A_{\hat{\alpha}}$. (Note that here the transformations g and h are also assumed to be linear.)

Consider the same situation but for square matrices $A_\alpha \in R^{n \times n}$, assuming $g = h$. In this case, show that the bifurcation points consist of all $\alpha \in R$ such that $\lambda_i(A_\alpha) \neq \lambda_i(A_{\hat{\alpha}})$ for *some* $i = 1, 2, \ldots, n$, where $\lambda_i(A) \doteq i$th characteristic value of the matrix A.

8. In the situation of the last problem with $g \neq h$ (necessarily), show that the *complexity* of the matrix A equals the rank of A. Thus, the complexity equals the maximum number of independent "pieces" of information contained in the description A.

Notes and References

§1. The quote involving the need for a philosophical attitude toward system modeling is taken from

Rosen, R., *Anticipatory Systems*, Pergamon, London, 1985.

Part of the thesis of this book is that there is a genuine need for developing an actual **theory** of models, employing a metalanguage that goes beyond the level of the phenomenon itself. It's just not possible to speak about the relationships between competing models of a given situation by using terms appropriate to the level of the model; we must go beyond the model to a meta-level in which the language is constructed to speak about issues involving properties of models, not properties of systems being represented by models. This idea has been a central theme in the philosophy of science for some time, and is well-summarized by the following statement of Wittgenstein's taken from the penultimate section of the *Tractatus*:

> My propositions serve as elucidations in the following way: anyone who understands me eventually recognizes them as nonsensical, when he has used them—as steps—to climb up beyond them. (He must, so to speak, throw away the ladder after he has climbed up it.) He must transcend these propositions, and then he will see the world aright.

For an account of these matters, see

Oldroyd, D., *The Arch of Knowledge,* Methuen, London, 1986,

Mosedale, F., ed., *Philosophy and Science,* Prentice-Hall, Englewood Cliffs, NJ, 1979,

Putnam, H., "Models and Reality," *J. Symbolic Logic,* 45 (1980), 464–482.

§2. The idea of linkage was introduced by Rosen to serve as a generalized version of the more traditional notion of the dependency of variables. The difference is that the linkage concept allows us to speak of various degrees of dependence, whereas the classical idea is an all-or-nothing concept. For a detailed account of the linkage concept and its use in describing the dynamics of interaction between systems, see

Rosen, R. *Fundamentals of Measurement and Representation of Natural Systems,* Elsevier, New York, 1978.

From the time of Newton, the concept of a system "state" has always been a somewhat murky affair, and it's only in recent years that the real role of the system state-space has been clarified. The modern view of the state is that it is purely a mathematical construction introduced to mediate between the system inputs and outputs, and that it is a mistake to attach any intrinsic system-theoretic significance to the notion of state. In particular, it is only a coincidence, and of no special consequence, that on occasion we can give a physical interpretation to the state. What's important is the physically observed quantities, the inputs and outputs, not the states. This point is made clear by the triangle example of the text, which was used originally in

Krohn, K., J. Rhodes and R. Langer, "Transformations, Semigroups and Metabolism," in *System Theory and Biology,* M. Mesarovic, ed., Springer, New York, 1968.

§3. Problems of causation have bedeviled philosophers from the time of Aristotle, who elucidated the four causal categories (material, efficient, formal, final) in his *Metaphysics.* The basic problem of causation as currently seen by philosophers is to understand whether there is any necessary connection between a cause and its effect. As noted in the text, this question is sometimes confused with the issue of determinism which, philosophically speaking, involves issues pertaining to free will and really has nothing to do with the much more restricted notion of causal connections. Good introductory accounts of the problems of causation are found in

Taylor, R., "Causation," in *The Encyclopedia of Philosophy,* P. Edwards, ed., Macmillan, New York, 1967,

Popper, K., *The Logic of Scientific Discovery,* Hutchinson, London, 1959,

Troxell, E., and W. Snyder, *Making Sense of Things: An Invitation to Philosophy,* St. Martin's Press, New York, 1976.

For an interesting account of the causality/determinism issue in the light of modern physics and computing, see

Manthey, M., "Non-Determinism Can be Causal," *Int. J. Theoretical Physics,* 23 (1984), 929–940.

The existence of a deterministic relationship between system observables also introduces the question of whether or not every such relationship constitutes a "law" of Nature, or whether there are additional conditions that must be imposed before we would dignify such a relationship by calling it a "law." A spectrum of treatments of this issue is given by

Ayer, A., "What is a Law of Nature?", in *The Concept of a Person,* A. Ayer, ed., London, 1963,

Lakatos, I., *Mathematics, Science and Epistemology,* Cambridge U. Press, Cambridge, 1978,

Eddington, A., *The Nature of the Physical World,* U. of Michigan Press, Ann Arbor, 1958,

Casti, J., "System Similarities and the Existence of Natural Laws," in *Differential Geometry, Topology, Geometry and Related Fields,* G. Rassias, ed., Teubner, Leipzig, 1985, pp. 51–74.

§4. The classic work outlining how biological species can be transformed into one another by smooth coordinate changes is

Thompson, d'Arcy, *On Growth and Form,* Cambridge U. Press, Cambridge, 1942, abridged ed., 1971.

See also

Rosen, R., "Dynamical Similarity and the Theory of Biological Transformations," *Bull. Math. Biology,* 40 (1978), 549–579.

It is impossible to overemphasize the importance of the concept of system equivalence in addressing basic questions of modeling. In this regard, the quantities that remain invariant under coordinate changes are the only aspects of the system that have any right to be termed *intrinsic* system-theoretic properties, and we should ideally only make use of the invariants in answering questions about the system. A formal account of some of these matters is found in

Kalman, R., "Identifiablity and Problems of Model Selection in Econometrics," *Fourth World Congress of the Econometric Society,* France, August 1980.

§5. The term "bifurcation" is used in several different ways in mathematics. In particular, in the theory of ordinary differential equations it's usually taken to mean the value of a parameter at which the solution to a set of ordinary differential equations becomes multiple-valued. Although this usage is consistent with that employed in the text, our interpretation of the term is of much broader currency, incorporating other types of "discontinuous" changes in system behavior as well.

The example for stellar stability is taken from

Thompson, J. M. T., *Instabilities and Catastrophes in Science and Engineering,* Wiley, Chichester, 1982.

The treatment of the so-called "nonelementary" catastrophes quickly leads one into the deep waters of modern singularity and bifurcation theory. For an account of some of this work, see Chapter 17 of

Gilmore, R., *Catastrophe Theory for Scientists and Engineers,* Wiley, New York, 1981,

as well as the work that sparked off the interest in catastrophe theory,

Thom, R., *Structural Stability and Morphogenesis,* Benjamin, Reading, MA, 1975.

§6. An introductory account of the "relativistic" view of system complexity is given in the articles

Casti, J., "On System Complexity: Identification, Measurement and Management," in *Complexity, Language and Life: Mathematical Approaches,* J. Casti and A. Karlqvist, eds., Springer, Berlin, 1986,

Rosen, R., "Information and Complexity," *op. cit.*

A variety of approaches to measuring system complexity, assuming that the complexity is an intrinsic property of the system independent of any other system, are described in

Casti, J., *Connectivity, Complexity and Catastrophe in Large-Scale Systems,* Wiley, New York, 1979.

The question of simple vs. complex has been one that has troubled system thinkers and philosophers for a number of years. A somewhat eclectic list giving a flavor of these musings is provided in

Quine, W. v. O., "On Simple Theories of a Complex World," in *Form and Strategy,* J. Gregg and F. Harris, eds., Reidel, Dordrecht, 1964,

Simon, H. A., "The Architecture of Complexity," in *Sciences of the Artificial,* MIT Press, Cambridge, 1969,

Beer, S., "Managing Modern Complexity," *Futures,* 2 (1970), 245–257,

Ashby, W. R., "Some Peculiarities of Complex Systems," *Cybernetic Medicine,* 9 (1973), 1–8,

Gottinger, H., *Coping with Complexity,* Reidel, Dordrecht, 1983,

Warsh, D., *The Idea of Economic Complexity,* Viking, New York, 1984.

§7. It can safely be asserted that the problem of identifying and controlling roundoff error is one of the principal pillars upon which much of modern numerical analysis is based. A good treatment of many of the problems encountered (and some of the remedies) can be found in the classic work

Wilkinson, J., *The Algebraic Eigenvalue Problem,* Oxford U. Press, Oxford, 1965.

The issue of "surprise" as a system property ultimately centers upon the discrepancy between one's model of reality and reality itself. In turn, this gap can be represented by those observables with which the system Σ is in interaction, but which are not accounted for in our description of Σ. In short, surprise comes about from interactions to which the system is open, but to which the model is closed. The traditional way to account for this type of uncertainty is by classical probability theory, assuming the unmodeled interactions to be represented by random variables having appropriate distribution functions. Such an approach has come under increasing attack, primarily because it is based upon the dubious assumption that randomness and uncertainty are somehow the same thing, or at least we can represent all types of uncertainty by various types of randomness. To address this difficulty, a variety of alternatives to probability theory have been advanced, each of which purports to represent uncertainty in a nonprobabilistic fashion. Some of these ideas are found in

Zadeh, L., "The Concept of a Linguistic Variable and its Application to Approximate Reasoning—I, II, III," *Information Sciences,* 8, 9 (1975), 199–249, 301–357, 43–80,

Kickert, W. J. M., *Fuzzy Theories on Decision-Making,* Nijhoff, Leiden, 1978,

Klir, G. J., and T. A. Folger, *Fuzzy Sets, Uncertainty and Information,* Prentice-Hall, Englewood Cliffs, NJ, 1988,

Klir, G. J., and E. C. Way, "Reconstructability Analysis: Aims, Results, Open Problems," *Systems Research,* 2 (1985), 141–163.

§8. An outstanding introduction to the idea of a formal system, Gödel's Theorem, the mechanization of thought (and much more) is the popular book

Hofstadter, D., *Gödel, Escher, Bach: An Eternal Golden Braid,* Basic Books, New York, 1979.

A somewhat more technical account of formal systems, but still at the university textbook level, is

Davis, M. and E. Weyuker, *Computability, Complexity and Languages,* Academic Press, Orlando, 1983.

There is a very deep and perplexing connection between the idea of a formal system and the problem of the mechanization of human thought processes. This question has a very extensive literature, much of which is summarized in the Hofstadter book cited above. The original paper in this area, and still required reading for anyone even remotely interested in the topic, is

Turing, A., "Computing Machinery and Intelligence," *Mind,* 59 (1950), (reprinted in *The Mind's I,* D. Hofstadter and D. Dennett, eds., Basic Books, New York, 1981, pp. 53–67).

Following Turing's paper, a virtual torrent of work has appeared in the scientific as well as philosophical literature arguing the pros and cons of the "can machines think?" question. A representative sampling of these arguments is provided in the following collection:

A. R. Anderson, ed., *Minds and Machines,* Prentice-Hall, Englewood Cliffs, NJ, 1964,

Gunderson, K., *Mentality and Machines,* 2d ed., U. of Minnesota Press, Minneapolis, 1985,

Webb, J., *Mechanism, Mentalism and Metamathematics,* Reidel, Dordrecht, 1980,

C. Hookway, ed., *Minds, Machines and Evolution,* Cambridge U. Press, Cambridge, 1984.

§9. Global modeling as a growth industry got its start with the famous works

Forrester, J. W., *World Dynamics,* Wright-Allen Press, Cambridge, MA, 1971,

Meadows, D., D. Meadows, J. Randers, and W. Behrens, *The Limits to Growth,* Universe Books, New York, 1972.

An excellent summary of the many global models that have been developed through the years, their strong and weak points, uses in policymaking and future prospects is provided in

Meadows, D., J. Richardson, and G. Bruckmann, *Groping in the Dark,* Wiley, Chichester, 1982.

From a system-theoretic perspective, the pathological instabilities inherent in the original Forrester-Meadows exercise was first pointed out in the paper

Vermuelen, P. J., and de Jongh, D. C. J., "Growth in a Finite World—A Comprehensive Sensitivity Analysis," *Automatica,* 13 (1977), 77–84.

Beginning with the work of Leontief on input/output models of national economies, there has been an ever-increasing interest in the problem of identification of input/output coefficients using tools and techniques from linear system theory and statistics. The ideas underlying these efforts are presented in the classic works

Gale, D., *The Theory of Linear Economic Models,* McGraw-Hill, New York, 1960,

Baumol, W., *Economic Theory and Operations Analysis,* 2d ed., Prentice-Hall, Englewood Cliffs, NJ, 1965.

Unfortunately, the past two decades of work on parameter identification in econometric models has not resulted in a completely satisfactory state of affairs. Some of the reasons why are detailed in

Kalman, R., "Identification from Real Data," in *Current Developments in the Interface: Economics, Econometrics, Mathematics,* M. Hazewinkel and A. H. G. Rinnooy Kan, eds., Reidel, Dordrecht, 1982, pp. 161–196,

Kalman, R., "Dynamic Econometric Models: A System-Theoretic Critique," in *New Quantitative Techniques for Economic Analysis,* G. Szegö, ed., Academic Press, New York, 1982, pp. 19–28.

The uses of modern tools from differential geometry, such as fiber bundles, in physics is of relatively recent vintage. For an account of how these ideas are employed, especially in the context of renormalization theory, see

Hermann, R., *Yang-Mills, Kaluza-Klein, and the Einstein Program,* Math Sci Press, Brookline, MA, 1978.

Figure 1.5 shows clearly the separation between the real world of natural systems on the left side of the diagram, and the purely imaginary world of mathematics on the right. In our view, the natural home of the system scientist, as opposed to the natural scientist or mathematician, is as the keeper of the encoding and decoding operations $\mathcal{E}$ and $\mathcal{D}$. Given the way in which universities and research institutes are administratively organized along traditional disciplinary lines, lines that place primary emphasis upon the two sides of Fig. 1.5, it's no surprise that the system scientist finds it difficult to find a natural professional home in modern institutions of higher learning. To function effectively, the system scientist must know a considerable amount about the natural world **and** about mathematics, without being an expert in either field. This is clearly a prescription for career disaster in today's world of specialization, as it almost invariably ensures comments like "this is interesting work, but it's not really mathematics," or "this is an intriguing idea, but it ignores the experimental literature." At the same time, we see an ever increasing move toward inter- and trans-disciplinary attacks upon problems in the real world, resulting in the emergence of fields such as the cognitive sciences, sociobiology, computer science, and other fields that cannot be comfortably encapsulated within traditional disciplinary boundaries. Our feeling is that the system scientist has a central role to play in this new order, and that role is made clear by the diagram presented in Fig. 1.5: to understand the ways and means of how to encode the natural world into "good" formal structures, and then to see how to use these structures to interpret the mathematics in terms of the questions of interest to the experimental scientist. The sooner this role is explicitly recognized by the international academic establishment, the sooner system science will be able to take its rightful place in the coming intellectual order.

DQ #3. The thesis of *reductionism* so pervades every area of science as the tool of choice for investigation of the properties of systems that it's of interest to examine the reductionist program in somewhat greater detail. Basically, the reductionist creed is based upon the following principles:

A. There exists a *universal* way $\mathcal{U}$ of encoding any natural system Σ, and every other encoding of Σ is equivalent to $\mathcal{U}$.

B. The encoding $\mathcal{U}$ is canonically determined from an appropriate series of *fractions* (proper subsystems) of Σ.

C. The fractions of Σ may be isolated from each other (i.e., be made independent) by the imposition of appropriate dynamics on Σ.

Thus we see that reductionism is really an *algorithm,* i.e., a prescription for finding the universal encoding $\mathcal{U}$ by determining the fractions of Σ using imposed dynamics. Needless to say, no such algorithm has ever been offered, and it's difficult to understand the degree of tenacity with which the scientific community clings to the fiction of the reductionist program in the face of the obvious impossibility of ever successfully seeing it through. A simple counterexample to the possibility of carrying out the steps outlined above is provided by the results of Problem 2 of this chapter, in which we saw that, in general, we can know the properties of individual subsystems characterized by the observables f and g, but not be able to predict from this knowledge what the composite system $h = g \circ f$ will be doing. Another way of seeing the difficulties involved is to note that the imposition of the dynamics on Σ involves a breaking of linkage relationships among the observables of Σ. These relationships are broken in order to simplifying the *state* description of the system, i.e., to "decompose" the original linkage relationships between the states. Unfortunately (for the reductionist), an operation geared to simplify state descriptions will usually do drastic and terrible things to the tangent vectors associated with the original system dynamics. This inability to simultaneously simplify *both* the state and velocity descriptions of Σ is the main reason why reductionism fails as a universal scheme for the analysis of natural systems. For a fuller account of these matters, see the Rosen book cited under §2 above.

CHAPTER TWO

PATTERN AND THE EMERGENCE OF LIVING FORMS: CELLULAR AUTOMATA AND DISCRETE DYNAMICS

1. *Discrete Dynamical Systems*

From the time of Heraclitus and the ancient Greeks, philosophers have contemplated the thesis that "all things are in flux." This is our leitmotif as well, and in this book we will encounter the mathematical manifestation of this concept in a variety of forms. All of these forms, however, can be subsumed under the general heading *dynamical system.* So in this chapter we examine the principal components of this most important of mathematical "gadgets," and explore in some detail the simplest class of dynamical systems—cellular automata. As we shall see, however, even this simplest type of dynamical process provides an astonishing variety of behavioral patterns and serves in some ways as a universal representative for other types of dynamical phenomena. We return to this point later, but for now let's look at the general features needed to mathematically characterize a dynamical system.

The essence of the idea of dynamics is the notion of a *change of state.* Assume that we have a set of abstract states X, as discussed in the previous chapter. Then a *dynamic* on X is simply a rule $T_t\colon X \to X$, specifying how a given state $x \in X$ transforms to another state $T_t(x)$ at time t. For example, if the set $X = \{\text{ON, OFF}\}$, and the rule T_t is that the switch should be turned to the ON position every other hour starting at midnight, and to the OFF position on every odd-numbered hour, then the dynamic defined on X would have the form

$$T_t(\text{ON}) = \begin{cases} \text{ON}, & t = \text{even-numbered hour} \\ \text{OFF}, & t = \text{odd-numbered hour}, \quad t = 0, 1, 2, \ldots 23, \end{cases}$$

and similarly for the situation when the system is in the state OFF. This simple rule defines a *flow,* or change of state, on the set X.

The first point to note about the definition of the dynamic T_t is that both the state set X and the time-set can be either continuous or discrete. In the simple example of the light switch they are both taken to be discrete, but this need not necessarily be the case. In fact, we most often encounter the situation when one or both of these sets is continuous. But for the remainder of this chapter we will usually assume that both the time-set and the state-set are discrete, although not necessarily finite. Thus, typical state-sets for us will be $X = \mathbf{Z}_+^n$, $n-$tuples of nonnegative integers, or X a finite subset of $\mathbf{Z}_+^n$. Our time-set in this chapter will always be the nonnegative integers $t = 0, 1, 2, \dots$.

It's customary to express the transformations $\{T_t\}$ in component form by the dynamical law

$$x_{t+1} = \mathcal{F}(x_t, t), \qquad x_0 = x_0, \qquad x_t \in X, \tag{\dagger}$$

indicating that the state at time $t + 1$ depends upon the previous state x_t, as well as upon the current time t itself. Sometimes there will also be dependencies upon past states $x_{t-1}, x_{t-2}, \dots$, but we will generally omit consideration of these situations in this chapter. As a typical example of this set-up, consider the case when $X = \mathbf{Z}_+^5 \mod 2$, i.e., each element of X is a 5-tuple whose entries are 0 or 1. Let x_t^i represent the value of the ith component of x_t at time t, and define the transition rule to be

$$x_{t+1}^i = \left(x_t^{i+1} + x_t^{i-1}\right) \mod 2.$$

In other words, the value at $t + 1$ is just the sum of the values on either side of x^i taken "mod 2." Here to avoid boundary condition difficulties, we identify the state components "mod 5," meaning that we think of the component x^6 as the same as x^1, x^7 as x^2, and so forth. We will see later that the behavior of even such a simple-looking dynamic conceals a surprising depth of structure leading to long-term behavior of a bewildering degree of complexity from initial states having as little as a single nonzero entry.

Given the dynamical rule of state-transition $\mathcal{F}$ and the initial state $x_0 \in X$, there is really only one fundamental question that we can ask about the system: what happens to the state x_t as $t \to \infty$? Of course, there are many subquestions lurking below the surface of this overarching problem, questions about what kind of patterns emerge in the long-term limit, how fast these limiting patterns are approached, the types of initial states giving rise to different classes of limiting patterns and so on, but all issues of this sort ultimately come down to the question of whether or not certain types of limiting behaviors are possible under the rule $\mathcal{F}$ from the initial state x_0.

Example: Linear Dynamics

Consider the simple linear system

$$x_{t+1} = a x_t, \quad x_0 = x_0, \quad x_t, a \in R.$$

We consider the long-term behavior of the state x_t as $t \to \infty$. It's evident that at any time t, we have $x_t = a^t x_0$. Thus, there are only three qualitatively different types of limiting behavior:

$$x_\infty = \begin{cases} 0, & \text{for } |a| < 1, \\ \infty, & \text{for } |a| > 1, \\ \pm x_0, & \text{for } |a| = 1. \end{cases}$$

Furthermore, the rate at which these limiting behaviors are approached depends upon the magnitude of a : the closer a is to 1 in absolute value, the slower the corresponding limit is reached. In addition, it's easy to see that the limiting behavior is reached monotonically, i.e., $|x_t - x_\infty| > |x_{t+1} - x_\infty|$ for all $t \geq 0$. Finally, we see that the limiting behavior is independent of the initial state x_0, except in the special case $|a| = 1$. This situation is exceptional and is a consequence of the system's linear structure. Usually there will be several different types of limiting behaviors, and which one the system displays will depend critically upon the particular starting state x_0. Nevertheless, the foregoing example, elementary as it is, displays in a very transparent form the type of information that we want to know for all such processes: what do they do as $t \to \infty$, how fast do they do it, and in what way does the limiting behavior depend on the starting point and the nature of the dynamical rule $\mathcal{F}$? These are the issues that we will explore in the remainder of this chapter within the context of Eq. (†).

Example: Rabbit Breeding and the Fibonacci Sequence

To illustrate a situation where the change of state depends not only upon the past state but also upon the state at earlier times, consider the famous problem of rabbit breeding. Assume that at time $t = 0$ we have a pair of rabbits (one male, one female) that breeds a second pair of rabbits in the next period, and thereafter these produce another pair in each period and each pair of rabbits produces another pair in the second period following birth and thereafter one pair per period. The problem is to find the number of pairs at the end of any given period.

With a little calculation, we easily see that if u_i represents the number of pairs at the end of period i, then the following recurrence relation governs the rabbit population,

$$u_i = u_{i-1} + u_{i-2}, \quad i \geq 2,$$

with the initial condition $u_0 = u_1 = 1$. This is a discrete dynamical system in which the state transition rule depends not only upon the immediate past state, but also upon the state one period earlier than the immediate past one. It's also easy to see that the limiting behavior of this system is the well-known population explosion of rabbits, $u_i \to \infty$.

If we compute the sequence $\{u_i\}$, we find the values $1, 1, 2, 3, 5, 8, 13, \ldots$, the well-known sequence of Fibonacci numbers. As we shall see throughout this book, the Fibonacci sequence crops up in a surprising array of settings ranging from plant phyllotaxis to cryptographic codes.

2. *Cellular Automata: The Basics*

Imagine an infinite checkerboard in which each square can assume the color black or white at each moment of time t. Further, assume that we have a rule that specifies the color of each square as a function of its four adjacent squares: above, below, left and right. Suppose that an initial pattern of black and white squares is given at time $t=0$, and we then let the rule of transition operate for all future times and examine the pattern that emerges as $t \to \infty$. This set-up describes the prototype for what is termed a *cellular automaton* or, more accurately, a two-dimensional cellular automaton. Let's look at the characteristic features of such an object.

- *Cellular State-Space*—The "backcloth" upon which the dynamics of the automaton unfold is a cellular grid of some kind, usually a rectangular partitioning of R^n. Our interest will focus upon the one- and two-dimensional cellular automata for which the grids forming the state-spaces will be either the checkerboard structure described above, or its one-dimensional counterpart consisting of lattice points ("nodes") on the infinite line R^1.

- *Finite States*—Each cell of the state-space can assume only a finite number of different values k. So if we have a finite grid of N cells, then the total number of possible states is also finite and equals k^N. Since the cellular grid is assumed to be at most countable, the total number of states is also countable, being equal to $k^{\aleph_0}$, where $\aleph_0$ represents the cardinality of the integers $\mathbb{Z}$.

- *Deterministic*—The rule of transition for any cell is a deterministic function of the current state of that cell and the cells in some local neighborhood of that cell. On occasion we will modify this feature and allow stochastic transitions, but for the most part we consider only deterministic transition rules.

- *Homogeneity*—Each cell of the system is the same as any other cell in that the cells can each take on exactly the same set of k possible values at any moment, and the same transition rule is applied to each cell.

- *Locality*—The state-transitions are local in both space and time. The next state of a given cell depends only upon the state of that cell and those in a local neighborhood at the previous time period. Thus we have no time-lag effects, nor do we have any nonlocal interactions affecting the state transition of any cell. For one-dimensional automata, the local neighborhood consists of a finite number of cells on either side of a given cell; for two-dimensional automata, there are two basic neighborhoods of interest: the *von Neumann neighborhood,* consisting of those cells vertically and horizontally immediately adjacent to the given cell and the *Moore neighborhood,* which also includes those cells that are diagonally immediately adjacent. These neighborhoods are illustrated in Fig. 2.1.

Example: Racial Segregation

To illustrate the use of cellular automata theory in a social setting, consider the problem of racial integration of an urban housing neighborhood.

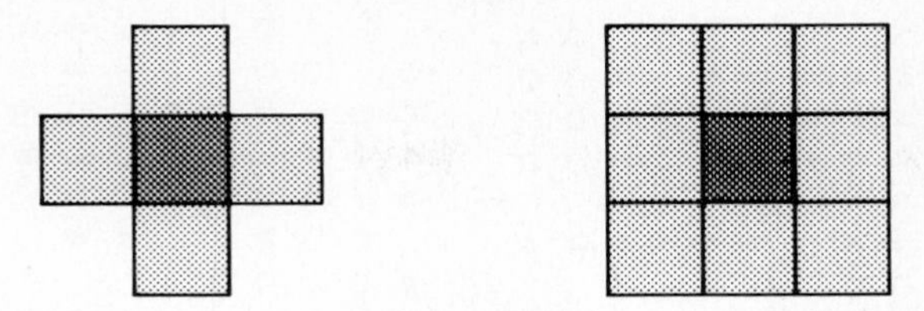

Figure 2.1 (a) Von Neumann Neighborhood, (b) Moore Neighborhood

Suppose we partition the urban region into a rectangular grid of size 16 by 13 cells, and assume that each cell represents a house in the region. Assume that each house can be occupied by a white or a black family, or is empty. Thus, here $k = 3$, and we have a total of $(16 \times 13)^3 = 8,998,912$ possible states of the region.

As the rule of transition, suppose we postulate that each racial group would prefer to have a certain percentage of its immediate neighbors being of the same group, and if this is not the case, then that party would move to the nearest grid location where the percentage of like neighbors was acceptable. In order to have a reasonable choice of where to move, it has been empirically observed that around 25%–30% of the housing locations should be vacant. If we use the Moore neighborhood and start with the initial state shown in Fig. 2.2 in which "o" denotes a white family, "#" denotes a black household and a blank space means the housing space is empty, we arrive at the steady-state distribution of Fig. 2.3(a) when we impose the condition that at least half of one's neighbors must be of the same color. Figure 2.3(b) shows the steady-state distribution from the same initial state when the requirement is that at least $\frac{1}{3}$ of one's neighbors must be of the same group.

```
o # # # # o o o o     # #   o o
o   # o o o   # #     #   # o
#   # o o # #   #     o #   # #
#     # #     o # # # o o
o   o o # # # #     #   #   o o
  # o #   o o #     o o     # #
# o o #         o o o # # #
o   # o   # #   # o o o       #
o   # o         # # o         #
o o       #     o # o o o o # #
  o # # o o o o   o # #   o # #
#   o # o #   o o # o # o   o
  o o     o # o # o o o     # #
```

Figure 2.2 Initial Housing Distribution

The results of the simulation of this vastly overly simplified version of urban housing patterns already suffice to illustrate some important features of cellular automata models. First of all, the rule of transition is one that has

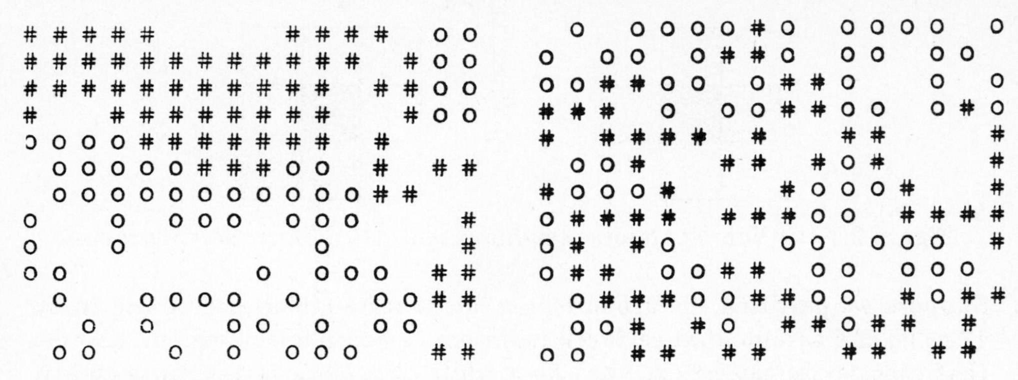

Figure 2.3 (a) $\frac{1}{2}$ of Neighbors the Same, (b) $\frac{1}{3}$ of Neighbors the Same

no special analytic structure: it is a linear threshold rule, with the threshold set by the parameter measuring the number of neighbors (in the Moore neighborhood) that are of the same color group. Furthermore, we see from Fig. 2.3 that the structure of the final distribution is critically dependent upon this parameter. Fig. 2.3(b) looks much like the initial distribution of Fig. 2.2, i.e., pretty unstructured. But Fig. 2.3(a) shows a considerable amount of structure emerging from the unstructured initial distribution. Thus, the passage from demanding $\frac{1}{3}$ of one's neighbors of the same color to $\frac{1}{2}$ induces a rather pronounced "phase transition" in the final housing pattern that emerges. We shall see more of this type of bifurcation behavior in other cellular automaton models later in the chapter.

Example: The Game of "Life"

Undoubtedly, the most famous and well-studied cellular automaton is the two-dimensional system described by J. H. Conway's board game "Life." In this metaphor for birth, growth, evolution and death, the state-space is taken to be an infinite checkerboard with each square capable of being either ON or OFF at each time period. Thus, for "Life" there are only $k = 2$ possible values for each cell at each period. The rule for state-transition is set up in order to mimic some basic features of living organisms. Life uses the Moore neighborhood to define the rules of transition which are explicitly given by:

A. If the number of ON neighbors is exactly 2, then the cell maintains its current state in the next period.

B. If the number of ON neighbors is exactly 3, then the cell will be ON in the next period, regardless of its current state.

C. If the number of ON neighbors is any other number, then the cell will be OFF in the next period.

There are no other rules. The rules basically say that the cell dies if it has either too few neighbors (isolation) or too many (crowding), but prospers if the number of neighbors is just right (Rules A and B). We should note that Life is not a game in the usual sense of the term, as the players make no decisions other than to decide what the initial configuration of ON and OFF cells will be. In a later section we will return to a detailed study of many of the features of this cellular automaton, so for now it's sufficient just to note that the types of questions of greatest interest in the Life universe center about the possibilities of self-reproducing patterns of various sorts.

3. *One-Dimensional Cellular Automata*

The Game of Life and the Racial Integration illustration of the preceding section are examples of two-dimensional cellular automata in which the system state-space has been the plane R^2 or a finite subset of it. But it turns out that even the simpler case of a one-dimensional automata already contains many of the characteristic features of the dynamical behavior of these objects, and in a setting in which it is far easier to analyze the behavior patterns that come about. So in this section we will focus our interest upon those cellular automata whose state-space is the infinite subset of the real line consisting of the points $\ldots, -3, -2, -1, 0, 1, 2, 3, \ldots$. Furthermore, we will make the situation even easier by assuming that each grid point can assume only $k = 2$ values, which we shall denote graphically by either a 0 or a blank space (OFF) or a 1 or the symbol $*$ (ON). Thus, if we let $a_t{}^i$ be the value of grid point i at time t, we can express the transition rule in the form

$$a_t{}^i = \mathcal{F}(a_{t-1}{}^{i\pm r}), \quad i = 0, \pm 1, \pm 2, \pm 3, \ldots; t = 0, 1, 2, \ldots\ ; r = 0, 1, 2, \ldots, R,$$

where $R \geq 0$ is an integer expressing the size of the neighborhood and $\mathcal{F}$ is the specific rule of transition. Usually we will take $R = 1$ or 2. Now let's consider the matter of the transition rule $\mathcal{F}$ in more detail for the case $k = 2$, $R = 1$.

Assume the rule of transition $\mathcal{F}$ is the so-called *mod 2* rule in which the value of the cell at time $t+1$ is just the sum of its two neighbors taken modulo 2, i.e., we have

$$a_{t+1}^i = (a_t^{i-1} + a_t^{i+1}) \mod 2,$$

for $i = 0, \pm 1, \pm 2, \ldots$. The local rule for the time evolution of this automaton is

$$\frac{111}{0} \quad \frac{110}{1} \quad \frac{101}{0} \quad \frac{100}{1} \quad \frac{011}{1} \quad \frac{010}{0} \quad \frac{001}{1} \quad \frac{000}{0}$$

In this diagram, the upper part shows one of the eight possible states that three adjacent cells can be in at time t, whereas the lower half shows the state that the central cell of the trio will assume at time $t+1$. We can compactly denote this rule by interpreting the eight binary digits composing the lower-half of the diagram as the binary representation of a decimal number and then call the rule by this number. Thus for the mod 2 rule illustrated in the diagram, we have the binary expression 01011010, which translates into the decimal number 90. Consequently, we call the mod 2 rule (Rule 90). It's clear that any eight binary digits defines a particular transition rule, so that there are a total of 2^8 possible rules when $k = 2$, $R = 1$. In general, there are $k^{k^{2R+1}}$ possible rules. The time evolution of the automaton is then obtained by simultaneous application of the given rule at each cell for each time-step. So, for example, using the preceding rule we obtain the following single-step transition from the initial state on the upper line to the next state given on the lower line:

$$\frac{0\,1\,0\,1\,1\,0\,1\,1\,0\,1\,0\,1\,0\,1\,1\,1\,0\,0\,0\,1\,0}{0\,0\,1\,1\,0\,1\,1\,0\,0\,0\,0\,0\,1\,0\,1\,1\,0\,1\,0}$$

Of the 256 theoretically possible local rules for a one-dimensional cellular automaton with $k = 2$, $R = 1$, we impose some restrictions on the set of rules in order to eliminate uninteresting cases from further examination. First, a rule will be considered "illegal" unless the zero-state 000 remains unchanged. Thus any rule whose binary specification ends in a 1 is forbidden. Second, we demand that rules be reflection symmetric, so that 100 and 001, as well as 110 and 011, yield identical values. Imposition of these restrictions leaves us with 32 legal rules of the form *abcdbed*0, where each letter can assume the value 0 or 1.

In Fig. 2.4 we show the evolution of some of the 32 legal rules, starting from an initial configuration consisting of a single nonzero cell. The evolution is shown until a particular configuration appears for the second time (a "cycle") or until 20 time-steps have been performed. We can see several classes of behavior in the above patterns. In one class, the initial 1 is erased (Rules 0 and 32) or maintained unchanged forever (Rules 4 and 36). Another class of behaviors copies the 1 to generate a uniform structure that expands by one cell in each direction on each time-step (Rules 50 and 122). These kinds of rules are termed *simple.* A third class of rules, termed *complex,* yields complicated, nontrivial patterns (e.g., Rules 18, 22 and 90).

The patterns of Fig. 2.4 were generated from a single nonzero initial cell. In Fig. 2.5 we show the results of the same experiment but now with an initial configuration selected randomly, with each cell independently having the state 0 or 1 with equal probability. Again we see a variety of behaviors with complex rules yielding complicated patterns. The most interesting feature of Fig. 2.5 is the fact that the independence of the initial cell states is totally destroyed with the dynamics generating correlations between values at separated cells. This is the phenomenon of *self-organization* in cellular

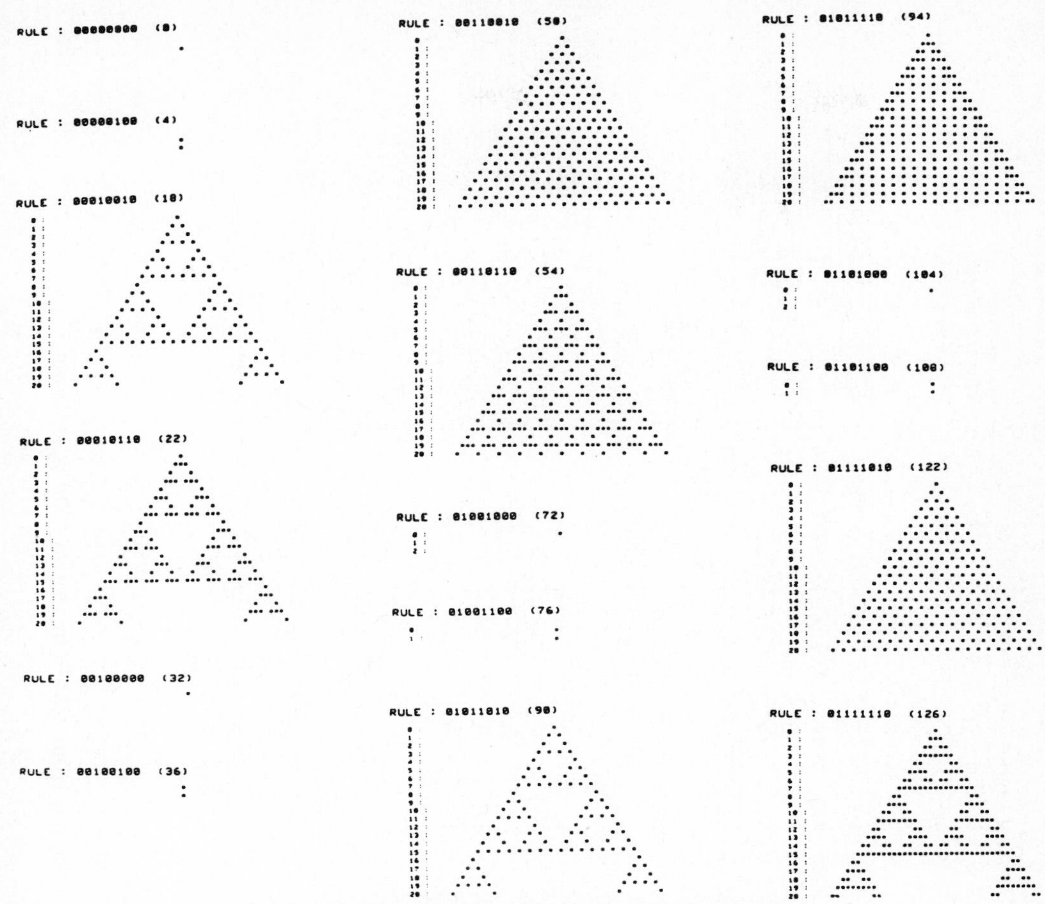

Figure 2.4 Evolution of Some Legal Rules

automata, in which an initially random state evolves to a state containing long-range correlations and nonlocal structure. The behavior of the automata of Fig. 2.5 is strongly reminiscent of the behavior of the dynamical systems that we will study in Chapter Four, with simple rules leading to steady-state behaviors consisting of fixed points or limit cycles, and complex rules giving rise to behavior analogous to more complicated types of "strange attractors."

In passing, it's worthwhile to note that the foregoing experiments were carried out under the assumption of periodic boundary conditions, i.e., instead of treating a genuinely infinite line of cells, the first and last cells are identified as if they lay on a circle of finite radius. It would have also been possible to impose null boundary conditions under which cells beyond each end are modified to maintain the value zero rather than evolve according to the local rule.

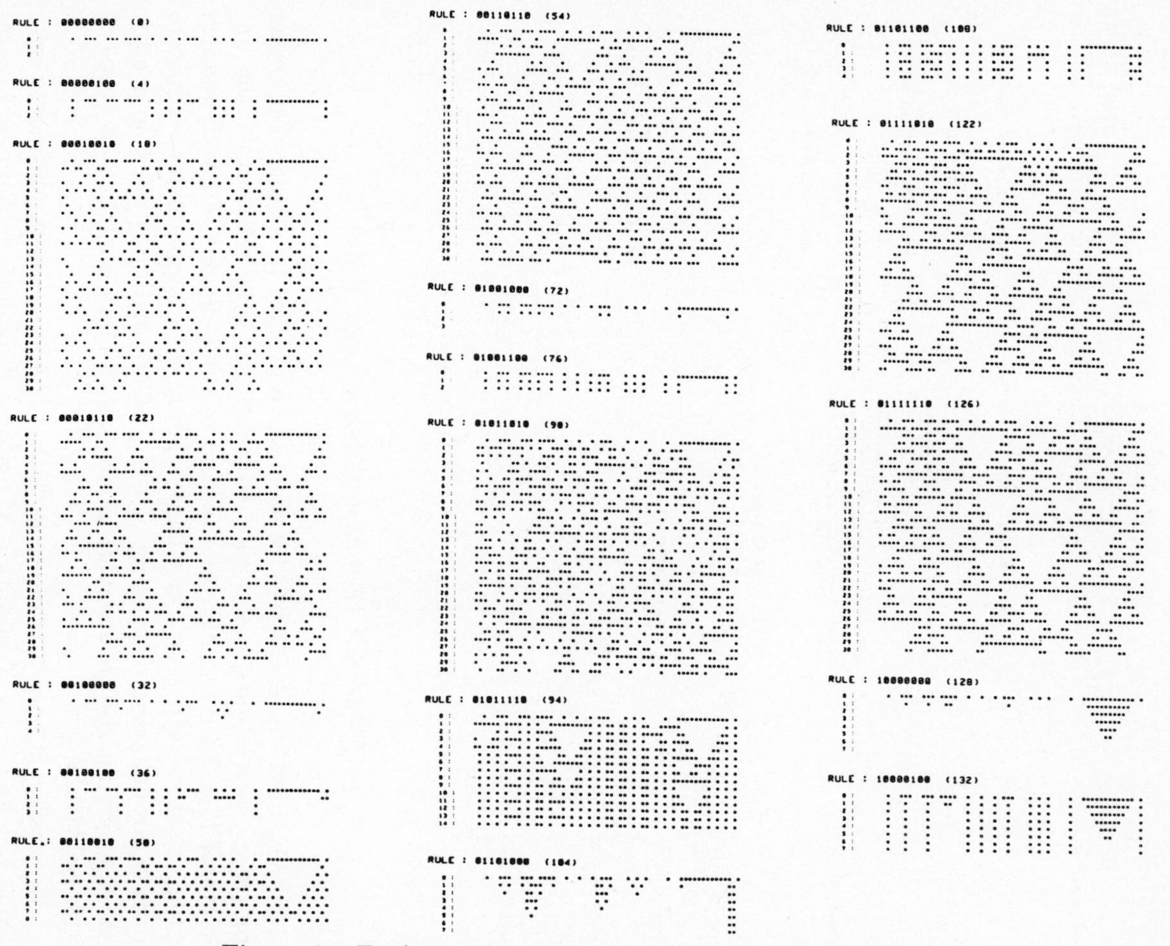

Figure 2.5 Evolution from Random Initial Configurations

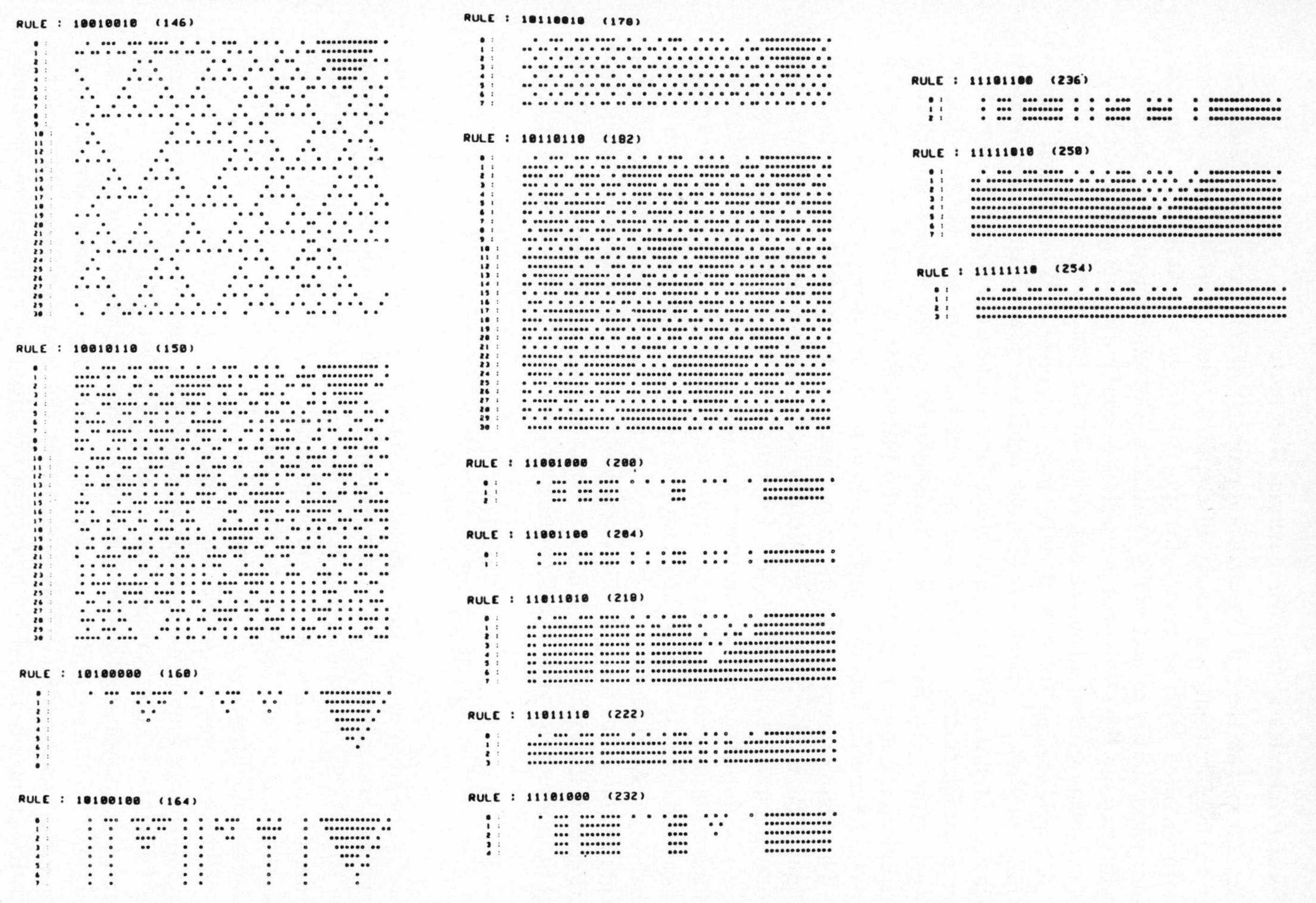

Figure 2.5(cont'd.) Evolution from Random Initial Configurations

4. *Local Properties of One-Dimensional Automata*

In Fig. 2.5 we saw the self-organization property of cellular automata emerge from the application of a variety of complex rules. Now we want to take a more detailed look at means for quantitatively characterizing the self-organization that is so pictorially evident in the experiments. In particular, we will examine the statistical properties of configurations generated by the time evolution of the automata described in the last section.

A configuration (or line of cells) may be considered disordered, or essentially random, if values of different cells are statistically uncorrelated. Deviations of statistical measures of configurations from their values for disordered configurations signify order and indicate the presence of correlations between values at different cells. A disordered configuration is specified by the single parameter p representing the probability of each cell taking the value 1. To specify ordered configurations requires more parameters.

The simplest statistical quantity that can be used to characterize a configuration is the average fraction of cells taking on the value 1. We denote this quantity by ρ. Clearly, for a disordered configuration $\rho = p$.

To begin with, consider the density ρ_1 obtained from a disordered initial configuration by evolution for one time-step. When $p = \rho = \frac{1}{2}$, a disordered configuration contains all eight possible three-cell neighborhoods with equal probability. Applying any rule, specified say by the binary sequence **R**, to this initial state for one time-step yields a configuration in which the fraction of cells with value 1 is given by the fraction of the eight possible neighborhoods that yield a 1 according to the rule **R**. This fraction is given by

$$\rho_1 = \frac{\#_1(\mathbf{R})}{(\#_0(\mathbf{R}) + \#_1(\mathbf{R}))} = \frac{\#_1(\mathbf{R})}{8},$$

where $\#_d(S)$ represents the number of occurrences of the digit d in the binary representation of S. So, for example, if we take cellular automaton Rule (01111110) = Rule 126 = **R**, we have $\#_1(\mathbf{R}) = 6$ and, consequently, the density $\rho_1 = 6/8 = 0.75$ if an infinite number of cells is included. The generalization of this formula to the case when $p \neq \frac{1}{2}$ is discussed in the Problems and Exercises at the end of the chapter.

When we come to consider the density ρ_τ for time-steps $\tau > 1$, we find that the simple argument used for ρ_1 cannot be used because correlations induced by the cellular automaton evolution interfere with the independence assumption underlying the derivation of ρ. However, exact results for ρ_τ can be obtained with the mod 2 rule (Rule 90), since this rule obeys the property of *additive superposition.* This means that the configurations obtained by evolution from any initial configuration are given by appropriate combinations of those for evolution from an initial configuration having only a single nonzero cell. Cellular automata satisfy additive superposition only if their rule is of the form $ab0cbac0$ with $c = (a + b) \mod 2$. Let's consider Rule 90 and calculate ρ_τ.

Let $N_\tau^{(1)}$ be the number of cells at time τ with value 1 obtained by evolution from an initial state containing a single nonzero cell. It can be shown that

$$N_\tau^{(1)} = 2^{\#_1(\tau)},$$

where $\#_1(\tau)$ is the number of ones in the binary representation of the integer τ. (Proof?)

The expression for $N_\tau^{(1)}$ shows that the density averaged over the region of nonzero cells (the "light cone") in the evolution according to Rule 90 is given by $\rho_\tau = N_\tau^{(1)}/(2\tau + 1)$, which does not tend to a definite limit as $\tau \to \infty$. However, the time-average density,

$$\bar{\rho}_T = \frac{1}{T}\sum_{\tau=0}^{T} \rho_\tau,$$

does tend to the expected limit 0 as T grows large, with the asymptotic behavior being like $T^{\log_2 3 - 2}$. In general, the value of a cell at time τ is a sum mod 2 of the initial values of $N_\tau^{(1)}$ cells, each of which have value 1 with probability ρ_0. If each of a set of k cells has value 1 with probability p, then the probability that the sum of the values of the cells will be odd (equal to 1 mod 2) is

$$\sum_{i \text{ odd}} \binom{k}{i} p^i (1-p)^{k-i} = \tfrac{1}{2}[1 - (1-2p)^k].$$

Hence, the density of cells with value 1 obtained by evolution for τ time-steps from an initial state with density ρ_0 using Rule 90 is given by

$$\rho_\tau = \tfrac{1}{2}[1 - (1 - 2\rho_0)^{2^{\#_1(\tau)}}].$$

This result is shown as a function of τ for the case $\rho_0 = 0.2$ for the Rules 90, 18 and 182 in Fig. 2.6. For large $\tau, \#_1(\tau) = O(\log_2 \tau)$ except at a set of points of measure zero, so that $\rho_\tau \to \frac{1}{2}$ as $\tau \to \infty$ for almost all τ. Note that in Fig. 2.6, for the nonadditive complex Rules 18 and 182, the values of cells at time τ depend on the values of $O(\tau)$ initial cells, and ρ_τ tends smoothly to a definite limit that is independent of the density of the initial disordered configuration.

The very essence of the idea of self-organization is that the evolution of a cellular automaton should generate correlations between cell values at different sites. The average density measure just discussed is much too coarse to give any information of this sort, so we consider the simplest type of correlation measure, the two-point correlation function

$$C^{(2)}(r) = \langle S(m)S(m+r)\rangle - \langle S(m)\rangle\langle S(m+r)\rangle,$$

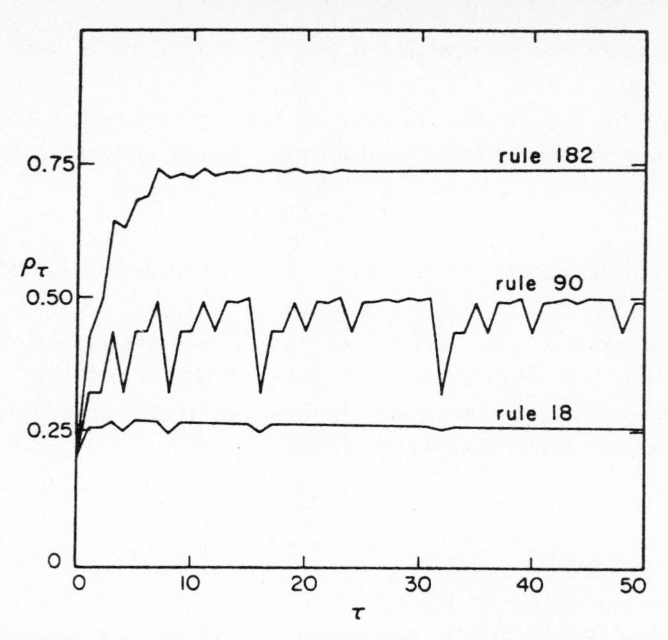

Figure 2.6 Average Density of Nonzero Cells as a Function of Time

where the average is taken over all possible positions m in the cellular automaton at a fixed time, with $S(k)$ taking on the values -1 and +1 when the cell at position k has values 0 and 1, respectively. As already noted, a disordered configuration gives $C^{(2)}(r) = 0$ for $r > 0$, with $C^{(2)}(0) = 1-(2\rho_0-1)^2$. With the single-cell initial states, the periodicities of the configurations that result from complex rules give rise to peaks in $C^{(2)}(r)$. At time-step t, the largest peaks occur when $r = 2^k$, with the digit corresponding to 2^k appearing in the binary decomposition of t. Smaller peaks occur when $r = 2^{k_1} \pm 2^{k_2}$, and so on. For the additive rules, the form of $C^{(2)}(r)$ is obtained by a convolution of the foregoing result with the correlation function for the particular initial configuration. Thus, with these rules we have $C^{(2)}(r) = 0$ for all disordered initial configurations. However, for nonadditive rules there exists the possibility of nonzero short-range correlations from disordered initial configurations. The existence of a nonzero correlation length is the first sign of the generation of order via the evolution of an automaton. In Fig. 2.7 we show the form of $C^{(2)}(r)$ for the complex Rule 18. Here we see that the correlation appears to fall off exponentially with a correlation length $\cong 2$.

The preceding measures of structure give some insight into the local patterns that may emerge during the evolution of the automaton's state. However, inspection of Figs. 2.4 and 2.5 shows that considerable additional structure is present that these measures are incapable of identifying. For example, the two-dimensional picture arising from the evolution of all legal rules displays many triangular figures of various sizes. These triangles are

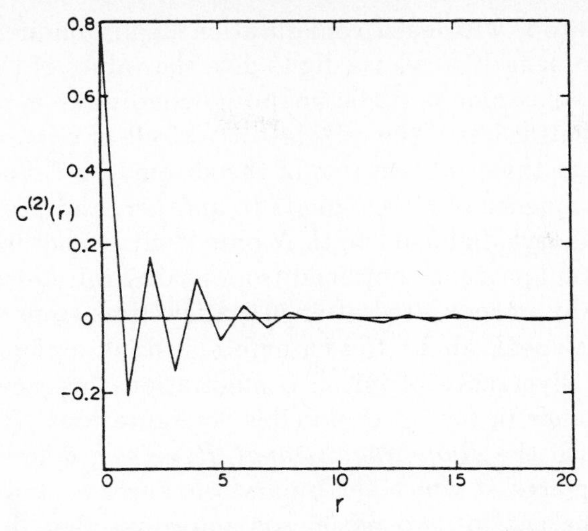

Figure 2.7 The Correlation Function $C^{(2)}(r)$ for Large t Using Rule 18

formed when a long sequence of cells, which suddenly all attain the same value 1, is progressively reduced in length by local "thermal" fluctuations. If we let $T(n)$ be the density of triangles of base length n, then with disordered initial states it can be shown that $T(n) \sim \lambda^{-n}$, where for additive rules $\lambda \sim 2$ and for nonadditive rules $\lambda \sim \frac{4}{3}$. Thus, the spectrum of triangles generated by complex rules is universal, independent of both the initial state as well as the precise rule being used. This result should be compared with the estimate $T(n) \sim n^{-\log_2 3} \approx n^{-1.59}$. Other measures like the sequence density can also be used to pinpoint additional local structure in the patterns emerging from the cellular automaton rules. We refer the reader to the Notes and References at the end of the chapter for further details of these tests and now turn our attention to problems of *global* structure.

5. *Global Properties of One-Dimensional Automata*

In the preceding section, we examined the behavior of cellular automata by considering the statistical properties of the set of values of cells in individual configurations. Now we want to take an alternate approach and consider the statistical properties of the set composed of all possible final configurations of an automaton. These are the configurations that the automaton reaches after the "transient" states have died out. This view of the the evolutionary behavior of an automaton allows us to make contact with issues in dynamical systems involving the structure of the attractor set for a dynamical process, as well as with certain questions centering upon self-organization that are in some sense complementary to those considered in the last section.

For the most part in this section, we shall consider "finite" automata having $N < \infty$ cells. Thus, there are a total of 2^N possible configurations for such an automaton, with each configuration being uniquely specified by a binary integer of length N whose digits give the values of the corresponding cells. Further, we assume periodic boundary conditions so that the first and last cells are identified as if the cells lay on a circle of circumference N. This set-up shows that the evolution rule of the automaton defines a transformation from one sequence of binary digits to another, and provides a mapping from the set of binary digits of length N onto itself. Experiments have shown that the final configurations obtained from nearby initial configurations are usually essentially uncorrelated after just a few time-steps.

In order to speak about the ensemble of final configurations and the convergence or divergence of initial configurations, we need to have some measure of distance in the set of possible configurations. It turns out to be convenient to use the *Hamming distance* $H(s_1, s_2)$, defined by measuring the number of places at which the binary sequences s_1 and s_2 differ. Let's first consider the case of two initial configurations that differ only by the value of a single cell, i.e., the Hamming distance between them is 1. After T time-steps, this initial difference may affect the values of at most $2T$ cells. However, for simple rules it is observed that the Hamming distance goes rapidly to a small constant value. The behavior of the Hamming distance for complex rules differs radically, depending upon whether the rule is additive or not. For additive rules (e.g., Rule 90 or Rule 150), the Hamming distance at time-step t is given by the number of nonzero cells in the configuration obtained by evolution from a single cell, and for Rule 90 has the form $H_t = 2^{\#_1(t)}$. For nonadditive rules, the difference between configurations obtained through evolution no longer depends just on the difference between the initial configurations. Generally, it is seen that in this case we have $H_t \sim t$, for large t. Thus, a bundle of initially close trajectories diverges with time into an exponentially expanding volume.

We can consider a statistical ensemble of states for a finite automaton by giving the probability for each of the 2^N possible configurations. A collection of states (configurations) such that each configuration appears with equal probability will be termed an "equiprobable ensemble." Such an ensemble may be thought of as representing the maximum degree of disorganization possible. Any automaton rule modifies the probabilities for states in an ensemble to occur, thereby generating organization. In Fig. 2.8 we show the probabilities for the 1024 possible configurations of a finite automaton with $N = 10$ obtained after 10 time-steps using Rule 126 from an initially equiprobable ensemble. The figure shows that the automaton evolution modifies the probabilities for the different configurations from the initially equally likely occurrence of all states to a final probability density that has some states' probabilities reduced to zero, while increasing some and decreasing others from their initial levels. Properties of the more probable configurations dominate the statistical averages over the ensemble, giving

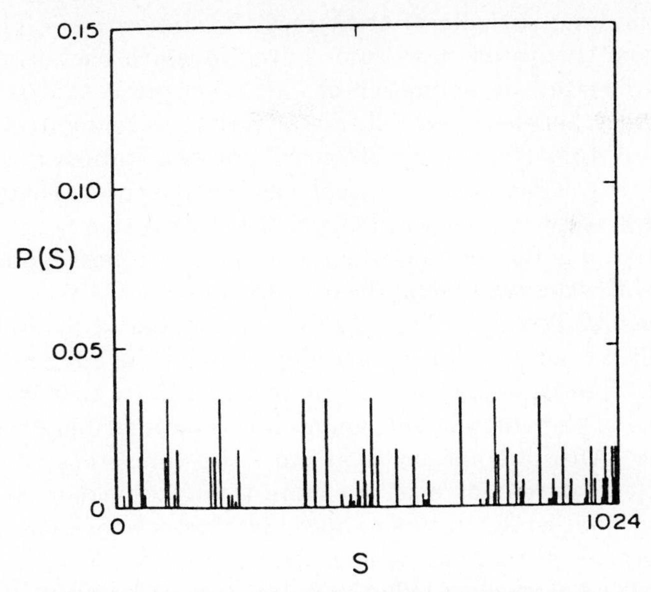

Figure 2.8 Configuration Probabilities with Rule 126

rise to the average local features of equilibrium configurations discussed in the last section.

It's of considerable importance in understanding the long-term behavior of cellular automata to examine the problem of *reversibility*. Most cellular automata rules can transform several different initial configurations into the same final configuration. Thus a particular configuration can have many ancestors, but only a single descendant, such as the trivial Rule 0. In a reversible system, each state has a unique ancestor and descendant so that the number of possible configurations must remain constant in time (Liouville's Theorem). But in an irreversible system the number of possible configurations may decrease with time, leading to the observation that final configurations of most cellular automata become concentrated in limited regions and do not fill the available volume of state-space uniformly and densely. This kind of behavior is exactly what leads to the possibility of self-organization by allowing some configurations to occur with larger probabilities than others, even in the steady-state limit. It turns out that as N becomes large, the fraction of possible configurations that can actually be reached by an irreversible rule tends to zero at the rate λ^N, where $\lambda \simeq 0.88$ for nonadditive rules. Such irreversible behavior can also be studied using ideas of entropy and information, but we leave a consideration of these notions to the chapter's Problems and Exercises.

For finite cellular automata with N cells, there are a total of only 2^N states meaning that the sequence of configurations reached from any initial state must become periodic after at most 2^N time-steps. Thus, after an initial transient the system must enter a cycle in which a set of configurations is cyclically repeated. Examination of Fig. 2.5 suggests that simple cellular automata give either short cycles involving just a few configurations or evolve after around N time-steps to a state consisting of a stationary configuration of cycle length 1. Complex rules yield much longer cycle lengths involving isolated independent regions, each of which follows a short cycle of length at most 2^p, where p is the number of regions. It is also possible that complex rules can yield cycles whose length increases unboundedly as N increases.

Experiments also show that of the total number of possible configurations 2^N, only a small fraction are actually visited from any particular initial configuration. For example, using Rule 126 with $N = 8$, it was found that at most 8 out of the possible 256 configurations are visited from any initial configuration. After a transient of at most two time-steps, the automaton enters a cycle that repeats after no more than six further steps. A total of 29 distinct final configurations appear in these cycles. For $N = 10$, the maximum number of configurations is 38, whereas for $N = 32$, the number is at least 1547. This type of behavior seems typical for most other complex rules.

Using additive rules, it's possible to give analytic results for transients and cycle lengths, as will be outlined in the next section. Using these techniques for Rule 90, if we let Π_N be the period obtained from evolution from a single nonzero cell, then it can be shown that

$$\Pi_N = 1, 1, 3, 2, 7, 1, 7, 6, 31, 4, 63, 14, 15, 1, 15, 14, 511, 12,$$

for $N = 3$ to 20. In connection with the length of the transient phase, if we let Φ_N be the length of the transient with a single nonzero initial cell, we find with Rule 90 that $\Phi_N = 1$ for N odd, and $\Phi_N = D(N)/2$ otherwise, where $D(j)$ is the largest power of 2 that divides j. Note that $\Phi_N = 1$ for N odd implies that in these cases exactly half of the 2^N possible configurations appear on cycles using Rule 90.

Finally, in passing, we should note that, in general, infinite cellular automata do not usually exhibit finite cycles except in exceptional circumstances. Any initial configuration with a finite number of nonzero cells either evolves to the zero state or yields a pattern whose size increases progressively with time. However, it can be shown that if the values of the initial cells form an infinite periodic sequence of period k, then the evolution of the infinite cellular automaton will be identical to that of a finite cellular automaton with $k = N$, and cycles with length much less than 2^k will be found.

We can summarize the foregoing discussion by observing that cellular automaton configurations can be divided into essentially three classes according to the circumstances under which they may be generated:

- *Gardens of Eden*—Some configurations can only appear as initial states of an automaton. These states can never be generated in the course of cellular automaton evolution.
- *Transient States*—These are configurations that can arise only within, say, the first t steps of an automaton's evolution. Such configurations leave no long-term "descendants."
- *Cyclic States*—Such configurations appear in cycles and are visited repeatedly. Cyclic configurations can be generated at any time-step and may be considered attractors to which any initial configuration ultimately leads.

For the most part, the discussion of the preceding sections has been based upon an empirical observation of computer-generated simulations of the legal rules as displayed in Figs. 2.4 and 2.5. Before looking at a few applications of these simplest of all cellular automata, we want to give some indication as to how *analytic* tools can be brought to bear upon the study of the behavior of cellular automata, at least in the case of additive rules like Rules 90 and 150.

6. *Algebraic Properties of Additive Cellular Automata*

We consider a one-dimensional cellular automaton consisting of N cells arranged in a circle (to give periodic boundary conditions). We denote the values taken on by the cells at time t by $a_0^{(t)}, a_1^{(t)}, \ldots, a_{N-1}^{(t)}$. As before, each $a_i^{(t)} = 0$ or 1.

The complete configuration of such an automaton can be represented by a *characteristic polynomial*

$$A^{(t)}(x) = \sum_{i=0}^{N-1} a_i^{(t)} x^i,$$

where the value of cell i is the coefficient of x^i, with all coefficients taking on the values 0 or 1.

It's convenient to consider generalized polynomials containing both positive and negative powers of x. These polynomials are usually termed *dipolynomials,* and possess the same divisibility and congruence properties as ordinary polynomials. It's clear from the above representation that multiplication of $A(x)$ by $\pm x^k$ results in a dipolynomial that represents a configuration obtained from $A(x)$ by shifting each cell k places to the left $(+x^k)$ or the right $(-x^k)$. Periodic boundary conditions are enforced by reducing the characteristic dipolynomial modulo the fixed polynomial $x^N - 1$ at every stage, i.e.,

$$\left(\sum a_i x^i\right) \mod (x^N - 1) = \sum_{i=0}^{N-1} \left(\sum_j a_{i+jN}\right) x^i.$$

For our mod 2 rule (Rule 90), we have the dynamics given by

$$a_i^{(t)} = \left(a_{i-1}^{(t-1)} + a_{i+1}^{(t-1)}\right) \quad \text{mod } 2.$$

The time evolution of this automaton is represented by multiplication of the characteristic polynomial specifying the initial state by the dipolynomial

$$\mathbf{T}(x) = x + x^{-1},$$

i.e., the state at time t emerging from the configuration $A(x)$ at time t–1, is given by

$$A^{(t)}(x) = \mathbf{T}(x)A^{(t-1)}(x) \quad \text{mod } (x^N - 1),$$

where all arithmetic is performed mod 2.

The foregoing representation shows that an initial configuration consisting of a single nonzero cell evolves after t time-steps to a configuration given by

$$\mathbf{T}(x)1 = (x + x^{-1})^t = \sum_{i=0}^{t} \binom{t}{i} x^{2i-t}.$$

For $t < N/2$, the region of nonzero cells grows linearly with time with the position of the nonzero cells given by $\pm 2^{j_1} \pm 2^{j_2} \pm \ldots$, where the j_i give the positions of nonzero digits in the binary decomposition of the integer t. The additive superposition property implies that patterns generated from initial states containing more than one nonzero cell may be obtained by addition mod 2 of the patterns generated from states with a single nonzero cell.

The first simple result that we can obtain from this machinery involves a characterization of "Garden-of-Eden" configurations. This is given in the following result:

LEMMA 2.1. *Configurations containing an odd number of cells with value 1 can never occur in the evolution of the cellular automaton defined by Rule 90, and can occur only as initial states.*

PROOF: Consider any initial configuration given by the polynomial $A^{(0)}(x)$. The next state of the system is given by the polynomial $A^{(1)}(x) = (x + x^{-1})A^{(0)}(x) \mod (x^N - 1)$. Thus,

$$A^{(1)}(x) = (x^2 + 1)B(x) + R(x)(x^N - 1)$$

for some dipolynomials $B(x)$ and $R(x)$. Since $x^2 + 1 = x^N - 1 = 0$ for $x = 1$, $A^{(1)}(1) = 0$. Thus $A^{(1)}(x)$ contains an even number of terms, and corresponds to a configuration with an even number of nonzero cells. Therefore, only such configurations can be reached from an arbitrary initial configuration $A^{(0)}(x)$.

The preceding lemma allows us to prove the following basic result governing the density of reachable configurations for the automaton of Rule 90.

THEOREM 2.1. *The fraction of the 2^N possible configurations of a cellular automaton with N cells evolving according to Rule 90 that can occur only as initial states is $\frac{1}{2}$ for N odd, and is $\frac{3}{4}$ for N even.*

PROOF:

Case A: N even. We have $(x^2 + 1) = (x + 1)^2 = (x - 1)^2$ where, as always, we work mod 2. Further, by number-theoretic results

$$(x - 1)^2 \,|\, (x^{N/2} - 1)^2 = (x^N - 1)$$

for N even, so we can conclude that

$$(x - 1)^2 | A^{(1)}(x),$$

where $A^{(1)}(x) = (x^2 + 1)B(x) + R(x)(x^N - 1)$, as in Lemma 2.1. But since $(x - 1)^2$ contains a constant term, $A^{(1)}(x)/(x - 1)^2$ is then an ordinary polynomial if $A^{(1)}(x)$ is also. Consequently, all reachable configurations represented by the polynomial $A^{(1)}(x)$ have the form

$$A^{(1)}(x) = (x - 1)^2 C(x)$$

for some polynomial $C(x)$. The predecessor of any such configuration is $xC(x)$, so any configuration of this form is in fact reachable. Since the $\deg A(x) < N$, $\deg C(x) < N - 2$. There are thus exactly 2^{N-2} reachable configurations, or $\frac{1}{4}$ of all the 2^N possible configurations.

Case B: N odd. Appealing to Lemma 2.1, the proof in this case reduces to showing that all configurations containing an even number of nonzero cells have predecessors. A configuration with an even number of nonzero cells can always be written in the form $(x + 1)D(x)$. However,

$$\begin{aligned} A^{(1)}(x) = (x + 1)D(x) &\equiv (x + x^{-1})(x^2 + \cdots + x^{N-1})D(x) \mod (x^N - 1) \\ &\equiv \mathbf{T}(x)(x^2 + x^4 + \cdots + x^{N-1})D(x) \mod (x^N - 1), \end{aligned}$$

giving an explicit predecessor for $A^{(1)}(x)$.

Now let's take a look at how the algebraic set-up enables us to speak precisely about the cyclic properties of the Rule 90 automaton.

To characterize the properties and density of cycles, we first need the following technical results:

LEMMA 2.2. *The lengths of all cycles in a cellular automaton of N cells evolving according to Rule 90 must divide the length Π_N of the cycle obtained with an initial configuration containing a single cell with a nonzero value.*

This result follows immediately from the additivity of the Rule 90, since any configuration is a superposition of configurations emerging from initial states with a single nonzero cell.

LEMMA 2.3. *For the Rule 90 cellular automaton, $\Pi_N = 1$ if N has the form $N = 2^j$ and $\Pi_N = 2\Pi_{N/2}$ if N is even but not of the form 2^j.*

The proof of this lemma is left as an exercise. However, note that the first part of Lemma 2.3 shows that when N has the form $N = 2^j$, then any initial configuration evolves ultimately to a fixed point consisting of all zeros, since

$$(x + x^{-1})^{2^j} 1 \equiv (x^{2^j} + x^{-2^j}) \equiv (x^N + x^{-N}) \equiv 0 \mod (x^N - 1).$$

The foregoing lemmas enable us to characterize cycle lengths for all automata having an even number of cells. The following result establishes the cyclic pattern when N is odd:

THEOREM 2.2. *For the Rule 90 with N odd,*

$$\Pi_N | \Pi_N^* = 2^{\mathrm{sord}_N(2)} - 1,$$

where $\mathrm{sord}_N(2)$ *is the multiplicative* suborder function, *defined as the least integer j such that* $2^j = \pm 1 \mod N$.

PROOF: By Lemma 2.1, an initial configuration containing a single nonzero cell cannot be reached at any time $t > 0$. Let $A^{(0)}(x)$ be such an initial configuration. Then we have

$$A^{(1)}(x) = (x + x^{-1})A^{(0)}(x) \mod (x^N - 1),$$

a configuration that in fact occurs again after $2^{\mathrm{sord}_N(2)} - 1$ time-steps, since

$$\begin{aligned} \mathbf{T}(x)^{2^{\mathrm{sord}_N(2)}} 1 \equiv (x + x^{-1})^{2^{\mathrm{sord}_N(2)}} &\equiv (x^{2^{\mathrm{sord}_N(2)}} + x^{-2^{\mathrm{sord}_N(2)}}) \\ &\equiv (x^{\pm 1} + x^{\mp 1}) \equiv (x + x^{-1}) \mod (x^N - 1) \end{aligned}$$

Using the above Lemmas 2.2, 2.3 and Theorem 2.2, one can compute directly the maximal cycle lengths that are given in Section 5 for the cases $N = 3$ to 20. Using similar arguments, it's also possible to determine the number and the length of the distinct cycles that emerge from Rule 90. We leave these matters, together with a consideration of extensions to more than two possible states for each cell, to the chapter's Problems and Exercises. Now we want to examine a few application areas in which one-dimensional cellular automata have proved to be useful.

7. *Languages: Formal and Natural*

Abstractly, human languages can be thought of as a set of rules for constructing sequences of symbols according to definite rules of composition (grammars) in order to convey semantic content (meaning) about some aspect of the world. It makes little difference in this view whether the symbols are actual marks on a piece of paper, sound patterns emitted by the vocal apparatus, or "bits" on a magnetic tape. It's a small leap from this vision of linguistic structure to the consideration of a cellular automaton as a metaphoric device for studying the abstract properties of languages, in general. The sequence of symbols that represent the state of a one-dimensional automaton at time t corresponds to one of the potentially infinite statements that can be made in a language, and the rule of state-transition in the automaton determines the allowable ways in which new, grammatically correct statements can be generated from an initial "seed." In this section, we will explore this interconnection between automata and languages as illustration of the power of even such a simple device as a one-dimensional automaton to shed light upon some of the deeper issues in human information processing.

To formalize the above considerations, we first need to define more precisely our notion of a finite automaton.

DEFINITION. *A finite automaton* $\mathcal{M}$ *is a composite consisting of a finite input alphabet* $A = \{a_1, \ldots, a_n\}$, *a finite set of states* $X = \{x_1, x_2, \ldots, x_m\}$, *and a state-transition function* $\delta\colon X \times A \to X$, *giving the next state in terms of the current state and current input symbol. In addition, there is a set* $F \subseteq X$, *termed the set of final or accepting states of the automaton.*

If $\mathcal{M}$ is an automaton that starts in state x_1, and if $u \in A^*$ is any finite input *string* formed as a sequence of elements from the alphabet A, then we will write $\delta^*(x_1, u)$ to represent the state that the automaton will enter if it begins in state x_1 and then moves across u from left to right, one symbol at a time, until the entire string has been processed. We then say that $\mathcal{M}$ accepts the word u if $\delta^*(x_1, u) \in F$, and $\mathcal{M}$ rejects the word u otherwise. Finally, the *language* $\mathrm{L}(\mathcal{M})$ accepted by $\mathcal{M}$ is the set of all words $u \in A^*$ that are accepted by $\mathcal{M}$. A language L is called *regular* if there exists a finite-state automaton that accepts it.

Example

Consider an automaton $\mathcal{M}$ with input alphabet $A = \{a_1, a_2\}$ and a state space given by $X = \{x_1, \ldots, x_4\}$, with the rule of transition being

$$\delta(x_1, a_1) = x_2, \quad \delta(x_1, a_2) = x_4, \quad \delta(x_2, a_1) = x_2, \quad \delta(x_2, a_2) = x_3,$$
$$\delta(x_3, a_1) = x_4, \quad \delta(x_3, a_2) = x_3, \quad \delta(x_4, a_1) = x_4, \quad \delta(x_4, a_2) = x_4.$$

Assume that the set of accepting states is $F = \{x_3\}$, with the initial state being x_1. It's now easily checked that the above automaton $\mathcal{M}$ will accept the input string $a_1a_1a_2a_2a_2$ but will reject the string of inputs $a_2a_1a_2a_1$, since in the first case the termination state is $x_3 \in F$, whereas for the second input string the system terminates in the state $x_4 \notin F$. It's easy to verify that this $\mathcal{M}$ accepts the language

$$L = \{a_1^{[n]} a_2^{[m]} : n, m > 0\}.$$

Thus, the above language L is a regular language.

It's sometimes convenient to represent the transition function δ of an automaton by a *directed graph,* in which the nodes of the graph represent the system states and the arcs represent the possible inputs that can be applied at each state. The direction of the arcs determines the next state that will be reached from a given state when the input on an arc emerging from that state is applied. Below we show the state-transition graph for the example given earlier:

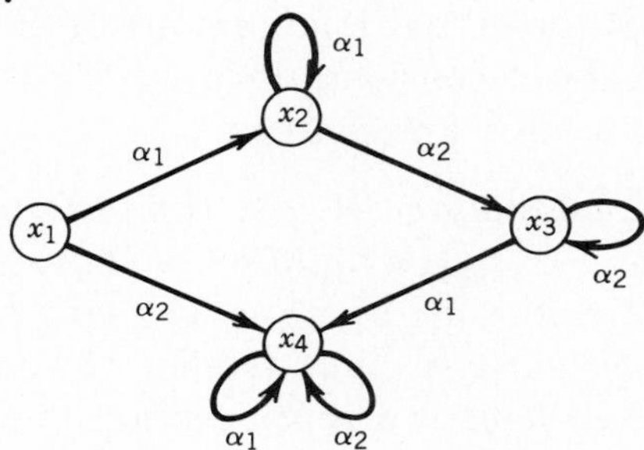

In the study of formal languages, it turns out to be convenient to extend the foregoing idea of a deterministic finite automata to allow for the idea of a nondeterministic (but not random) state-transition. This leads to the notion of a *nondeterministic automaton,* obtained by allowing the next state to be a *subset* of X, rather than just a single element. Thus in calculating $\delta^*(x_1, u)$, we accumulate *all* states that the automaton can enter when it reaches the end of the word u when starting in the state x_1. We then say that $\mathcal{M}$ accepts u if at least one of these states that could be entered is in F. It can be proved that a language is regular if and only if it is accepted by a nondeterministic finite automaton. In view of the above definition of a regular language, this shows that a nondeterministic automaton is equivalent to a deterministic one, at least insofar as accepting languages is concerned. So why introduce the nondeterministic automata? Basically the reason is that it is usually easier to design a nondeterministic automaton to accept a particular language, but any actual machine that is built to accept this language must be a deterministic one. Thus, the equivalence theorem guarantees that the design can be physically implemented.

There are usually taken to be four types of languages distinguished by the amount of memory needed in an automaton that accepts them. These language types are:

0. *Unrestricted* languages—indefinitely large memories.
1. *Context-sensitive* languages—memory proportional to word length.

2. *Context-free* languages—memory arranged in a stack, with a fixed number of elements available at any moment.

3. *Regular* languages—no memory.

The above types of languages form a hierarchy with the unrestricted languages being the most general. Each of the types is associated with a particular class of automata that accepts it, with Type 0 being the only language that requires the power of a universal computer (Turing machine) for its implementation. We have already seen that a finite, deterministic automaton suffices to implement a regular language, with context-free and context-sensitive languages requiring what are termed *pushdown automata* and *linear-bounded automata,* respectively.

This classification of languages bears a striking similarity to the classification of one-dimensional cellular automata according to their limiting behaviors. In the preceding sections, we have seen that all such automata ultimately settle into long-term behaviors of one of the following sorts:

A. The initial pattern disappears.

B. The initial pattern evolves to a fixed finite size.

C. The initial pattern grows indefinitely at a fixed rate.

D. The initial pattern grows and contracts with time.

It's tempting to try to identify the types of automata A–D with the language classes 0–3. To follow up the identification here would require more time and space than we have available in a book of this sort, but it's worthwhile pursuing a few of the simpler aspects of the language $\leftrightarrow$ automata duality just to see how far one can go toward establishing the conjecture.

Let $\Omega^{(t)}$ represent the set of configurations generated by a cellular automaton at time t, i.e., the set of configurations that can be reached at time t when any one of the admissible initial configurations is used. It's been shown that after any finite number of time-steps t, the set $\Omega^{(t)}$ forms a regular language. Thus, the real case of interest is the nature of the set $\lim_{t\to\infty}\Omega^{(t)}$. One approach to getting a handle on this object is to define the *entropy* of the set $\Omega^{(t)}$.

If we have an automaton with k possible cell values, then the total number of possible sequences of X symbols is k^X. In general, only some number $N(X)$ of these sequences occur in nontrivial regular languages. These sequences correspond to the distinct paths through the transition graph of the automaton. The *topological entropy* of the set $\Omega^{(t)}$ is then defined to be

$$s_t = \lim_{X\to\infty} \frac{1}{X} \log_k N(X).$$

For any regular language, this number is also given in terms of the largest characteristic value of the adjacency matrix of the state-transition graph as $s_t = \log_k \lambda_{\max}$. By the locality assumption on the state-transition rule in the automaton, it follows that the set $\Omega^{(t)}$ always contracts or remains unchanged in time. Consequently, the entropy of $\Omega^{(t)}$ is nonincreasing with t. Further, it can be shown that Class A cellular automata have $s_t \to 0$ as $t \to \infty$. Class B–D automata have nonzero limiting entropies, although Class B automata yield periodic limiting patterns and, hence, generate zero *temporal* entropy. The entropies of regular languages are always logarithms of algebraic integers, i.e., the root of a monic polynomial having integer coefficients. Context-free languages, however, have entropies given by logarithms of general algebraic numbers but can still be computed by finite procedures. The entropies for context-sensitive and unrestricted languages are, in general, noncomputable numbers. So if Class C and D automata do indeed yield limit sets corresponding to context-sensitive or unrestricted languages, then the entropies of these sets are generally not computable. However, we can *estimate* the entropy of any automata from experimental data by fitting parameters in various types of models that reproduce the data.

Another approach to the language $\leftrightarrow$ automata question is through the idea of *complexity*. We define the complexity of the set $\Omega^{(t)}$ to be the number of states in the smallest automata that generates the set $\Omega^{(t)}$. Call this number $\Xi^{(t)}$. It's experimentally observed that $\Xi^{(t)}$ is nondecreasing in time and that Class A and B automata appear to give complexities that tend to constants after just one or two time-steps, or increase at most quadratically with time. On the other hand, Class C and D automata usually give complexities that increase exponentially with time. In all cases we have the bound

$$1 \leq \Xi^{(t)} \leq 2^{k^{2rt}} - 1,$$

where r is the size of the neighborhood in the automaton transition rule.

To give some indication of the nature and magnitude of the entropies and complexities, we give the values of these quantities for some typical one-dimensional cellular automata below:

Rule	$\Xi^{(0)}$	$\Xi^{(1)}$	$\Xi^{(2)}$	$\lambda_{\max}$
0	1	1	1	1.0
90	1	1	1	2.0
128	1	4	6	1.62
182	1	15	92	1.89

We can summarize this very brief excursion into formal language theory and cellular automata in the following compact form:

THEOREM. *Class A and B cellular automata have limit sets that form regular languages. For most Class C and D automata, the regular language complexity increases monotonically with time, so that the configurations obtained in the limit usually do not form a regular language. Instead, the limit sets for Class C automata appear to form context-sensitive languages, whereas those for Class D automata seem to correspond to general, unrestricted languages.*

At this juncture, the obvious question that arises is what about natural languages like French, Russian, English, and Chinese. To what degree do any of the formal languages discussed above adequately mirror the structure of these most human of all communication modes? This question lies at the very heart of Chomsky's "generative/transformational" theory of grammars which, at the very least, has disposed of the view that any normal human language can be modeled as a regular language and, hence, generated by a finite-state automaton. It would take another volume the size of this one to even begin to enter into the details of Chomsky's theories, but one of the major conclusions that seems to have emerged is that human languages are also probably not "computable" by any structure less powerful than a full unrestricted language of Type 0. In any case, certainly no automaton with finite memory can account for the myriad statements possible in any human tongue. It seems very likely that, at best, only a genuine Turing machine (a universal computer) can even begin to approach the complexity and structure that we see in even the simplest of human languages. The interested reader is invited to consult the chapter's Notes and References for a wealth of detail on the linguistic revolution inspired by Chomsky's work and the deep connections of this work to problems of automata theory.

8. *DNA Sequences and Cellular Automata*

While natural human languages are almost unimaginably complex, there is another type of language employed by nature that offers many possibilities for analysis by the kind of automata-theoretic ideas we have been discussing. This is the language of DNA sequences, the so-called Language of Life.

At the simplest possible level, the DNA molecule can be viewed as a one-dimensional lattice with four states per cellular site. Of course such a view is hopelessly incomplete, as the DNA molecule is much more than just a linear string of base pairs, having many other mechanisms for information storage. Nonetheless, as a first approximation we can consider the DNA molecule as a one-dimensional cellular automaton capable of self-reproduction having four possible states per cell, which we shall label A, C, G and T, representing the bases Adenine, Cytosine, Guanine and Thymine. Structurally, the pattern present on the DNA molecule is composed of a sequential arrangement of A, C, G and T along the strand which is twisted into an antiparallel double-stranded helix, with the alignment of the two strands mediated through hydrogen bonding of the base pairs. Furthermore, the base pairing is quite specific: A always pairs with T, and G always pairs with C. As a consequence, the base sequence of one strand completely determines the base sequence on the complementary strand. The celebrated *double helical* structure of the DNA molecule is depicted in Fig. 2.9. The first part of the figure gives a linear representation of the base-pairing of the two strands, together with the sugar and phosphate bonds on the strands themselves, and the second part displays the way the strands are twisted to form the characteristic double helix geometry. Now let's begin by briefly reviewing the functioning of the DNA molecule in the day-to-day chemical business of the cell.

While simple organisms have only a single DNA strand, higher organisms contain several separate bundles of DNA called *chromosomes.* The number of chromosome strands varies from species to species, being 46 for humans, 16 for onions and 60 for cattle. Each of these strands is further subdivided into sections called *genes,* where each such gene is by definition the amount of information needed to code for either one protein (a *structural* gene), or to either induce or repress certain chemical operations of the cell (*regulatory* gene). The Central Dogma of Molecular Biology, due to Francis Crick, is a statement about the flow of information in the cell. Compactly, it states that

$$\text{DNA} \longrightarrow \text{RNA} \longrightarrow \text{Protein}.$$

The essential message of the Central Dogma is that there can be no Lamarckian-type of information flow back from the protein (phenotype) to the cellular nucleus (genotype). The Central Dogma is a description of the way in which a DNA sequence specifies both its own reproduction (replication) and the synthesis of proteins (transcription and translation). Replication is carried out by means of the base-pairing complementarity of the two DNA strands, with the cellular DNA-synthesizing machinery reading each strand to form its complementary strand. The protein-synthesizing procedure proceeds in two steps: first of all, in the *transcription* step a messenger RNA (mRNA) strand is formed, again as a complement to one of the DNA strands, but with the base element U (Uracil) taking on the role of T (thymine). Note that usually RNA is a single-stranded molecule. In the second step, the cellular protein-manufacturing machinery (the ribosomes) *translates* a section of the mRNA strand using the genetic code in order to form a protein strand. The mRNA base sequence is read as a nonoverlapping set of contiguous triplets called *codons.* Proteins are composed of amino acids of which there are 20 different types. Since there are $4^3 = 64$ possible codons, there is some redundancy in the translation of specific codons into amino acids.

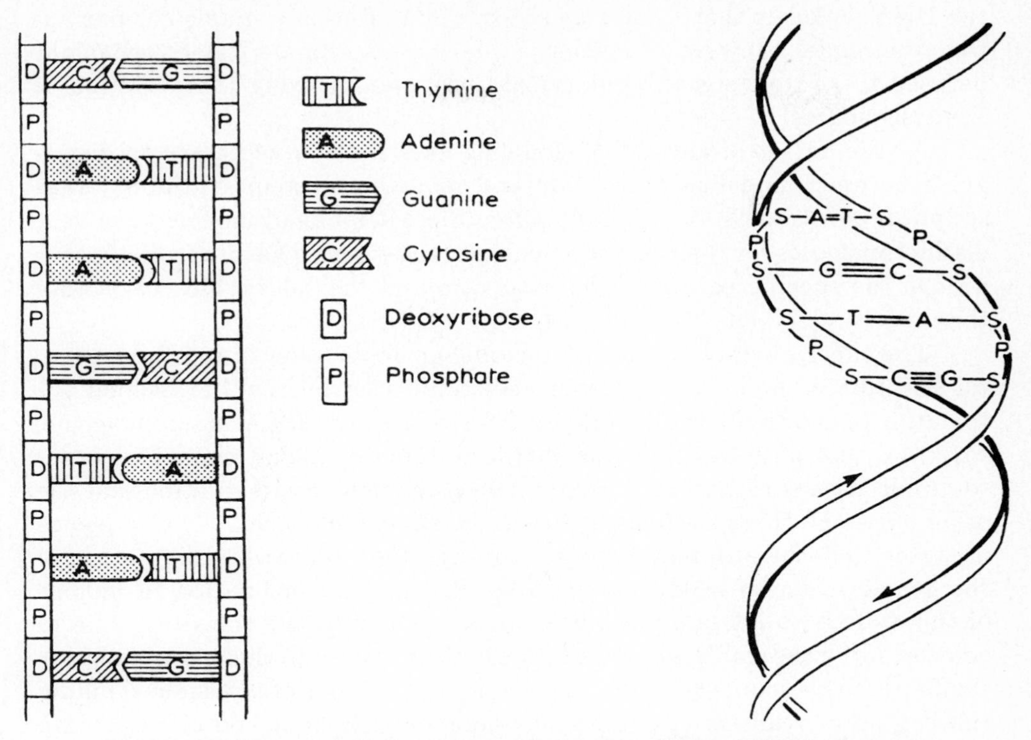

Figure 2.9 Form and Geometry of the Two Helical DNA Strands

The foregoing description of the Central Dogma, oversimplified as it is, already suggests the utility of modeling the evolution of DNA sequences within the framework of one-dimensional cellular automata of the type we have been considering. The sugar and phosphate groups form a one-dimensional lattice (strand) of cellular sites, and the state of any given cell in the automaton is determined by one of the four elements from the nucleotide alphabet. Changes in the configuration of the sequence (mutations) are seen after replication, which we may think of as an iteration of the automaton according to the particular rule that has been programmed. The nature of the transition rule will be determined, of course, by the biological and chemical constraints of the situation. Before going on, we should caution that the above description of the functional activity of DNA is incomplete in several ways, two of the most important being that not all DNA sequences code for protein, and that the three-dimensional configuration of the DNA molecule also acts to store dynamic information about the role of the DNA sequence in the functioning of the cell. Thus, our idea of using the elementary one-dimensional cellular automata set-up to model the action of DNA sequences can be at most only a starting point for much more elaborate structures capable of taking into account these additional aspects of the DNA molecule that cannot be accommodated in any simple cellular automaton model. Now let's consider just how the automata framework might be used to address some foundational questions involving the evolution of DNA sequences.

At the outset of any DNA modeling exercise, it's necessary to distinguish between modeling replication and modeling protein synthesis (transcription and translation). These quite different kinds of models have very distinct purposes for their development, purposes whose nature strongly conditions the specific nature of the automata and the rules which constitute the model.

The most obvious purpose for modeling replication is to study natural selection or prebiotic evolution. We could use the one-dimensional automaton to represent the germ-line DNA or the pre-DNA macromolecule, with the rule of state-transition in the automaton being chosen in order to mimic base substitution, base insertion/deletion, recombination and the other types of changes that can occur in DNA over time. The sequence of states that the automaton passes through then represents the nature of the genetic changes in the population. The natural inclination in models of this type is to introduce a stochastic element into the transition rule to account for "random" genetic mutations. One way to do this is to randomly shuffle the base sequence according to one of a number of plausible shuffling rules. In more realistic situations, this random element should also take into account chemical constraints as well.

A simplified version of the above scheme was carried out by Holland, who produced an elementary one-dimensional chemistry involving only two base elements, two amino acids, two types of bonds, the above simple form of mutation involving shuffling operators and a primitive notion of catalysis. Starting with a random ensemble of initial states, these experiments showed that catalysis eventually dominates, and the arrangement of amino acids always clusters into groupings that are functionally equivalent to enzymes. Such behavior is strongly reminiscent of the Class D automata discussed in earlier sections. These were the most complex one-dimensional automata, displaying limiting behavior patterns that corresponded to the most general type of unrestricted language. Given the evident complexity of living organisms, the result of this experiment with "simplified" DNA is reassuring and tends to validate the use of automata-theoretic ideas in the analysis of DNA evolution.

One of the difficulties in dealing with DNA replication by cellular automata models is the problem of selection. In the prebiotic stages, selection can be handled simply in terms of competition for raw materials among the competing enzymes. However, at later stages of evolution this is an inadequate criterion since more highly developed organisms have a phenotype that is quite different in form than the genotype, and many evolutionary criteria act on the phenotype rather than the genotype. Development of an appropriate notion of selection remains one of the major stumbling blocks to the effective use of cellular automata for modeling the process of DNA replication and evolution.

We can also use automata to model the processes of transcription and translation. The purpose of a model of this type is to attempt to understand the global mechanisms underlying the functioning of the somatic DNA, as well as a variety of problems underlying development. In such models, it's essential to incorporate some means of dynamical information storage. The transition rule of the automaton corresponds to the conformational changes that take place in the organism, as well as possible replications. The need to deal with conformational changes introduces an unusual aspect into the cellular automaton construction since, although the automaton is basically a one-dimensional object, it must also have some idea of its own geometry in three dimensions. Changes in this geometry are one of the main processes taking place during the state transition of the automaton. Examples of how this can be done are reported in the papers cited in the chapter's Notes and References.

Our discussion of the uses of cellular automata for studying the dynamics of DNA sequence development, sketchy as it is, shows clearly the value of the theoretical work noted in the earlier sections on the long-term behavior patterns that can emerge from various types of one-dimensional automata. Now let's take a look at another way such mathematical machines are being used to study living processes—the growth of plants.

9. *L–Systems and the Growth of Plants*

To complete our discussion of one-dimensional automata, we briefly consider a model for the development of filamentous plants originally proposed by Lindenmayer in the late 1960s. This model contains the novel feature that the number of cells in the automaton is allowed to increase with time according to the prescription laid down by the automaton state-transition rule. In this way the model "grows" in a manner mimicking the growth of a filamentous plant like the blue-green algae *Anabaena*.

The simplest version of the *L-system* of Lindenmayer assumes the form of a one-dimensional lattice that starts at time $t = 0$ with a single active cell in state 1. The rule of transition is as follows: the next state at cell i is given by the current state at i, together with the state of the cell immediately to the left of cell i, i.e., cell i–1. The state transition is given by the rule

$$(i = 0, i - 1 = 0) \Rightarrow i \to 0, \qquad (i = 0, i - 1 = 1) \Rightarrow i \to 1,$$
$$(i = 1, i - 1 = 0) \Rightarrow i \to 11, \qquad (i = 1, i - 1 = 1) \Rightarrow i \to 0.$$

The interesting feature of this rule is the cell division given in the case when $i = 1, i - 1 = 0$. If we set the initial state with the first cell at the left being

0 and the cell next to it being in the state 1, then the time behavior of the automaton is [0 1], [0 1 1], [0 1 1 0], [0 1 1 0 1 1], ..., for the times $t = 1, 2, 3$ and 4, respectively. So here we already see the growth of the initial "seed" at cell 1 to several seeds at the cells 2, 3, 5 and 6.

In this example, we could change the division rule to yield a [0 1] progeny pattern instead of the above [1 1] pattern when cell i is in state 1 with i–1 being in state 0. Such a change gives a banded pattern of the following sort emerging from the seed at cell 1: [0 1], [1 0 1], [0 1 0 1], [1 0 1 1 0 1], [0 1 0 1 0 1 0 1], ... In a different direction, we could think of the cell at site i–1 as constituting the "environment" for the cell at site i, and keeping the new progeny rule, consider the case when the environment is constant; i.e., we fix cell i–1=1. In this situation, site i yields a sequence of values that constitute an increasing string of 0s always with a 1 at the right-hand end.

Interesting as the above ideas are in certain special situations, for the consideration of most realistic plant growth situations it's necessary to account for more extended neighborhoods of cell i, as well as more complicated transition rules. In addition, it is also desirable to allow for higher-dimensional geometries to account for the plant growth patterns. With this as our cue, we now leave the study of one-dimensional automata and turn our attention to the case of two-dimensional automata.

10. *Planar Automata and the Game of Life*

In Section 2, we introduced the notion of a two-dimensional cellular automaton which we can think of as having a state-space consisting of an infinite checkerboard, each of whose squares can assume one of k symbols from a finite alphabet at any moment t. In contrast to the case of one-dimensional automata, where the local neighborhood of a cell i is simply a line of cells on either side of site i, for two-dimensional automata there are several possible inequivalent neighborhoods, depending upon whether or not we allow cells diagonal to site i in the neighborhood. Earlier we have spoken about the two most important such neighborhoods, the von Neumann neighborhood consisting only of the cells adjacent to cell i horizontally and vertically, and the Moore neighborhood consisting of *all* cells immediately adjacent to cell i. The rule for a two-dimensional automaton using the von Neumann neighborhood is given by

$$a_{i,j}^{(t+1)} = \phi[a_{i,j}^{(t)}, a_{i,j+1}^{(t)}, a_{i+1,j}^{(t)}, a_{i,j-1}^{(t)}, a_{i-1,j}^{(t))}].$$

It's often suitable to use what are termed "totalistic" rules in which the value of a cell at time $t+1$ depends only on the sum of the values of the cells in the neighborhood at time t. To give some indication of the spectrum of possibilities for two-dimensional rules, we note that for $k = 2$ states per cell, there are a total of $2^{32} \approx 4 \times 10^9$ rules for the von Neumann neighborhood, although this number drops dramatically to $2^5 = 32$ rules of the totalistic type. For the Moore neighborhood, the corresponding number of rules are $2^{512} \approx 10^{154}$ general rules, and $2^9 = 512$ totalistic rules.

In view of the magnitude of these numbers, it's clearly impossible to explore the implications of all rules, even with the fastest computers. However, computer experiments have been carried out that sample various rules. These experiments indicate that the long-term behavior of two-dimensional automata tends to fall into the same basic categories that we observed for one-dimensional automata. Thus, for the two-dimensional situation we have the analogues of the Class A–D behavior discussed earlier for the one-dimensional geometry. But there are some differences to note. First of all, the two-dimensional processes often generate "dendritic" patterns characterized by noninteger growth dimensions. In addition, for the one-dimensional automata we saw that the patterns formed after any finite number of time-steps formed a regular language; in the two-dimensional case this result no longer holds. As a consequence, there are questions about configurations generated by two-dimensional automata after a finite number of steps that can be posed but cannot generally be answered by any finite computational process, and are therefore formally undecidable. Examples of such questions are whether a particular configuration can be generated after one time-step from any initial configuration, or whether there exist configurations that have a particular period in time and are thus invariant under some number of iterations of the automaton rule. To explore just a few of the kinds of things that can happen with two-dimensional automata, we now consider the most well-studied such process, namely, that generated using Conway's rules for playing "The Game of Life."

Recall from Section 2 that Life uses the Moore neighborhood with the following simple rule:

$$a_{i,j}^{(t+1)} = \begin{cases} a_{i,j}^{(t)}, & \text{if } \sum_n a_n = 2, \\ 1, & \text{if } \sum_n a_n = 3, \\ 0, & \text{otherwise,} \end{cases}$$

where we have used the notation $\sum_n a_n$ to represent the sum of all the cells in the Moore neighborhood of site (i,j). Of course, for Life we have only the values ON or OFF for the cells, i.e., $k = 2$. Using the foregoing rule, we display the fate of some initial triplets in Fig. 2.10. We see that the first three triplets die out after the second generation, whereas the fourth triplet forms a stable block, and the fifth oscillates indefinitely.

Of considerable interest in this automaton is whether there are initial configurations that will eventually reproduce themselves. The first example of such a configuration is the famous "Glider," which is displayed in Fig. 2.11.

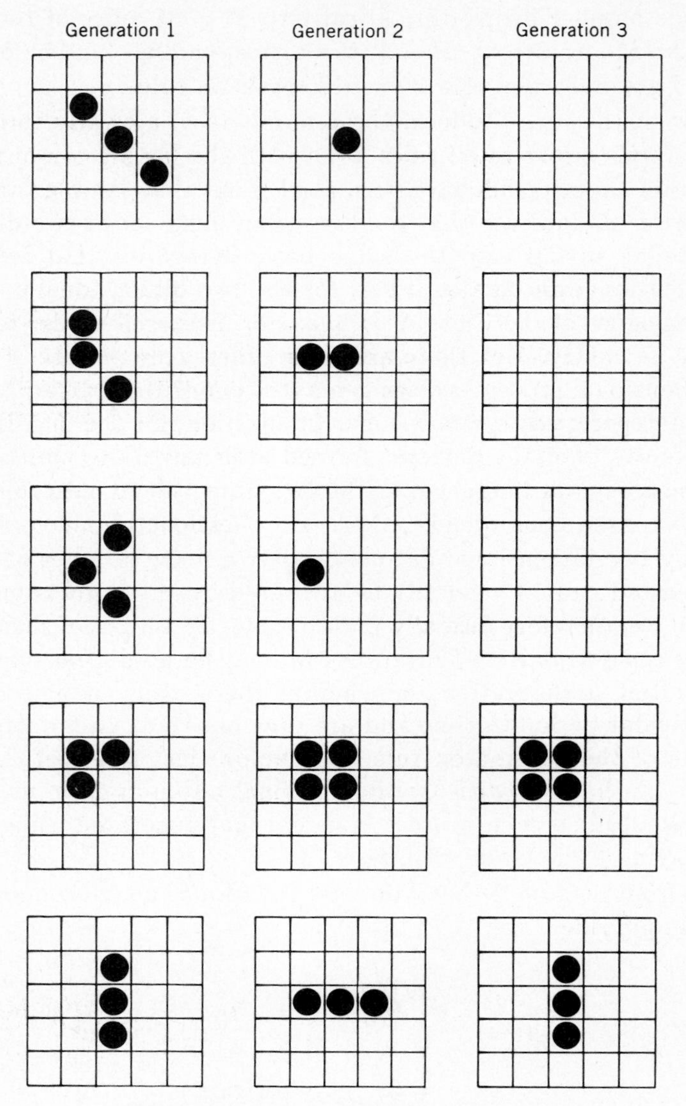

Figure 2.10 Triplet Histories in Life

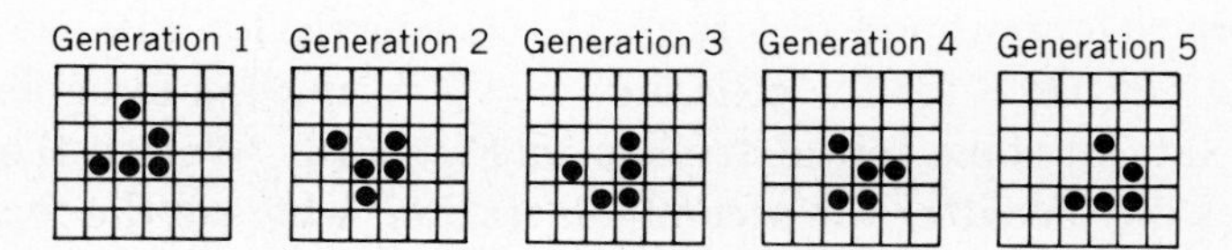

Figure 2.11 The Glider

Originally it was conjectured that because of the overpopulation constraint built into the Life rule, there could be no configurations that could grow indefinitely. J. H. Conway, the inventor of the Life rule, offered a $50 reward to anyone who could produce such a configuration. The configuration that won the prize, termed a "Glider Gun," is depicted in Fig. 2.12. Here the "Gun" shown at the lower-left part of the figure is a spatially fixed oscillator that resumes its original shape after 30 generations. Within this period, the Gun emits a Glider that wanders across the grid and encounters

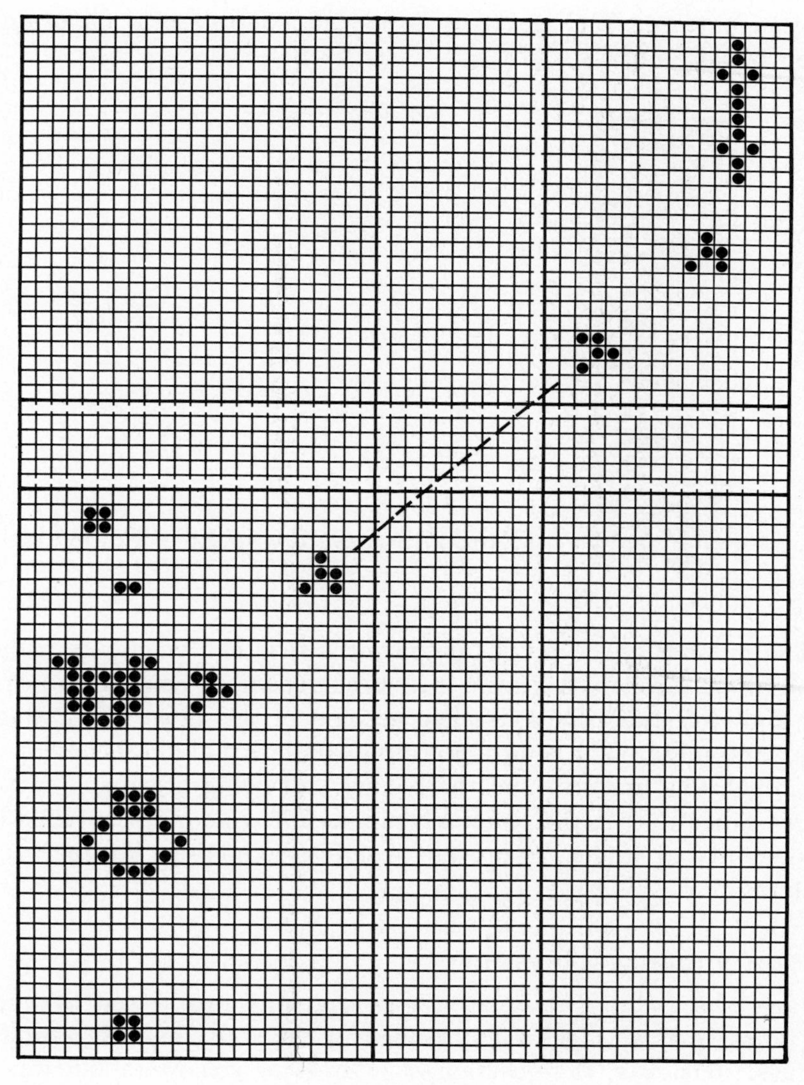

Figure 2.12 The Glider Gun

an "Eater" shown at the top right of the figure. The Eater, a 15-generation oscillator, swallows up the Glider without undergoing any irreversible change itself. Since the Gun oscillates indefinitely, it can produce an infinite number of Gliders, which implies that there do exist configurations that can grow indefinitely. Another type of unlimited growth pattern can also emerge from what is termed a "Puffer Train." This is a moving configuration like a Glider that leaves behind a trail of stable debris-like objects (see Problem 16).

Another question of interest for the Life automaton is whether there exist "Garden-of-Eden" configurations. Such configurations could arise *only* as initial configurations and never as the result of applying the Life rule to any other pattern. Such patterns do indeed exist as seen, for example, in the configuration of Fig. 2.13.

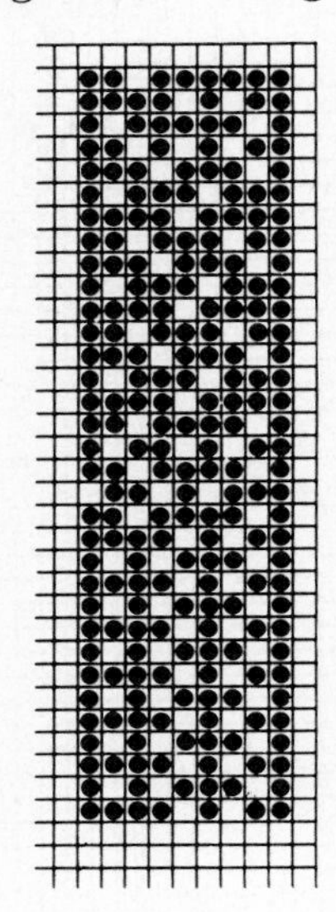

Figure 2.13 A Garden-of-Eden Configuration

The number and type of Life patterns that can emerge is truly enormous and we strongly urge the interested reader to consult the chapter's Notes and References for a detailed account of this most fascinating of all cellular automata. The almost uncanny similarity of the behavior of patterns in the Life universe and the behavior of living organisms in the real universe suggests that the seemingly simple Life rule may contain within it the possibility to model the process of self-reproduction. We turn now to a consideration of this issue.

11. *Self-Reproducing Automata*

The idea of a cellular automaton was originally introduced by von Neumann to serve as an idealized structure for the study of "self-reproducing" machines. Basically, the question of interest to von Neumann was to examine the *kind* of logical organization that suffices for an automaton to be able to reproduce itself. Thus, von Neumann wanted to abstract from natural biological self-reproduction the *logical* form of the reproduction process, independent of its material realization by physico-chemical processes. Von Neumann was able to exhibit a universal Turing machine consisting of a two-dimensional automaton (with $k = 29$ states per cell) that was capable of universal *construction* using the 5-cell von Neumann neighborhood. Self-reproduction then followed as a special case when the machine described on the constructor's input was the constructor itself.

The heart of the self-reproduction problem is in the way we handle the issue of reproducing the "blueprint" of the machine. Imagine that we have succeeded in building a universal constructor, and we then feed the plans for the constructor back into it as input. The constructor will then reproduce itself, but without the instructions that describe how to build itself. This is a trivial and nonperpetuating type of reproduction, and not at all what we have in mind when we speak of self-reproducing machines. So how do we arrange it so that the blueprint, as well as the constructor, are reproduced? This was the big difficulty that von Neumann had to surmount and, as we shall soon see, his solution involved using the information on the blueprint in both a *syntactic* and *semantic* mode, so that the same information could serve two very distinct purposes.

The resolution of the reproduction dilemma was to build into the constructor a "supervisory unit" which was to function in the following manner. Initially the blueprint is fed into the constructor as before, and the constructor reproduces itself. At this point, the supervisory unit switches its state from "construction-mode" to "copy-mode," and then copies the blueprint as raw, uninterpreted data. This copy of the blueprint is then appended to the previously produced constructor (which includes a supervisory unit), and the self-reproducing cycle is complete. The key element in this scheme is to prevent the description of the constructor from being a part of the constructor itself, i.e., the blueprint is located outside the machine and is then appended to the machine at the end of the construction phase by the copying operation of the supervisory unit.

The crucial point to note about von Neumann's solution is the way in which information on the input blueprint is used in two fundamentally different ways. First, it's treated as a set of instructions to be *interpreted* which, when executed, cause the construction of a machine somewhere else in the automaton array. Second, the information is treated as *uninterpreted* data, which must be copied and attached to the new machine. These two different uses of information are also found in natural reproduction, the interpreted instructions corresponding to the process of *translation,* and the uninterpreted data corresponding to the process of *transcription.* These are exactly the processes we discussed above in connection with the action of cellular DNA, and it's worth noting that von Neumann came to discover the need for these two different uses of information long before their discovery in the unravelling of the genetic code. The only difference between the way von Neumann arranged things and the way Nature does it is that von Neumann arbitrarily chose to have the copying process carried out after the construction phase, whereas Nature engages in copying of the DNA early in the replication process.

Simpler machines than von Neumann's can be shown to be capable of self-reproduction, so the question arises of *how simple* a self-reproducing machine can be. This is the converse of the question originally posed by von Neumann, who considered what would be *sufficient* for self-reproduction. Now we are concerned with what is *necessary.* As noted above, there are many types of "pseudo–self-reproduction." A simple example is the automaton defined by the mod 2 addition rule using the von Neumann 5-cell neighborhood in the plane. In this case, starting with a single ON cell, a little later we will see 5 isolated cells that are ON. This clearly doesn't constitute self-reproduction since the initial cell was "reproduced" by the transition rule, rather than having reproduced itself. Any notion of reproduction here is due entirely to the transition rule and in no way resides within the configuration.

To rule out the kind of pseudo–self-reproduction described above, it is now customary to require that any configuration that is genuinely self-reproducing must have its reproduction actively directed by the configuration itself. Thus, we want to require that responsibility for reproduction reside *primarily* with the parent structure, but not *totally.* This means that the structure may take advantage of certain features of the transition "physics," but not to the extent that the structure is merely passively copied by mechanisms built into the transition rule. Von Neumann's requirement that the configuration make use of its stored information as both instructions to be interpreted and as data to be copied provides an appropriate criterion for distinguishing real from pseudo–self-reproduction. It's appropriate to close this discussion of self-reproduction by outlining the manner in which Conway showed that Life rules admit configurations that are capable of self-reproduction in exactly the sense just described. With this bit of mathematical wizardry, Conway showed that the simple rules of Life are actually complicated enough to include the computation of **any** quantity that can be computed.

Conway's self-reproduction proof is based on the observation that Glider Guns, as well as many other Life objects, can be produced in Glider collisions. He then shows that large constellations of Glider Guns and Eaters can produce and manipulate Gliders to force them to collide in just the right way to form a copy of the original constellation. The proof begins not by considering reproduction *per se,* but by showing how the Life rule allows one to construct a universal computer. Since the Life universe consists of an array of ON–OFF pixels on a sufficiently large video screen, what this amounts to is showing that one can construct a Life pattern that *acts* like a computer in the sense that we start with a pattern representing the computer and a pattern representing its programming, and the computer calculates any desired result that would itself have to be expressed as a Life pattern. For numerical computations this could involve the Life computer emitting the requisite number of figures or, perhaps, arranging the required number of figures in some prespecified display area.

The basis of the Life computer is the demonstration that any binary number can be represented by a Glider stream and that other Life patterns could be arranged to function as AND, OR and NOT gates, the necessary building blocks of any computer. The biggest hurdle in these constructions is to show how the various streams representing the "wires" of the Life computer can be made to interpenetrate without losing their original structure. Wires and logic gates are all that any finite computer needs, but a universal computer needs something more: a potentially infinite memory! Conway's solution is to use the Life configuration termed a "Block" to serve as an external memory element. The Block, consisting of a 2×2 array of ON cells, has the property that it is a so-called "still life," i.e., it remains invariant under the Life rule. The idea is to use the Block as a memory element outside the computer pattern, and to use the distance of the Block from the computer to represent the number being stored. To make this scheme work, it's necessary to devise a procedure to move the Block even though it's not in the computer. This can be accomplished by a tricky set of Glider–Block collision mechanisms. The final result of all these maneuvers is Conway's proof that the circuitry of *any* possible computer can be translated into an appropriate Life pattern consisting only of Guns, Gliders, Eaters and Blocks. But what about the other part of the self-reproduction process, the universal constructor?

The second part of the Conway proof is to show that any conceivable Life pattern can be obtained by crashing together streams of Gliders in just the right way. The crucial step in this demonstration is to show how it's possible to arrange to have Gliders converge from four directions at once in order to properly represent the circuits of the computer. The ingenious solution to this seemingly insoluble problem, termed "side tracking," is much too complicated to describe here, but it provides the last step needed to complete Conway's translation of von Neumann's self-reproduction proof into the language of Life.

Now what would a self-reproducing Life pattern look like? For one thing, it would be **BIG**. Certainly it would be bigger than any computer or video screen in existence could possibly display. Moreover, it would consist mostly of empty space since the design considerations require the use of extremely sparse Glider streams. The overall shape of the pattern could vary considerably depending upon design considerations. However, it would have to have an external projection representing the computer memory. This projection would be a set of Blocks residing at various distances outside the pattern's computer. Further, at least one of these blocks would be special in that it would represent the blueprint of the self-reproducing pattern. (In actuality, it is the *number* represented by this block that is the blueprint.) For a detailed description of how the reproduction process actually works, we refer to the Notes and References. We close this discussion of Life by noting the sobering estimate for how big such a self-reproducing Life pattern would be. Rough estimates indicate that such a pattern would probably require a grid of around 10^{13} cells. By comparison, a high-resolution graphics terminal for a home computer can display around 10^5 cells (pixels). To get some feel for the magnitude of this difference, to display a 10^{13} cell pattern, assuming that the pixels are 1 mm^2, would require a screen about 3 km ($\approx$ 2 miles) across. This is an area about six times greater than Monaco! Thus, we can safely conclude that it is unlikely in the extreme that Conway's vision of living Life patterns will ever be realized on any real-world computer likely to emerge from even the rosiest of current estimates of future computer technology. From this sad state we now pass to a consideration of two-dimensional cellular automata that are realized in nature on a regular basis, the mosaic patterns seen on numerous vertebrates such as zebras and tropical fish.

12. *A Cellular Automaton Model of Vertebrate Skin Patterns*

From an evolutionary standpoint, it's of considerable interest to study how vertebrates acquire skin patterns like the characteristic geometric shapes of zebra stripes and leopards' spots because these patterns are important to the survival of the animal. Most such patterns are spots or stripes that are formed by specialized pigment cells (melanocytes), and the problem of how skin patterns are formed can be formulated as describing how the colored pigment cells come to be distributed on the embryo's skin.

Currently, it's thought that the pattern is formed by a reaction-diffusion process in which uniformly distributed pigment cells produce two or more species of morphogen molecules that react with each other and diffuse in space to produce a pattern of concentrations with a characteristic wavelength. The morphogen concentrations form a preliminary pattern that then induces a differentiation of the pigment cells resulting in the final skin pattern.

Here we consider only the simplest process of the above type in which there are just two pigment cell types: a colored cell and an uncolored cell, labeled C and U, respectively. We assume that each C cell produces an inhibitor morphogen that stimulates the change of other nearby C cells to the U state; the C cells also produce an activator morphogen that stimulates the change of nearby U cells to the C state. The two morphogens are diffusable, with the inhibitor having the greater range. The U cells are passive and produce no active substances. Thus, the fate of each pigment cell will be determined by the sum of the influences on it from all neighboring C cells.

We can describe the production, diffusion and decay of morphogens by the generalized diffusion equation

$$\frac{\partial M}{\partial t} = \nabla \cdot \mathbf{D} \cdot \nabla M - KM + Q,$$

where $M = M(x,t)$ is the morphogen concentration (either activator or inhibitor), and the other terms represent the diffusion, chemical transformation and production, respectively, with x being the spatial position in the plane. To eliminate having to deal with this complicated partial differential equation, assume each C cell produces two morphogens at a *constant* rate that diffuse away from their source and are uniformly degraded by the neighboring cells. The two morphogens together constitute a "morphogenetic field" $w(R)$, where R is the distance form the C cell that is producing the morphogens. Assume the net activation effect close to the C cell is a positive constant, but the net inhibition effect farther from the cell is a constant negative value. Thus, the activation-inhibition field has the shape of an annulus with the C cell at the center and radii $R_1 < R_2$, with $w(R) = w_1 > 0$ if $0 \leq R < R_1$, and $w(R) = w_2 < 0$ if $R_1 \leq R \leq R_2$.

To model the development of various types of skin patterns using cellular automata, we begin by randomly distributing C cells on a rectangular grid of lattice points in R^2, the lattice points representing pigment cells. Then for a grid point at position x, the field values due to all neighboring C cells are summed up. If $\sum_i w(|x - x_i|) > 0$, then the cell at x becomes (or remains) a C cell; if the sum is negative, the cell becomes (or remains) a U cell. But if the sum is zero, the cell remains unchanged. Thus, the continuous problem represented by the diffusion equation above is replaced by the discrete two-dimensional cellular automaton with the above simple additive rule. The dynamics are carried out until the resulting pattern stabilizes, which is empirically seen to occur after about five iterations. The only remaining issue is what neighborhood is being used to carry out the above sum.

In experiments by Young, the radii of the activation-inhibition annulus were taken to be $R_1 = 2.30$, $R_2 = 6.01$, with the activation field value $w_1 = 1.0$, and the inhibition field value w_2 ranging over a variety of negative values. Fig. 2.14 displays the results of some of these experiments. We note that as inhibition is decreased, the spot pattern connects up into a pattern of stripes. In closing, we note that by using an anisotropic neighborhood (ellipses rather than circles), it's possible to obtain patterns exhibiting *both* spots and stripes. Details can be found in the Notes and References.

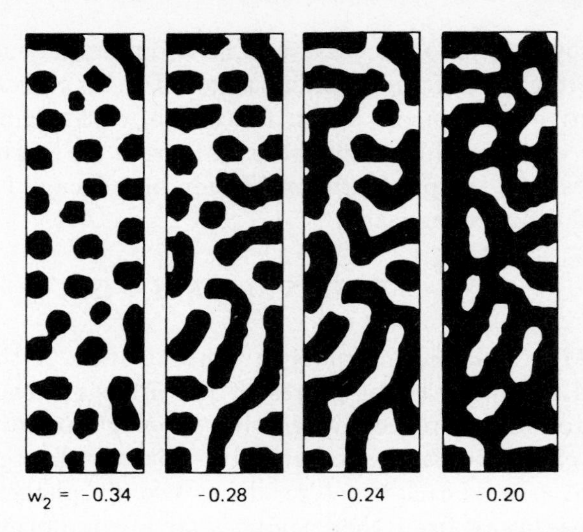

Figure 2.14 Skin Patterns Emerging from Different Inhibitor Values

13. *Cellular Automata—Extensions and Generalizations*

In the limited space available, we have only begun to scratch the surface of the potential for cellular automata to capture the underlying structure of patterns that emerge in the physical, life, social and behavioral sciences. By way of completeness, let's now very briefly indicate some directions in which the basic ideas developed above can be extended to more complicated types of automata.

• *More States*—For the most part, we have confined our attention to the case when each cell can assume only $k = 2$ states. There is no reason why each cell cannot assume any number of different values, an uncountable number even. The problem, of course, is that by allowing a large number of possible values for each cell, it's difficult to interpret whatever structure may be present in the patterns that emerge. One way out of this difficulty is to make use of the increasingly sophisticated color graphics that are becoming ever more widespread and assign a different color to each of the possible cellular values k. This procedure works well in a number of cases and, in fact, could be the basis for a scheme in which the cells could assume a continuum of values by allowing a spectrum of colors instead of just a discrete distribution. It's tempting to imagine what kind of cellular "Picassos" or "Chagalls" might emerge from such experiments, as well as the features of the rules that lead to such computational "masterpieces" (see Discussion Question #5).

• *Higher Dimensions*—We have shown the type of increased pattern complexity that can arise when one passes from one- to two-dimensional cellular automata, but there is no intrinsic reason to stop at planar grids.

We can easily imagine defining automata on higher-dimensional grids as well, as was done in the very infancy of cellular automata studies by Ulam and his co-workers. The problem with passing to dimensions higher than two is again the difficulty in discerning the patterns that may be present in the automaton's behavior as well as the geometric increase in the number of rules, not to mention the number of different types of neighborhoods that can occur. Here again modern computer graphics may come to the rescue, at least in the sense of being able to display the output of such a higher-dimensional automaton in a form that would enable one to identify interesting structures by looking at lower-dimensional sections of the overall configuration.

• *Complex Rules*—We have imposed severe restrictions on the types of transition rules in order to keep the set of possible automata to a manageable level. However, there are many natural phenomena that seem to call for more elaborate state-transitions if a cellular automaton is to serve as a credible metaphor for the system. In one direction, we could allow the transition rule to be *nonlocal* so that the value of a given cell is determined not just by those cells adjacent to it (in any sense of adjacency) but also by cells that lie in some other part of the array. Rules of this type might be appropriate for certain classes of problems in modern quantum physics and quantum electronics.

In a somewhat different vein, we might want to allow rules that are *time-* or *space-dependent.* Such a rule would actually represent many rules at once, as the change of cell value would then depend upon the spatial location of the cell in the grid as well as upon the particular time-step. Clearly, there are an astronomical number of such rules, and the choice of one to explore would have to be narrowed down on physical and/or aesthetic grounds. However, it's certainly plausible that many phenomena in the life and social sciences might evolve according to rules of this sort which would change dramatically as environmental, economic and/or social circumstances shift.

Finally, we may want to consider state-transition rules that are *stochastic* rather than deterministic. With such rules, the actual value of a cell would be determined not just by the values of neighboring cells but also by the values of random variables evolving according to one or another probability distribution. Such a rule might be quite appropriate for a cellular automaton model of, say, genetic mutations or various types of population migration processes. Of course, with such a rule we would have to consider *ensembles* of automaton trajectories and study the patterns that emerge in some statistical sense.

• *Boundary Conditions*—Cellular automata are assumed to "live" on an infinite grid; computer experiments must be carried out on some finite subset of this infinite space. This introduces the question of how to deal with the behavior of cells that are on the boundary of the finite region. The simplest solution is to adopt the periodic boundary assumption that we have used throughout this chapter. But there are many other possibilities, especially when one considers automata in dimensions higher than one. It's of considerable interest to know the degree to which the patterns that emerge are artifacts of the particular choice of boundary conditions or are genuine properties of the automaton itself. Even in low dimensions, examples can be given that show that the boundary conditions **do** matter. A systematic investigation of this phenomenon would be of considerable value.

• *Control*—The cellular automata we have been considering are discrete versions of the classical dynamical systems of Newton, Laplace, and Lagrange. From the standpoint of an observer, this is a very passive kind of dynamics: turn the system on and see what happens (if you live long enough!). A major extension to this Newtonian view is the introduction of *active* control. In cellular automaton terms, this means that the cell's value would depend not only upon the value of its neighbors but also upon the value of a *control* element chosen by the system controller (manager, decision maker, etc.).

The idea of active control drastically changes many important features and introduces an added level of complexity into the analysis of the behavior of the controlled system. For instance, we might now want to ask whether it's possible to arrange for the automaton pattern to arrive at some prespecified configuration after some predetermined number of time-steps by choosing some suitable sequence of controls. This is a typical problem of *reachability,* which is now well understood for continuous-state, continuous-time dynamical systems (see Chapter Three). The extension of these results to the cellular automaton world should prove to be interesting.

Discussion Questions

1. We have defined the spatial topological entropy of a sequence of cells as a measure of the likelihood of particular sequences of X cells appearing from random initial configurations. We can also define a spatial *measure* entropy formed from the probabilities of possible sequences, as well as temporal entropies, to count the number of sequences that occur in the time-series of values taken on by each cell. Topological entropies reflect the possible configurations of a system; measure entropies reflect those that are probable. Can you connect up the properties of topological/measure, spatial/temporal entropies with the four classes of one-dimensional cellular automata discussed in the text? That is, can you classify the various types of automata if you know the numerical values of the various entropies?

2. For Class C one-dimensional automata, we can see some cases of patterns with large triangular regions and low entropies; others give highly irregular patterns with no long-range structure. There appears to be no statistical test to distinguish between these Class C structures. How would you go about developing some measure that would identify such subclasses of the four main types of one-dimensional automata?

3. Consider a two-dimensional automaton with a finite state-space consisting of the 64 cells of a checkerboard. Each cell can assume one of $k = 4$ values, which we label red, yellow, blue and green. We begin with an initial configuration consisting of a random distribution of the four colors in equal proportions on the cellular space; i.e., initially there are 16 cells of each color randomly distributed on the board. At each time-step, *one* of the cells is selected at random and the following rule is applied to determine the next state of that cell:

A. The color on the cell is removed and the color is noted.

B. Another cell is selected at random, and if one of the cells in the von Neumann neighborhood of this cell is occupied by a color *preceding* the color removed from the first cell, then the color from the second cell is placed in the first cell; otherwise, the first cell remains empty. The cyclical color precedence ordering is red $>$ yellow $>$ green $>$ blue $>$ red.

The steps A and B are repeated alternately until the pattern stabilizes.

Describe what you think will be the likely patterns that will emerge during the course of time from this automaton. For example, do you think there is an appreciable chance that any color will get "wiped out" as the process unfolds? Or do you imagine that the relative proportions of the colors will oscillate? (*Remark:* The rules of this automaton are set up to mimic the development of self-reproducing "hypercycles," a popular model for various sorts of autocatalytic chemical reactions that are considered further in Chapter Six.)

4. Kolmogorov and Chaitin have developed the idea that the "complexity" of a sequence of numbers is equivalent to the size of the minimal program needed to reproduce the sequence. This idea turns out to be identical in many cases to the notion of entropy, which measures the exponential growth rate of the number of distinct orbits as a function of time in a dynamical system. The set of Lyapunov exponents are a closely related collection of numbers that measure the time-averaged local asymptotic divergence rate of the orbits. The averaged sum of all the positive Lyapunov exponents is always less than or equal to the metric entropy of the system. To what degree do you think either the Kolmogorov-Chaitin or the Lyapunov exponent concept serves as a good measure of the complexity of a cellular automaton?

(Note that the Lyapunov exponents measure only the *local* divergence of trajectories, and are inherently incapable of ascertaining any *global* macroscopic divergence.)

5. The Dutch artist Piet Mondrian was famous for developing a cubist style of painting that emphasized checkerboard patterns in various colors. Displayed below is a black-and-white version of his work, *Checkerboard, Bright Colors,* which involves a seemingly random scattering of eight colors on a rectangular grid of 256 cells, random, that is, to the untutored eye. Imagine a celluar automaton with this grid as its state-space and the $k = 8$ colors as the allowable "values" of each cell.

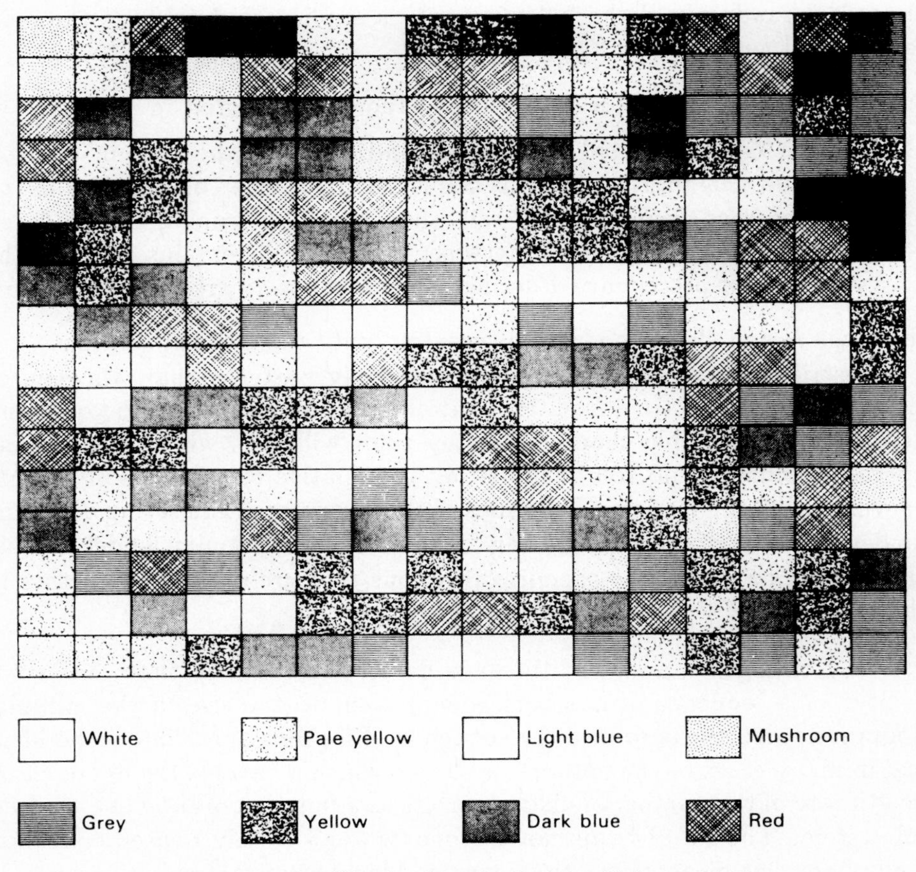

Mondrian's *Checkerboard, Bright Colors,* 1919

a) What kind of state-transition rule could you devise that would lead from any initial configuration to that of *Checkerboard, Bright Colors?*

b) Can you think of any criterion that could be used to judge an artistically aesthetic pattern that might emerge from applying any transition rule to the above system?

c) Imagine that you were given a Mondrian painting and a Picasso, both projected onto the same grid using the same set of basic cell values (colors). Do you think it would be possible to develop a "filter" that would enable you to distinguish the Mondrian from the Picasso? That is, is it possible to produce a test to determine the identity of the artist of any painting? (*Remark:* We shall return to a deeper consideration of these matters in Chapter Eight.)

6. Apply now the method of Problem 5, which used a two-dimensional state-space to study paintings, to the the case of musical works. In music we have a linear, sequential structure, so it seems plausible to use a one-dimensional automaton to represent the unfolding of a musical piece, say, a Mozart symphony or a Strauss waltz. Each state of the automaton would represent the notes that are being played at that moment in time, with the state-transition being given by the musical score prescribing the notes that are to be sounded at the next moment. Do you think that it's possible to develop this idea into a mechanistic procedure for musical works? How would you account for the basic aspects of rhythm and intensity in such a scheme? Presumably, each composer's style would be represented by a set of invariant features of the state-transition rule for the automata that represent this composer's work. How would you go about identifying the specific properties of a transition rule that would stamp it as belonging to a particular composer? Consider the entire issue of computer-generated music in the context of one-dimensional cellular automata and examine the issue of how to represent a musical composition in cellular automata terms.

7. The preceding questions on artistic forms raise the issue of how different behaviors are distributed within the space of cellular automaton rules. We have seen that empirical studies suggest that for symmetric one-dimensional rules, Class A and B automata appear to become progressively less common as the number of state values k and the size of the neighborhood r increase. Class C automata become more common, and Class D slowly less common. For two-dimensional automata, Class C (chaotic) is overwhelmingly the most common, with Class D being very rare.

Consider the above observations in the context of the artistic identifications discussed above. Since it may be assumed that an aesthetically pleasing form does not arise as the result of a random or chaotic configuration, it would appear that genuine works of art would be represented by Class A, B or D automata, although Class D also seems unlikely as it would involve an infinite time pattern that shows no periodicity. Discuss the contention that artistic "signatures" would be represented by only Class A or

B rules. How would you establish a metric on the space of rules so that the "nearness" of one rule to another could be measured?

8. We have seen that for one-dimensional automata, the set of configurations reached after a finite time constitutes a formal language, which can be represented by the paths through a finite-directed graph. The smallest such graph provides a canonical description of the set of all words in the language, and the number of its nodes represents the *complexity* Ξ of the language.

a) In general, the value of Ξ is bounded above by the quantity $2^{k^{2rt}} - 1$, and appears to be nondecreasing with time. Discuss the claim that the increase of complexity in time is a principle for self-organizing systems analogous in generality, but complementary in content, to the law of entropy increase in thermodynamic systems (the famous Second Law of Thermodynamics).

b) Consider the thesis that the complexity Ξ would be a good measure for the artistic merit of a work described by a one-dimensional automaton, in the sense that works of higher complexity could be considered to be "better" in an artistic sense than those of lower complexity. Examine this contention by comparing the complexities of a rock or C&W tune with a Beethoven or Haydn symphony.

c) How would you go about developing a procedure for calculating the complexity Ξ *directly* from the transition rule and the initial configuration, rather than computing it in "real time" from the unfolding of the automaton's state?

d) It's often held that Darwinian evolution involves a passage from a state of lower to increasingly higher complexity. Does the foregoing notion of complexity seem to you to be consistent with this conventional wisdom? Can you tie together this idea of biological evolution and complexity with evolution and complexity of patterns in the two-dimensional automaton Life?

9. The text describes an L-system for modeling one-dimensional plant growth. How could you extend the set-up of the text to two- and three-dimensional L-systems that could account for spatial features of real plant growth such as sprouting of leaves, reproduction by airborne transmission of seeds, and budding of flowers?

10. If a system is capable of universal computation, then with appropriate initial conditions its evolution can carry out any finite computational process. Any predictions about the behavior of a cellular automaton must be made, of course, by performing some computation. However, if the automaton is capable of universal computation, then this computation generally must reduce to a direct simulation of the automaton's evolution.

Consequently, questions about the infinite time-limiting behavior of such automata may require infinite computations, and therefore would be formally undecidable.

a) Universal computation can be used to establish the undecidability of questions about the behavior of a system. What about the converse? That is, can undecidability occur even in systems *not* capable of universal computation? Can you construct an example of such a case? (*Hint:* Find another system that is capable of universal computation, and show that a reduction of its capabilities does not affect undecidability.)

b) Rice's Theorem states that almost all questions about arbitrary recursively enumerable sets are undecidable. What do you think about the possibility that simple questions that can be stated in a few logical symbols usually *are* decidable for any particular automaton? That is, we are not usually interested in *all* propositions that can or cannot be decided about a given automaton but only in those "natural" questions that can be stated in a relatively compact way.

11. We can think of the evolution of a cellular automaton as a *pattern-recognition* process, in which all initial configurations in the basin of attraction of a particular attractor are thought of as instances of some pattern with the attractor being the "archetype" of this pattern. Thus, the evolution of the different state-trajectories toward this attractor would constitute recognition of the pattern. How could you formalize this idea into a practical pattern-recognition device for, say, image processing?

12. In our discussion of self-reproduction, we saw that any such process must contain four components:

A. A *blueprint* for constructing offspring.

B. A *factory* to carry out the construction.

C. A *controller* to make sure the construction follows the plan.

D. A *duplicating machine* to transmit a copy of the blueprint to the offspring.

We also saw how these components entered into biological replication in a living cell. Do you think the same four basic components appear in other "living" organisms such as a society? A manufacturing enterprise? An ecological network?

13. The actual emergence of structure in an automaton's evolution pattern is a consequence of a *shift of level* of description. Even though the initial configuration *looks* random, it must fully encode (and hence "know") the final structure that the computation unveils. But it takes an external observer to appreciate any pattern that may be present in this structure;

the emergence of structure decreases the number of available *microstates* but may increase the number of observed *macrostates.* The ordered states coming about from an irreversible computation are few, but *recognizable.* Discuss how a mathematical procedure could be developed to characterize this recognizability.

14. The propagation of an infectious disease has many striking similarities with the way two-dimensional automata patterns form from initial "cases." How would you go about constructing an automaton rule that would mimic the way diseases spread in a planar population? (Note that here it seems likely that the rule would have to involve *stochastic* state-transitions to account for the fact that not all people exposed to a disease actually contract it.) Do you think a model for disease propagation could be modified to suit a situation involving the migration of human or animal populations from one geographic region to another? How?

15. In Ulam's Coral Reef automaton, the rule of transition is that a cell is ON in the next generation if the following three conditions are met:

A. It borders (in the von Neumann neighborhood) one and only one newly occupied cell, where newly occupied means the cell came ON only in the current generation.

B. It must not touch (in the Moore neighborhood) an old occupied cell, where old means a cell that came ON prior to the current generation.

C. Among all cells that would qualify to come ON by conditions A and B, eliminate those that would touch each other (in the Moore neighborhood).

a) What kind of patterns do you think would emerge from a single ON seed cell?

b) How would you build in an extension to the above rule to account for deaths of cells?

c) What kind of "pseudo–self-reproducing" patterns do you think could come out of the above rule (with death included)?

d) Show by computer experiment that if two or more patterns interpenetrate, they may become involved in a struggle in which the winning pattern destroys the rest. How would you interpret this in terms of competition among living organisms?

16. The Life automaton rule is carefully balanced between allowing too many births (and a consequent population explosion) and too many deaths (and an extinction of the "species"). Consider other Life rules that would also balance life and death in order to create nontrivial population patterns

over long time periods. For example, a popular variant is *3–4 Life,* in which the state-transition rule is quite simple: a cell will be ON (alive) in the next generation if and only if it is surrounded (in the Moore neighborhood) by exactly three or four ON neighbors in the current generation. Discuss the implications of the fact that 3–4 Life is more pro-birth, but at the same time deaths by isolation are more common. Do you think this would lead to more or less the same types of patterns as in Life, or would the two universes be radically different?

17. We have seen that self-reproducing patterns are possible in Life, as well as with other types of automata. This leaves open the question of whether or not a machine can make another machine more complicated than itself. Do you think this is possible? How might it be done? What is the connection between a "complexity-increasing" construction and the idea of natural evolution for living organisms?

18. Most of our discussion in the text has been for *irreversible* state-transition rules, i.e., rules for which the state at time t cannot be uniquely determined from the state at time $t+1$.

a) Show that *reversible* rules exist by explicitly displaying one.

b) If the initial configuration for a one-dimensional automaton is selected at random and you apply a reversible rule, what kind of long-term behavior would you expect to see?

c) If you choose an initial configuration and a reversible rule that allow information to propagate, what kind of temporal behavior patterns will emerge?

d) Interpret your answer to part (c) by considering each state of the automaton to be a "message" with the contents of the cells being the characters of the message, assuming only local measures of correlation.

e) Consider the possibility of using cellular automata with reversible rules to model the processes of classical Newtonian physics (space, time, locality, microscopic reversibility, conservation laws, etc.)

19. Cellular automata often serve as discrete versions of partial differential equations in one or more spatial variables. What value do you see to the "inverse problem," i.e., developement of a partial differential equation whose space-time behavior exactly matches a given cellular automaton at the discrete cellular grid points? Could such an equation be used to predict the long-term behavior of a Class C or D automaton without having to engage in direct simulation? Why or why not?

20. Comment on the following argument often given to support (or justify) cellular automata research:

A. There are now available novel computational resources that may, on a given task, outperform conventional resources by a great many orders of magnitude.

B. The conceptual developments of mathematical physics must have been strongly influenced by the nature of the computational tools available.

C. Therefore, the new resources suggest new approaches to the modeling and simulation of physical systems, and it should be possible to replace conventional formulations involving real variables, continuity, etc., by more constructive and "physically real" counterparts.

21. In our discussion of self-reproduction, we have assumed that the blueprint (or instructions) for the automaton to be produced was given in advance. Consider the possibility of self-reproduction by means of *self-inspection,* i.e., the machine to be reproduced **creates** the blueprint by means of self-inspection. Do you see any logical contradictions in such a scheme? If such a reproduction process were possible, what would the implications be for Lamarckian inheritance in biological reproduction, i.e., the inheritance by the offspring of characteristics acquired by the parent(s)?

22. Current DNA research suggests that many subsequences of the DNA molecule are there solely "by chance," and serve no intrinsic biological function. Assuming this to be true, how could it come about in the cellular automata models of DNA outlined in the text?

Problems and Exercises

1. a) For a one-dimensional cellular automata with $k = 2$ and a single nonzero initial cell, show that the mod 2 rule (Rule 90) leads to patterns in which the value of a site at a given time-step is just the value modulo 2 of the corresponding coefficient in Pascal's triangle, with the initial nonzero cell forming the apex of the triangle. That is, the values are the coefficients in the expansion $(1 + x)^n \mod 2$.

b) For Rule 150 in which the value of each cell is the sum of its own value and the values of its nearest neighbors mod 2, show that the sequence of binary digits obtained from a single initial nonzero cell for n time-steps is the sequence of coefficients of $(1 + x + x^2)^n \mod 2$.

2. Feedback shift-registers consist of a sequence of "sites" carrying the value $\alpha(i)$ at site i. At each time-step, the site values evolve by a shift $\alpha(i) = \alpha(i-1)$ and feedback $\alpha(0) = \mathbf{F}[\alpha(j_1), \alpha(j_2), \ldots]$, where the j_i give the positions of "taps" on the shift-register. Show that a one-dimensional cellular automaton of N cells corresponds to a feedback shift-register of N

sites with site values 0 and 1 and taps at positions $N-2, N-1$, and N, by the automaton rule given by the Boolean function $\mathbf{F}$. Show that for one time-step of the automaton, it requires N time-steps in the shift-register.

3. The average fraction of cells with value 1 emerging over one time-step from a disordered configuration has been given in the text by

$$\rho_1 = \frac{\#_1(\mathbf{R})}{(\#_0(\mathbf{R}) + \#_1(\mathbf{R}))} = \frac{\#_1(\mathbf{R})}{8},$$

where $\#_d(\mathbf{R})$ represents the number of occurrences of the digit d in the binary representation of $\mathbf{R}$. This formula holds only for the case when each cell can take on the initial value 1 or 0 with equal probablility $p = \frac{1}{2}$. Show that the generalization of this formula to the case when $p \neq \frac{1}{2}$ is given by weighting each of the eight possible three-cell neighborhoods σ by the weight $p(\sigma) = p^{\#_1(\sigma)}(1-p)^{\#_0(\sigma)}$, and then adding the probabilities for those σ which yield 1 upon application of the automaton rule.

4. Self-similar figures on a plane may be characterized in the following manner. Find the minimum number $N(a)$ of squares with side a necessary to cover all parts of the figure (all cells with nonzero values in the cellular automaton case). The figure is *self-similar* or scale invariant if rescaling a changes $N(a)$ by a constant factor independent of the absolute size of a. If this is the case, $N(a) \sim a^{-D}$, where D is defined to be the Hausdorff-Besicovitch or *fractal* dimension of the figure.

a) Show that a figure that fills the plane would have $D = 2$, but a line would give $D = 1$.

b) In Fig. 2.4 we showed the triangular patterns that emerge from various initial configurations. Show that the pattern for Rule 90 is self-similar with $D = \log_2 3 \cong 1.59$.

c) For Rule 150, the density $T(n)$ of triangles of base length n satisfies the recurrence relation

$$T(n = 2^k) = 2T(2^{k+1}) + 4T(2^{k+2}), \quad T(1) = 0, \quad T(2) = 2.$$

Show that for large k this yields

$$T(n) \sim n^{-\log_2(2\varphi)} = n^{-\log_2(1+\sqrt{5})} \sim n^{1.69},$$

and, hence, conclude that the limiting fractal dimension of the pattern generated by Rule 150 is $D = \log_2(2\varphi) \cong 1.69$.

d) Show that when the number of cell values is k = prime, the fractal dimension of the emerging triangle pattern is given by

$$D_k = \log_k \sum_{i=1}^{k} i = 1 + \log_k \left(\frac{k+1}{2}\right).$$

5. Show that irreversibility of the one-dimensional cellular automata rules stems from the condition that the next state depends only upon the state at the previous time-step, and that reversible rules can be obtained by allowing the next configuration to depend upon the *two* previous configurations.

6. A *random mapping* between k elements is defined by the rule that each element is mapped to any one of the k elements with equal probability $p = 1/k$. Thus, all k^k possible mappings are generated with equal probability.

a) Show that the probability of a particular sequence of k elements having no predecessor under such a random mapping is $(k-1)^k/k^k = (1 - 1/k)^k$. Thus, this probability approaches $1/e \simeq 0.37$ as $k \to \infty$. That is, a fraction $1/e$ of the possible states are not reached by iteration of a random mapping.

b) Show that the probability of a cycle of length r appearing in the iteration of such a random mapping is

$$\sum_{i=r}^{k} \frac{(k-1)!}{(k-i)!\,k^i}.$$

c) For complex nonadditive cellular automata rules, almost all configurations become unreachable as $N \to \infty$. With this fact in mind, consider the degree to which cellular automata behave like random mappings for large N.

7. If the general rule of transition for a one-dimensional automaton is given by

$$a_i^{t+1} = \mathbf{F}[a_{i-r}^t, a_{i-r+1}^t, \ldots, a_{i+r-1}^t, a_{i+r}^t],$$

show that a necessary (but not sufficient) condition for unbounded growth of the initial configuration is either one of the following:

A. $\mathbf{F}[a_{i-r}, a_{i-r+1}, \ldots, 0, 0, \ldots, 0] \neq 0$,

B. $\mathbf{F}[0, \ldots, 0, a_{i+1}, \ldots, a_{i+r}] \neq 0$.

8. Prove the following "Malthusian" limit regarding the growth of self-reproducing planar configurations: if a configuration is capable of reproducing $f(T)$ offspring by time T, then there exists a positive constant k such that $f(T) \leq kT^2$. How would you extend this result to n-dimensional configurations?

9. Prove the following assertion: For any two-dimensional automaton with an irreversible rule, there must exist "Garden-of-Eden" configurations, i.e., configurations that have no predecessor.

10. Consider the following one-dimensional version of Life using a neighborhood size of n cells. The state-transition rule is: A cell is ALIVE at time $t+1$ if it is DEAD at time t and the number of ALIVE neighbors is greater than or equal to $n+1$ and less than or equal to $n + k_0$ (birth), or the cell is ALIVE at time t and the number of neighbors that are ALIVE is greater than or equal to $n + 1$ and less than or equal to $n + k_1$ (survival). Here k_0 and k_1 are positive integers. Let's agree to identify any finite configuration by its *support* (i.e., the cells that are ALIVE) and denote the configuration by

$$c = \bigcup_{1 \leq i \leq p} [a_i, b_i],$$

where $a_i \leq b_i$ and $b_i + 1 < a_{i+1}$. The intervals $\{[a_i, b_i]\}$ are called the *components* of the configuration c. Assume that the above Life game evolves on a finite interval $[a, b]$, i.e., cells x for which $x < a$ or $x > b$ are always DEAD.

a) Show that if $k_0 \geq k_1$, then any finite configuration evolves in a finite number of time-steps toward a stable configuration c such that $n \leq b_i - a_i \leq n + k_1 - 1$ and $n + 2 \leq a_{i+1} - b_i$ for all i.

b) Prove that if $n - k_1 + 1 < k_0 < 2k_1 - n$, then one-dimensional Life admits cycles of length 2.

(*Remark:* These results show that Life rules that exhibit more complex behavior are those for which birth conditions are more restrictive than conditions for survival.)

11. Consider a three-cell one-dimensional automaton with the property that each cell can assume a *continuum* of values in the range 0 to 1. Let the rule of the automaton be given by

$$\begin{aligned} x_1(t+1) &= x_2^2 + x_3^2 + 2x_1x_2, \\ x_2(t+1) &= 2x_1x_3 + 2x_2x_3, \\ x_3(t+1) &= x_1^2, \end{aligned}$$

where the three cells x_i, $i = 1, 2, 3$, satisfy the local and global constraints

$$0 \leq x_i \leq 1, \qquad x_1 + x_2 + x_3 = 1. \tag{$*$}$$

a) Show that the state-transition rule maps the region $(*)$ to itself.

b) Refute or verify the conjecture that for *almost every* initial configuration p, the limit set of this automaton is given by a point $p \in R^3$ satisfying the periodicity condition

$$p_2 = T(p_1),$$
$$p_3 = T(p_2),$$
$$p_1 = T(p_3),$$

where T is the rule of transition given above. Show by counterexample that this property cannot hold for *all* initial configurations.

c) If the cells x_i represent the relative population of species i, describe in words the interpretation of the transition rule, the region $(*)$ and the limiting point p.

12. Consider the generalization of Problem 11 in which the transition rule is

$$x_i(t+1) = \sum_{k,m=1}^{N} \gamma_i^{km} x_k x_m, \qquad 1 \leq i \leq N,$$

with the coefficients satisfying

$$\gamma_i^{km} = \begin{cases} \gamma_i^{mk} > 0 & \text{if } \min(k,m) \leq i \leq \max(k,m), \\ \gamma_i^{mk} = 0 & \text{otherwise,} \end{cases}$$

$$\sum_{i=m}^{k} i\gamma_i^{km} = \frac{m+k}{2}, \qquad \sum_{i=m}^{k} \gamma_i^{km} = 1.$$

We normalize the set $\{x_i\}$ by

$$0 \leq x_i \leq 1, \qquad \sum_{i=1}^{N} x_i = 1.$$

a) Show that the quantity

$$\sigma \equiv \sum_{i=1}^{N-1} (N-i) x_i,$$

is invariant for the above transformation.

b) Given an initial vector $\{x_i^0\}$, show that there is exactly one value of the index j such that

$$N - j \geq \sigma \geq N - j - 1,$$

and that in terms of j, every initial vector converges to the vector whose coordinates are

$$x_j = \sigma - (N - j - 1), \quad x_{j+1} = N - j - \sigma, \quad x_i = 0, \text{ for all other } i.$$

Note that this fixed point is *independent* of the coefficients γ_i^{km}.

13. Show that each of the following languages with the alphabet A forms a regular language by constructing a finite automaton that accepts it:

a) $A = \{a, b\}$, $L =$ words whose final four symbols form the string *abab*,

b) $A = \{a, b, c\}$, $L =$ all *palindromes* of length 6 or less. (*Recall:* A palindrome is a word that reads the same backward or forward.)

c) $A = \{0, 1\}$, $L =$ all binary number strings that are integral multiples of 5.

14. What are the complexities of the regular languages of the previous problem?

15. Let a measure μ be defined on the space of configurations of an $n-$dimensional automaton by specifying that the symbols that can occur at each cell are independent random variables, with the probability of a given symbol occurring being $1/k$. Prove that a cellular automaton preserves the measure μ in the sense that $\mu(S) = \mu[f^{-1}(S)]$ for any measurable configuration S, if and only if the automaton rule f is onto, i.e., if and only if for every configuration $\hat{c}$, there exists a configuration c such that $f(c) = \hat{c}$.

16. In Life we discussed the configuration termed the Glider Gun, which periodically gives off Gliders and, hence, serves as a generator of Life configurations of unbounded growth. There are other such configurations called Puffer Trains, that are moving patterns producing debris as they sweep across the plane, with the debris tail growing indefinitely. Construct a specific example of a Puffer Train. (*Warning:* You will need a computer for this Exercise!)

Notes and References

§1. Accounts of dynamical system theory easily accessible to undergraduates can be found in the well-known texts

Hirsch, M., and S. Smale, *Differential Equations, Dynamical Systems and Linear Algebra,* Academic Press, New York, 1974,

Arnold, V. I., *Ordinary Differential Equations,* MIT Press, Cambridge, MA 1973.

For a more advanced account, see

Guckenheimer, J., "A Brief Introduction to Dynamical Systems," in *Nonlinear Oscillations in Biology,* F. Hoppenstadt, ed., Amer. Math. Soc., Providence, RI, 1979,

Guckenheimer, J., and P. Holmes, *Nonlinear Oscillations, Dynamical Systems, and Bifurcations of Vector Fields,* Springer, New York, 1983.

The Fibonacci sequence is one of the most ubiquitous and important in all of mathematics, arising in a bewildering variety of settings ranging from the spiral pattern of pinecones and sunflower seeds to branching patterns on trees and on to the spiral pattern of the DNA molecule. Much of this universality ultimately derives from the relationship between the sequence of Fibonacci numbers and the "golden ratio" ϕ of the ancient Greek geometers given by

$$\phi = \lim_{n\to\infty} \frac{u_n}{u_{n-1}},$$

where u_i = ith Fibonacci number. For more information on this sequence and its connections to the morphology of living objects, see

Huntley, H., *The Divine Proportion,* Dover, New York, 1970,

Cook, T., *The Curves of Life,* Dover, New York, 1979,

Stevens, P., *Patterns in Nature,* Penguin, London, 1976.

§2. The example of urban housing patterns in a racially mixed neighborhood is adapted from

Schelling, T., "Dynamic Models of Segregation," *J. Math. Socio.,* 1 (1971), 143–186.

Conway's Life game was brought to the attention of the general public in a series of articles by Martin Gardner in *Scientific American.* The complete set of Life articles as well as the treatment of a number of related topics can be found in

Gardner, M., *Wheels, Life and Other Mathematical Amusements,* Freeman, San Francisco, 1983.

An extensive popular account of the Life game, together with computer programs for playing it, is given in

Poundstone, W., *The Recursive Universe,* Morrow, New York, 1985.

§§3–5. The results of this and the next few sections are due primarily to Stephen Wolfram, who is almost single-handedly responsible for the current revival of mathematical and computational interest in cellular automata. The material of this section has been abstracted from his paper

Wolfram, S., "Statistical Mechanics of Cellular Automata," *Rev. Mod. Physics,* 55 (1983), 601–644.

This paper, along with many others, is reprinted in the following work which is essential reading for all *aficionados* of cellular automata:

Wolfram, S., ed., *Theory and Applications of Cellular Automata,* World Scientific, Singapore, 1986.

§6. For these results, as well as many more, see

Martin, O., A. Odlyzko and S. Wolfram, "Algebraic Properties of Cellular Automata," *Comm. Math. Phys.,* 93 (1984), 219–258.

§7. A good introductory treatment of the relationship between formal languages and automata is

Davis, M., and E. Weyuker, *Computability, Complexity and Languages,* Academic Press, New York, 1983.

A somewhat more mathematically detailed account is given in

Eilenberg, S. *Automata, Languages and Machines,* Vols. A and B, Academic Press, New York, 1974 and 1976.

The treatment of cellular automata, languages and complexity follows

Wolfram, S., "Computation Theory of Cellular Automata," *Comm. Math. Phys.,* 96 (1984), 15–57.

The pioneering work responsible for taking the study of natural languages from the realm of descriptive analysis to that of a formal scientific discipline is due to Noam Chomsky. Introductory accounts of his life and work are found in

Leiber, J. *Noam Chomsky: A Philosophic Overview,* Twayne Pub. Co., Boston, 1975,

Lyons, J., *Noam Chomsky,* rev. ed., Penguin, London, 1977.

For the ideas of the master himself, see

Chomsky, N., *Language and Mind,* Harcourt, Brace, Jovanovich, New York, 1972,

Chomsky, N., *Reflections on Language,* Pantheon, New York, 1975.

Chomsky's ideas have by no means met with universal love and admiration. An assessment of some of the shortcomings of his views, as well as some of the strengths, is given in the collection

Harman, G., ed., *On Noam Chomsky: Critical Essays,* U. of Massachusetts Press, Amherst, MA 1982.

§8. The classic popular account of the discovery of the double helix structure of the DNA molecule is provided in

Watson, J., *The Double Helix,* Atheneum, New York, 1968.

The complete text of Watson's book, as well as commentaries upon it's role in the sociology of science, together with reprints of the original papers on the double helix structure may be found in

G. S. Stent, ed., *The Double Helix: Text, Commentary, Reviews, Original Papers,* Norton, New York, 1980.

Good references for understanding the operation of the DNA in both protein production and replication are

Hofstadter, D., "The Genetic Code: Arbitrary?" *Scientific American,* March 1982 (reprinted in Hofstadter, D., *Metamagical Themas,* Basic Books, New York, 1985),

Rose, S., *The Chemistry of Life,* 2d ed., Penguin, London, 1979,

Rosenfield, I., E. Ziff and B. van Loon, *DNA for Beginners,* Norton, New York, 1982.

Our discussion of DNA modeling via cellular automata follows that given in

Burks, C., and D. Farmer, "Towards Modeling DNA Sequences as Automata," *Physica D,* 10D (1984), 157–167.

§9. Some of Lindenmayer's work on modeling plant growth using cellular automata is given in his paper

Lindenmayer, A., "Mathematical Models for Cellular Interactions in Development, Parts I and II," *J. Theor. Biol.,* 30 (1967), 455–484.

Other work along the same lines is reported in

Ransom, R., *Computers and Embryos: Models in Developmental Biology,* Wiley, Chichester, 1981.

§10. In addition to the popular accounts of Life cited under §2 above, a more technical account is given in Chapter 25 of

Berlekamp, E., J. H. Conway and R. Guy, *Winning Ways for Your Mathematical Plays,* Vol.2, Academic Press, London, 1982.

§11. Von Neumann's proof of the possibility of self-reproducing automata is given in

von Neumann, J., *Theory of Self-Reproducing Automata,* U. of Illinois Press, Urbana, IL 1966.

Von Neumann's original work is a bit difficult to follow. A simpler account of his ideas given from several perspectives is found in the collection

Burks, A., ed., *Essays on Cellular Automata,* U. of Illinois Press, Urbana, IL 1970.

The idea of a living organism as a machine has proven irresistably attractive to scientists and philosophers since the time of Aristotle. For some modern perspectives on this eternal question, see

Laing, R., "Machines as Organisms: An Exploration of the Relevance of Recent Results," *Biosystems,* 11 (1979), 201–215,

Laing, R., "Anomalies of Self-Description," *Synthese,* 38 (1978), 373–387.

§12. For a much more detailed discussion of skin pigmentation modeling problem using cellular automata, as well as further computational results, see

Young, D., "A Local Activator-Inhibitor Model of Vertebrate Skin Patterns," *Math. Biosciences,* 72 (1984), 51–58.

Other work along the same lines is reported in

Bard, J., "A Model for Generating Aspects of Zebra and Other Mammalian Coat Patterns," *J. Theor. Biol.,* 93 (1981), 363–385,

Murray, J., "A Pre-Pattern Formation Mechanism for Animal Coat Markings," *J. Theor. Biol.,* 88 (1981), 161–199,

Swindale, N., "A Model for the Formation of Ocular Dominance Stripes," *Proc. Roy. Soc. London, Ser. B,* 208 (1980), 243–264.

§13. For a fuller account of the various intracacies and subtleties of cellular automata, see the Wolfram volume cited under §3 above as well as

Farmer, D., T. Toffoli and S. Wolfram, eds., *Cellular Automata,* North-Holland, Amsterdam, 1984,

Demongeot, J., E. Golès and M. Tchuente, eds., *Dynamical Systems and Cellular Automata,* Academic Press, London, 1985.

DQ #3. This Hypercycle game, as well as many other simple cellular automata illustrating a wide variety of prototypical situations in biology, language and life, can be found in

Eigen, M., and R. Winkler, *The Laws of the Game,* Knopf, New York, 1981.

DQ #4. The idea of describing the randomness of a sequence by the length of the shortest program needed to reproduce the sequence seems to have been hit upon independently by the great Russian mathematician A. N. Kolmogorov and by G. Chaitin (while he was still a graduate student!). The original papers are

Kolmogorov, A. N., "Three Approaches to the Quantitative Definition of Information," *Prob. Info. Transmission,* 1 (1965), 1–7,

Chaitin, G., "Information-Theoretic Limitations of Formal Systems," *J. Assn. Comp. Mach.,* 21 (1974), 403–424,

Chaitin, G., "A Theory of Program Size Formally Identical to Information Theory," *J. Assn. Comp. Mach.,* 22 (1975), 329–340.

DQ #15. The Coral Reef automaton is discussed in a paper by Ulam given in the Burks book cited under §11 above.

DQ #18. The idea of a reversible rule leads immediately to the possibility of an information-lossless computer, i.e., one that would dissipate no heat during the course of its computations. For a discussion of this possibility, see

Bennett, C., and R. Landauer, "The Fundamental Physical Limits of Computation," *Scientific American,* July 1985,

Bennett, C., "The Thermodynamics of Computation," *Int. J. Theor. Physics,* 21 (1982), 905–940,

Landauer, R., "Irreversibility and Heat Generation in the Computing Process," *IBM J. Res. Dev.,* 5 (1961), 183–191.

PE #4. An extensive treatment of the theory of fractals is given in the well-known book

Mandelbrot, B., *The Fractal Geometry of Nature,* Freeman, New York, 1983.

Other accounts of a somewhat more technical nature are found in

Falconer, K., *The Geometry of Fractal Sets,* Cambridge U. Press, Cambridge, 1985,

Fischer, P., and W. Smith, eds., *Chaos, Fractals, and Dynamics,* Marcel Dekker, New York, 1985.

PE #11. For a more extensive discussion of this problem as well as many other problems of a similar nature, see

Stein, P. R., and S. M. Ulam, "Nonlinear Transformation Studies on Electronic Computers," *Rozprawy Matematyczne,* 39 (1964), 1–66 (This paper is reprinted in W. A. Beyer, J. Mycielski and G. C. Rota, eds., *Stanislaw Ulam: Sets, Numbers, and Universes,* MIT Press, Cambridge, MA 1974.)

CHAPTER THREE

The Analytical Engine: A Newtonian View of Brains, Minds and Mechanisms

1. *Brains and Minds*

One of the standard, if not *the* standard, conundrums of philosophy is the teaser, "What is mind?" Descartes' view that the brain and the mind are totally separate entities is aptly summed up in the epigram from Philosophy 101: "What is mind? No matter; what is matter? Never mind." This *dualist* interpretation of the mind-body problem stood more-or-less unchallenged until rather recently when neurophysiological advances, coupled with the arrival of the digital computer, brought forth a host of fresh insights and competing views.

The "brain as computer" metaphor leads to the intuitively appealing idea that somehow the brain is a piece of hardware acting as an information-processing device, while the mind is naturally associated with the software. In other words, all the functions and aspects we usually associate with mind, like creative thought, emotions, pain and so forth are just programs being run in the neurophysiological hardware of the brain. Roughly speaking, this is a weak form of what philosophers of mind call the *central-state identity* hypothesis, i.e., that all mental events are identical with neurophysiological events in a material brain. There are many minor variations upon this basic theme, but they are all united in the claim that mind resides primarily in the nonmaterial "algorithms" for carrying out functional activities rather than in the material pieces and connections in a physical object, be it a computer or an organic brain.

In this chapter we take no position on the brain-mind problem; rather, our goal is to indicate how a mathematical machine can serve to illuminate some aspects of how **any** type of information processing object like a brain would carry out functional activities such as remembering external stimuli, engaging in cognitive introspection, and executing instructions for bodily functions. We will also touch upon one of the most heated psychological debates of this century involving the competing claims of the behaviorists and cognitivists. This debate has by now been pretty much played out, but we shall see that rather simple system-theoretic arguments provide additional ammunition to sink the already floundering ship of behaviorism. These matters will be taken up in somewhat more detail later; for now let us take a longer look at the kind of mathematical machine that will provide the formal system for a model that links the brain and the mind.

2. *Input/Output Relations and Dynamical Control Systems*

The idea of a mechanism, or a machine, as a means to formalize our notion of a causal, dynamic relationship between observables was introduced in the last chapter. The basic ingredients consisted of inputs, outputs and states, together with two maps expressing how inputs act to change states and how states generate outputs. In the automata considered in Chapter Two, the characterizing feature of the machines was that all three sets were assumed to be *finite;* no assumptions were made about the nature of the state-transition and output maps other than that they be well-defined.

In this chapter, we want to drop the finiteness assumption in favor of another type of hypothesis concerning the nature of the input and output sets. In place of finiteness, we shall assume that the input and output sets are vector spaces. Furthermore, we will make no *a priori* assumptions about the state-space, but rather take the position that the states are to be constructed in a "natural" way from the system's input/output description. These states are then used to generate the system's output in a manner to be described later.

Thus, let's assume we are given the input/output relation $f : \Omega \to \Gamma$, where Ω and Γ are vector spaces, and let our task be to construct a state-space X, an initial state x_0, and maps $g : \Omega \to X$, and $h : X \to \Gamma$ so that the diagram

$$\begin{array}{ccc} \Omega & \xrightarrow{f} & \Gamma \\ g \searrow & & \nearrow h \\ & X & \end{array}$$

commutes. Furthermore, for technical reasons that will become apparent later, we require that the map g be onto X, while demanding that h be one-to-one. The commutativity of the diagram, together with the conditions on g and h, comprise the requirements for $\Sigma = (X, g, h, x_0)$ to constitute what we call a *canonical model,* or a *realization,* of the input/output relation $\mathcal{I} = (\Omega, \Gamma, f)$.

As we will see below, the abstract model Σ is equivalent to the dynamical system

$$\begin{aligned} x_{t+1} &= \phi(x_t, u_t), \quad x|_{t=0} = x_0, \\ y_t &= h(x_t), \end{aligned}$$

which formally has the same appearance as the automata studied in Chapter Two; only the interpretations of the quantities x, u, y differ. Part of our story now will be to see how this change in setting from one formal system (finite-state automata) to another (dynamical control systems) changes the nature of the questions that can be asked and the answers that can be obtained about the natural system N which the dynamical process models, and which is represented here by the input/output relation (Ω, Γ, f). To make our treatment as painless as possible, we will confine attention to the case of **linear** control systems, i.e., the case when the maps f, g, h are linear and the spaces Ω, Γ and X are vector spaces. As will be seen, even this simple setting contains plenty of surprises and is rich enough to supply us with a considerable wealth of ideas about matters of current concern in neurophysiology and psychology. But before embarking upon the analytical pyrotechnics, let us pause for a moment to review some matters in psychology that serve to motivate our later development.

3. *Behavioral vs. Cognitive Psychology*

In the early 1920s, John Watson made the radical suggestion that behavior does not have mental causes. Stimulated by the general idea of logical positivism, this thesis was further developed and modified by Hull, Skinner and others, and has come to be termed *psychological behaviorism.* A principal motivation for adoption of the behaviorist view was to rid psychology of the dualist attitude that mind is a nonphysical entity, somehow disjoint from the physical brain. The behaviorist solution is to eliminate all notions of mind, mental states and mental representation from psychological investigation, concentrating solely upon externally observable *stimulus-response* behavior patterns.

By the early 1960s, it was recognized that both Descartes' dualist position and the behaviorist approach to human behavior were unattractive, and effort was focused upon developing a materialist theory of mind that allowed for mental causes. One such theory, termed *logical behaviorism,* was quite similar to classical behaviorism and is really just classical behaviorism in a semantic form. Another theory, *central-state identity,* postulates that mental events, states and processes are identical with neurophysiological events in the brain. Thus, under the central-state identity theory, a behavioral effect is the result of a causal pattern of physical events in the brain. The problem with the central-state identity notion is that in either its weak or strong form, *token* and *type physicalism,* respectively, it asserts that all mental particulars that exist or could ever exist are neurophysiological. Thus, the logical possibility of machines or various types of disembodied spirits having mental properties is ruled out because they are not composed of neurons.

During the last decade or so, a way out of these dilemmas has been provided by the theory of *functionalism,* an outgrowth of that amalgam of physics, linguistics, neurophysiology, computer science and psychology loosely labeled *cognitive science.* Functionalism is based upon the idea that a mental state can be defined by its causal relations to other mental states, and that such mental states can be realized by many systems. In essence, behavior is driven by software, not hardware. An account of these various notions is given in the popular articles and books cited in the Notes and References. Since it will not be necessary for us to distinguish between the central-state identity theory and functionalism, we adopt the generic term *structuralism* to represent any theory of the mind that involves physical mental states, be they manifested in a human brain, a disembodied cloud from space or a collection of silicon wafers in a machine.

One of the main results of this chapter is to display a precise, system-theoretic argument for asserting the *abstract* equivalence of behaviorism and cognitivism, while showing that operationally only the cognitive view offers the basis for a predictive, causal theory of human behavior. Such a conclusion is a natural consequence of the so-called Realization Theorem of mathematical system theory. Following the path laid out by the cognitive framework, we then provide a fairly detailed mathematical description of the way in which a "brain" would process and store external stimuli in order to generate observed behavioral responses. At the end of the chapter, we give some speculations based upon the theory of system invariants for how thoughts are generated as consequences of internal system dynamics.

4. *Stimulus-Response Patterns and External System Models*

Suppose we have an information-processing object $\mathcal{I}$ (human being, machine, cloud, etc.) consisting of the proverbial "black box" connected to its environment by certain input and output channels (Fig. 3.1). Assume that at any given moment t, the stimulus u_t is selected from some set of symbols U (not necessarily numbers), while the observed response y_t belongs to another set of symbols Y. To simplify the exposition, assume that t takes on only the discrete values $t = 0, 1, 2, \ldots$. Then a given stimulus-response pattern of $\mathcal{I}$ is represented by the sequence $B_{\mathcal{I}} = (u_t, y_{t+1})$, $t = 0, 1, 2, \ldots$.

$$u \longrightarrow \boxed{\mathcal{I}} \longrightarrow y$$

Figure 3.1 An Information-Processing Object

If we let Ω denote the set of all possible stimuli sequences, with Γ representing the set of all response sequences, then the overall *external behavior* of the object $\mathcal{I}$ can be denoted by a stimulus-response map

$$f : \Omega \to \Gamma,$$
$$\omega \mapsto \gamma,$$

where

$$\omega = \{u_0, u_1, u_2, \ldots\}, \qquad \omega \in \Omega, \quad u_i \in U,$$

and

$$\gamma = \{y_1, y_2, \dots\}, \qquad \gamma \in \Gamma, \quad y_i \in Y.$$

(Note that to respect causality, the first output appears one time unit after application of the first input. Here Ω and Γ are infinite-dimensional vector spaces of sequences.)

According to the behaviorists, all that can ever be known about $\mathcal{I}$ are the map f, together with the sets Ω and Γ. Or, put another way, a behaviorist would claim that to be given f would be to be given everything that could be known about the disposition of the object $\mathcal{I}$ to behave in a certain way, and that it would be *nonscientific* to assert the existence of any unobservable internal mechanism generating f. Mathematical system theory provides an honest, true, clear and direct refutation of this claim.

5. *Cognitive States and Internal Models*

An internal model Σ of the behavioral pattern f involves postulating the existence of a set X of internal *state variables,* an initial state $x_0 \in X$, and a dynamic relationship g linking the stimuli u and the states, as well as a rule h specifying how internal states combine to generate the response y. More compactly, we have

$$\begin{aligned} x_{t+1} &= \phi(x_t, u_t), \quad x_0 = x_0 \\ y_t &= h(x_t), \end{aligned} \tag{Σ}$$

$x_t \in X$, $u_t \in U$, $y_t \in Y$. We then say that Σ is an internal model of the observed behavior f if the stimulus-response pattern of Σ, i.e., the observed input/output behavior of Σ, agrees with that of $\mathcal{I}$. Note that in order for this to happen, it's necessary to construct an appropriate set X, together with appropriate maps

$$\begin{aligned} \phi &: X \times U \to X, \\ h &: X \to Y. \end{aligned}$$

From an abstract point of view, the first step in the structuralist program is to ensure that for any given external model $\mathcal{I} = (\Omega, \Gamma, f)$, a corresponding internal model $\Sigma = (X, g, h, x_0)$ exists. If this is the case, it would then be natural to associate the abstract states X with the postulated physical states of the brain in some fashion, while interpreting the maps g and h as means for encoding and decoding external stimuli and mental states, respectively. It's one of the great triumphs of mathematical system theory to have been able to provide a rather definitive resolution of this question, happily in the affirmative. This chapter gives a detailed account of this solution as well as an exposition of how the encoding/decoding operations are explicitly carried out.

6. *Realizations and Canonical Models*

Loosely speaking, we can phrase the behavioral-structuralist problem as follows:

> *Given a stimulus-response pattern $B_{\mathcal{I}}$ for the external description $\mathcal{I}$, find an internal model Σ such that $B_{\mathcal{I}} = B_{\Sigma}$, where B_{Σ} is the stimulus-response pattern of Σ.*

It turns out that the solution to this problem is trivially easy: there are an infinite number of models $\Sigma = (X, g, h, x_0)$ such that $B_{\mathcal{I}} = B_{\Sigma}$. So let's make the problem even more interesting by asking: how can we identify a *good* model from this infinitude of candidates? The answer hinges upon invoking a system-theoretic translation of Occam's Razor, i.e., a "good" model is one that is "compact," or "minimal," in some well-defined sense. Now let us make this idea more precise.

Assume we are given *any* model $\Sigma = (X, g, h, x_0)$. Then we say that Σ is *completely reachable* if for any state $x^* \in X$, there exists an input sequence $\omega \in \Omega$ and a time T such that $x_T = x^*$, i.e., the input ω transfers the system state from x_0 at time $t = 0$ to x^* at time T. Notice that the property of complete reachability depends upon Ω, T, and g but is independent of the output function h.

Example

Assume we have the scalar system given by the dynamics

$$x_{t+1} = x_t^2 + u_t, \quad x_0 = 0, \qquad x_t \in X = R^1,$$

where the inputs $u_t \geq 0$, i.e., the inputs consist of the set

$$\Omega = \{(u_0, u_1, u_2, \ldots,) : \; u_i \geq 0\}.$$

In this case, it's clear that $x_t \geq 0$ whatever input sequence $\omega \in \Omega$ is applied. Thus no states $x_T < 0$ can ever be reached, and the system is **not** completely reachable. On the other hand, it's intuitively clear that if we drop the non-negativity constraint on u_t, then any state x_T can be reached from the origin by application of a suitable input sequence (proof?).

Now let's focus upon the output of Σ. We call Σ *completely observable* if any initial state x_0 can be uniquely determined from knowledge of the system input sequence, together with observation of the system output y_t over an interval $0 < t \leq T$. Note that observability depends upon Ω, T and g, as well as h.

Putting the two concepts together, we call Σ *canonical* if it is both completely reachable and completely observable. The minimality criterion is now clear: the state-space X of a canonical model is minimal in the sense that there are no elements in X that cannot be accessed by application of some stimulus, and no two distinct initial states give rise to the same behavior sequence. Thus, a canonical model is characterized by a state-space containing no elements "extraneous" to its input/output behavior B_Σ.

7. *Linear Systems and Canonical Realizations*

In order to give concrete mathematical meaning to the sets Ω, Γ and X and the maps f, g, and h comprising the "data" of the Realization Problem, we consider the case when the sets are vector spaces and the maps are all linear.

Assume our information-processing object $\mathcal{I}$ has m independent input channels and p output channels, and that the inputs and outputs are real numbers. Then for each moment $t = 0, 1, 2, \ldots$, we have $u_t \in R^m$ and $y_t \in R^p$. Thus, the input space Ω consists of *sequences* of elements from R^m, whereas the output space Γ is comprised of sequences of vectors from R^p. Further, for technical reasons, let's assume that there are only a finite number of inputs applied, i.e., there exists an $N < \infty$ such that $u_t \equiv 0$ for all $t > N$. Hence, we have

$$\Omega = \{(u_0, u_1, u_2, \ldots, u_N) : u_i \in R^m,\ N < \infty\},$$

$$\Gamma = \{(y_1, y_2, y_3, \ldots) : y_i \in R^p\}.$$

(*Note:* We impose no finiteness condition on Γ.)

Turning now to the maps (f, g, h), if we assume f is linear, the abstract map $f : \Omega \to \Gamma$ can be represented by the linear input/output relation

$$y_t = \sum_{i=0}^{t-1} A_{t-i}\, u_i, \qquad t = 1, 2, \ldots, \tag{I/O}$$

where the matrices $A_j \in R^{p\times m}$. Another way of looking at this relation is to observe that an input sequence $\omega = (u_0, u_1, u_2, \ldots, u_N)$ is transformed into an output sequence $\gamma = (y_1, y_2, y_3, \ldots)$ by the lower-triangular block Toeplitz matrix

$$\mathcal{F} = \begin{pmatrix} A_1 & 0 & 0 & 0 & \ldots \\ A_2 & A_1 & 0 & 0 & \ldots \\ A_3 & A_2 & A_1 & 0 & \ldots \\ \vdots & & & & \end{pmatrix}, \qquad A_i \in R^{p\times m}.$$

Thus, the *abstract* map f is represented by the *concrete* infinite block matrix $\mathcal{F}$. The above representations make it evident that when we say that f is given, we mean the same thing as saying that we are given the behavior sequence $B_{\mathcal{I}} = \{A_1, A_2, A_3, \ldots\}$, i.e.,

$$f \leftrightarrow \{A_1, A_2, A_3, \ldots\}.$$

There are several points to note about the foregoing set-up:

1) For each t, the relation (I/O) specifies a set of p equations in the pm unknowns of the matrix A_t. In general, unless $m = 1$ (a single-input system), this set of equations is underdetermined and there will be many matrices A_t that will serve to generate the behavior sequence $B_{\mathcal{I}}$; however, this nonuniqueness disappears under the finiteness hypothesis made above on the input sequence ω (see Chapter Seven, Problem 4 for additional details).

2) If there are only a finite number of outputs, i.e., if after some time $T > 1$, we have $y_t = 0$ for all $t \geq T$, then there are only a finite number of nonzero A_i and the matrix $\mathcal{F}$ is effectively finite.

3) Tacitly assumed in the above set-up is that the initial internal state of the system is $x_0 = 0$, i.e., the system starts in an equilibrium state and will not depart from this state unless a nonzero input is applied. If this is not the case, we have a problem of observability which will be discussed later.

Now we turn to the matter of generating a canonical realization given the input/output description f, i.e., given a behavior sequence B.

In view of the linearity assumption, a realization of f will consist of the construction of an n-dimensional vector space X, together with three real matrices F, G, and H of sizes $n \times n, n \times m$, and $p \times n$, respectively. We *assume* for the moment that $\dim X = n < \infty$. Later we'll return to a discussion of the nature and implications of this assumption. The space X represents the *state-space* of our system, and it is related to the given input and output spaces via the dynamical relations

$$\begin{aligned} x_{t+1} &= Fx_t + Gu_t, \qquad x_0 = 0, \\ y_t &= Hx_t, \end{aligned} \tag{Σ}$$

$x_t \in X$, $u_t \in R^m$, $y_t \in R^p$. In view of the assumption that X is finite-dimensional, there is no loss of generality in taking $X = R^n$. Given an input sequence $\omega \in \Omega$, it's clear that Σ generates an output $\gamma \in \Gamma$. If the input/output pair $(u_t, y_{t+1})_\Sigma$ from Σ agrees with the pair $(u_t, y_{t+1})_{I/O}$ given by the relation (I/O) above, then we call Σ an *internal* model of the input/output description f. It's a simple exercise to see that this will be the case if and only if

$$A_t = HF^{t-1}G,$$

for all $t = 1, 2, 3, \ldots$. This relation links the input/output description given by the sequence $\{A_1, A_2, \ldots\}$, with the state-variable description Σ given by the matrices F, G, and H. The above condition assures that the behavior of the two objects f and Σ are identical; however, there may be many systems $\Sigma = (F, G, H)$ satisfying this relation, so we need additional conditions to single out a canonical model from this set of candidates.

We spoke earlier of the concepts of reachability and observability. These notions involved the state-space X and, hence, are properties of the system Σ. Consequently, the reachability and observability of states of Σ must be expressible by means of the matrices F, G, and H. Let us see how this can be done.

From the dynamical relation defining Σ, at time $t = 1$ we have the system state

$$x_1 = Gu_0,$$

i.e., all states that can be "reached" at time $t = 1$ are given as linear combinations of the columns of G. Similarly, at time $t = 2$ the dynamical equations yield

$$x_2 = FGu_0 + Gu_1.$$

Thus, all states reachable at time $t = 2$ are linear combinations of the columns of FG and the columns of G. Carrying on this process, we see that at time $t = k$ the reachable states consist of linear combinations of the columns of the matrices $\{G, FG, F^2G, \ldots, F^{k-1}G\}$. But by our finiteness assumption on $\dim X = n$, the Cayley-Hamilton Theorem implies that

$$F^r = \sum_{i=0}^{n-1} \alpha_{ri} F^i, \qquad \alpha_{ri} \in R,$$

for all $r \geq n$. Consequently, no new states can appear in the above list after the term $F^{n-1}G$. Putting all these observations together, we have the following:

REACHABILITY THEOREM. *A state $x \in R^n$ is reachable from the origin for the system Σ if and only if x is a linear combination of the columns of the matrices $G, FG, F^2G, \ldots, F^{n-1}G$.*

COROLLARY 1. *Every state $x \in R^n$ is reachable (i.e., Σ is completely reachable) if and only if the $n \times nm$ matrix*

$$\mathcal{C} = [G|FG|F^2G|\ldots|F^{n-1}G]$$

has rank n.

COROLLARY 2. *If a state $x \in R^n$ is reachable, then it is reachable in no more than n time-steps.*

COROLLARY 3. *The reachable states form a subspace of R^n, i.e., if $x, \bar{x} \in R^n$ are reachable, the states $\alpha x + \beta \bar{x}$ are also reachable for all real α, β.*

Example: The Controlled Pendulum

Consider the dynamics describing the (small) oscillations of a pendulum of unit length and mass with controllable velocity (in discrete-time)

$$x_{t+1} = Fx_t + Gu_t, \qquad x_0 = 0,$$

where

$$x = \begin{pmatrix} x_1 \\ x_2 \end{pmatrix}, \qquad F = \begin{pmatrix} 0 & 1 \\ -1 & 0 \end{pmatrix}, \qquad G = \begin{pmatrix} 1 \\ 0 \end{pmatrix}.$$

Here x_1 is the position and x_2 the velocity of the pendulum. Imagine we want to move the pendulum to the zero position, but with unit velocity. Can this be done using control on the velocity alone? Intuitively the answer is yes, since we have free control over the velocity. Let's verify this conclusion using the Reachability Theorem.

The state we wish to reach is $x^* = (0,1)'$. According to the dynamics, we have

$$x_1 = Gu_0 = u_0 \begin{pmatrix} 1 \\ 0 \end{pmatrix}, \qquad u_0 \in R.$$

Thus, after one time-step we can reach any nonzero position, but with zero velocity. Consequently, x^* is not reachable at the first step. At time $t = 2$ we have

$$x_2 = FGu_0 + Gu_1 = u_0 \begin{pmatrix} 0 \\ -1 \end{pmatrix} + u_1 \begin{pmatrix} 1 \\ 0 \end{pmatrix}.$$

Here we see that the state x^* can be reached at time $t = 2$ simply by choosing $u_0 = -1, u_1 = 0$. In fact, we have

$$\mathcal{C} = [G|FG] = \begin{pmatrix} 1 & 0 \\ 0 & -1 \end{pmatrix} = \text{rank } 2,$$

implying that all states $x \in R^2$ can be reached by time $t = 2$.

Now let's turn to the question of observations. So far, we have assumed the system Σ started off in the state $x_0 = 0$. What if we cannot verify this assumption? Do we have any means for identifying the actual initial state solely from knowledge of the system inputs and observation of the output sequence $(y_1, y_2, y_3, \ldots)$? If so, then we say that x_0 is *observable;* if we can identify any initial state x_0 from the inputs and outputs, then we call Σ *completely observable.*

At first glance, it would appear from the dynamical equations for Σ that the determination of x_0 from the output would depend upon the particular input sequence, since the input generates the output. In general, this is true; however, for a large class of systems, including linear systems, we can skirt this difficulty by observing that the output is given explicitly by the formula

$$y_t = HF^t x_0 + \sum_{i=0}^{t-1} HF^{t-i-1} G u_i, \qquad t = 1, 2, \ldots .$$

Here we see that the effect of the initial state x_0 upon the output is separated from the effect of the input sequence. Thus, without loss of generality, we can use the standard input sequence $u_i \equiv 0$, for all $i = 0, 1, 2, \ldots$, since if this is not the case, we consider the output sequence

$$\bar{y}_t = y_t - \sum_{i=0}^{t-1} HF^{t-i-1} G u_i,$$

instead of the given output sequence. So for all questions of observability, we assume $u_i \equiv 0$.

Now consider what it would be like for a state x_0 *not* to be observable. This would mean that there is another state $\bar{x}_0 \neq x_0$ such that $\bar{x}_0$ and x_0 yield identical outputs for all t, i.e., x_0 and $\bar{x}_0$ are indistinguishable on the basis of the output. Mathematically, this means

$$HF^t x_0 = HF^t \bar{x}_0,$$

or

$$HF^t (x_0 - \bar{x}_0) = 0, \qquad t = 0, 1, 2, \ldots .$$

However, again we can appeal to the Cayley-Hamilton Theorem to see that the above condition need be checked only for a finite number of values of t. More precisely, if $\dim X = n < \infty$, we will have x_0 indistinguishable from $\bar{x}_0$ for all t, if and only if

$$HF^t (x_0 - \bar{x}_0) = 0. \qquad t = 0, 1, 2, \ldots, n-1.$$

Another way of saying this is that the state $x_0 - \bar{x}_0$ yields the same output as would the initial state $x_0 \equiv 0$. Putting all these remarks together, we obtain the following result:

OBSERVABILITY THEOREM. *An initial state $x_0 \in R^n$ is unobservable for the system Σ if and only if x_0 is contained in the kernel of the matrix*

$$\theta = \begin{pmatrix} H \\ HF \\ HF^2 \\ \vdots \\ HF^{n-1} \end{pmatrix}.$$

(*Recall:* If A is an $n \times m$ matrix, $\ker A = \{x \in R^m : Ax = 0\}$.)

COROLLARY 1. *Σ is completely observable if and only if rank $\theta = n$, i.e., $\ker\theta = \{0\}$.*

COROLLARY 2. *The unobservable states form a subspace of R^n.*

The Reachability and Observability Theorems enable us to split up the state-space X into four disjoint pieces: states that are reachable/unreachable combined with those that are observable/unobservable. This is the content of the so-called Canonical Decomposition Theorem for linear systems. As noted earlier, our interest is only in the piece of X corresponding to those states that are *both* reachable and observable, since they are the only states that can play a role in generating the input/output map f. For this reason, we call an internal model $\Sigma = (F, G, H)$ a *canonical model* of f if and only if Σ is both completely reachable and completely observable. Summarizing our development thus far, we are given the external behavior

$$B = \{A_1, A_2, A_3, \ldots\}, \qquad A_i \in R^{p\times m},$$

and we call a system $\Sigma = (F, G, H)$ a canonical model for B if

1) $A_i = HF^{i-1}G, \quad i = 1, 2, 3, \ldots$

2) Σ is completely reachable and completely observable.

Example: The Fibonacci Numbers

Suppose we apply the input sequence

$$\omega = \{1, 0, 0, \ldots, 0\}$$

and observe the output

$$\gamma = \{1, 1, 2, 3, 5, 8, 13, \ldots\}.$$

Here we recognize γ as the sequence of Fibonacci numbers generated by the recurrence relation

$$y_t = y_{t-1} + y_{t-2}, \qquad y_1 = y_2 = 1.$$

Since the inputs and outputs are scalars, we see that any linear system with this input/output behavior corresponds to a system with $m = p = 1$, i.e., a single-input/single-output system. Furthermore, it's easy to see that the system's behavior sequence is given by

$$B = \{1, 1, 2, 3, 5, 8, 13, \ldots\},$$

i.e., $A_i \in R^{1\times 1}$ and $A_i \equiv y_i$, $i = 1, 2, 3, \ldots$.

Consider the dynamical model

$$x_{t+1} = Fx_t + Gu_t, \qquad x_0 = 0,$$
$$y_t = Hx_t,$$

with $x_t \in R^2$ and

$$F = \begin{pmatrix} 0 & 1 \\ 1 & 1 \end{pmatrix}, \quad G = \begin{pmatrix} 1 \\ 1 \end{pmatrix}, \quad H = (1 \quad 0).$$

A little algebra soon shows that

$$A_i = y_i = HF^{i-1}G,$$

for this system. Furthermore, both the reachability matrix

$$\mathcal{C} = [G|FG] = \begin{pmatrix} 1 & 1 \\ 1 & 2 \end{pmatrix},$$

and the observability matrix

$$\mathcal{O} = \begin{pmatrix} H \\ HF \end{pmatrix} = \begin{pmatrix} 1 & 0 \\ 0 & 1 \end{pmatrix},$$

have rank $2 = n = \dim X$. Thus, the system $\Sigma = (F, G, H)$ constitutes a canonical model for the Fibonacci sequence.

The foregoing example leaves open the question of just where the system $\Sigma = (F, G, H)$ came from. How did we *know* that $X = R^2$? And how can we compute n and Σ directly from the behavior sequence B? Once we have a candidate model, it's relatively easy to check whether it satisfies the conditions to be a canonical model, but this is far from having a procedure to calculate the model from B alone. We turn now to the solution of this central problem.

8. *A Realization Algorithm*

We begin by writing the infinite behavior sequence B in the so-called Hankel form

$$\mathcal{H} = \begin{pmatrix} A_1 & A_2 & A_3 & A_4 & \cdots \\ A_2 & A_3 & A_4 & A_5 & \cdots \\ A_3 & A_4 & A_5 & A_6 & \cdots \\ \vdots & & & & \end{pmatrix},$$

where $\mathcal{H}_{ij} = A_{i+j-1}$. Let's *assume* the infinite Hankel array $\mathcal{H}$ has finite rank n, i.e., all $r \times r$ submatrices of $\mathcal{H}$ have determinant zero for $r > n$. We will return below to the question of what to do if this assumption cannot be verified. The simplest case where it is easily verified is when A_k is a fixed constant matrix for all $k \geq N$, for some $N \geq 1$.

Under the finite rank hypothesis, there exists an $r < \infty$ such that

$$A_{r+j+1} = -\sum_{i=1}^{r} \beta_i A_{i+j},$$

for some $\beta_1, \beta_2, \ldots, \beta_r$, $\beta_i \in R$. This also implies that there exist matrices P and Q such that

$$P\mathcal{H}_{rr}Q = \begin{bmatrix} I_n & 0 \\ 0 & 0 \end{bmatrix},$$

where $\mathcal{H}_{rr}$ is the $r \times r$ submatrix of $\mathcal{H}$ consisting of its first r rows and columns, and $I_n = n \times n$ identity matrix. A canonical realization of the behavior sequence B can now be obtained using P and Q according to the following prescription.

Ho Realization Algorithm. *Assume the Hankel array $\mathcal{H}$ has finite rank n. Then a canonical realization of the behavior sequence*

$$B = \{A_1, A_2, A_3, \ldots\}$$

is given by the system $\Sigma = (F, G, H)$, where the state-space X has dimension n and

$$F = \mathcal{R}_n P \sigma(\mathcal{H}_{rr}) Q \mathcal{C}^n, \quad G = \mathcal{R}_n P \mathcal{H}_{rr} \mathcal{C}^m, \quad H = \mathcal{R}_p \mathcal{H}_{rr} Q \mathcal{C}^n.$$

(Here the term $\sigma(\mathcal{H}_{rr})$ means "left-shift" the matrix $\mathcal{H}_{rr}$, i.e., move all rows of $\mathcal{H}_{rr}$ one position to the left, and the operators $\mathcal{R}_n$ and $\mathcal{C}^m$ have the effect of "keep the first n rows" and "keep the first m columns," respectively.)

Example: Fibonacci Numbers (Cont'd.)

To illustrate Ho's algorithm, we return to the behavior sequence given in the preceding section consisting of the Fibonacci numbers $B = \{1, 1, 2, \ldots\}$. Forming the Hankel array, we have

$$\mathcal{H} = \begin{pmatrix} 1 & 1 & 2 & 3 & 5 & 8 & \cdots \\ 1 & 2 & 3 & 5 & 8 & 13 & \cdots \\ 2 & 3 & 5 & 8 & 13 & 21 & \cdots \\ \vdots & & & & & & \end{pmatrix}.$$

Because we *know* that B consists of the Fibonacci sequence generated by the recurrence relation $A_i = A_{i-1} + A_{i-2}$, it's easy to see that each column of $\mathcal{H}$ is the sum of the two preceding columns; hence, we can verify the finite rank assumption and obtain rank $\mathcal{H} = n = 2$.

To apply Ho's algorithm, all we need to do now is find matrices P and Q, reducing $\mathcal{H}$ to the standard form. However, given that rank $\mathcal{H} = 2$, together with the action of the operators $\mathcal{R}_2$ and $\mathcal{C}^2$, it suffices to confine our attention to the 2×2 submatrix of $\mathcal{H}$ given by

$$\mathcal{H}_2 = \begin{pmatrix} 1 & 1 \\ 1 & 2 \end{pmatrix} ,$$

and reduce $\mathcal{H}_2$ to standard form, i.e., find matrices P_2 and Q_2 such that

$$P_2 \mathcal{H}_2 Q_2 = \begin{pmatrix} 1 & 0 \\ 0 & 1 \end{pmatrix} = I_2 .$$

It's easy to see that such a pair is given by

$$P_2 = \mathcal{H}_2^{-1} = \begin{pmatrix} 2 & -1 \\ -1 & 1 \end{pmatrix},$$

$$Q_2 = I_2 = \begin{pmatrix} 1 & 0 \\ 0 & 1 \end{pmatrix}.$$

Furthermore, the left-shift of $\mathcal{H}_2$ is given by

$$\sigma(\mathcal{H}_2) = \begin{pmatrix} 1 & 2 \\ 2 & 3 \end{pmatrix}.$$

Applying the prescription of the Ho Realization Algorithm yields the canonical system

$$\begin{aligned} F &= \mathcal{R}_2 P_2 \sigma(\mathcal{H}_2) Q_2 \mathcal{C}^2, \\ &= \mathcal{R}_2 \mathcal{H}_2^{-1} \sigma(\mathcal{H}_2) I_2 \mathcal{C}^2, \\ &= \begin{pmatrix} 0 & 1 \\ 1 & 1 \end{pmatrix}, \end{aligned}$$

$$\begin{aligned} G &= \mathcal{R}_2 P_2 \mathcal{H}_2 \mathcal{C}^1, \\ &= \mathcal{R}_2 \mathcal{H}_2^{-1} \mathcal{H}_2 \mathcal{C}^1, \\ &= \begin{pmatrix} 1 \\ 0 \end{pmatrix}, \end{aligned}$$

$$\begin{aligned} H &= \mathcal{R}_1 \mathcal{H}_2 Q_2 \mathcal{C}^2, \\ &= \mathcal{R}_1 \mathcal{H}_2 I_2 \mathcal{C}^2, \\ &= (1 \ \ 1). \end{aligned}$$

It's now a simple matter to verify that the system

$$x_{t+1} = \begin{pmatrix} 0 & 1 \\ 1 & 1 \end{pmatrix} x_t + \begin{pmatrix} 1 \\ 0 \end{pmatrix} u_t, \qquad x_0 = 0,$$

$$y_t = (1 \quad 1)\, x_t,$$

produces the Fibonacci sequence using the standard input $u_0 = 1$, $u_t = 0$, $t > 0$. Further, a simple computation shows that the system $\Sigma = (F, G, H)$ is both completely reachable and completely observable.

Remarks

1) Comparing the canonical model for B given in the last section with the model Σ given above, we see they involve different matrices G and H. This is to be expected since the matrices P and Q are not unique; we can always introduce a nonsingular matrix T and use the matrices $\widehat{P} = PT$, $\widehat{Q} = T^{-1}Q$, which will, in general, yield the system matrices $\widehat{F} = TFT^{-1}$, $\widehat{G} = TG$, and $\widehat{H} = HT^{-1}$. From elementary linear algebra, we know that such a transformation T represents a linear change of variables in the state-space X. The choice of coordinate system in X is the only arbitrariness in the canonical realization of B; once the coordinate system is fixed, i.e., T is chosen, then the realization of B is unique.

2) Even though the actual matrices F, G, H may not be unique, the dimension of X remains the same for *all* canonical realizations of B. Thus, all canonical models for the Fibonacci sequence have state-spaces X with $\dim X = 2$.

3) Ho's Algorithm is far from the most computationally efficient method for finding a canonical realization. It has been singled out for attention here for two reasons: i) historically, it was the first such procedure, and ii) it's easy to explain and see its tie-in with the classical invariant factor algorithm of linear algebra. For pointers to its computationally more efficient relatives, the reader is urged to consult the Notes and References.

9. *Partial Realizations*

The use of Ho's Algorithm to construct a canonical model of the Fibonacci sequence hinged critically upon the knowledge that $\dim X = n < \infty$; moreover, we made explicit use of the precise value $n = 2$, although it's clear that once n is known to be finite, then a finite, though perhaps tedious, calculation will produce the exact value. But what's to be done if we are denied this *a priori* information? In other words, what if we don't know how to "fill-in the dots" in the sequence $B = \{A_1, A_2, A_3, \ldots\}$? This means, in effect, that we have only a *finite* sequence of data available and we wish to develop a procedure for canonically realizing it. This is the so-called *partial realization problem,* and leads to some of the deepest and most difficult waters in mathematical system theory. Here we will only touch upon one or two aspects of particular interest.

Define the $n \times m$ principal minor of the Hankel array $\mathcal{H}$ as

$$\mathcal{H}_{n,m} = \begin{pmatrix} A_1 & A_2 & A_3 & \cdots & A_m \\ A_2 & A_3 & A_4 & \cdots & A_{m+1} \\ \vdots & & & & \\ A_n & A_{n+1} & A_{n+2} & \cdots & A_{n+m-1} \end{pmatrix}.$$

We say that $\mathcal{H}$ satisfies the Rank Condition if there exist integers n, m such that

$$\operatorname{rank} \mathcal{H}_{n,m} = \operatorname{rank} \mathcal{H}_{n+1,m} = \operatorname{rank} \mathcal{H}_{n,m+1}, \tag{RC}$$

i.e., the rank $\mathcal{H}_{n,m}$ remains unchanged by the addition of one block row or one block column. Whenever the Rank Condition holds, we can appeal to the following result as justification for the use of Ho's Algorithm.

PARTIAL REALIZATION THEOREM. *Let n, m be integers such that the Rank Condition (RC) is satisfied. Then a system Σ given by Ho's Algorithm applied to $\mathcal{H}_{n,m}$ is a canonical model of the behavior sequence $B_{n,m} = \{A_1, \ldots, A_{n+m}\}$.*

Furthermore, if the Rank Condition is not satisfied for n, m, then every partial realization of $B_{n,m}$ has dimension greater than rank $\mathcal{H}_{n,m}$.

Example: The Fibonacci Numbers (Cont'd.)

For the Fibonacci sequence $B = \{1, 1, 2, 3, \ldots\}$, it's easily checked that the Rank Condition is first satisfied for $n = m = 2$, i.e.,

$$\operatorname{rank} \mathcal{H}_{22} = \operatorname{rank} \mathcal{H}_{32} = \operatorname{rank} \mathcal{H}_{23} = 2.$$

Thus, using Ho's Algorithm with $\mathcal{H} = \mathcal{H}_{22}$, we obtain the same canonical realization as in the last section. The Partial Realization Theorem then ensures that this model will canonically realize the finite sequence

$$B_{2,2} = \{A_1, A_2, A_3, A_4\} = \{1, 1, 2, 3\}.$$

Remark

It can be shown that for *any* infinite sequence $\{A_1, A_2, A_3, \ldots\}$, the Rank Condition will be satisfied infinitely often. Hence, every finite sequence of data can be canonically realized as a subsequence of some finite behavior sequence.

10. *The State-Space X*

For our goal in discussing the role of linear systems in modeling cognitive thought processes in the brain, it turns out to be useful to give a somewhat more explicit characterization of the elements of the canonical state-space X. So far we have just thought of the elements of X, the states, as points of the space R^n. In the advanced theory of linear systems, a much deeper interpretation is attached to the states: a state x represents an "encoding" of an input ω in the most compact form possible consistent with the production of the output γ associated with ω by the input/output map f. In more technical terms, each state in X is an *equivalence class of inputs,* where we think of two inputs ω, ω' as being equivalent if they generate the same output under f, i.e., $\omega \approx \omega'$ if and only if $f(\omega) = f(\omega')$. Thus, somehow the encoding of $\omega \to x = [\omega]_f$ represents the way in which the system Σ "remembers" the input ω. But what can such an encoding operation look like?

The simplest way to see explicitly how x codes the input ω is to *formally* associate each input ω with a vector of polynomials. Since, by definition,

$$\omega = (u_0, u_1, u_2, \ldots, u_N),$$

with each $u_i \in R^m$, $N < \infty$, we can formally identify ω with the vector polynomial

$$\omega \longleftrightarrow u_0 + u_1 z + u_2 z^2 + \cdots + u_N z^N \doteq \omega(z),$$

where z is an indeterminate symbol. It's important to note here that we are *not* thinking of z as any sort of number; it is just an indeterminate symbol whose only role is to act as a "time-marker" for the input ω, i.e., the element z^t corresponds to the "time t" at which the input u_t is applied to the system.

Now we turn to the internal dynamics matrix F. With the above identification of input sequences with polynomials, we have the isomorphism $\Omega \approx R^m[z]$, the set of polynomials with coefficients in R^m. Let $\psi_F(z)$ be the minimal polynomial of F, i.e., $\psi_F(z)$ is the nonzero polynomial of least degree such that $\psi_F(F) = 0$. As is well-known from matrix theory, $\psi_F(z)$ is a divisor of the *characteristic polynomial* of F which is always of degree n if $F \in R^{n\times n}$. Hence, $\deg \psi_F(z) \leq n < \infty$.

For a variety of technical reasons that would take us too far afield to elaborate here, it turns out that

$$x \doteq [\omega]_f \in \Omega \bmod \psi_F(z),$$

i.e., x is the remainder after division of the vector polynomial ω by the scalar polynomial $\psi_F(z)$.

Example

Let us return to the Fibonacci sequence realization given above. Just as $\dim X = 2$ for any realization, it is also true that $\psi_F(z)$ is the same for each canonical model. So let's take F as

$$F = \begin{pmatrix} 0 & 1 \\ 1 & 1 \end{pmatrix},$$

which gives $\psi_F(z) = z^2 - z - 1$. For a general scalar input

$$\omega = u_0 + u_1 z + u_2 z^2 + \cdots + u_N z^N,$$

we compute

$$x = \text{remainder } \left[\omega / \psi_F(z)\right].$$

As ω ranges over all of $R[z]$, it's an easy matter to see that x ranges over the set of all *linear* polynomials, i.e.,

$$X \approx \{\text{polynomials } p(z) : \deg p \leq 1\}.$$

Hence, $\dim X = 2 = \deg \psi_F(z)$, a result true for this case since the minimal and characteristic polynomials of F coincide. In general,

$$\dim X = \deg\{\text{minimal polynomial of } F\}.$$

To see how a *specific* input is coded into an element of X, suppose we apply the standard input $u_0 = 1, u_t = 0$, $t > 0$. Then $\omega \approx 1$ and we have

$$x = \omega / \psi_F(z) = \frac{1}{(z^2 - z - 1)},$$

implying that $x = 1$, i.e., $[\omega]_f = 1$. If we had chosen instead the input

$$u_0 = 1,\ u_1 = 1,\ u_2 = 1,\ u_t = 0, \quad t > 2,$$

then $\omega \approx 1 + z + z^2$, and we would have

$$\begin{aligned} x = [\omega]_f &= \text{remainder } [\omega / \psi_F(z)] \\ &= \text{remainder } \left[\frac{(z^2 + z + 1)}{(z^2 - z - 1)}\right] \\ &= 2z + 2 \end{aligned}$$

The chapter's Problems and Exercises give additional examples of how various systems code inputs as states. The important point to note is that

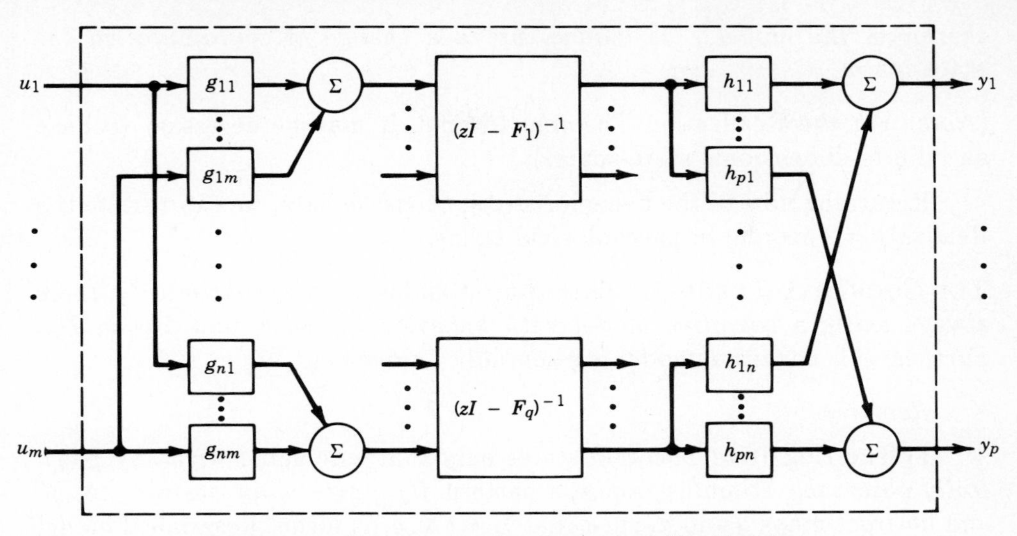

Figure 3.2 Structure of a Canonical Realization

the coding is determined by the system matrix F alone; thus, different systems will encode the inputs differently and, hence, we will see different state-spaces X. This observation provides a precise confirmation of our remarks in Chapter One to the effect that the state-space is **not** an intrinsic part of the system's observed behavior; it is a mathematical construction whose sole purpose is to provide a means to "remember" the inputs in a form convenient to produce the associated outputs.

To conclude our discussion of the Realization Problem, it's instructive to examine graphically the structure of a canonical realization. In Fig. 3.2 we display the general picture of a canonical model. The blocks labeled $(zI - F_i)^{-1}$ correspond to a decomposition of F into Jordan normal form, i.e.,

$$F \cong \operatorname{diag}(F_1, F_2, \ldots, F_q),$$

and the boxes labeled g_{ij} and h_{ij} represent the components of the matrices G and H, respectively. What's significant about this picture is its high degree of internal connectivity: there are a large number of small pieces linked together in a tightly woven web of interconnections. This is a point we shall return to below, as we now leave linear system theory and return to the brain-mind question.

11. *The Cognitive Theorem*

The collection of results given in the preceding sections can compactly be summarized in the following result:

REALIZATION THEOREM. *Given an input/output map $f\colon \Omega \to \Gamma$, there always exists a canonical model $\Sigma = (F, G, H)$ such that $B_I = B_\Sigma$. Fur-*

thermore, the model Σ is unique, up to a change of coordinates in the state-space.

(*Note:* For the Realization Theorem to hold, it may be necessary to have an infinite-dimensional state-space.)

Returning now to the behaviorist-cognitivist debate, we can restate the Realization Theorem in psychological terms.

THE COGNITIVE THEOREM. *Given any stimulus-response pattern $B_{\mathcal{I}}$, there always exists a cognitive model with behavior B_Σ such that $B_\Sigma = B_{\mathcal{I}}$. Further, this cognitive model is essentially unique.*

Remarks

1) The Cognitive Theorem states only that associated with any *physically observable* stimulus-response pattern $B_{\mathcal{I}}$, there is an *abstract* set X and *abstract* maps g and h, such that $\Sigma = (X, g, h)$ forms a canonical model with $B_{\mathcal{I}} = B_\Sigma$. In order for the Cognitive Theorem to form the basis for a cognitive (or materialist) theory of behavior, it's necessary for these abstract objects to somehow be related to *actual* mental states. Later we shall examine just how this might be done.

2) At one level, the Cognitive Theorem says that there is no essential difference between the behaviorist and the cognitive schools of thought: they are *abstractly equivalent.* On another level, the two theories are worlds apart; it all depends upon your point of view. The cognitive model offers an *explanatory mechanism* (the states X and the maps g and h) for behavior that also has a built-in *predictive capacity,* as well (the dynamics $x_{t+1} = \phi(x_t, u_t)$). The behaviorist model provides neither; it offers only a catalog of experimental observations; the raw data, so to speak.

12. *Memory and Behavior*

The diagrammatic representation of the realization problem given in Section 2 makes it clear that the role of the state-space X is to "mediate" in some way between the external inputs from Ω and the observed outputs from Γ. But what kind of psychological interpretation can we attach to this process of mediation? Or, more precisely, what functional interpretation can we attach to the maps g and h? As already discussed, the only consistent answer to this question is to assert that the role of g is to "encode" an external stimuli ω and represent ω internally as a state, whereas the role of h is to "decode" a state and thereby produce an observable output γ. We have already seen that this encoding operation can be explicitly represented by operations involving polynomials, at least for the case of linear systems, so that we could think of Σ as a kind of "pattern recognition" device: the input pattern ω is "remembered" as the state $[\omega]_f$, which is represented by any polynomial ω^* such that $\omega - \omega^* \equiv 0 \mod \psi_F$. The simplest such polynomial ω^* is obtained by dividing ω by ψ_F and designating the remainder as ω^*.

Let's imagine that the stimulus-response pattern $f\colon \Omega \to \Gamma$ corresponds to some sort of elementary behavior (raising your arm, blinking your eye, scratching your nose, etc.). Then the Cognitive Theorem tells us that however complicated the behavior of f may be, it is composed of a combination of elementary behavioral "atoms," interconnected as in Fig. 3.2. It is the high level of interconnectivity that enables an arbitrarily complicated behavior f to be composed of such elementary behaviors. From an evolutionary standpoint, we would not expect to see a noncanonical realization of f because it's just too large and unwieldy (too many complicated components substituting for the high level of connectivity). In short, it's more efficient and reliable to interconnect many simple behavioral modes than to rely on fewer, more complicated types. And this is exactly what we do see when we look at a brain: the elementary components, the neurons, have extremely simple behavior, but the density of interconnections is overwhelmingly large. Such experimental observations strongly argue for the view of a brain as a canonical realization of observed behavior.

Now let's turn to the problem of *pattern recognition.* It's clear that one of the characteristic features of intelligence is the ability to learn and respond to a wide variety of external stimuli (patterns). Once a pattern is learned, in some fashion the brain must be able to recognize the pattern again among the myriad patterns presented by the external world. The set-up outlined above provides a simple criterion for how this can be accomplished.

Imagine the pattern we want to recognize is represented by the input $\Phi \in \Omega$, and we want to build a system that fails to react to any other input $\pi \neq \Phi$. By what has gone before, we know that the system is totally characterized by its minimal polynomial ψ, so our problem is to find a system Σ that recognizes Φ, but fails to respond to an input $\pi \neq \Phi$. In the language of polynomials, the solution is trivial: we need to find a ψ such that $\psi | \pi$ but $\psi \not| \Phi$. A solution is possible if and only if $\pi \not| \Phi$. (*Interpretation:* The pattern discrimination problem can be solved as long as the pattern to be recognized Φ is not a multiple of the recognition circuit *and* every pattern we want to reject is such a multiple. In order to carry out such a discrimination, a brain must clearly have many such elementary "circuits" wired-up in various series-parallel combinations. Referring to Fig. 3.2, we could imagine each of the blocks in the canonical realization as being one such elementary circuit, the entire circuit being devoted to recognition of a *single* such pattern Φ. Then a brain would consist of an unimaginably large number of copies of Fig. 3.2.)

To summarize our position so far, each compound behavior $(\omega, f(\omega))$ is remembered as a state of the "machine" Σ that cognitively represents f.

The system Σ is explicitly generated so that it will produce the "response" $\gamma = f(\omega)$ when the input ω is applied. However, if some other input $\pi \neq \omega$ is given, then Σ will, in general, produce a response $f(\pi) \neq \gamma$. Thus, Σ is capable of generating many behaviors besides the specific pattern $(\omega, f(\omega))$ for which it was developed. This observation allows us to present a theory for how *internal* (i.e., unobservable) stimulus-response patterns may arise, and their connection to the issue of thoughts, feelings and emotions. This theory is based upon the idea that internal concepts like thoughts and feelings arise as a result of "parasitic" impulses arising out of the brain's *normal* functions associated with the processing of sensory inputs into behaviors. In the next section, we examine how such an idea can be given mathematical form within the context of our dynamical system metaphor.

13. *Thoughts and Group Invariants*

So far, we have created a mechanism at the functional level whereby external behavioral modes of an organism can be coded and decoded via the internal mental states of some kind of "brain." Now we wish to explore the manner in which this mechanism might give rise to what we ordinarily regard as internal "thoughts," as distinct from external behavioral activities. Any decent model for a brain must account for subjective emotional experiences like pain, love, jealousy and pleasure, and not just externally observed actions like motion, sleep and talking.

According to recent work in brain physiology, the central cortex of the brain consists of around 4 million neuronal modules, each composed of a few thousand nerve cells. Each module is a column that is vertically oriented across the cerebral cortex, and is about 0.25 mm across and 2 to 3 mm long. These modules are now recognized to be the functional units of communication throughout the association cortex which forms about 95% of the human neocortex. Thus, a human brain can be thought of as something like a piano with 4 million keys. Carrying the musical analogy a step further, let us postulate four parameters that the cortical modules utilize in generating the virtually infinite number of spatiotemporal patterns that constitute the conscious experiences that can be derived from the brain. These parameters are *intensity* (the integral of the impulse firing in the particular module's output lines), the *duration* of the impulse firing from the module, the *rhythm* (or temporal pattern of modular firings) and the *simultaneity* of activation of several modules.

As a working hypothesis, we have associated each neuronal module with a state-variable model $\Sigma_f = (F, G, H)$ of a particular behavioral pattern f. Thus, even though a module Σ_f is *originally* needed to account for the pattern f, once the mechanism (wiring diagram) corresponding to Fig. 3.2 is physically implemented in the neuronal hardware, the module Σ_f may generate many other behavioral responses as well. We have already seen that the output from such a system Σ_f is

$$y_t = \sum_{i=0}^{t-1} HF^{t-i-1}Gu_i .$$

Thus, Σ_f will reproduce the input/output behavior f as long as the input sequence ω is that originally given as part of the description of f; however, if a different input sequence ω^* is given, then Σ_f will, in general, produce an output sequence $\gamma^* \neq \gamma$. As a consequence, each of our neuronal modules Σ_f will correspond to a particular "learned" behavioral mode f, but it can also produce an infinite variety of other modes $f^* \neq f$, once the neuronal pathways (essentially the connective structure of F, together with the connections G and H linking the inputs and outputs to the states) have been laid down.

The foregoing type of ambiguity (or lack of one-to-one correspondence) between f and Σ_f can be eliminated by employing the tacit assumption that a standard input sequence is used, generally $u_0 = 1, u_t = 0, t \neq 0$. It's tempting to conjecture that much of the processing of stimuli carried out by the body's receptor organs is arranged to implement such a normalization prior to the input reaching the neuronal module. We shall assume that this is the case and that there is a one-to-one match between cortical modules and behaviors.

Under the foregoing hypotheses, there are on the order of 4 million or so "elementary" behaviors, one for each cortical module. These "atoms"of cognitive life correspond to the keys on the piano. The intermodular connections, coupled with the four parameters of intensity, duration, rhythm and simultaneity, then generate all behavioral modes. Let's now take a look at how these elementary behavioral modules could be stored in a brain.

First of all, each module consists of about 2500 neurons capable, therefore, of storing 2500 bits of information. If we assume that a single real number requires 25 bits, then this means that a given module can store around 100 real numbers. If the system $\Sigma_f = (F, G, H)$ corresponding to the module has a state-space of dimension n, and the number of input channels m and output channels p are such that $p, m \leq n$, then to store Σ_f requires $n(p + m + n)$, i.e., $O(n^2)$, numbers. With a brute force storage arrangement of this type, each module can only have a system Σ_f such that $\dim \Sigma_f \leq 10$. But this seems much too small to be able to account for even reasonably complex "elementary" behaviors.

A way out of the foregoing dilemma is to recall that the canonical model Σ_f is determined only up to a change of coordinates in the state-space. Thus, any other model $\widehat{\Sigma}_f = (\widehat{F}, \widehat{G}, \widehat{H}) = (TFT^{-1}, TG, HT^{-1})$, $\det T \neq 0$, will display exactly the same elementary behavior. By standard arguments in linear system theory, it can be shown that as T ranges through the group of nonsingular $n \times n$ matrices, there exists a representative of the behavior class of f, call it $\tilde{\Sigma}_f = (\tilde{F}, \tilde{G}, \tilde{H})$, such that the number of nonfixed elements in $\tilde{\Sigma}_f$ is $O(n)$, i.e., by viewing the states in an appropriate basis, it's possible to represent the behavior f by storing only $O(n)$ numbers. These numbers form what are called the *Kronecker invariants* of the group action given by letting the nonsingular $n \times n$ matrices act upon the state-space X. A reasonable conjecture is that evolutionary adaptation has arranged matters so that the "hard-wired" neuronal connections in the cortex are such that the brain represents each learned behavior in something close to this optimal "Kronecker" coordinate system in the state-space. Thus, with the same 2500 neurons in each module, it's possible to accommodate elementary behavioral modes f requiring canonical realizations Σ_f such that $\dim \Sigma_f$ can be on the order of 100 or so, an order of magnitude increase over the brute-force storage scheme.

Up to now we have considered each cortical module Σ_f as a means for representing a given *observed* behavioral pattern f. But what about internal *thoughts?* How can we account for aspects of consciousness involving notions like hope, fear, pain, hunger, thirst and other nonbehavioral, but nonetheless real, mental phenomena? Is there any way to accommodate these aspects of consciousness within the systems framework developed above? Any decent theory of the brain should be able to account for these phenomena, so let's see how it might be done using the framework outlined above.

To make progress on the problems of emotional states and thoughts, let's reconsider the diagram of Σ given in Fig. 3.2 and examine the meaning of the blocks denoted there as g_{ij}, h_{ij} and $(zI - F_i)^{-1}$. Our contention is that the elements g_{ij} and h_{ij} are just pre- and post-processors linking the module to sensory effectors/affectors, *as well as to parts of the brain and to other modules,* while the elements $(zI - F_i)^{-1}$ and the lines into and out of these blocks represent the internal workings of the cortical module itself. With this picture in mind, we consider separately the question of emotional states and cognitive states.

There is now a great deal of experimental evidence suggesting that most emotional states (hunger, pain, taste, etc.) have their origin in the *limbic system,* that collection of nuclei and connecting pathways at the base of the brain. If this is indeed the case, then as far as cortical modules are concerned it makes little difference whether the inputs come from external sensory stimuli or from another part of the brain such as the limbic system, the cerebellum or the optical cortex.

From the perspective of the cerebral cortex, where our modules Σ_f "live," inputs from the sense organs and inputs from the limbic system are treated equally, and appropriate cortical modules are developed early on to handle each. In terms of Fig. 3.2, some of the input channels to the g_{ij} come from sensory receptors, and others from the limbic system. The emotional states arising in the limbic system may or may not evoke observable outputs depending upon the post-processors h_{ij}, since, as we know, sometimes emotional states generate observable responses (crying, hunger pangs, violent movements) and sometimes not. In any case, in our set-up there is no need to distinguish emotional states from sensory stimuli, other than that one comes from the outside world while the other comes only from outside the neocortex. To the "inner world" of the neocortex, they are completely indistinguishable from each other and are processed in exactly the same manner (at least in our theory).

Accounting for cognitive thoughts poses a somewhat more delicate task, since such thoughts are assumed to be self-generated within the cortex itself, quite independently of stimuli from the sense organs or other parts of the brain. Our somewhat speculative approach to this problem is to contend that such thoughts are by-products of primary cortical stimulation through the external input channels g_{ij}. We have already asserted that each module Σ_f is established by a particular behavioral mode f, with the g_{ij} conditioned to pre-process the appropriate stimuli, transforming it into a standard form. But it's also the case that each such cortical module shares connections with 10 or so neighboring modules, which may generate stimuli that feed *directly* into the internal blocks $(zI - F_i)^{-1}$, bypassing the pre-processors. Such inputs would, in general, cause the module Σ_f to emit outputs to the h_{ij} that may even result in a behavioral output different from f if the threshold of the h_{ij} is attained.

In general, we may assume that such direct stimuli from the other modules is weak compared to that from the pre-processors, so when the "real" input signal for f is present the "noise" from the other modules is too feeble to influence Σ_f. Note also that in order for Σ_f to be ready to function properly when the right stimuli for f are applied, it must be the case that the matrix F is stable with a rather quick damping back to the zero state. Otherwise Σ_f would not be in a position to respond properly to rapid repetition of the same stimuli. Just how this condition is translated into the actual physical and chemical properties of the neurons, axons and synapses of the brain is a matter for experimental investigation.

So we conclude that thoughts are generated only when the module Σ_f is in it quiescent state waiting to perform its main function, and such thoughts are generated by the noise present in Σ_f from other modules. At first hearing, this may seem like a very bizarre notion of how thoughts arise since conventional wisdom dictates that thoughts are somehow voluntary creations of the human mind and, as a result, stand above the merely involuntary activities associated with maintainence of the various bodily functions. However, upon further examination the basic idea seems not so outlandish after all, since there is no *a priori* reason to imagine that the brain itself can distinguish between a creative thought and the impulses that generate the bat of an eyelash—they are both electrochemical impulses in the neural circuitry, and it is only the external world that *interprets* them differently. It's at exactly this stage that the idea of *consciousness* enters the picture, an idea that we prefer to steer away from at this stage of our elementary exposition.

To summarize, the brain's cortical modules correspond to elementary behaviors f that are represented internally by the objects Σ_f. For compactness and efficiency, we further contend that nature has arranged things so that the objects Σ_f are stored by the Kronecker invariants of Σ_f under the group of state coordinate changes,

$$F \to TFT^{-1}, \quad G \to TG, \quad H \to HT^{-1}, \quad \det T \neq 0.$$

Each such collection of numbers characterizes an entire *class* of systems Σ_f, all of which canonically represent the same external behavior f. The simplest such element $\widetilde{\Sigma}_f$ of each class contains $O(n)$ parameters, enabling the brain to efficiently reproduce elementary behaviors involving state-spaces of dimension on the order of 100. Since there are around 4 million such cortical modules in the human brain, various series-parallel connections of those elementary behavioral/cognitive "atoms" provide ample material for the almost unlimited variety of thoughts, emotions and experiences of human life. This view of the brain is entirely *functional,* of course, omitting all material aspects associated with the actual physicochemical activities of any real brain. The big experimental challenge would be to bring the functional view presented above into congruence with the known experimental facts surrounding the brain's actual physical structure and behavior.

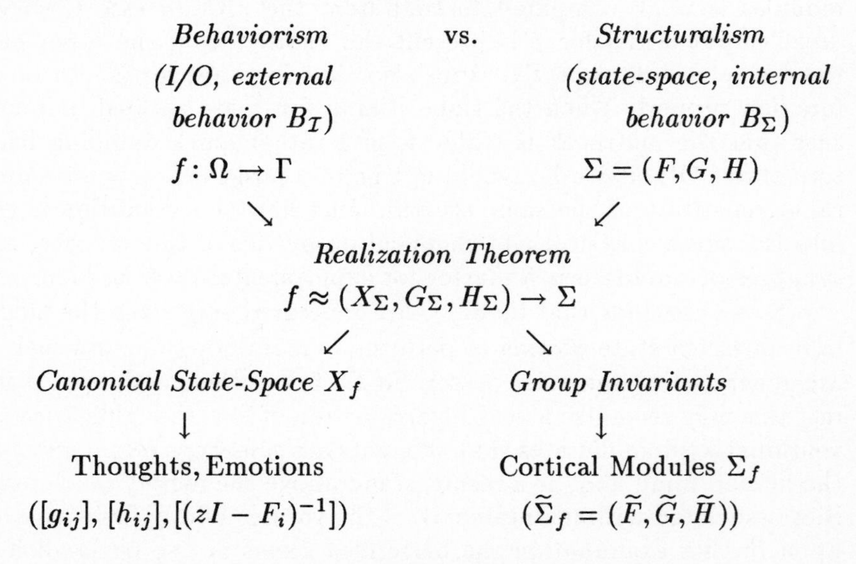

Figure 3.3 A System-Theoretic View of the Brain

Pictorially, we can schematically represent this system-theoretic view of the brain as shown in Fig. 3.3 in which we depict the relationship between the Behavioral and Cognitive (Structural) approaches to brain modeling. This figure shows clearly the central role played by the Realization Theorem in serving as a bridge to illuminate the interrelationships between the two schools of thought in brain modeling. Here we also see the important function of the group invariants in determining the activities of the individual cortical modules, as well as the functional role of the canonical state-space in its relation to the generation of thoughts and emotions in the brain. Of course, the diagram of Fig. 3.3 is only a sketch for a system-theoretic view of the brain and virtually everything remains to be done to turn it into an actual theory as noted above. However, we contend that there is the germ of a major research program encapsulated in this figure for an enterprising cognitive scientist.

14. *Linearity?*

Let's conclude our speculations with a few remarks concerning the emphasis upon *linear* structures in this chapter. After all, given the enormous complexity of the functions the brain clearly performs, on what grounds can we justify the arguments given here which seem to be highly dependent upon linearity? There are several answers to this objection, depending upon the level at which the question is considered. We examine a number of them in turn.

- *Realization Level*—Our main tool has been the Canonical Realization Theorem, and as already noted, the equivalence between a behavior f and a canonical internal state model Σ_f is in no way dependent upon the linearity of f. The theorem is true under very weak hypotheses on f, Ω and Γ. So at this level there is no objection.

We have focused most of our specifics upon the case of linear f because it is the situation in which the algebraic ideas can most easily be made explicit and accessible to nonmathematicians. An important aspect of our development was the description of the canonical state-space as an equivalence class of inputs. For linear processes, this space can be described explicitly by simple mathematical gadgets—polynomials. For more general f, an explicit characterization is either impossible, or at least algebraically much more complicated as, for instance, when f is bilinear, in which case the state-space is an algebraic variety.

Thus, we don't necessarily claim that the intrinsic brain modules are actually linear; only that they are based upon the same *concepts* as given here explicitly for the linear case.

- *Dynamical Level*—At another level, one might object that the dynamical processes of observed neural phenomena are so complicated that there *must* be complicated nonlinearities at work. Perhaps so, but the recent work in cellular automata theory outlined in the last chapter, in chaos theory and fractals detailed in Chapter Five, and other recent work in dynamical system theory seems to indicate otherwise. The main message of all of this work is that very complicated behavior can and does emerge from simple (even linear) local interactions when the number of interconnected subsystems is large enough. And with at least 4 million or so cortical modules, it's not unreasonable to suppose that almost arbitrarily complicated patterns might arise in the brain from linear or almost linear building blocks.

• *Approximation Level*—As discussed, we do not claim that the cortical modules are necessarily linear; however, if they are truly nonlinear, we have the comforting system-theoretic fact that *any* reasonably smooth behavior f can be arbitrarily closely approximated by a *bi*linear process, and such processes are amenable to the same sort of algebraic treatment we have presented for linear processes.

So, in summary, it's not the linearity of f that is important; it's the concept of a canonical realization and the algebraic structure of its associated state-space. These are the ingredients that make the magic work.

Discussion Questions

1. Discuss the thesis that it's possible for artificial devices like computers to have actual mental states. To what degree do you believe that such mental states are uniquely characteristic of human brains? (The *central-state identity* theory asserts that they are; i.e., mental events, states and processes are identical with neurophysiological events in a biological brain.)

2. The theory of *functionalism* is based upon the idea that a mental state can be defined by its causal relations to other mental states and that such states can be realized by many types of systems. What do you think about the prospects for a workable theory of cognitive behavior based upon this type of "bootstrap" theory?

3. One often sees the analogy postulated between the brain/mind and the digital computer's hardware/software, with brain being hardware and mind the software. Does this analogy make any sense to you? If so, how do you think the software gets written? Does the brain's hardware offer any insight into the design of new computer architectures? Can computer software advances give us any help in better understanding of the mind? Do you think the "operating system" is wired in at birth?

4. We have associated the *abstract* states X of a canonical realization with the *physical* states of a brain. In electrical engineering, the states X can

also be associated with the resistors, capacitors and/or inductors of an *RLC* electrical circuit. Consider whether or not the elements of such a circuit and groups of neurons in a brain bear any relationship to each other.

5. Can you think of any experiments that would test our contention that elementary behaviors f are "coded" into the connectivity pattern of a neuronal module?

6. Behavior patterns in real life are generally learned responses; i.e., the behavior f is the result of positive and negative reinforcements (at least this is what the behaviorists claim). How would you modify our system-theoretic model to include learning?

7. The state-space of a canonical realization contains no unreachable or unobservable states. If the abstract states are identified with physical brain states, how would you interpret unreachable and/or unobservable states of the brain?

8. The philosopher Karl Popper and the neurophysiologist J. C. Eccles have postulated a brain-mind theory termed *dual-interactionism,* in which brain and mind are two separate entities interacting through some sort of "liason" modules in the physical brain. Discuss the degree to which such a picture is compatible (or can be made compatible) with the system-theoretic framework developed in this chapter.

9. In the *holographic memory* theory of brain operation as developed by Karl Pribram and David Bohm, the memory function of the brain is postulated to function like a hologram in the sense that memory is not localized in the brain. Roughly, the idea is that the physical brain "tunes in" or somehow "reads" a holographic universe that exists on some plane transcending space and time. Consider the manner in which the "cortical module" model developed here can account for this holographic aspect of brain function.

10. The discussion of this chapter carefully omits any considerations of intelligence. What additions, if any, do you think would be needed to our cognitive model of behavior in order to have it shed light on the question of *intelligent* behavior? In principle, could a computer display intelligent behavior? Why?

11. One of the original and still bench-mark tests for intelligence is the *Turing test,* in which an interrogator (you) is able to pose questions of any sort via teletype to two respondents whose physical nature is hidden from the interrogator. One of the respondents is a computer; the other a human. If, after an extended series of questions and answers to and from both respondents, the interrogator is unable to distinguish the computer from the human, then the computer is deemed to possess intelligence.

Discuss the adequacies of this sort of operational definition of intelligence. In particular, consider whether the passing of the Turing test constitutes "true" intelligence or is only a *simulation* of an intelligent being. Is there any real difference?

12. The computer metaphor for cognition assumes that we are speaking of a *digital* computer. Do you see any difference if we would consider an *analog* computer instead?

13. Gödel's Theorem says that in any consistent axiom system rich enough to generate the arithmetic of the natural numbers, there are statements we cannot prove within this system, but that can be seen to be true by other means. In computer science, it is established that any type of computing machine can be represented by a special type of primitive computer, a *Turing machine.* Virtually all arguments against the mechanization of intelligence have the following common skeleton: a man cannot be represented by a Turing machine (computer) since, if Joe were a realization of some Turing machine T, then (by Gödel) there would be something A that Joe could not do, but Joe *can* do A; therefore, Joe is not a realization of T and this is true of all possible Turing machines T. Hence, Joe transcends the limits of mechanism. (*Remark:* Here the activity A could be taken to be A = prove T's Gödel sentence, i.e., the description of T in Gödel's coding scheme.)

Does this antimechanism argument seem convincing to you? Can you offer any counterarguments or pick out any hidden assumptions underlying the argument?

14. Imagine that a friend told you he had constructed a machine that could feel pain, love, jealousy, fear, and so forth. Suppose further that the machine could communicate to you in ordinary, everyday language, and you asked it to describe what its pains, loves, etc., *felt* like. What do you imagine the response would be? Now imagine that you were of a mischievious nature and wanted to cause this machine some grief, say, to feel jealous of your new girlfriend that the machine just met. What kind of actions/statements would you take to bring out this emotion in the machine?

Problems and Exercises

1. Instead of describing a linear system's external behavior by the sequence $B = \{A_1, A_2, \dots\}$, it's often alternately characterized by the $p \times m$ *transfer matrix*

$$W(z) = \sum_{i=1}^{\infty} A_i z^{-i},$$

where z is a complex parameter. (*Remark:* The matrix $W(z)$ can also be thought of as the linear operator relating the discrete Laplace transform of the system's input and output.)

a) Show that if $\Sigma = (F, G, H)$, then $W(z) = H(zI - F)^{-1}G$.

b) Show that the blocks "$(zI - F_i)^{-1}$" in Fig. 3.2 correspond to a decomposition of the proper rational matrix $W(z)$ into matrix partial fractions.

c) Prove that the complete reachability of Σ corresponds to the condition that the matrix $W(z)$ is *irreducible;* i.e., there exist polynomial matrices $S(z)$ and $T(z)$ of sizes $p \times m$ and $m \times m$, respectively, such that $W(z) = S(z)T^{-1}(z)$, where $\det T(z) \neq 0$, and

$$\text{rank} \begin{pmatrix} S(z) \\ T(z) \end{pmatrix} = m,$$

for every complex z.

2. Consider the behavior sequence

$$B = \{1, 2, 5, 12, 29, 70, \dots\} = \text{ Pell numbers},$$

defined by the recurrence relation $A_{i+2} = 2A_{i+1} + A_i$.

a) Show that B has a two-dimensional canonical realization. Compute it using Ho's Algorithm.

b) Show that the behavior sequence

$$B = \{1, 2, 3, 4, 5, \dots\} = \text{natural numbers},$$

also has a two-dimensional canonical realization. Calculate it.

c) Is there any connection between the canonical system Σ_P of the Pell numbers and the system Σ_N that realizes the natural numbers?

3. Prove that the sequence of *prime numbers*

$$B = \{2, 3, 5, 7, 11, \dots\},$$

has no finite-dimensional realization (difficult!).

4. Consider the input sequence $\omega = \alpha_0 + \alpha_1 z + \alpha_2 z^2 + \cdots + \alpha_N z^N$. Compute the "remembered" state $[\omega]_f$ for the following systems:

a) $\psi_F(z) = z$,

b) $\psi_F(z) = z - 1$ ("integrator"),

c) $\psi_F(z) = z^k$ ("truncator"),

d) $\psi_F(z) = z^k - \alpha$ ("spectrum analyzer"),

e) $\psi_F(z) = z^k + \sum_{i=0}^{k-1} \beta_i z^i, \quad \beta_i \in R$.

5. Let Σ_i, $i = 1,2,3,4$ be linear systems connected in the following series-parallel network

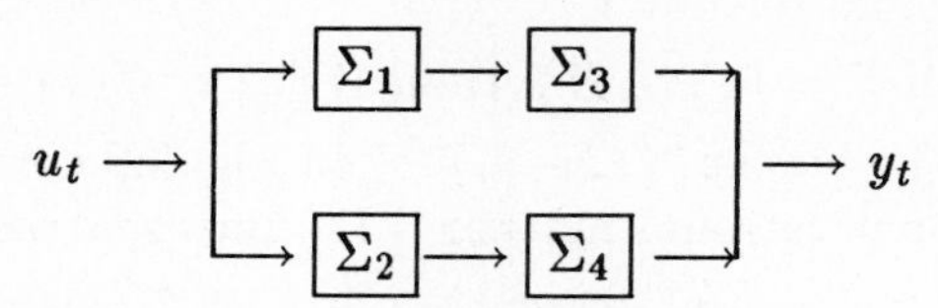

Compute the input/output relation of this system in terms of the Σ_i. (*Remark:* According to Fig. 3.2, the brain consists of a staggeringly large number of copies of this kind of network.)

6. Prove that the properties of complete reachability and complete observability are *generic,* i.e., in the space of all linear systems, the reachable and observable systems form an open, dense set. (*Remark:* What this means is that for each reachable and observable system, there is an open neighborhood of the system containing only reachable and observable systems. Further, every neighborhood of an unreachable and/or unobservable system contains reachable and observable systems.)

7. (Duality Principle) Given the *control* system $\Sigma_C = (F, G, -)$, show that any properties of Σ_C can be expressed as equivalent properties of the *observation* system $\Sigma_O = (F', -, G')$, i.e., the two systems Σ_C and Σ_O are *dual* under the transformation

$$F \longrightarrow F', \qquad G \longrightarrow H'.$$

(*Remark:* This pivotal result enables us to halve the work in system analyses and to consider *only* reachability or *only* observability properties of the system. The Duality Principle also holds for nonlinear systems, at the expense of a more elaborate definition of Σ_C and Σ_O.)

8. (Kalman) Let W_1 and W_2 be two transfer matrices. We say that W_1 *divides* W_2 if there exist polynomial matrices V and U such that $W_1 = VW_2U$. We say that the system Σ_1 *simulates* Σ_2 if the state-space X_1 of Σ_1 is isomorphic to a subspace of X_2.

a) If $\{\psi_i(W)\}$ denote the *invariant factors* of W, $i = 1,2,\ldots r$, show that Σ_1 can be simulated by Σ_2 if and only if $\psi_i(W_1)$ divides $\psi_i(W_2)$ for all i. (*Remark:* This means that a computer with transfer matrix W_2 can simulate a brain with transfer matrix W_1 if and only if each invariant factor of the brain is a divisor of the corresponding invariant factor of the computer.)

b) Show that from a dynamical point of view, the condition $W_1|W_2$ means that the inputs and outputs of the machine having transfer matrix W_2 are re-coded by replacing the original input ω_2 by $U(z)\omega_2$, whereas the output γ_2 is replaced by $V(z)\gamma_2$. Such a change involves a delay $d = \deg \psi_{W_2}$, where ψ_{W_2} is the denominator polynomial of W_2.

9. Determine whether or not the following systems $\Sigma = (F, G, H)$ are completely reachable and/or completely observable:

a) $$F = \begin{pmatrix} 0 & 1 & 0 & \dots & 0 \\ 0 & 0 & 1 & \dots & 0 \\ 0 & 0 & 0 & \dots & 0 \\ \vdots & & & & \\ 0 & 0 & 0 & \dots & 1 \\ \alpha_1 & \alpha_2 & \alpha_3 & \dots & \alpha_n \end{pmatrix}, \quad G = \begin{pmatrix} 0 \\ 0 \\ \vdots \\ 0 \\ 1 \end{pmatrix},$$

$$H = (\beta_1 \quad \beta_2 \quad \dots \quad \beta_n),$$

b) $$F = \begin{pmatrix} \alpha_1 & 0 & \dots & 0 \\ 0 & \alpha_2 & \dots & 0 \\ \vdots & & & \\ 0 & 0 & \dots & \alpha_n \end{pmatrix}, \quad G = \begin{pmatrix} 1 \\ 1 \\ \vdots \\ 1 \end{pmatrix}, \quad H = (\beta_1 \quad \beta_2 \quad \dots \quad \beta_n),$$

c) $$F = \begin{pmatrix} 0 & 0 & \dots & 0 & \alpha_1 \\ 1 & 0 & \dots & 0 & \alpha_2 \\ 0 & 1 & \dots & 0 & \alpha_3 \\ \vdots & \vdots & \dots & \vdots & \vdots \\ 0 & 0 & \dots & 1 & \alpha_n \end{pmatrix}, \quad G = \begin{pmatrix} \beta_1 \\ \beta_2 \\ \vdots \\ \beta_n \end{pmatrix}, \quad H = (0 \quad \dots \quad 0 \quad 1),$$

d) $$F = \begin{pmatrix} 1 & 3 & 4 \\ 0 & 1 & 1 \\ 2 & 3 & 2 \end{pmatrix}, \quad G = \begin{pmatrix} 1 & 1 \\ 0 & 1 \\ 2 & 3 \end{pmatrix}, \quad H = \begin{pmatrix} 1 & 1 & 0 \\ 2 & 2 & 3 \end{pmatrix},$$

e) $$F = \begin{pmatrix} 5 & 9 & 1 & 0 \\ 4 & 12 & 1 & 0 \\ 3 & 0 & 0 & 1 \\ 2 & 1 & 2 & 0 \end{pmatrix}, \quad G = \begin{pmatrix} 1 & 0 & 0 \\ 1 & 2 & 0 \\ 1 & 5 & 1 \\ 2 & 3 & 2 \end{pmatrix},$$

$$H = \begin{pmatrix} 5 & 0 & 6 & 7 \\ 1 & 1 & 2 & 3 \\ 4 & 2 & 1 & 2 \end{pmatrix}.$$

10. Prove that the constant linear system $\Sigma = (F, G, -)$ can be completely reachable if and only if the number of inputs (the number of columns of G) is greater than or equal to the number of nontrivial invariant factors of F.

11. (Kalman) Consider the partial behavior sequence

$$B_6 = \{1, 1, 1, 2, 1, 3\}.$$

a) Show that the system $\Sigma = (1,1,1)$, i.e., $F = [1]$, $G = [1]$, $H = [1]$ realizes the first three terms of B_6.

b) Using the Partial Realization Theorem, show that the "natural" continuation of B_6 is

$$B_{11} = \{1,1,1,2,1,3,2,3,5,2,9\},$$

the behavior for the canonical system $\Sigma = (F,G,H)$, where

$$F = \begin{pmatrix} 1 & 1 & 0 \\ 0 & -1 & 1 \\ 1 & 0 & -1 \end{pmatrix}, \quad G = \begin{pmatrix} 1 \\ 0 \\ 0 \end{pmatrix}, \quad H = (1 \quad 0 \quad 0).$$

c) Show that the transfer matrix for the above system is

$$W(z) = \frac{z^2 + 2z + 1}{z^3 + z^2 - z - 2}.$$

12. Consider the scalar transfer matrix

$$W(z) = \frac{c_{n-1}z^{n-1} + \cdots + c_1 z + c_0}{z^n + d_{n-1}z^{n-1} + \cdots + d_1 z + d_0}$$

a) Show that the system $\Sigma = (F,G,H)$ given by

$$F = \begin{pmatrix} 0 & 1 & \cdots & 0 \\ 0 & 0 & \cdots & 0 \\ \vdots & & & \\ 0 & 0 & \cdots & 1 \\ -d_0 & -d_1 & \cdots & -d_{n-1} \end{pmatrix}, \quad G = \begin{pmatrix} 0 \\ \vdots \\ 0 \\ 1 \end{pmatrix}, \quad H = (c_0 \quad \cdots \quad c_{n-1}),$$

realizes $W(z)$, i.e., $W(z) = H(zI - F)^{-1}G$.

b) Is this realization canonical, in general?

c) If not, now could you make it so?

13. Consider the transfer matrix

$$W(z) = \frac{z^2 + 2z + 1}{z^3 + z^2 - z - 2}.$$

a) Show that the behavior sequence associated with W is

$$B = \{1,1,2,1,3,\ldots\}.$$

b) Compute a canonical realization of $W(z)$.

14. Show that a single-input system $\Sigma = (F,g,-)$ is completely reachable if and only if in the transfer matrix $W(z) = (zI - F)^{-1}g$, the numerator and denominator of $W(z)$ have no common factors.

Notes and References

§1. The brains-minds-machines issue has never been more actively pursued than today, providing the groundwork for that growing academic discipline cognitive science. For an enlightening and entertaining introduction to these matters, see

Flanagan, O., *The Science of the Mind,* MIT Press, Cambridge, 1984,

Churchland, P. M., *Matter and Consciousness,* MIT Press, Cambridge, 1984,

Gunderson, K., *Mentality and Machines,* 2d ed., U. of Minnesota Press, Minneapolis, 1985.

For an account of the brain-mind issue by two of the twentieth century's most prominent thinkers, see

Popper, K., and J. Eccles, *The Self and Its Brain,* Springer, Berlin, 1977.

§2. Mathematical system theory appears to have gotten its initial push from electrical engineering with the problem of circuit design. The fundamental question spawning the subject was the design of a lumped-parameter (RLC) circuit displaying a given input/output voltage pattern with a minimal number of elements (resistors, inductors and capacitors). The generalization of this problem has led to the mathematical theory of linear systems considered in this chapter.

§3. An excellent summary of the development of the behaviorist and cognitive schools of psychology is given in the Flanagan book cited under §1. Further consideration of some of the issues involved, especially from the philosophical standpoint, is given in the popular article

Fodor, J., "The Mind-Body Problem," *Scientific American,* 244 (1981), 114–124.

Important critiques of the behavioral school are

Chomsky, N., "Review of Skinner's *Verbal Behavior,*" in *Readings in the Philosophy of Language,* L. Jakobovits and M. Miron, eds., Prentice-Hall, Englewood Cliffs, NJ, 1967,

Dennett, D., "Skinner Skinned," in Dennett, D., *Brainstorms: Philosophical Essays on Mind and Psychology,* Bradford Books, Montgomery, Vermont, 1978.

The point of view espoused in this chapter is very close to that associated with the Artificial Intelligence (AI) school of thought on the mind-body problem. For introductory accounts of this position, see

Boden, M., *Artificial Intelligence and Natural Man,* Basic Books, New York, 1977,

Hunt, M., *The Universe Within: A New Science Explores the Human Mind,* Simon & Schuster, New York, 1982.

For some adverse views on the AI school of thought, see

Searle, J., "Minds, Brains and Programs," *The Behavioral and Brain Sciences,* 3 (1982), 417–457,

Weizenbaum, J., *Computer Power and Human Reason,* Freeman, San Francisco, 1976,

Dreyfus, H., and S. Dreyfus, *Mind Over Machine,* Free Press, New York, 1986.

§4–5. Textbook accounts of the external vs. internal model view of dynamical processes may be found in

Casti, J., *Linear Dynamical Systems,* Academic Press, Orlando, 1987,

Brockett, R., *Finite-Dimensional Linear Systems,* Wiley, New York, 1970,

Fortmann, T., and K. Hitz, *An Introduction to Linear Control Systems,* Dekker, New York, 1977,

Rosenbrock, H. *State-Space and Multivariable Theory,* Wiley, London, 1970.

§6. The recognition of reachability and observability as the central system properties involved in determining canonical models is traceable to the work of Kalman in the 1960s. The basic references are

Kalman, R., "On the General Theory of Control Systems," *Proc. 1st IFAC Congress, Moscow,* Butterworths, London, 1960,

Kalman, R., "Canonical Structure of Linear Systems," *Proc. Nat. Acad. Sci. USA,* 48 (1962), 596–600.

It's also of interest to note the relationship between the reachability of a single-input linear system and the nth-order equation

$$\frac{d^n x}{dt^n} + a_1 \frac{d^{n-1} x}{dt^{n-1}} + \cdots + a_n x = u(t),$$

usually seen in classical engineering and physics texts. Rewriting this nth-order system in matrix form, we come directly to the *control canonical* form given in Problem 6(a). It's then an easy matter to verify that any such system is completely reachable. Hence, this vital system-theoretic property is "built in" by the manner in which the equation is written. This observation underscores the point that the first step to a good theory is often a good notation!

§7. The Realization Problem, as we have defined it, is a noise-free version of the *statistical* problem of identification of the parameters of a linear model when the observations are subject to various types of stochastic perturbations. A simple version of the statistical problem is when we are given the scalar input/output relation

$$y_t = \sum_{i=1}^{n} A_{t-i}u_i + \epsilon_t, \qquad t = 1, 2, \ldots ,$$

where the observations $\{y_t\}$ and the inputs $\{u_t\}$ are known, and the quantities $\{\epsilon_t\}$ are assumed to be normally distributed random variables representing the uncertainty in the model. The problem is to develop a "good" internal model of the form

$$\begin{aligned} x_{t+1} &= Fx_t + Gu_t, \\ y_t &= Hx_t + \epsilon_t, \end{aligned}$$

so that the behavior of the model agrees with that of the input/output relation in some sense. For various approaches to this kind of "stochastic" realization problem, see the works

Hazewinkel, M., and A. H. G. Rinnooy Kan, eds., *Current Developments in the Interface: Economics, Econometrics, Mathematics,* Reidel, Dordrecht, 1982,

Kalman, R., "Identification of Linear Relations from Noisy Data," in *Developments in Statistics,* P. R. Krishnaiah, ed., Vol. 4, Academic Press, New York, 1982,

Deistler, M., and B. Pötscher, eds., *Modeling Problems in Econometrics, Applied Mathematics & Computation,* Vol. 20, Nos. 3 and 4, 1986.

§8. The Ho Realization Algorithm was first presented in

Ho, B. L., and R. E. Kalman, "Effective Construction of Linear State-Variable Models from Input/Output Functions," *Regelungstech.,* 14 (1966), 545–548.

Other algorithms are given in the books by Casti and Brockett cited under §4–5 above. Of special interest are the computational aspects of the various realization methods, some of which are notoriously numerically unstable. An especially detailed study of the numerical aspects of a well-known recursive algorithm for computing minimal realizations is presented in

de Jong, L. S., "Numerical Aspects of Recursive Realization Algorithms," *SIAM J. Cont. Optim.*, 16 (1978), 646–659.

§9. The partial realization question has been extensively explored by R. Kalman in work summarized in

Kalman, R., "On Partial Realizations, Transfer Functions and Canonical Forms," *Acta Poly. Scand.*, 31 (1979), 9–32.

For other important work along the same lines, see

Kalman, R., "On Minimal Partial Realizations of a Linear Input/Output Map," in *Aspects of Network and System Theory,* R. Kalman and N. de Claris, eds., Holt, New York, 1971,

Rissanen, J., "Recursive Identification of Linear Systems," *SIAM J. Cont. Optim.,* 9 (1971), 420–430.

An account of some of the relationships between the Partial Realization Problem for linear systems and the properties of Hankel and Toeplitz matrices is given in

Fuhrmann, P., "On the Partial Realization Problem and the Recursive Inversion of Hankel and Toeplitz Matrices," in *Linear Algebra and Its Role in Systems Theory,* R. Brualdi, D. Carlson, B. Datta, C. Johnson, and R. Plemmons, eds., Amer. Math. Soc., Providence, RI, 1985.

§10. The treatment given here of the concept of state may seem needlessly abstract, or even eccentric, to those schooled in the more traditional views of classical physics and the idea of a *phase* space composed of the position and velocity vectors of the various components of the system. This view is a very special case of the far more general notion of state, and it's a major conceptual mistake to think of the state as being comprised of elements to which we can attach any sort of intuitive, physical interpretation. Our presentation makes clear the primary role of the state-space: a purely mathematical construction introduced to mediate between the measured inputs and observed outputs. Any physical interpretation that one can attach to the states is purely coincidental and independent of their primary function. The modern algebraic view of linear systems further amplifies and clarifies this point. For an account of this algebraic point of view, see the Casti book cited under §4–5 or

Tannenbaum, A., *Invariance and System Theory: Algebraic and Geometric Aspects,* Springer Lect. Notes in Math., Vol. 845, Berlin, 1981.

§11–13. A much more detailed account of the arguments presented in these sections is found in

Casti, J., "Behaviorism to Cognition: A System-Theoretic Inquiry into Brains, Minds and Mechanisms," in *Real Brains, Artificial Minds,* J. Casti and A. Karlqvist, eds., Elsevier, New York, 1987.

Mathematical models of various brain and mind functions have appeared with great regularity in the literature over the past couple of decades. For instance, Hoffman has developed a model of brain function that is based upon the premise that the neuron is an infinitesimal generator of perceptions, cognition and emotion. These ideas make extensive use of the correspondence between a Lie group germ and neuron morphology to give a very stimulating account of many aspects of form memory and vision. Hoffman uses the usual mathematical structure governing invariance in the presence of an infinitesimal generator, namely, Lie transformation groups together with their prolongations, to establish higher-order differential invariants. These invariants then show how the memory engram is stored within the brain. For an account of these ideas, see

Hoffman, W., "The Neuron as a Lie Group Germ and a Lie Product," *Q. Applied Math.,* 25 (1968), 423–441,

Hoffman, W., "Memory Grows," *Kybernetik,* 8 (1971), 151–157,

Hoffman, W., "Subjective Geometry and Geometric Psychology," *Math. Modelling,* 3 (1981), 349–367.

In a quite different direction, the neurophysiologist Karl Pribram and the theoretical physicist David Bohm have jointly proposed a model of holographic memory storage in the brain, which was briefly considered in Discussion Question 9. Their holographic view has the great merit that it enables one to account for the observed fact that patients with brain damage do not seem to lose their memory function, suggesting that memory is stored in a distributed manner throughout the brain, rather than being localized to particular regions. The Bohm-Pribram theory is discussed more fully in

Bohm, D., *Wholeness and the Implicate Order,* Routledge and Kegan Paul, London, 1980,

Pribram, K., "Towards a Holonomic Theory of Perception," in *Gestalttheorie in der modernen Psychologie,* S. Ertel, ed., Steinkopff, Durnstadt, 1975.

In a long series of papers over the past two decades, Stephen Grossberg has presented a theory of learning, perception, cognition, development and motor control that involves a rather elaborate theory of nonlinear processes, emphasizing the role of "adaptive resonances" in neural circuitry to explain behavioral phenomena. The essence of this work is summarized in

Grossberg, S., *Studies of Mind and Brain,* Reidel, Dordrecht, 1982.

§14. Most of the results that we've presented within the context of linear systems can be extended to a wide class of nonlinear systems, but at the price of a somewhat more elaborate mathematical formalism and machinery. For an account of these matters, see

Casti, J., *Nonlinear System Theory,* Academic Press, Orlando, 1985.

CHAPTER FOUR

Catastrophes and Life: The Singularities of Ecological and Natural Resource Systems

1. *Smoothness and Discontinuity*

Most models of natural phenomena make use of the elementary functions of analysis such as $\exp x$, $\sin x$, $\cos x$, x^k, as the raw material from which the equations of state are formed. These are the smoothest of smooth functions (analytic, even) and, consequently, can display no discontinuous changes of any sort: what happens at a point $x + \Delta x$ is pretty much the same as what's happening at x, if Δx is small enough. Nevertheless, a substantial number of physical procsses *do* display discontinuities, with the onset of turbulence, cell differentiation, beam buckling and stellar collapse being but a miniscule sample of such situations. So how can we effectively capture such discontinuous behavior in mathematical models using smooth components? This is the applied problem that modern bifurcation theory and its close relatives, singularity theory and catastrophe theory, strive to address.

The key to making discontinuity emerge from smoothness is the observation that the overall behavior of both static and dynamical systems is governed by their behavior near their singular, or critical, points. These are the points at which the first derivative of the system's equation of state vanishes. Away from the critical points, the Implicit Function Theorem tells us that the behavior is boring and predictable, linear, in fact; only at the critical points does the system have the possibility of breaking out of this mold and establishing a new mode of operation. It's at the critical points that we have the opportunity to affect dramatically the nature of the system's future behavior by making small "nudges" in the system dynamics, with one type of nudge leading to a limit cycle, another type to a stable equilibrium, and yet a third type resulting in the system moving from its current attractor region into the domain of another. It's in these nudges in the equations of motion that the germ of the idea of discontinuity from smoothness blossoms forth into the modern theory of singularities, catastrophes and bifurcations, and we see how to make discontinuous behavior come out of smooth descriptions.

We can formalize this idea of "nudging" the system by recognizing that virtually all system descriptions, static or dynamic, contain constituent parameters whose values specify the operating conditions of the problem. Be they chemical reaction rates, technological coefficients, the Reynolds number of a fluid or elasticity constants, such parameters define a *class* of systems, and we can regard a "small nudge" as being the same thing as a "small" change of such parameters, i.e., a change from one system to another nearby system. It then makes sense to ask: If we make a small change in parameters, does this result in a *substantial* change in the system's equilibrium (or steady-state) behavior? In this way, we can envision smooth changes in parameters resulting in discontinuous changes in the equilibrium behavior of systems whose descriptions are given in terms of smooth functions.

Example

To fix the foregoing ideas, consider a system described by the one-parameter family of quartics

$$f(x, a) = \frac{x^4}{4} + \frac{ax^2}{2} + x.$$

The singular points, where $\partial f/\partial x = 0$, are given by the set

$$\{x : x^3 + ax + 1 = 0\}.$$

From elementary algebra, it's easy to see that we have three singular points when $|a| < 2/(3\sqrt{3})$, one singular point when $|a| > 2/(3\sqrt{3})$ and a repeated (multiple) singular point when $|a| = 2/(3\sqrt{3})$. Thus, we have the structure shown in Fig. 4.1.

Here we have designated the roots as $x^*(a)$ to explicitly indicate their dependence upon the parameter a. In terms of the root structure, it's evident that the system undergoes a dramatic change at the points $a = \pm 2/(3\sqrt{3})$, shifting from three roots to one. Thus, the map $a \mapsto x^*(a)$ is discontinuous (and multiple-valued) at these values of a. Even though $f(x, a)$ is as smooth as it can be in both x and a, the behavior of the equilibria of f can be highly discontinuous, even with smooth changes in a. It is this kind of discontinuity that catastrophe theory and its parents, singularity theory and bifurcation theory, have been created to study.

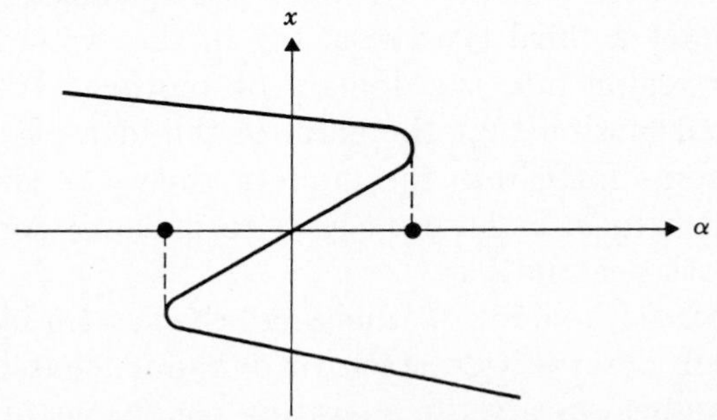

Figure 4.1 Solution Set for $x^3 + ax + 1 = 0$

2. *Smooth Functions and Critical Points*

From now on we shall be concerned only with *smooth* maps and functions, i.e., maps $f\colon R^n \to R^m$ such that f has derivatives of all orders for all $x \in R^n$. Typical examples of such objects are

$$\begin{aligned} f(x_1, x_2) &= x_1^2 + 2x_1x_2^3, \\ f(x_1) &= \exp x_1, \\ f(x_1, x_2) &= \sin x_1 + \cos x_2. \end{aligned}$$

Nonsmooth maps would be things like

$$f(x_1, x_2) = 1/\sqrt{x_1x_2}, \quad \text{or} \quad f(x_1) = \log x_1.$$

We denote such maps by $C^\infty(R^n, R^m)$, and we call them smooth *functions* if $m = 1$. Most of the relationships, models and equations of state of mathematical physics, engineering and the life sciences are expressed using such functions, and our concern will be to look at the behavior of such smooth maps in the neighborhood of a *critical point* x^*, i.e., a point at which we have $\partial f/\partial x = 0$. In component form, if $f(x_1, x_2, \ldots, x_n) = \{f_1(x_1, \ldots, x_n), \ldots, f_m(x_1, \ldots, x_n)\}$, then

$$\frac{\partial f}{\partial x} \doteq \begin{pmatrix} \frac{\partial f_1}{\partial x_1} & \frac{\partial f_1}{\partial x_2} & \cdots & \frac{\partial f_1}{\partial x_n} \\ \frac{\partial f_2}{\partial x_1} & \frac{\partial f_2}{\partial x_2} & \cdots & \frac{\partial f_2}{\partial x_n} \\ \vdots & & & \\ \frac{\partial f_n}{\partial x_1} & \frac{\partial f_n}{\partial x_2} & \cdots & \frac{\partial f_n}{\partial x_m} \end{pmatrix}$$

The reader will recognize this as the usual *Jacobian* matrix of the map f. For notational simplicity, we shall always assume that a preliminary coordinate change in R^n has been carried out so that the critical point of interest is $x^* = 0$.

Consider, for a moment, the situation from elementary calculus when f is a smooth function of a single variable ($n = m = 1$). If $f'(0) = 0$ and we scale the axes so that $f(0) = 0$, then the graph of f might look like Fig. 4.2, while the Taylor series expansion near the origin would be

$$f(x) = f''(0)\frac{x^2}{2!} + f'''(0)\frac{x^3}{3!} + \cdots .$$

Locally, the sign of $f''(0)$ determines the character of $f(x)$ near the origin. If $f''(0) > 0$, we have the situation of Fig. 4.2; if $f''(0) < 0$, then f has a local maximum at the origin, while $f''(0) = 0$ means the origin is an inflection

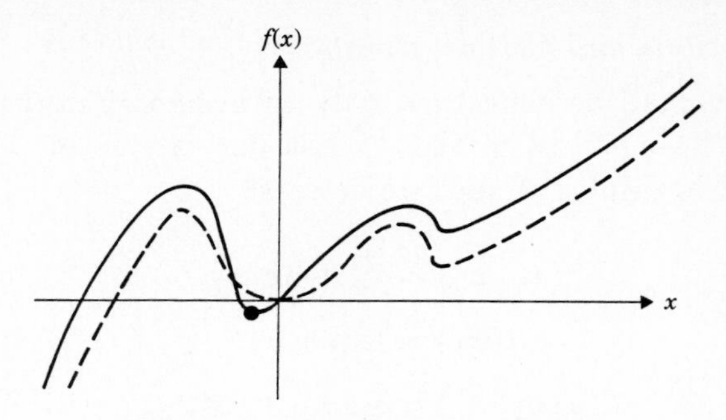

Figure 4.2 The Graph of $f(x)$

point and we need to consider higher-order derivatives to determine the local character of the critical point.

What's important here is that the entire local behavior of f is governed by the first nonvanishing term in the Taylor series expansion, in the above case the term $\frac{1}{2}f''(0)x^2$ (assuming $f''(0) \neq 0$). Thus, locally f is a quadratic, and as we will show later, there is a local coordinate change $x \to p(x) = x'$ such that in the x' variable we have the *exact* local representation $f(x') = \pm x'^2$ (Morse's Lemma). Furthermore, a small perturbation of f to $f + \epsilon g$ will preserve this situation, with only a small shift in the critical point from 0 to a point x_ϵ^* near 0, the quadratic nature of f being unchanged by the perturbation. In singularity theory we try to generalize this type of result by addressing the following basic questions:

1) What happens when n and/or $m > 1$?

2) What about the case of degenerate critical points, i.e., those for which $f''(0) = 0$?

3) What does a "typical" $f \in C^\infty(R^n, R^m)$ look like near the origin?

4) If f is *not* typical, what is the simplest family that f can belong to such that the *family* is typical in the space of families of smooth functions?

We have fairly complete answers to these questions when $m = 1$ (functions) and some information when $m > 1$ (maps). A detailed mathematical account of the methods behind the answers would take us well beyond the scope of this book, but we shall try to convey as much of the underlying flavor of the results as we can during the course of our subsequent exposition. But first a few preliminary ideas about what we mean when we speak of a function being "like" another function, and what qualifies a function to be termed "typical."

3. *Structural Stability and Genericity*

For our purposes, a function f being "like" another function $\hat{f}$ means that if $\hat{f}$ is "near" f by some measure of closeness, then the critical points of f

and $\hat{f}$ share the same topological structure (local min, max, saddle point, etc.). Looked at from another vantage point, we would say that the two functions are alike if by a coordinate change in the domain and range of f we can transform f into $\hat{f}$, and conversly. Thus, if $f, \hat{f}: R^n \to R^m$, then f would be *equivalent* to $\hat{f}$ if there exist coordinate changes

$$g: R^n \to R^n, \quad h: R^m \to R^m,$$

such that the following diagram is commutative:

$$\begin{array}{ccc} R^n & \xrightarrow{f} & R^m \\ {\scriptstyle g}\downarrow & & \downarrow{\scriptstyle h} \\ R^n & \xrightarrow[\hat{f}]{} & R^m \end{array}$$

We will always assume that $f, \hat{f} \in C^\infty(R^n, R^m)$, with g and h being origin-preserving diffeomorphisms (i.e., one-to-one, onto and smooth, having the normalization $g(0) = h(0) = 0$).

One natural way of obtaining a function $\hat{f}$ from a given function f is when f is a member of a *parameterized family* of functions. If we let $\alpha \in R^k$ be a vector of parameters, then we would have $f_\alpha: R^n \to R^m$ for some fixed value of α. If we then let $\alpha \to \hat{\alpha}$, we obtain a new member of the family, $f_{\hat{\alpha}}$. In this situation we would say that $f_{\hat{\alpha}}$ is equivalent to f_α if we can find coordinate changes g and h as before, as well as a smooth change $a: R^k \to R^k$, such that the earlier diagram commutes taking $f = f_\alpha$ and $\hat{f} = f_{\hat{\alpha}}$. In general, the critical points of f_α will also depend upon the parameter value α, so there is an induced map

$$\begin{aligned} \chi: R^k &\to R^n \\ \alpha &\mapsto x^*(\alpha), \end{aligned}$$

where x^* is a critical point of f. Of great theoretical and applied concern are those values of α where the map χ is discontinuous.

Example

Consider the parameterized family of functions

$$f_\alpha(x) = \frac{x^3}{3} + \alpha x, \qquad \alpha \in R^1.$$

The critical points $x^*(\alpha)$ are the real roots of the equation

$$f'_\alpha(x) = x^2 + \alpha = 0.$$

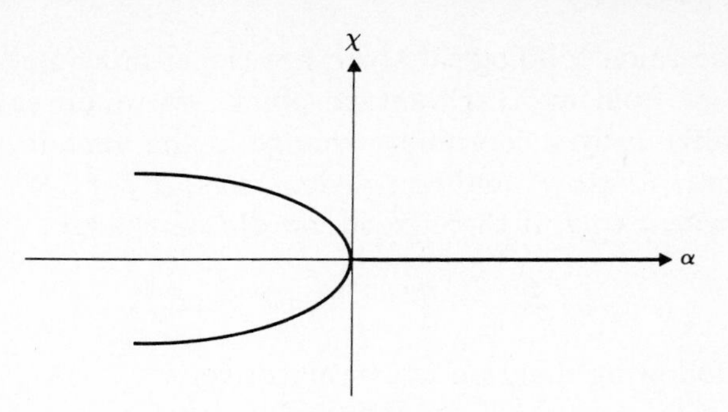

Figure 4.3 The Map χ

Consequently, the graph of the map χ is given by the diagram in Fig. 4.3. Here we see the discontinuity (nonexistence, actually) of χ as we pass smoothly through the point $\alpha = 0$. For $\alpha < 0$, the family has two critical points of opposite sign (one a local max, the other a local min), and these two points coalesce at $\alpha = 0$ to an inflection point, which then gives way to no critical points for $\alpha > 0$.

By the methods we will develop later on, it can also be shown that any function $f_{\hat{\alpha}}$ in this family is equivalent to f_α for $\hat{\alpha}$ in a local neighborhood of α as long as that neighborhood does not include the point $\alpha = 0$. Thus, the bifurcation point $\alpha = 0$ splits the family $f_\alpha = x^3 + \alpha x$ into three disjoint classes: functions having two, one or no critical points, and these classes are represented by the values $\alpha < 0$, $\alpha = 0$, and $\alpha > 0$, respectively.

The preceding considerations enable us to formulate precisely the notion of a function being "stable" relative to smooth perturbations. Let f and p be in $C^\infty(R^n, R^m)$. Then we say that f is *stable* with respect to smooth perturbations p if f and $f + p$ are equivalent, i.e., if there exist coordinate changes g and h such that the diagram

$$\begin{array}{ccc} R^n & \xrightarrow{\;f\;} & R^m \\ g\downarrow & & \downarrow h \\ R^n & \xrightarrow[f+p]{} & R^m \end{array}$$

commutes for all smooth p sufficiently close to f. For this notion to make sense, we need to impose a topology on the space $C^\infty(R^n, R^m)$ in order to be able to say when two maps are "close." The usual topology employed is the *Whitney topology,* which defines an ϵ-neighborhood of f by demanding that all functions in such a neighborhood agree with f and all of its derivatives up to accuracy $\epsilon > 0$. Details can be found in the chapter Notes and References.

If we pass to a parameterized family of functions, the preceding definition of stability must be modified to take into account variation not only in

x, but also in the parameters α. Let $f_\alpha(x) \in C^\infty(R^n, R^m)$ be such a smooth family. Then we say that $f_\alpha(x)$ is equivalent to $q_\beta(x)$, α, $\beta \in R^k$ if there exists a *family* of diffeomorphisms $g_\alpha\colon R^n \to R^n, \alpha \in R^k$, a diffeomorphism $e\colon R^k \to R^k$, and a smooth map $h_\alpha\colon R^m \times R^k \to R^m$ such that the diagram

$$\begin{array}{ccc} R^n \times R^k & \xrightarrow{\ f_\alpha\ } & R^m \\ {\scriptstyle g_\alpha}\big\downarrow & & \big\downarrow{\scriptstyle h_\alpha} \\ R^n \times R^k & \xrightarrow[\ q_\beta\ ]{} & R^m \end{array}$$

commutes. If f and q are equivalent, then

$$q_\beta(x) = f_{e(\alpha)}\big(g_\alpha(x)\big) + h_\alpha(x)$$

for all $(x, \alpha) \in R^n \times R^k$ in a neighborhood of $(0, 0)$.

The concept of structural stability now extends to families by saying that f_α is structurally stable *as a family* if f_α is equivalent to $f_\alpha + q_\alpha$, where q_α is a sufficiently small (i.e., nearby) *family* of functions.

Now let's turn to the question of what we mean by "typical," in a mathematical sense. Imagine we are given a set of elements S (any kind of set in which a topology has been defined so that we can speak of an element being "close" to another). Further, let there be some property P that the elements of S each may or may not possess. Then we say that the property P is *generic* relative to S under the following circumstances:

1) If $s \in S$ possesses property P, then there exists *some* neighborhood N of s such that every $s \in N$ also possesses property P (openness),

AND

2) If $s \in S$ does not possess property P, then *every* neighborhood of s contains an element possessing property P (denseness).

In loose terms we can say that if P is *generic* for the set S, "almost every" $s \in S$ possesses the property P.

Example 1

Let S = all quadratic functions of a single variable x, i.e.,

$$S = \{f(x) : f = x^2 + bx + c,\ b, c \in R^1\},$$

and let P be the property that the roots of f are distinct. Thus, $f \in S$ possesses property P if and only if $b^2 - 4c \neq 0$. It's clear that P is generic for the set S, since we can represent S with the plane R^2 by identifying the coefficients of any $f \in S$ with the point $(b, c) \in R^2$. Those $f \in S$ not possessing P lie on the curve $b^2 = 4c$ (see Fig. 4.4), and it is evident that :

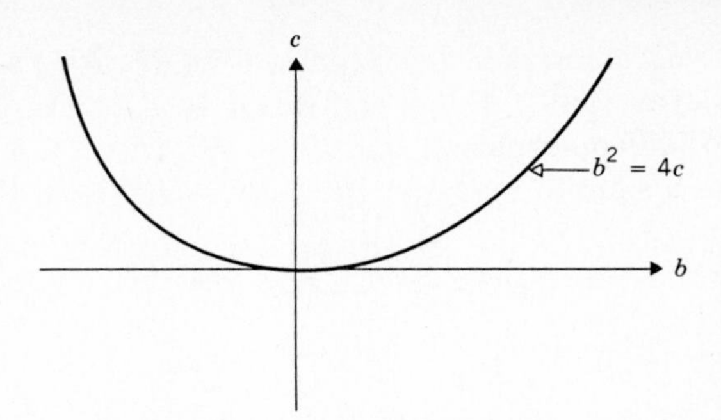

Figure 4.4 Genericity for Quadratic Polynomials

1) Every f such that $b^2 \neq 4c$ has some neighborhood in which all points (b', c') are such that $b'^2 \neq 4c'$.

2) In every neighborhood of every point (b^*, c^*) such that $b^{*^2} = 4c^*$, there exist points (b, c) such that $b^2 \neq 4c$.

Consequently, the property of having distinct roots is generic for quadratic polynomials. This argument can be extended to nth-degree polynomials in a fairly straightforward manner.

Example 2

Let $S = \{f \in C^\infty(R^n, R)\colon f(0) = \text{grad } f(0) = 0\}$, and let the property P be that f has a nondegenerate critical point at the origin; i.e.,

$$\mathcal{H}(0) \doteq \det \left[\frac{\partial^2 f}{\partial x^2}\right](0) \neq 0.$$

We can see that nondegeneracy is generic by the following argument.

First of all, the characteristic polynomial of $\mathcal{H}$ is an nth-degree polynomial $p(\lambda)$, and the requirement that $\det \mathcal{H}(0) \doteq p(0) = 0$ means that $p(\lambda)$ must have no constant term, i.e.,

$$p(\lambda) = \lambda^n + a_{n-1}\lambda^{n-1} + \cdots + a_1\lambda.$$

But the set of all nth-degree polynomials can be identified with the points of R^n as in the previous example, and it is then easy to verify that points of the form $(a_{n-1}, a_{n-2}, \ldots, a_1, 0)$ are nongeneric in R^n. Hence, nondegeneracy of critical points is a generic property for smooth functions on R^n.

4. *Morse's Lemma and Theorem*

The natural starting point for our study of the critical points of smooth functions is to consider functions that have a *nondegenerate* or *regular* critical point at the origin. Thus, we have a smooth function $f\colon R^n \to R$, with $f(0) = 0$, grad $f(0) = 0$, such that $\det \mathcal{H}(f)(0) \neq 0$, where

$$\mathcal{H}(f) \doteq \frac{\partial^2 f}{\partial x^2}\,,$$

is the Hessian matrix of f. If the critical values are distinct, i.e., $f(x^*) \neq f(\hat{x})$, for $x^*, \hat{x}$ critical points, then f is termed a *Morse function.*

If f is a Morse function, the Taylor series of f begins with quadratic terms, i.e.,

$$f(x) = \sum_{i,j=1}^{n} \mathcal{H}_{ij}(f)(0)x_i x_j + \ldots,$$

and we investigate the degree to which the Hessian matrix $\mathcal{H}(f)$ determines the local structure of f near the origin. Clearly, if we can find a coordinate transformation $x \rightarrow y(x)$ such that in the y variables we have

$$f(y) = \sum_{i,j=1}^{n} \widehat{\mathcal{H}}_{ij}(0)y_i y_j,$$

where $\widehat{\mathcal{H}}$ is the transformed Hessian matrix, then the Hessian matrix would *entirely* determine the behavior of f near the critical point at the origin. Morse's Theorem asserts that this is indeed the case if the origin is a nondegenerate critical point and, moreover, that Morse functions are both stable and generic in $C^\infty(R^n, R)$. Thus, the "typical" kind of smooth function we meet is a Morse function, and if we have a non-Morse function g at hand, an arbitrarily small perturbation of g will make it Morse. Furthermore, if f is Morse, then all functions sufficiently close to f are also Morse, where nearness is measured in the Whitney topology discussed above. Before entering into the proof of this key result, it's worthwhile to reflect for a moment upon its message.

At first glance, it might appear that Morse's Theorem covers virtually all cases of practical interest. After all, if we have the bad luck to be looking at a non-Morse function, we can always perturb it by a small amount and find ourselves back in the class of "good" functions. This is very reminiscent of the situation in matrix theory where if we have a singular matrix A, we can always slightly perturb some of the elements and thereby make it nonsingular. But we know that in many important situations this procedure is not acceptable, and we must deal directly with the singular matrix A itself (e.g., in studying the properties of unreachable and/or unobservable linear systems). So it is also with smooth functions. Sometimes the essence of the problem is the degeneracy in the critical point, as in phase transitions, and it's necessary to deal directly with non-Morse functions. Such situations arise naturally when we have a parameterized family of functions, since then there will always be some member of the family that is non-Morse, and it's important to know how the character of the critical point changes as we pass through these "atypical" functions in the family.

As an aside, it's tempting to speculate that the generic quadratic behavior of smooth functions ensured by Morse's Theorem accounts for the fact that almost all the laws of basic physics are expressed in terms of quadratic forms (conservation of energy, Fermat's principle, the principle of least action, etc.). Since it's both stable and "typical" to be locally quadratic, it's reasonable to suppose that the local laws of classical physics will also be expressible in terms of quadratic forms, as indeed they are.

As the first step toward Morse's Theorem, let us consider the following basic lemma.

MORSE'S LEMMA. *Let $f\colon R^n \to R$ be a smooth function with a nondegenerate critical point at the origin. Then there is a local coordinate system $(y_1, y_2, \ldots, y_n)$ in a neighborhood Y of the origin, with $y_i(0) = 0$, $i = 1, 2, \ldots, n$, such that*

$$f(y) = -y_1^2 - y_2^2 - \cdots - y_l^2 + y_{l+1}^2 + \cdots + y_n^2$$

for all $y \in Y$.

PROOF: We have

$$\begin{aligned} f(x_1, x_2, \ldots, x_n) &= \int_0^1 \frac{d}{dt} f(tx_1, \ldots, tx_n)\, dt \\ &= \int_0^1 \sum_{i=1}^n \frac{\partial f}{\partial x_i}\Big|_{(tx_1,\ldots,tx_n)} x_i\, dt. \end{aligned}$$

Hence, we can write

$$f(x) = \sum_{j=1}^n x_j g_j(x),$$

in some neighborhood of the origin by taking

$$g_i(x) = \int_0^1 \frac{\partial f}{\partial x_i}\Big|_{(tx_1,\ldots,tx_n)}\, dt.$$

Furthermore,

$$g_i(0) = \frac{\partial f}{\partial x_i}\Big|_0 .$$

Since the origin is a critical point, we have $g_i(0) = 0$. Hence, applying the above representation "trick" again, there exist smooth functions h_{ij} such that

$$g_j(x) = \sum_{i=1}^n x_i h_{ij}(x),$$

and we can write

$$f(x) = \sum_{i,j=1}^n x_i x_j h_{ij}(x).$$

If we symmetrize the h_{ij} by taking

$$\widehat{h}_{ij} = \frac{1}{2}(h_{ij} + h_{ji}),$$

the equation for $f(x)$ still holds, and we have $\widehat{h}_{ij} = \widehat{h}_{ji}$.

Differentiating the representation for $f(x)$ twice, we obtain

$$\frac{\partial^2 f(0)}{\partial x^2} = 2\widehat{h}_{ij}(0),$$

which shows that the matrix

$$\left[\widehat{h}_{ij}(0)\right] = \frac{1}{2}\left[\frac{\partial^2 f}{\partial x^2}\Big|_0\right],$$

is nonsingular, since the origin is a nondegenerate critical point.

Now suppose that there exist local coordinates $(y_1, y_2, \ldots, y_n)$ in a neighborhood Y of the origin such that

$$f = \pm y_1^2 \pm y_2^2 \pm \cdots \pm y_{r-1}^2 + \sum_{i,j \geq r} y_i y_j H_{ij}(y_1, \ldots, y_n),$$

where $H_{ij} = H_{ji}$. By a relabeling of the last r coordinates, we may assume $H_{rr}(0) \neq 0$. Let

$$g(y_1, y_2, \ldots, y_n) = \sqrt{|H_{rr}(y_1, y_2, \ldots, y_n)|}\ .$$

By the Inverse Function Theorem, g is smooth in some neighborhood V of the origin, with $V \subset Y$. We again change coordinates to $(v_1, v_2, \ldots, v_n)$ defined by

$$v_i = y_i, \quad i \neq r,$$

$$v_r = g(y_1, y_2, \ldots, y_n)\left(y_r + \sum_{i>r} \frac{y_i H_{ir}(y_1, \ldots, y_n)}{H_{rr}(y_1, \ldots, y_n)}\right),$$

which is also a local diffeomorphism (again by the Inverse Function Theorem). Now we have

$$f(v_1, v_2, \ldots, v_n) = \pm v_1^2 \pm v_2^2 \pm \cdots \pm v_r^2 + \sum_{i,j \geq r+1} v_i v_j H'_{ij}(v_1, \ldots, v_n).$$

Hence, by induction, we have established that locally f is quadratic.

Remarks

1) The above proof should be compared with the procedure for reducing a symmetric matrix to diagonal form.

2) A function of the form

$$f(z) = z_1^2 + z_2^2 + \cdots + z_{n-l}^2 - z_{n-l+1}^2 - \cdots - z_n^2,$$

is called a *Morse l-saddle.* Thus, Morse's Lemma asserts that every smooth f can be transformed by a smooth coordinate change to an l-saddle in the neighborhood of a nondegenerate critical point.

3) The number l is a topological invariant describing the *type* of the critical point at the origin, in the sense that l remains unchanged when we employ a smooth coordinate change in the variables $x_1, x_2, \ldots, x_n$.

Now let's complete the discussion of Morse functions by giving a complete statement of Morse's Theorem.

MORSE'S THEOREM. *A smooth function $f\colon R^n \to R$ is stable if and only if the critical points of f are nondegenerate and the critical values are distinct (i.e., if x^* and $\hat{x}$ are distinct nondegenerate critical points of f, then $f(x^*) \neq f(\hat{x})$).*

Furthermore, the Morse functions form an open, dense set in the space of smooth functions $C^\infty(R^n, R)$.

Finally, in the neighborhood of any critical point of f there exists an integer l, $0 \leq l \leq n$, and a smooth coordinate change $g\colon R^n \to R^n$, such that

$$(f \circ g)(x_1, x_2, \ldots, x_n) = y_1^2 + y_2^2 + \cdots + y_l^2 - y_{l+1}^2 - \cdots - y_n^2.$$

We will see many examples of the use of Morse's Lemma and Theorem later on, so for the moment let's pass directly to a consideration of what can happen if the critical point is *degenerate.*

5. *The Splitting Lemma and Corank*

We have already seen functions like $f(x) = x^3$ which have degenerate critical points. It's clear from an examination of the graph of such functions that near such a critical point, it's not possible to make them "look quadratic" by any kind of smooth change of coordinates. This raises the question: If they are not locally quadratic, what *do* such functions look like near a critical point? To address this matter, we need to develop some additional concepts, starting with the idea of a way to measure the degeneracy of the critical point.

The most natural approach toward measuring the degree of degeneracy of a critical point is to recall that nondegeneracy is defined solely in terms of the nonsingularity of the Hessian matrix

$$H(f) = \frac{\partial^2 f}{\partial x^2},$$

evaluated at the critical point in question. Thus, if the critical point is at the origin and $\det H(f)(0) \neq 0$, then the origin is nondegnerate. This means that $H(f)(0)$ is of full rank n, i.e., rank $H(f)(0) = n$ if and only if the origin is a nondegnerate critical point for f. So we can define the degree of degeneracy by the *rank deficiency* of H. This number, termed the *corank* of f, is given by

$$r \doteq \text{corank } f = n - \text{ rank } H(f)(0).$$

From what has gone before, it's evident that r is invariant under smooth coordinate changes.

Using the corank r and arguments similar to those employed in the preceding section, it's possible to prove the following key extension of Morse's Lemma:

THE SPLITTING LEMMA. *Let f have a degenerate critical point at the origin and assume corank $f = r$, i.e., $n -$ rank $H(f)(0) = r$. Then there exists a smooth coordinate change $x \to y$ such that near the origin f assumes the form*

$$f(y) = g(y_1, \ldots, y_r) + Q(y_{r+1}, \ldots, y_n),$$

where g is a smooth function of order cubic and Q is a nondegenerate quadratic form.

Remarks

1) The term "Splitting Lemma" follows from the fact that we are able to "split" the new coordinate variables y into two *disjoint* classes: $\{y_1, \ldots, y_r\}$ that belong to the higher-order function g, and $\{y_{r+1}, \ldots, y_n\}$ which are part of the nondegenerate quadratic form Q. The number of variables in each class is determined solely by the integer r, the corank of f.

2) In practice, the total number of variables n may be very large with the corank of f still very small (often only 1 or 2). In such situations, almost all of the variables of the problem are involved with the quadratic form Q, an object with very nice analytic properties, especially for questions of optimization. Only a relatively small number of variables remain in the "bad" part of the decomposition of f. Consequently, the Splitting Lemma allows us to separate the analysis of f into a large, "nice" part, and a small, "bad" part.

3) The Splitting Lemma is a natural generalization of Morse's Lemma as can be seen from the fact that if the origin is a nondegenerate critical point, $r = 0$ and the Splitting Lemma reduces to Morse's Lemma.

4) At this stage, we can say nothing more specific about the function g other than that it begins with terms of order cubic or higher. Later, by imposing additional conditions upon f, we will be able to assert that not only is $g \in O(|y|^3)$, but that g is actually a *polynomial.* The precise nature of these additional conditions is an integral part of the celebrated Classification Theorem of elementary catastrophe theory, to which we shall devote considerable attention later in the chapter.

6. *Determinacy and Codimension*

Our main problem is to decide when two smooth functions f and g are equivalent, i.e., differ only by a smooth local change of variables. Let $j^k f$ denote the terms up through degree k in the Taylor series expansion of f about the origin. We call $j^k f$ the *k-jet* of f. The k-jet of f enables us to direct our attention to tests for the equivalence of f and g using finite segments of the Taylor series via the concept of *determinacy.*

We say that f is *k-determined* if for all smooth g such that $j^k f = j^k g$, we have that f and g are equivalent. In other words, f is *k-determinate* if f is equivalent to *every* g whose Taylor series agrees with that of f through terms of degree k. The smallest value of k for which this holds is called the *determinacy* of f, denoted $\sigma(f)$.

Notice that to calculate $\sigma(f)$ by the foregoing definition involves an infinite calculation since we must, in principle, test *every* smooth g whose k-jet agrees with that of f. For this reason, we seek an alternate test that can be computed in a finite number of steps, and that implies finite determinacy. The basis of such a test is provided by the notion of *k-completeness.*

Let us define

$$m_n \doteq \{f \in C^\infty(R^n, R)\colon f(0) = 0\}.$$

We say that a function f is $O(|x|^k)$ if the Taylor series expansion of f begins with terms of degree k. Further, let $f_{,i}$ denote the partial derivative of f with respect to x_i, i.e., $f_{,i} \doteq \frac{\partial f}{\partial x_i}$. Then we say that f is *k-complete* if every $\phi(x) \in O(|x|^k)$ can be written as

$$\phi(x) = \sum_{i=1}^{n} \psi_i(x) f_{,i}(x) + O(|x|^{k+1}),$$

where $\psi_i(x) \in m_n$.

Example

Let $f(x_1, x_2) = \frac{1}{4}(x_1^4 + x_2^4)$. The partial derivatives are

$$f_{,1} = x_1^3, \qquad f_{,2} = x_2^3.$$

Using these elements, we can generate any function of the form

$$\phi(x_1, x_2) = \psi_1(x_1, x_2)x_1^3 + \psi_2(x_1, x_2)x_2^3,$$

$\psi_1, \psi_2 \in m_n$. It's an easy exercise to see that no choice of ψ_1 and ψ_2 enables us to generate mixed fourth-order terms like $x_1^2 x_2^2$, although pure quartics like x_1^4 and x_2^4 are possible. However, all fifth-order terms, pure and mixed, can be generated through appropriate selection of $\psi_1, \psi_2 \in m_n$. Thus f is 5-complete, but not 4-complete.

The determination of exactly what kinds of functions are generated by the elements $\{f_{,i}\}$ leads to the final concept that we need, the notion of *codimension.* The space m_n is an infinite-dimensional real vector space (a ring, actually), with basis elements

$$\{x_1^{\alpha_1} x_2^{\alpha_2} \dots x_n^{\alpha_n} : \sum_i \alpha_i \geq 0\}.$$

Geometrically, any term in this list that *cannot* be generated using the elements $\{f_{,i}\}$ represents a "missing" direction in the space m_n. Loosely speaking, the number of such missing directions is termed the *codimension* of f, and we denote this number by codim f. More formally, if we define the *Jacobian ideal* of f as

$$\Delta(f) = \left\{ \phi \in m_n : \phi = \sum_{i=1}^{n} \psi_i(x) \frac{\partial f}{\partial x_i}, \quad \psi_i \in m_n \right\},$$

then

$$\text{codim } f = \dim_R [m_n / \Delta(f)] .$$

The importance of the codimension will emerge as we proceed, but for now it's sufficient just to note that even though both of the spaces m_n and $\Delta(f)$ are infinite-dimensional, the quotient space $m_n/\Delta(f)$ can be (and usually is) finite-dimensional.

Example

Returning to the example considered earlier with $f(x_1, x_2) = \frac{1}{4}(x_1^4 + x_2^4)$, we have already seen that all terms of order 5 and higher are generated by $f_{,1}$ and $f_{,2}$, and the only "bad" fourth-order term is $x_1^2 x_2^2$. However, no terms of order 3 or lower are contained in $\Delta(f)$. Thus, the "missing" directions in m_n are those generated by the following elements:

$$\{x_1, x_2, x_1^2, x_1 x_2, x_2^2, x_1^3, x_1 x_2^2, x_1^2 x_2, x_2^3, x_1^2 x_2^2\}.$$

Hence, codim $f = 10$.

The relations between determinacy, completeness and codimension are given in the following deep results of Thom, Mather, Arnold and Zeeman.

THEOREM 4.1.

i) f k-complete implies f is k-determinate.

ii) f k-determinate implies f is (k+1)-complete.

iii) codim $f < \infty$ if and only if f is finitely determined.

Remarks

1) If f is finitely determined, it means $f \sim j^k f$ (since we can always take $g = j^k f$). But $j^k f$ is just a polynomial of degree k. Thus, finite determinacy means that there is some local coordinate system in which f looks like a polynomial.

2) The codimension of f and the determinacy $\sigma(f)$ are jointly infinite or finite, but they are not equal. It can be shown that codim $f \geq \sigma(f) - 2$. There are similar inequalities relating the corank r with both $\sigma(f)$ and codim f. These relations are of considerable applied significance, as we shall see later on.

3) It's quite possible for very simple-looking functions to have infinite codimension (e.g., take $f(x_1, x_2) = x_1 x_2^2$), the moral being that even simple models may have rather complicated behavior. We shall return to this point in the examples as well as in the Problems and Exercises.

4) Just like corank f, the codimension is a topological invariant, thereby providing a tool for classifying smooth functions near a critical point. We shall see that in some cases these are the only invariants needed; i.e., they form an independent set of *arithmetic* invariants for classifying elements of m_n.

At this stage, it's natural to inquire to what degree a small perturbation of f introduces any change into the picture presented above. In other words, what kinds of perturbations can be removed by a suitable change of coordinates? This question leads to the idea of a *universal unfolding* of f.

7. *Unfoldings*

We have seen that the basis elements in the factor space $m_n/\Delta(f)$ represent "directions" in m_n that cannot be generated using the first partial derivatives of f. Intuitively, we might suspect that if $p(x)$ is a smooth perturbation of f, and if p contains components in these "bad" directions, it would not be possible to remove these components from p by a smooth change of variables. On the other hand, all the components of p in the "good" directions get intermixed with those same components of f, and we would suspect that these components of p could be transformed away by a coordinate change. The idea of an *unfolding* of f formalizes this intuitive notion.

Roughly speaking, an unfolding of f is a k-parameter family of functions into which we can embed f and which is stable *as a family*. If k is the smallest integer for which we can find such a family, then the unfolding is called *universal*. Let $\mathcal{F} = \{f_\alpha(x)\}$ be such a universal unfolding of f, with $f_0(x) = f(x)$. Then the family $\mathcal{F}$ represents all of the perturbations of f that cannot be transformed to zero by means of a smooth coordinate change. The codimension of f provides us with a systematic procedure for constructing such a universal unfolding $\mathcal{F}$.

Let f be a function with codimension $c < \infty$, and let $\{u_i(x)\}$, $i = 1, 2, \ldots, c$ be a basis for the space $m_n/\Delta(f)$. Thus, the elements $\{u_i(x)\}$ are the directions in m_n that are *not* generated by $f_{,i} = \frac{\partial f}{\partial x_i}$. Then it can be shown that a universal unfolding of f is given by

$$f(x) + \sum_{i=1}^{c} \alpha_i u_i(x), \qquad \alpha_i \in R.$$

In other words, the smallest parameterized family that f can be embedded within in a stable way is a c-parameter family determined by the basis elements of $m_n/\Delta(f)$.

In terms of perturbations to f, what this result means is that if $\hat{f} = f + p$ is a perturbed version of f with $p = \sum_{i=1}^{\infty} \beta_i \phi_i(x)$, where the set $\{\phi_i(x)\}$ spans $C^\infty(R^n, R)$, then we can write

$$\hat{f} = f + \sum_{i=1}^{c} \alpha_i u_i(x) + z(x),$$

where

$$z(x) = \sum_{i=1}^{\infty} \beta_i \phi_i(x) - \sum_{i=1}^{c} \alpha_i u_i(x).$$

The Unfolding Theorem implies that there exists a smooth change of variables $x \to T(x)$, such that $z \equiv 0$ in the new coordinates. Consequently, the only part of the perturbation p that cannot be made to vanish by looking at $\hat{f}$ in the proper coordinates is the part corresponding to the missing directions in m_n, i.e., those directions not generated by $\frac{\partial f}{\partial x_i}$, $i = 1, 2, \ldots, n$. These are the so-called essential perturbations; all other perturbations can be made to disappear if we look at $\hat{f}$ through the right pair of spectacles.

Often we are already given a k-parameter family of functions to which our function of interest f belongs, and we would like to know how the parameterized family $\{f_\beta(x)\}$, $\beta = (\beta_1, \beta_2, \ldots, \beta_k)$ relates to the universal unfolding of f discussed above, an unfolding that depends upon the integer c = codim f. Without loss of generality, assume $k > c$, since otherwise $\{f_\beta\}$ would be equivalent to the universal unfolding $\{f_\alpha\}$. The Universal Unfolding Theorem guarantees the existence of the following:

1) A map

$$e\colon R^k \to R^c$$
$$\beta \mapsto \alpha,$$

where e is smooth but, of course, not a diffeomorphism since $k > c$.

2) A mapping

$$y\colon R^{n+k} \to R^n,$$

such that for each β the map

$$y_\beta\colon R^n \to R^n$$
$$x \mapsto y(x, \beta),$$

is a local diffeomorphism at the origin.

3) A smooth map

$$\gamma\colon R^k \to R,$$

ensuring that the unfolding

$$\{f_\alpha(x)\} \doteq F(x, \alpha) = F(y_\beta(x), e(\beta)) + \gamma(\beta).$$

The maps e, y and γ smoothly reparameterize the unfolding $F(x, \beta)$ into the universal unfolding $F(x, \alpha)$. Note that if $k = c$ and e is a diffeomorphism, then the above situation is just the condition for equivalence of families of functions discussed in Section 3.

Example

To pull together all of the material on codimension and unfoldings, consider the function

$$f(x_1, x_2) = x_1^2 x_2 + x_2^3.$$

The quantities $f_{,1}$ and $f_{,2}$ are given by

$$\frac{\partial f}{\partial x_1} = 2x_1x_2, \qquad \frac{\partial f}{\partial x_2} = x_1^2 + 3x_2^2.$$

The space $\Delta(f)$ is

$$\Delta(f) = \{g\colon g = \phi_1(x_1, x_2)x_1x_2 + \phi(x_1, x_2)(x_1^2 + 3x_2^2), \quad \phi_1, \phi_2 \in m_n\}.$$

After a bit of algebra, it's easy to see that no terms of the form

$$u_1 = x_1, \quad u_2 = x_2, \quad u_3 = x_1^2 + x_2^2,$$

belong to $\Delta(f)$, but all other quadratics as well as all higher-order terms are obtainable. Thus, codim $f = c = 3$ and the elements $\{x_1, x_2, x_1^2 + x_2^2\}$ form a basis for $m_n/\Delta(f)$. A universal unfolding for f is given by

$$F(x, \alpha) = x_1^2 x_2 + x_2^3 + \alpha_1(x_1^2 + x_2^2) + \alpha_2 x_2 + \alpha_3 x_1.$$

Now consider the function $g(\hat{x}_1, \hat{x}_2) = \hat{x}_1^3 + \hat{x}_2^3$. It can be shown that g is also a cubic of codimension 3. Going through the above arguments for g, we are led to the unfolding

$$G(\hat{x}, \beta) = \hat{x}_1^3 + \hat{x}_2^3 + \beta_1 \hat{x}_1 + \beta_2 \hat{x}_2 + \beta_3 \hat{x}_1 \hat{x}_2.$$

Since f and g are equivalent via a linear change of coordinates, the two unfoldings $F(x, \alpha)$ and $G(\hat{x}, \beta)$ must also be equivalent.

To see this equivalence, set

$$e\colon R^3 \to R^3$$
$$(\beta_1, \beta_2, \beta_3) \mapsto (\alpha_1, \alpha_2, \alpha_3),$$

where

$$\begin{aligned}
\beta_1 &= -2^{2/3}\alpha_1, \\
\beta_2 &= 2^{-2/3}(\alpha_2 + \alpha_3\sqrt{3}), \\
\beta_3 &= 2^{-2/3}(\alpha_2 - \alpha_3\sqrt{3}).
\end{aligned}$$

Further, we choose the maps y and γ as

$$y\colon R^5 \to R^2$$
$$(x_1, x_2, \alpha_1, \alpha_2, \alpha_3) \mapsto (\hat{x}_1, \hat{x}_2),$$

with

$$\begin{aligned}
\hat{x}_1 &= 2^{-1/3}\left(\frac{x_2 + x_1}{\sqrt{3}} + \frac{2\alpha_1}{3}\right) \\
\hat{x}_2 &= 2^{-1/3}\left(\frac{x_2 - x_1}{\sqrt{3}} + \frac{2\alpha_1}{3}\right)
\end{aligned}$$

and

$$\gamma\colon R^3 \to R$$
$$(\alpha_1, \alpha_2, \alpha_3) \mapsto \frac{-4\alpha_1^3}{27} + \frac{2\alpha_1\alpha_2}{3}.$$

Using these transformations and multiplying out, we find that

$$\begin{aligned}
\hat{x}_1^3 + \hat{x}_2^3 + \beta_1\hat{x}_1\hat{x}_2 + \beta_2\hat{x}_1 + \beta_3\hat{x}_2 = x_1^2 x_2 + x_2^3 + \alpha_1(x_1^2 + x_2^2) \\
+ \alpha_2 x_2 + \alpha_3 x_1 + \gamma(\alpha_1, \alpha_2, \alpha_3),
\end{aligned}$$

as was to be shown.

The importance of this example is that it is far from immediately evident that the two functions f and g, together with their unfoldings F and G, are equivalent. The transformations e, y and γ given above are not difficult to find once we know they exist, but it requires the machinery of singularity theory to decide the existence issue.

All the concepts presented thus far—determinacy, codimension, corank, etc.—are drawn together in order to prove one of the major results of recent applied mathematics, the Thom Classification Theorem for elementary catastrophes.

8. *The Thom Classification Theorem*

The condition of equivalence under smooth coordinate changes defines an equivalence relation on the set of smooth functions $C^\infty(R^n, R)$: two functions $f, g \in C^\infty(R^n, R)$ are equivalent if they can be transformed into each other (up to an irrelevant constant) by a locally smooth change of variables $y: R^n \to R^n$. The big question is: How many equivalence *classes* are there, and what does a representative of each class look like? This question is answered by the Thom Classification Theorem, at least under certain circumstances.

It should be noted at the outset that the question as stated is still not entirely settled. What the Classification Theorem does is show that the question is mathematically meaningful by imposing natural conditions on the elements of C^∞ that allow us to obtain a *finite* classification. This pivotal result was first conjectured by Thom in the 1960s, with a complete proof given later by Mather, based upon key work of Malgrange. Arnold substantially extended the original results and the implications of the Classification Theorem for applications were extensively pursued by Zeeman, Poston, Stewart, Berry and others. So in a real sense the Classification Theorem, though attributed to Thom, is really the product of a long list of mathematicians and represents a classic example of how fundamental mathematical advances are put together, piece by piece, as a collective international effort. Now let's turn to a statement of the theorem.

Recall that Morse's Theorem told us that all functions were equivalent to quadratics near a nondegenerate critical point. Furthermore, quadratic functions themselves certainly fall into this category, so only functions of order 3 and higher can have degenerate critical points. Thom discovered that if we restrict attention to functions $f \in O(|x|^3)$ of codimension less than 6, then we could finitely classify this subset of $C^\infty(R^n, R)$. In addition, the codimension c and the corank r parametrize the equivalence classes, i.e., given a function f, if we calculate c and r, then these two integers determine the set of all functions that are locally equivalent to f at the origin, provided $c \leq 5$. A full statement of the result is as follows:

THOM CLASSIFICATION THEOREM. *Up to multiplication by a constant and addition of a nondegenerate quadratic form, every $f \in C^\infty(R^n, R)$ of codimension $c \leq 5$ is smoothly equivalent to one of the standard forms listed in Table 4.1.*

Table 4.1 Equivalence Classes of Codimension ≤ 5

Corank/Codim	Function	Universal Unfolding
1/1	x^3	$x^3 + a_1 x$
1/2	x^4	$x^4 + a_1 x^2 + a_2 x$
1/3	x^5	$x^5 + a_1 x^3 + a_2 x^2 + a_3 x$
1/4	x^6	$x^6 + a_1 x^4 + a_2 x^3 + a_3 x^2 + a_4 x$
1/5	x^7	$x^7 + a_1 x^5 + a_2 x^4$ $+ a_3 x^3 + a_4 x^2 + a_5 x$
2/3	$x_1^3 - 3x_1 x_2^2$	$x_1^3 - 3x_1^2 + a_1(x_1^2 + x_2^2)$ $+ a_2 x_1 + a_3 x_2$
2/3	$x_1^3 + x_2^3$	$x_1^3 + x_2^3 + a_1 x_1 x_2 + a_2 x_1$ $+ a_3 x_2$
2/4	$x_1^2 x_2 + x_2^4$	$x_1^2 x_2 + x_2^4 + a_1 x_1^2 + a_2 x_2^2$ $+ a_3 x_1 + a_4 x_2$
2/5	$x_1^2 x_2 + x_2^5$	$x_1^2 x_2 + x_2^5 + a_1 x_1^2 + a_2 x_2^2 + a_3 x_1$ $+ a_4 x_2 + a_5 x_2^3$
2/5	$x_1^2 x_2 - x_2^5$	$x_1^2 x_2 - x_2^5 + a_1 x_1^2 + a_2 x_2^2 + a_3 x_1$ $+ a_4 x_2 + a_5 x_2^3$
2/5	$x_1^3 + x_2^4$	$x_1^3 + x_2^4 + a_1 x_1 + a_2 x_2 + a_3 x_1 x_2$ $+ a_4 x_2^2 + a_5 x_1 x_2^2$

The Thom Theorem is of great theoretical and practical importance in just the same way that the Jordan Canonical Form Theorem is important in matrix theory. Both results consider a broad class of important mathematical objects, and show how we can constructively determine whether or not any two elements from the class can be transformed into each other by means of an appropriate change of coordinates in the underlying space. In the case of matrices, the coordinate changes are linear transformations themselves (i.e., matrices), while in the C^∞ case the coordinate transformations are origin-preserving diffeomorphisms. Furthermore, each theorem provides us with a canonical representative from each class of equivalent objects, a representative that contains all the "structure " present in each element of its class, where by structure we mean all those properties that are left unchanged by coordinate transformations, i.e., the invariant, or coordinate-free, properties. In a system-theoretic sense, these are the only properties that can be considered properties of the system itself; hence, the importance of such theorems.

There are a number of amplifying remarks needed to interpret properly the meaning of this fundamental result.

1) Table 4.1 omits functions that either have no critical point at the origin or have a nondegenerate critical point. In the first case, f is equivalent to any one of its linear coordinates, and Morse's Theorem covers the second case.

2) In the Splitting Lemma, it was seen that every f of corank r could be written as $f = g(x_1, x_2, \ldots, x_r) + Q(x_{r+1}, \ldots, x_n)$, where $g \in O(|x|^3)$ and Q is a nondegenerate quadratic form. In Table 4.1, the column labeled "Function" displays only the first part of this decomposition and omits the quadratic form.

3) The terms in the universal unfolding are not unique; what is unique is the integer c, the codimension of f, which represents the *number* of such terms.

4) For functions of corank 1, the codimension suffices to determine the class of f; for corank 2, the codimension plus additional information is needed, the details of which would take us too far off the track of this book to go into here. The standard theoretical catastrophe theory literature gives an account of how these corank 2 cases are obtained.

5) In codimensions greater than 5, the standard forms are no longer finite in number but can contain parameters (moduli). The classification in these cases bears analogy with the classification of matrices by congruence or similarity. In the first case, a single integer (the rank) suffices to determine the class of congruent matrices, and we have a finite classification. Under similarity, the characteristic values and the size of the Jordan blocks are the similarity invariants, implying an infinite classification. The situation is much the same in spirit, but far more difficult technically, for functions of higher codimension.

For applications, the most important parts of the Classification Theorem are the canonical unfoldings and the behavior of the critical points of f as we vary the unfolding parameters $\{\alpha_i\}$. This circle of questions leads to what is usually termed (elementary) *catastrophe theory,* which we shall take up in a later section. For now, let's consider an example from the field of electrical power engineering which involves the computation of corank, codimension and unfoldings, thereby illustrating most of the theoretical machinery developed thus far.

9. Electric Power Generation

Consider an electric power supply network with n generators. If we assume no transfer conductance between generators, the equations governing such a network are

$$M_i \frac{d\omega_i}{dt} + d_i\omega_i = \sum_{\substack{j=1 \\ j\neq i}}^{n} E_i E_j B_{ij} \times [\sin\delta_{ij}^* - \sin\delta_{ij}],$$

$$\frac{d\delta_i}{dt} = \omega_i, \qquad i = 1, 2, \ldots, n.$$

Here

$\omega_i =$ angular speed of rotor i,

$\delta_i =$ electrical torque angle of rotor i,

$M_i =$ angular momentum of rotor i,

$d_i =$ damping factor for rotor i,

$E_i =$ voltage of generator i,

$B_{ij} =$ short circuit admittance between generators i and j,

$\delta_{ij} = \delta_i - \delta_j$,

$\delta ij^* =$ the stable steady-state value of δ_{ij},

$\omega_{ij} = \omega_i - \omega_j$.

Our interest is in studying the behavior of the equilibrium values of ω_i and δ_i as a function of the parameters M_i, E_i, B_{ij}, and d_i.

If we define $a_{ij} = d_i - d_j$, $b_{ij} = E_i E_j B_{ij}$, then it can be shown that the function

$$V(\omega_{ij}, \delta_{ij}) = \sum_{i=1}^{n-1} \sum_{k=i+1}^{n} [\tfrac{1}{2} M_i M_k \omega_{ik}^2 - a_{ik}\delta_{ik} - b_{ik}(M_i + M_k)\cos\delta_{ik}]$$
$$- \sum_{i=1}^{n} \sum_{\substack{j=1 \\ j\neq i}}^{n-1} \sum_{\substack{k=j+1 \\ k\neq i}}^{n} M_i b_{jk} \cos\delta_{jk},$$

is a Lyapunov function for the above dynamics. We can now use the function V to study the behavior of the network near an equilibrium employing techniques of modern singularity theory.

To illustrate the basic ideas, consider the simplest case of $n = 2$ generators. In this situation, we have only the two basic variables $\omega_{12} \doteq x_1$, $\delta_{12} \doteq x_2$. The function V then becomes

$$V(x_1, x_2) = \tfrac{1}{2} M_1 M_2 x_1^2 - a_{12} x_2 - b_{12}(M_1 + M_2)\cos x_2 + K$$
$$\doteq \tfrac{1}{2}\alpha x_1^2 - \beta x_2 - \gamma \cos x_2 + K.$$

The local behavior of V near an equilibrium is not affected by the constant K, so we set $K = 0$. The critical points of V are

$$x_1^* = 0, \quad x_2^* = \arcsin\left(\frac{\beta}{\gamma}\right).$$

Using the rules of singularity theory, it can be shown that V is 2-determinate with codimension 0 if $\alpha \neq 0$ and $\gamma \neq \pm\beta$. The condition on α is necessary for the problem to make sense, so the only interesting possibility for a degeneracy in V occurs when we have $\gamma = \pm\beta$. If $\gamma \neq \pm\beta$, then V is equivalent to a Morse function in a neighborhood of its critical point and can be replaced by its 2-jet

$$\tfrac{1}{2}\alpha x_1^2 - \sqrt{\gamma^2 - \beta^2}\, x_2^2,$$

a simple Morse saddle. So let's assume $\gamma = \beta$. (The case $\gamma = -\beta$ is similar.)

The function V now assumes the form

$$V(x_1, x_2) = \tfrac{1}{2}\alpha x_1^2 - \beta(x_2 + \cos x_2).$$

Some additional calculations with V show that the corank and codimension are both 1, indicating that the canonical form for V is the so-called fold catastrophe. Thus, a universal unfolding of V in this critical case is

$$\tilde{V} = \frac{x_2^3}{3} + t x_2,$$

where t is the unfolding parameter.

From the above analysis, we conclude that an abrupt change in the stability of the power network can only be expected in the case when $\gamma = \beta$, in which case the system's potential function is locally equivalent to $\tilde{V}$. The stability properties of the system will change abruptly when the mathematical parameter t, which is dependent upon the physical parameters α, β and γ, passes through the value $t = 0$. Results of this sort are routine to obtain by standard arguments in singularity theory, and will play an increasingly important role in engineering calculations in the future as the methods become more well-known and widespread.

10. *Bifurcations and Catastrophes*

Each of the unfoldings in Table 4.1 represents a c-parameter family of functions in the variable(s) x. For most values of the parameters, the corresponding function $F(x, \alpha)$ has a nondegenerate critical point at the origin and the local behavior is governed by Morse's Theorem. Of considerable theoretical and applied interest are those values of the parameters for which the origin is a degenerate critical point. The set of such parameter values is called the *bifurcation set* of the function $f(x)$, and it plays an important role in applications of singularity and catastrophe theory.

Let $\mathcal{B}$ denote the bifurcation set. Then $\mathcal{B}$ is given by

$$\mathcal{B} = \left\{ \alpha_k \colon \frac{\partial F}{\partial x} = 0, \det \left[\frac{\partial^2 F}{\partial x^2} \right] = 0, \quad k = 1, 2, \ldots, c \right\}.$$

Note that the definition of $\mathcal{B}$ involves two conditions on the function family $F(x, \alpha)$, which contains $r+c$ variables, where $r = \text{corank } f$, and $c = \text{codim } f$. So as long as $r \leq 2$, we can always use these conditions to eliminate the x-dependence and express the set $\mathcal{B}$ solely in terms of the parameters $\{\alpha_i\}$. For example, if we take the famous "cusp" situation ($r = 1$, $c = 2$), the universal unfolding is

$$F(x, \alpha_1, \alpha_2) = x^4 + \alpha_1 x^2 + \alpha_2 x.$$

Consequently, the bifurcation conditions are

$$\frac{\partial F}{\partial x} = 0, \quad \frac{\partial^2 F}{\partial x^2} = 0,$$

leading to

$$4x^3 + 2\alpha_1 x + \alpha_2 = 0,$$

and

$$12x^2 + 2\alpha_1 = 0.$$

Eliminating x, we find that

$$\mathcal{B} = \{(\alpha_1, \alpha_2) \colon 8\alpha_1^3 + 27\alpha_2^2 = 0\}.$$

The geometrical picture of $\mathcal{B}$ in the (α_1, α_2)-plane is shown in Fig. 4.5. Here we see the two branches of the solution set of the equation $8\alpha_1^3 + 27\alpha_2^2 = 0$ coming together at the "cusp" point at the origin. This type of geometrical structure in $\mathcal{B}$ serves to motivate the names like "fold," "cusp," and so forth usually attached to the various classes in Table 4.1.

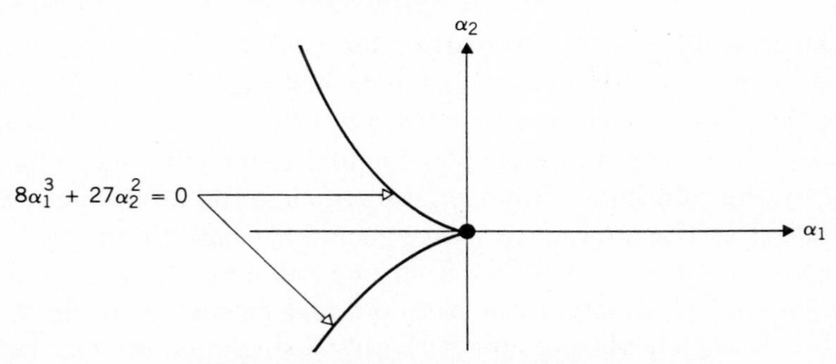

Figure 4.5 The Cusp Geometry

If we look at the solutions of the equation $\partial F/\partial x = 0$ as a function of x *and* the parameters α, then we obtain a surface M in R^{r+c}, which can be shown to be a smooth manifold. Each of the classes of functions in Table 4.1 has its own characteristic geometry for the surface M. But these geometries are related within the family of cuspoids or the family of umbilics in that as we pass from lower to higher codimension within a family, each manifold M is a proper submanifold of the higher-order singularity. The simplest version of this situation is for the fold and cusp geometries displayed in Figs. 4.6–4.7.

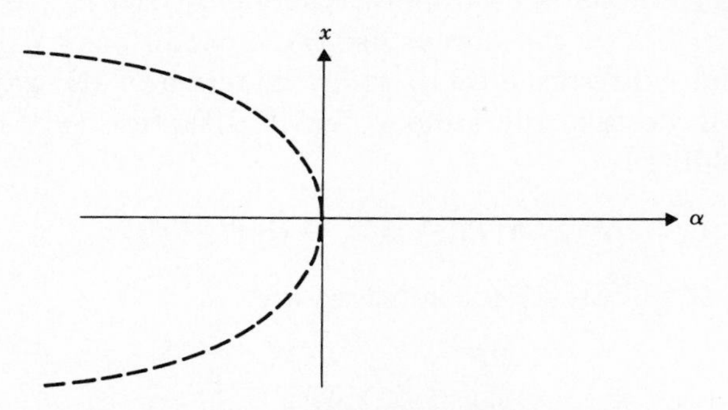

Figure 4.6 The Manifold M for the Fold

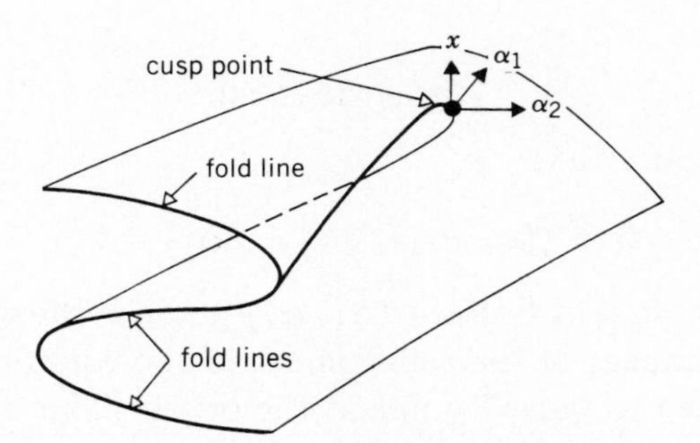

Figure 4.7 The Cusp Manifold M

In the case of the fold, the equation for M is $3x^2 + \alpha = 0$, leading to the solution $x^2 = \pm\sqrt{-\alpha/3}$ displayed as the dashed curve in Fig. 4.6. Passing to the cusp, the relevant equation for M is $4x^3 + 2\alpha_1 x + \alpha_2 = 0$. The geometry is shown in Fig. 4.7, where we see how the fold geometry is embedded within the cusp by the fold lines. However, we see that the additional parameter in the cusp gives rise to a much richer geometry than for the fold enabling us, for instance, to pass smoothly from one value of $x^*(\alpha_1, \alpha_2)$ to another on M when $\alpha_1 > 0$. Comparison with the fold geometry in Fig. 4.6 shows that such a passage is not possible with only a single parameter. In passing, we note that the bifurcation set $\mathcal{B}$ for the cusp as shown in Fig. 4.5 can be obtained by projecting the manifold M of Fig. 4.7 onto the (α_1, α_2)-plane. The bifurcation sets for the other singularity classes in Table 4.1 can also be obtained by projecting their characteristic manifold M onto the parameter space in the same manner.

An alternate way of viewing the preceding set-up is to regard the determination of the manifold M as that of finding a map

$$\chi: R^c \rightarrow R^r,$$
$$(\alpha_1, \alpha_2, \ldots, \alpha_c) \mapsto (x_1, x_2),$$

defined by the equation(s) grad $F(x, \alpha) = 0$. The values of $\alpha \in R^c$ where the map χ is discontinuous are exactly the points where the solution x^* acquires a new degree of degeneracy. Thom has termed these critical parameter values *catastrophes,* since they usually represent places where the physical process described by $F(x, \alpha)$ undergoes some sort of dramatic change in its behavior. Now let's leave the theory for awhile and examine some illustrative uses of the preceding machinery for addressing questions of interest in natural resource systems.

11. *Harvesting Processes*

There are two qualitatively different ways of making use of the singularity and catastrophe theory methods in applications, depending upon the assumptions we make concerning knowledge of the equation of state governing the system's behavior. The first, which we might term the *physical way,* assumes that a function $F(x, \alpha)$, $x \in R^n$, $\alpha \in R^k$, is known, where x represents the system's observed behavioral states and α is a vector of parameters (or inputs) that determine the states. Using the known analytic character of the function F, the mathematical machinery developed above can be employed to study the behavior of the critical points as functions of the inputs, and to identify those special values or regions in the parameter space where the system undergoes various types of significant transitions.

However, it's often the case that the function F is not known and we have knowledge only of the observed pattern of parameter values and the corresponding system behavior, i.e., we know the input/output behavior of the system but *not* the actual input/output map. Such situations are especially common in the social and behavioral sciences, where there are no "natural laws" of the scope and majesty of those found in physics and engineering upon which to base an equation of state F. In these cases we employ the *metaphysical way* of catastrophe theory, blithely *assuming* that the system variables x and α are given in the canonical coordinate system, the number of x-variables equaling the corank of the unknown equation of state and the number of input parameters equaling the codimension. Furthermore, the metaphysical way also assumes that we can ignore the nondegenerate quadratic form Q from the Splitting Lemma, so that the essential system behavior is carried only in the order cubic part of the decomposition. The underlying justification for these assumptions is the invariance under smooth coordinate changes of most of the important properties associated with the system's critical points. Consequently, the behavior in the canonical model is *qualitatively* the same as the behavior in whatever coordinate system is used to naturally describe the problem. The example given earlier on electric power generation networks illustrates the physical way for applying the catastrophe ideas; in this section, we give an example of the metaphysical way.

A commonly occurring situation in the management of natural resource systems is when we have a resource of some kind—fish, trees, grass, etc.—that, if left to itself, will grow at some natural rate dependent upon the environmental carrying capacity and natural fecundity. However, this resource has some economic value and is regularly harvested for human use. Due to technological advances, as well as laws of supply and demand, there are incentives for many harvesters to get into the business with high-tech equipment to reap the rewards of high harvesting yields. On the other hand, if too many harvesters, or too efficient harvesters, are allowed unrestricted access to the resource, the stock sizes decline to the point where the resource is not able to sustain itself and the whole industry becomes in danger of collapse. This situation is well documented, for example, in the harvesting of whales in the Pacific or anchovies off Peru. The first question of interest is how to balance the short- and long-term economics of the harvesting situation in order to construct a policy that serves both the economic needs of the harvesters and also preserves the resource. A related question, and the one that we will address here, is how the equilibrium levels of the resource are affected by both the number of harvesters and the technical efficiency of their equipment. For sake of definiteness, we formulate our discussion in terms of fishery harvests, although it should be clear that the arguments apply to many other types of natural resource processes as well.

Since we generally don't know the precise mathematical relationship between the equilibrium stock level and the levels of fishing effort and efficiency, we will invoke the metaphysical way of catastrophe theory and just assume that the fishing situation of interest can be described by the following input/output variables:

- *Input 1*: The level of fishing investment as measured by fleet size (the number of fishing boats or harvesting units).

- *Input 2*: The technological and economic efficiency measured in terms of the catchability coefficient and the capacity per boat, and in terms of the lowest stock size for which it would still be economical to continue fishing.

- *Output*: The stock size, which may be measured in many ways, and may consist of several geographical substocks or even a mixture of many species.

Using the above variables and the metaphysical way, we assume that the unknown relationship between the stock size and the inputs is a function of codimension 2 and corank 1, i.e., if we let the inputs be denoted by α_1,α_2, where the output is x, then there is a relationship $f(x, \alpha_1, \alpha_2) = 0$, such that codim $f = 2$ and corank $f = 1$. Since we are already being munificent with our assumptions, we'll also assume that $f \in C^\infty$. Then we are in the situation covered by the Thom Classification Theorem and can assert that whatever relationship f may happen to be, there exists a smooth change of variables so that in the new variables $f \rightarrow \hat{f} = \frac{1}{4}\hat{x}^4 + \frac{1}{2}\hat{\alpha}_1\hat{x}^2 + \hat{\alpha}_2\hat{x}$. Consequently, if we're only interested in the properties of f near the equilibrium stock levels, we can use the function $\hat{f}$. However, since we're making life easy anyway, we might as well assume that the natural coordinate system of the problem for x, α_1 and α_2 is also the canonical coordinate system, implying that we can take $f = \hat{f}$.

In view of the above considerations, we know that the cusp geometry governs the stock size compared to the harvesting effort/efficiency relationship. Thus we can invoke the Classification Theorem to produce a geometrical depiction of the relationship as shown in Fig. 4.8, which is taken from real data on the antarctic fin and blue whale stocks in the period 1920–1975. The reader will recognize this figure as a relabeled version of the canonical geometry given earlier in Fig. 4.7. The arrows from the origin show the equilibrium stock level trajectory for different fleet sizes and catching efficiency levels. It is imperative to note that this trajectory is most definitely *not* the usual time-history of the stock levels that emerges as the inputs are varied. Rather, the trajectory shown in Fig. 4.8 should be interpreted in the following sense: we first fix the fleet size and efficiency at given levels and then observe what equilibrium stock level emerges; this gives *one point* on the cusp manifold. We next change the inputs to new levels and wait for the stock level to stabilize; this gives a second point on the manifold, and so on. Thus, the cusp manifold is a collection of all the possible *equilibrium* levels that the stock can assume. So the curve shown is a curve of equilibrium positions for stock size and has nothing to do with the behavior of the system in its transient states on the way to equilibrium. To deal with this kind of transient behavior, we need the much deeper theory of bifurcation of *vector fields,* some of which we shall discuss later in the chapter.

We can gain considerable understanding of fishery management practices and policies from an interpretation of the behavior shown in Fig. 4.8. The period prior to 1930 was characterized by expanding fishing effort, but relatively low technological efficiency (e.g., shore-based processing plants); there is no clear evidence of catastrophic stock declines until the late 1930s.

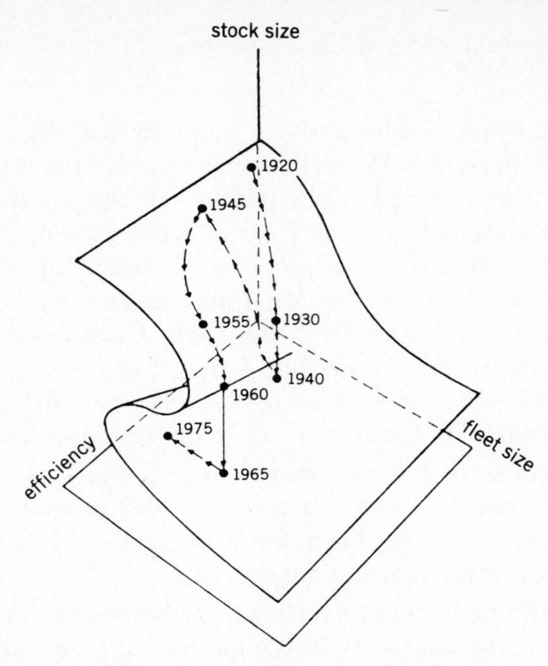

Figure 4.8 Stock Size Equilibria for the Antarctic Fin and Blue Whales

When the fishing industry came back to life after World War II, technological change was rapid, involving factory ships and fleets of highly mobile catcher boats. The catastrophic stock declines during the 1950s and 1960s were characterized by catches that remained very high until stocks became quite low. In this period, the high catches were limited by quotas rather than vessel capacities.

The implications of this sort of analysis for fishery harvesting policy is quite clear. If catastrophic behavior is to be avoided in the face of increasingly sophisticated technological advances and continued government subsidies to fleets, then precise monitoring and control policies designed to avoid the fold lines in Fig. 4.8 will become of prime importance. Since this seems to be difficult for a variety of technical and economic reasons, it may be necessary to consider alternate schemes arising from taking a hard look at the cusp geometry.

Such policies might be to avoid the cusp region by forcing a reduction in fishing efficiency, or broadening the cusp region by, for example, fleet size regulation. A third possibility is to induce deliberately a sequence of small catastrophes by invoking a policy that involves many crossings of the fold lines, each crossing taking the system from a region of high stocks to lower, or vice versa. Such a scheme would still preserve viability of the fishery system if the drops from the upper to the lower sheet were arranged to take place rather near the cusp point in the input space. One way to accomplish this would be to increase the harvesting capacity per vessel, while decreasing the number of boats. Another would be to increase the stock size below which fishing is uneconomical. This is a very novel idea to most fishery managers since it involves deliberately inducing a discontinuity in the equilibrium stock levels, a discontinuity that oscillates between low and high populations. The political, social and psychological consequences of such actions go far beyond historical mangement wisdom; nevertheless, such deliberately induced "booms and busts" may ultimately turn out to be the only way to avoid the kind of catastrophic collapse of the fishing grounds that managers (and fishermen) want to avoid above all else. This situation is somewhat reminiscent of that encountered in earthquake protection, in which it is often desirable to pump water into the fault lines to precipitate minor tremors and relieve the fault tension before it suddenly releases itself in the form of a devastating quake.

The above fishery problem shows how we can use the metaphysical way of catastrophe theory to address some real policy issues in the natural resource management area. Now let's take a look at how knowledge of the system's underlying equation of state can enable us to employ the physical way to determine effective management policies for another kind of natural resource system, a forest.

12. *Forest Insect Pest Control*

To illustrate how much more detailed our analysis can be when we actually do know the relationship between the system control inputs and the outputs, in this section we consider a problem involving the growth and death of a forest insect pest, the spruce budworm, that infests the spruce forests of eastern North America. Budworm outbreaks occur irregularly at 40 to 80-year intervals, and may expand to cover many million hectares of forest in a few years. We will investigate the feasibility of generating controlling actions that could permanently suppress these outbreaks and their consequent ecological and economic disruptions.

Neglecting spatial dispersion factors, the dynamical equations governing the birth, growth and death of the spruce budworm at a given geographical site have been postulated to be

$$\frac{dB}{dt} = \alpha_1 B\left[1 - B\frac{(\alpha_3 + E^2)}{\alpha_2 S E^2}\right] - \frac{\alpha_4 B^2}{(\alpha_5 S^2 + B^2)},$$

$$\frac{dS}{dt} = \alpha_6 S\left[1 - \frac{\alpha_7 S}{\alpha_8 E}\right],$$

$$\frac{dE}{dt} = \alpha_9 E\left[1 - \frac{E}{\alpha_7}\right] - \frac{\alpha_{10} B E^2}{S(\alpha_3 + E^2)}.$$

Here $B(t)$ is the budworm density in the site, $S(t)$ is the total surface area of the branches in the stand of trees in the site, and $E(t)$ represents the stand "energy reserve," a measure of the stand foliage condition and health. These quantities satisfy the natural physical constraints

$$B(t),\ S(t) \geq 0, \quad 0 \leq E(t) \leq 1.$$

The parameters α_1–α_{10} represent various intrinsic growth rates, predation rates, and so forth. Our interest will be in the manner in which the equilibrium (steady-state) values of B, S and E vary with changes in these parameters.

Since our primary concern is with $\bar{B}$, the equilibrium value of B, if we set the right side of the above dynamical equation equal to zero, and perform the necessary algebra, the equation for $\bar{B}$ turns out to be

$$\alpha_1\alpha_7^3(\alpha_3 + \bar{E}^2)\bar{B}^3 - \alpha_1\alpha_2\alpha_7^2\alpha_8\bar{E}^3\bar{B}^2 + \alpha_1\alpha_5\alpha_7\alpha_8^2(\alpha_3 + \bar{E}^2)\bar{B} + \alpha_2\alpha_4\alpha_7^2\alpha_8\bar{E}^3\bar{B} - \alpha_1\alpha_2\alpha_5\alpha_8^3\bar{E}^5 = 0.$$

In obtaining this equation, we have assumed that the only physically interesting equilibrium points of the system are those with $\bar{S} \neq 0$.

In the equation for $\bar{B}$, the quantity $\bar{E}$ has been regarded as an 11th parameter in the problem. This approach can be justified by noting that the dynamical equations imply that the relationship between $\bar{E}$ and $\bar{B}$ is

$$\bar{B} = \frac{-\alpha_8\alpha_9}{\alpha_7^2\alpha_{10}(\bar{E}^3 - \alpha_7\bar{E}^2 + \alpha_3\bar{E} + \alpha_3\alpha_7)}\ .$$

Substituting this relation into the equation above for $\bar{B}$ yields an eleventh-degree polynomial equation for $\bar{E}$ alone. The only physically relevant values for $\bar{E}$ are in the range $0 \leq \bar{E} \leq 1$, and it can be shown that there is a *unique* root of this equation in the unit interval for each set of values for the parameters $\{\alpha_i\}$. Thus, in what follows, we shall always use the symbol $\bar{E}$ to denote this unique root of the foregoing equation for $\bar{E}$.

To reduce the equation for $\bar{B}$ to canonical form (i.e., monic with no quadratic term), we introduce the new variable

$$y = \bar{B} - \frac{\alpha_2\alpha_8\bar{E}^3}{3\alpha_7(\alpha_3 + \bar{E}^2)}\ .$$

After a bit of manipulation, it turns out that y satisfies the cubic equation

$$-(y^3 + t_1 y + t_2) = 0,$$

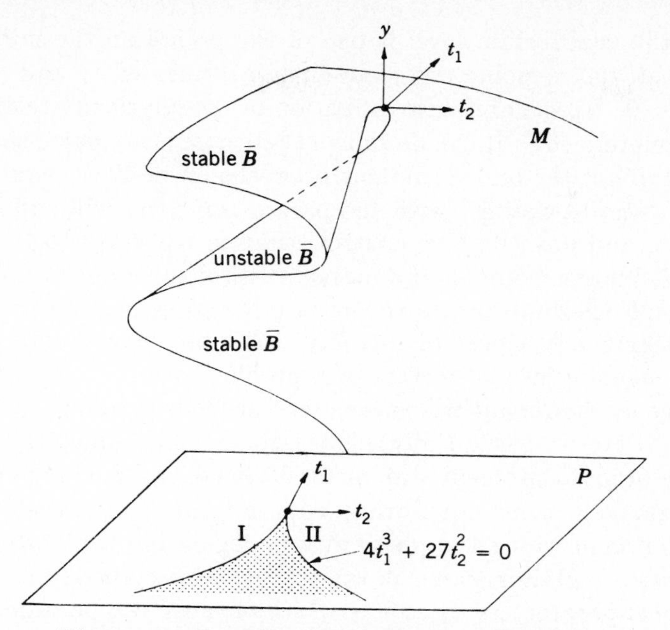

Figure 4.9 Equilibrium Budworm Densities

where the parameters t_1 and t_2 are given in terms of the original system parameters as

$$t_1 = \frac{-\alpha_8 \bar{E}^2}{\alpha_7^2}\left[\frac{\alpha_2^2\alpha_8\bar{E}^4}{3(\alpha_3+\bar{E}^2)^2} - \frac{\alpha_2\alpha_4\alpha_7\bar{E}}{\alpha_1(\alpha_3+\bar{E}^2)} - \alpha_5\alpha_8\right],$$

$$t_2 = \frac{-\alpha_2\alpha_8^2\bar{E}^5}{9\alpha_7^3(\alpha_3+\bar{E}^2)}\left[\frac{2\alpha_2^2\alpha_8\bar{E}^4}{3(\alpha_3+\bar{E}^2)^2} - \frac{3\alpha_2\alpha_4\alpha_7\bar{E}}{\alpha_1(\alpha_3+\bar{E}^2)} + 6\alpha_5\alpha_8\right].$$

Catastrophe theorists will recognize the y-equation as the equilibrium equation for the standard form of the cusp catastrophe of Table 4.1, where now the two mathematically meaningful parameters are t_1 and t_2. (In actuality, since further analysis shows that the equilibria on the upper and lower sheets of the cusp manifold for this problem are locally stable, this is called the *dual* cusp geometry rather than the cusp. This does not affect the geometry of the manifold but is of critical importance for applications, as we shall see shortly.) The geometry of this situation for the (y, t_1, t_2) interaction is shown in Fig. 4.9.

By examination of Fig. 4.9, it's evident that if we want to manipulate the parameters α_1–α_{10} to stabilize the budworm density on the lower sheet of the manifold, it will be necessary to be able to select these parameter values so that the quantity $4t_1^3 + 27t_2^2 \geq 0$. This is a necessary condition for being able to stabilize the budworm densities at a low level. If this cannot be done, then it can be expected that outbreaks will always occur whenever the equilibrium level is one of the points on the middle sheet of the manifold, i.e., a point corresponding to values of t_1 and t_2 such that $4t_1^3 + 27t_2^2 < 0$. Upon careful investigation of the physically realizable values of the parameters $\{\alpha_i\}$, it can be shown that there is no magic range of values that will stabilize the budworm density on the lower sheet; consequently, no amount of "knob-twisting" with the parameters $\{\alpha_i\}$ will suffice to control this system, and any effective control schemes will have to be based upon more sophisticated methods of dynamic control theory.

Although the budworm problem is interesting in its own right, by far the most significant aspect of this exercise is to show that the number of physically meaningful parameters in a problem may be very different from the number of *mathematical* parameters needed to address the questions of interest. Here we saw 10 physically important parameters (the $\{\alpha_i\}$) as part of the original problem statement; however, upon carrying out the elementary analysis of the equilibrium equation for $\bar{B}$, it turned out that the real question concerning the possibility of regulating the budworm density by parametric variation came down to the interrelationship between two mathematical parameters, t_1 and t_2. Each of these two parameters is a very complicated algebraic combination of all 10 of the physical parameters, and it's unlikely in the extreme that it could be guessed beforehand that this combination of the α-parameters and no other is the relevant combination for answering the question about control of budworm outbreaks. This illustration is the strongest possible testimony to the advantage of the physical way of catastrophe theory in allowing us to gain a deep understanding about what does and doesn't count in the analysis of a given system.

We have repeatedly emphasized that the ideas and techniques developed thus far in this chapter apply only to situations in which we can postulate a *static* function that expresses the equation of state for the problem at hand. Even in the foregoing budworm example, where we started with a dynamical description given in terms of a set of differential equations, our analysis was carried out using the algebraic equation given by the equilibrium values of the dynamical variables. At this stage it's natural to ask whether we can extend the ideas of singularity and catastrophe theory to deal directly with the changes in *transient* behavior of a dynamical process. It turns out that this is an issue of far greater subtlety and depth than for the case of static functions and maps, and we can only touch upon a few of the high points of this kind of dynamic bifurcation analysis in a book of this sort. We shall consider these matters in Section 14, but first let's look at the relatively simpler extension from functions to maps.

13. *Mappings*

For sake of exposition, we have concentrated our attention in this chapter on the case of bifurcations and singularities of smooth functions, i.e., mappings whose range is just the real numbers. Ironically enough, the setting that stimulated the development of the entire field of singularity theory involved the case of smooth *mappings* from the plane to the plane, i.e., mappings $f\colon R^2 \to R^2$. Allowing the range of our maps to be a space bigger than the real line opens up the possiblity of a host of additional types of behavior that the maps can canonically display, and this added freedom greatly complicates the classification of behavioral modes. Nevertheless, a substantial body of results now exists for these more general situations, and in this section we want to give a brief taste of what can be done by considering the classical case of smooth maps of the plane as originally developed by Whitney.

Whitney's work on planar maps was focused on the question of to what degree can the results of Morse's Theorem for smooth functions be carried over to smooth *maps.* Recall that the main features of Morse's Theorem are the genericity result showing that Morse functions form an open, dense set in $C^\infty(R^n, R)$, and the explicit classification of smooth functions by the eigenvalue structure of the Hessian matrix of f. This classification also provided a detailed characterization of the canonical representative of each class as a nondegenerate quadratic form. It's these features of Morse's Theorem that Whitney was interested in trying to extend to the more general setting of planar maps. Perhaps surprisingly, Whitney was able to provide a total and complete analogue of Morse's results, at least for the case of planar maps. Later we shall see that this kind of total generalization is not possible for general maps of $R^n \to R^m$, for the simple reason that there exist counterexamples to the genericity requirement, e.g., if $n = m = k^2$, Thom has shown that the stable maps don't form an open set for $k = 3$. Furthermore, not all combinations of dimensions n, m are such that the smooth maps $f\colon R^n \to R^m$ are dense in $C^\infty(R^n, R^m)$. But before taking up these matters, let us give a complete statement of Whitney's classic result.

WHITNEY'S THEOREM. *Let M be a compact subset of R^2, and suppose that $\phi\colon M \to R^2$ with $\phi \in C^\infty(R^2, R^2)$. Denote ϕ componentwise as*

$$\phi(x, y) = \big(u(x, y), v(x, y)\big).$$

Then

i) ϕ is stable at $(x_0, y_0) \in M$ if and only if near (x_0, y_0), ϕ is equivalent to one of the three mappings

$$u = x, \quad v = y \quad \text{(regular point)},$$

$$u = x^2, \quad v = y \quad \text{(fold point)},$$

$$u = xy - x^3, \quad v = y \quad \text{(cusp point)}.$$

ii) The stable maps form an open and dense set in $C^\infty(R^2, R^2)$.

iii) ϕ is globally stable if and only if ϕ is stable at each point of M, and the images of folds intersect only pairwise and at nonzero angles, and the images of folds do not intersect images of cusps.

There are several comments in order regarding this basic result:

1) Compared with Morse's Theorem for smooth functions, Whitney's Theorem says that near a critical point maps of the plane look like either the fold or the cusp, whose canonical forms are described in the second part of the theorem. This result is to be compared with Morse's Theorem which states that near a critical point of a smooth function, the behavior is quadratic.

2) The genericity conclusion is the same for both theorems: stable smooth maps of the plane are generic in $C^\infty(R^2, R^2)$, just as stable smooth functions are generic in $C^\infty(R^n, R)$.

3) The last part of Whitney's Theorem is a technical condition to ensure that we are dealing with planar maps that are "typical" in the space of such maps. The intersection condition given for global stability is the analogue for planar maps of the condition in Morse's Theorem involving the critical *values* of a smooth function being distinct.

The foregoing discussion has demonstrated the possibility of extending the fundamental aspects of Morse's Theorem to smooth planar maps. But what about *general* smooth maps of $R^n \to R^m$? By a result due to Thom, a general extension is not possible since there exist unstable maps from $R^9 \to R^9$ such that every map in an open neighborhood is also unstable. Thus, there is no possibility of extending the genericity result for smooth maps to this case, raising the question of whether there are certain "nice" combinations of n and m such that the stable maps from $R^n \to R^m$ form an open and dense set. If you'll pardon the pun, openness remains an open question, but we do have a complete resolution of the matter with regard to denseness as the following theorem shows.

DENSITY THEOREM. *Let $q = n - m$. Then the stable smooth maps $f\colon R^n \to R^m$ are dense if and only if*

i) $q \geq 4, \quad m < 7q + 8.$

ii) $q = 0, 1, 2, 3, \quad m < 7q + 9.$

iii) $q = -1, \quad m < 8.$

iv) $q = -2, \quad m < 6.$

v) $q \leq -3, \quad m < 7.$

COROLLARY. *Stable smooth maps are always dense if $n \leq 7$ or $m \leq 5$.*

Example: Urban Spatial Structure

To illustrate the Density Theorem, we consider an urban economic situation involving the flow of money from residents of one urban region to another. Let the zones of the region be labeled $i = 1, 2, \ldots, K$, and define

$$\begin{aligned} s_{ij} &= \text{flow of money from region } i \text{ to region } j, \\ e_i &= \text{per capita expenditure on shopping goods by residents of zone } i, \\ P_i &= \text{population of zone } i, \\ W_i &= \text{size of the "center" represented by zone } i, \\ c_{ij} &= \text{cost of travel from zone } i \text{ to zone } j. \end{aligned}$$

The standard aggregate model for s_{ij} is given by

$$s_{ij} = \frac{e_i P_i W_j^\alpha \exp(-\beta c_{ij})}{\sum_{k=1}^K W_k^\alpha \exp(-\beta c_{ik})}, \qquad i, j = 1, 2, \ldots, K,$$

where α and β are parameters representing consumer economies and "ease" of travel, respectively. Our interest here is with the map

$$\begin{gathered} S \colon R^K \to R^{K^2} \\ (W_1, W_2, \ldots, W_K) \mapsto (s_{11}, s_{12}, \ldots, s_{KK}). \end{gathered}$$

For clarity, consider the case of two regions ($K = 2$). The Jacobian matrix of the map S is given by

$$\mathcal{J}(W) = \alpha \begin{pmatrix} \frac{W_1^{\alpha-1} W_2^\alpha}{D_1^2} e^{-\beta(c_{11}+c_{12})} & \frac{-W_1^\alpha W_2^{\alpha-1}}{D_1^2} e^{-\beta(c_{11}+c_{12})} \\ \frac{-W_1^{\alpha-1} W_2^\alpha}{D_1^2} e^{-\beta(c_{11}+c_{12})} & \frac{W_1^\alpha W_2^{\alpha-1}}{D_1^2} e^{-\beta(c_{11}+c_{12})} \\ \frac{W_1^{\alpha-1} W_2^\alpha}{D_2^2} e^{-\beta(c_{21}+c_{22})} & \frac{-W_1^\alpha W_2^{\alpha-1}}{D_2^2} e^{-\beta(c_{21}+c_{22})} \\ \frac{-W_1^{\alpha-1} W_2^\alpha}{D_2^2} e^{-\beta(c_{21}+c_{22})} & \frac{W_1^\alpha W_2^{\alpha-1}}{D_2^2} e^{-\beta(c_{21}+c_{22})} \end{pmatrix},$$

where the D_i are distance parameters, and we have taken $e_i = P_i = 1$ for this analysis. It's easy to see that rank $\mathcal{J} = 1$ for all values of α and β, so that every point $W = (W_1, W_2)$ is a singular point for the map S. This means that S is an unstable map, implying that there exist arbitrarily close maps that are not equivalent to S. But the Density Theorem tells us that stable maps of $R^2 \to R^4$ are dense so, though there are maps close to S that are not equivalent to it, there are other maps close to S that are stable; i.e., S can be arbitrarily closely approximated by a map that *is* stable.

14. *Bifurcation of Vector Fields*

The natural bridge between the singularities of smooth *functions* considered earlier, and the bifurcations of *dynamical* systems is provided by the so-called gradient or gradient-like processes. We call a dynamical system

$$\frac{dx}{dt} = f(x, \mu), \qquad x \in R^n, \quad \mu \in R^k, \tag{Σ}$$

a *gradient* system if there exists a function $V\colon R^n \to R$, such that $f(x, \mu) = -\text{grad } V$ for all $\mu \in R^k$. It's easy to check that a necessary and sufficient condition for (Σ) to be a gradient system is that the Jacobian matrix $\mathcal{J}$ be symmetric, i.e.,

$$\mathcal{J}_{ij} \doteq \left[\frac{\partial f_i}{\partial x_j}\right] = \mathcal{J}_{ji}.$$

In this case, a suitable function V is given by

$$V = \sum_{i=1}^{n} \int_0^1 f_i(tx_1, tx_2, \ldots, tx_n) x_i \, dt.$$

We call f a *gradient-like* system provided there exists a smooth function $V\colon R^n \to R$, such that

$$(\text{grad } V(x), f(x, \mu)) \geq 0 \text{ for all } x \in R^n.$$

In either of the above cases, the equilibrium states of the system (Σ) coincide with the critical points of the function V, and we can directly employ the arguments outlined earlier to study the critical points of V in order to shed light upon the behavior of the dynamical process (Σ) in a neighborhood of an attractor. Gradient and gradient-like systems form the basis for the theory of *elementary* catastrophes, providing the mathematical starting point for most of the applications displayed in the literature. More complex types of dynamical systems, which have equilibrium sets more complicated than simple point equilibria or periodic trajectories, involve analyses that go beyond the simple elementary catastrophes discussed in earlier sections. We shall give indications of where some of the difficulties lie in the next chapter, but first let's take a look at an example of how to use gradient behavior to categorize the behavioral modes for the growth of blue-green algae in a shallow water pond.

Example: Phytoplankton Dynamics

Many shallow water ponds experience the phenomenon of phytoplankton "bloom," in which there is a rapid increase (or die-off) of blue-green algae, such as *Anabaena,* during certain seasons of the year. Typically, this bloom process occurs according to the following sequence:

1) A superphosphate fertilizer is applied to the pond while the algae concentration is low.

2) An algal bloom occurs during which *Anabaena* increase rapidly, with maximal algal concentration depending upon the time of the year when the pond is fertilized.

3) Die-off of algal bloom takes place without immediate remineralization of phosphate.

The above scheme indicates that the die-off is part of the overall dynamics of the system and is not triggered by other environmental factors. Our goal is to model this process using elementary catastrophe theory.

Denote the total algal concentration by $x(t)$, with the concentration of the blue-green nitrogen fixer *Anabaena* represented by $b(t)$. Since the algal bloom seems to be triggered by the amount of phosphate available, we let $a(t)$ represent the phosphate concentration of the pond. Using arguments based upon properties of the logistic equation, it has been argued that these quantities are governed by the following dynamics:

$$\frac{dx}{dt} = -(C_1x^3 - C_2ax + C_3b),$$

$$\frac{da}{dt} = -C_4x(a - a_0),$$

$$\frac{db}{dt} = C_5ab - C_6bx,$$

where the parameters C_1–C_6 represent various conversion factors and birth and death rates, and a_0 is the pond equilibrium concentration of phosphate.

The equation for x can be thought of as a gradient system for the function

$$V(x, a, b) = \tfrac{1}{4}C_1x^4 + \tfrac{1}{2}C_2ax^2 + C_3bx,$$

with a and b considered as time-varying parameters. To bring this function into the canonical form of the cusp geometry, we introduce the coordinate change

$$x = C_1^{-1/4}z,$$

$$a = \left(\frac{C_1^{1/2}}{C_2}\right)p_1,$$

$$b = \left(\frac{C_1^{1/4}}{C_3}\right)p_2,$$

so that the dynamics for x are governed by the scalar potential function

$$V(z, p_1, p_2) = \tfrac{1}{4}z^4 - \tfrac{1}{2}p_1z^2 + p_2z.$$

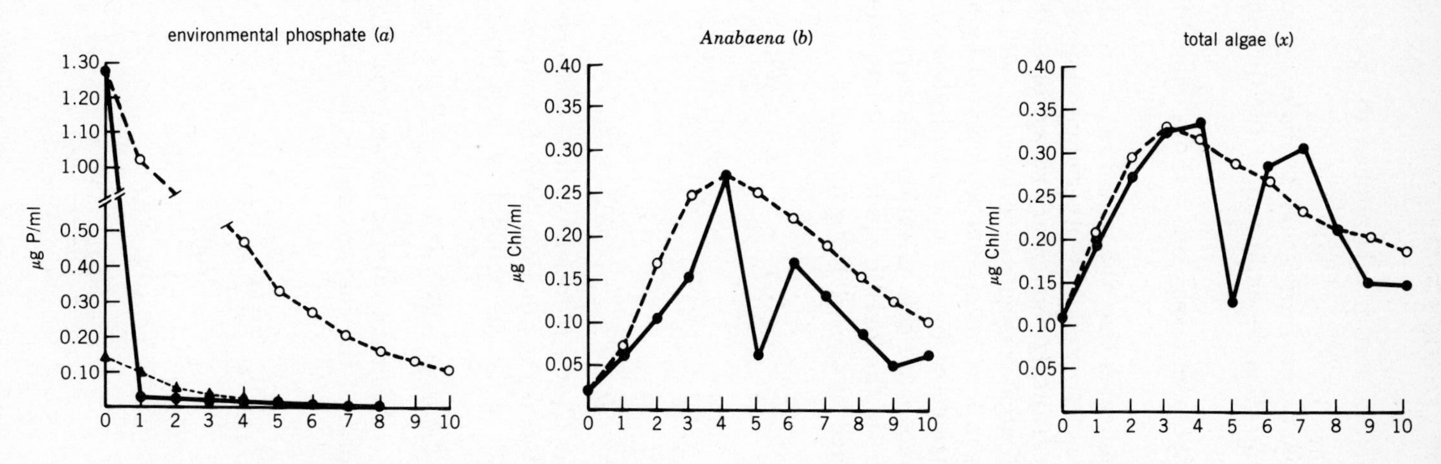

Figure 4.10 Simulated and Experimental Values of Pond Variables

Using typical values of the parameters C_1–C_6 and a_0, trajectories of the total algae, *Anabaena* and phosphate concentrations from the model versus experimental observations are depicted in Fig. 4.10. In general, the agreement is good with the discrepancy in the phosphate concentration explainable by a variety of factors that are discussed in the chapter Notes and References. The main point of this example is not so much to show the agreement with the actual experimental data, but rather to demonstrate how a gradient dynamical system can naturally arise in the process of investigating a realistic natural resource situation. It should be noted, of course, that every scalar dynamical process *is* a gradient system and, consequently, whenever we are interested in the behavior of a process whose output is a scalar quantity as in the lake pollution example, it's possible to employ elementary catastrophe theory analysis along the lines outlined above.

From the symmetry condition required for gradient systems, it's evident that gradient, and even gradient-like, systems do not form a dense set in the space of all smooth dynamical processes. It is exactly at this point that the real troubles begin when we attempt to classify the types of behaviors that can arise in the structure of the attractor set of the system (Σ). Even in the simplest-looking one- and two-dimensional systems, a bewildering array of behavioral modes can arise as we change system parameters, and only the most elementary types of dynamics can really be said to be understood. Since the next chapter considers such systems that display so-called chaotic behavior associated with aperiodic attractors, here we only look at the simplest type of nongradient system, the *codimension one* bifurcations, as an indication of what can be said about the classification of parameterized vector fields. This setting is an example of the situation in which we still have a scalar output, but with the focus of interest being on the transient as well as equilibrium behavior of the process.

The most important cases to consider in the single-parameter case are when for a certain value of the parameter, say $\mu = \mu_0$, the system Jacobian matrix $\mathcal{J}$ has either a pair of characteristic values on the imaginary axis (the Hopf bifurcation), or when $\mathcal{J}$ has a simple characteristic value $\lambda = 0$ (the saddle-node bifurcation). Thus, we assume that the system dynamics are such that $f(0, \mu) = 0$ for all real μ, and that as μ passes through μ_0, either a single root of $\mathcal{J}$, or a complex conjugate pair of roots, crosses the imaginary axis (with nonzero speed). It can be shown that if the Jacobian matrix $\mathcal{J}$ at $(0, \mu_0)$ has k, ℓ and m characteristic values with positive, negative and zero real parts, respectively, then locally (in (x, μ)-space) near $(0, \mu_0)$, the saddle-node dynamics can be brought to the form

$$\frac{dx}{dt} = \mu^* - x^2, \qquad \frac{dy}{dt} = A_+ y, \qquad \frac{dz}{dt} = A_- z,$$

where $A_\pm$ are matrices having k and ℓ roots in the right and left half-planes,

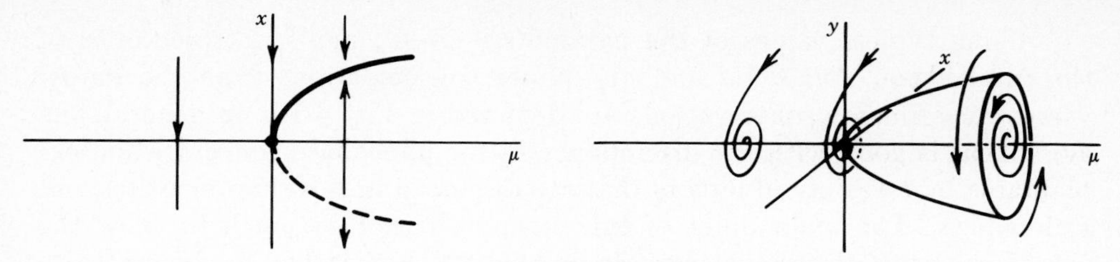

Figure 4.11 (a) Saddle-Node and (b) Hopf Bifurcations

respectively, and where $\mu^* = \mu - \mu_0$. Thus, the only locally nonlinear behavior comes from the one-dimensional "x-part" of the dynamics. Similarly, if $\mathcal{J}$ has a complex conjugate pair cross the imaginary axis at $(0, \mu_0)$, then a smooth coordinate change will produce the standard form

$$\frac{dx}{dt} = -y + x\left(\mu^* - (x^2 + y^2)\right),$$
$$\frac{dy}{dt} = x + y\left(\mu^* - (x^2 + y^2)\right),$$
$$\frac{dw}{dt} = B_+ w,$$
$$\frac{dv}{dt} = B_- v,$$

where again the matrices $B_\pm$ represent the linear parts of the dynamics coming from the eigenspaces corresponding to the roots of $\mathcal{J}$ in the left and right half-planes.

The importance of the saddle-node bifurcation is that all bifurcations of one-parameter families at an equilibrium with a zero characteristic value can be perturbed to saddle-node bifurcations. So we can expect that the zero characteristic value bifurcations that we meet in practice will be saddle-nodes, and if they are not, then there is probably something special about the problem that restricts the context to prevent the saddle-node from occurring. A typical type of constraint of this sort would be when there is some sort of symmetry in the problem that forces the roots of $\mathcal{J}$ to appear symmetric with respect to the imaginary axis. If the entries of $\mathcal{J}$ are not constrained by such considerations, then the *generic* way that a root can cross the imaginary axis is either for a real root to go through the origin, or for a complex conjugate pair to cross the imaginary axis. The first case leads to the saddle-node; the second to the Hopf bifurcation.

As far as the qualitative structure of the trajectories goes, the stable equilibrium at the origin becomes unstable at $\mu = \mu_0$, and a stable limit cycle is born whose radius is proportional to $\sqrt{\mu - \mu_0}$ in the case of the Hopf bifurcation; for the saddle-node, we have either the appearance or disappearance of a stable and unstable equilibrium, depending upon the direction in which the parameter μ passes through the critical value μ_0. The bifurcation diagrams for these two cases are displayed in Fig. 4.11.

Discussion Questions

1. Starting with Hadamard, in mathematical modeling of physical phenomena it has often been claimed that in order for a model to be a credible representation of the physical phenomenon, the behavior of the model should not change dramatically if we perturb the mathematical description a little bit. In other words, only descriptions that are structurally stable can serve as candidates for the representation of real-world processes. Discuss the pros and cons of this contention. Can you give examples of structurally *unstable* descriptions that, nevertheless, serve as "good" representations of physical processes?

2. Certain mathematical relationships between variables describing physical situations have been elevated to the status of a "natural law," while other seemingly similar types of dependencies are merely termed "empirical" relationships. The law of Conservation of Energy is an example of the former, and Ohm's "Law" illustrates the latter. Can you identify any criteria by which we could distinguish a natural law from an empirical relationship? Should invariance under coordinate transformations enter into the requirements for a natural law? If so, what does such a requirement have to do with our classification of smooth maps?

3. It has been suggested that the *complexity* of a system Σ can be measured by the number of inequivalent descriptions of Σ that can be constructed by a given observer $\mathcal{O}$. Clearly, this is a "relativistic" measure, dependent upon the particular observer. If we take "equivalence" to mean that the descriptions differ only by a smooth change of variables, this view of complexity would lead us to conclude that the maximum complexity of Σ as seen by *any* observer would be the number of equivalence classes that that observer could distinguish in the space of smooth descriptions of Σ. By the Thom Classification Theorem, this number is finite only for descriptions of codimension six or less. In higher codimensions, each class is represented by a canonical description containing parameters called *moduli;* hence, there are an infinite number of classes.

a) Discuss this relativistic view of system complexity. How would you use it to measure the complexity of a stone? An electrical circuit? An automobile? A corporation? A political system?

b) Consider how you would arrange to compare the complexities of two systems even in the case when the codimensions may be greater than six.

c) We could term the above measure of complexity, the complexity of Σ as seen by $\mathcal{O}$, and denote it as $\mathcal{C}_{\mathcal{O}}(\Sigma)$. However, the observer $\mathcal{O}$ is also a system that is in interaction with Σ, and hence, the system Σ can itself form a measure of complexity for $\mathcal{O}$. Call this the complexity of the observer as seen by the system, and term it $\mathcal{C}_{\Sigma}(\mathcal{O})$. In the physical sciences, it's usually tacitly assumed that the interaction between the system and the observer is highly asymmetric, with the system's impact upon the observer being negligible. Hence, $\mathcal{C}_{\Sigma}(\mathcal{O}) \approx 0$. Discuss this assumption within the context of social and behavioral systems. In cases where the system's impact upon the observer is not negligible, consider how you might use the difference $\mathcal{C}_{\mathcal{O}}(\Sigma) - \mathcal{C}_{\Sigma}(\mathcal{O})$ as a tool for policymaking on the part of the observer and/or the system.

4. Intuitively, our notion of "surprise" involves the discrepancy between the predictions we make on the basis of models (mathematical, mental or otherwise) and the actual observed behaviors of the systems being modeled. In the mathematical context, how could the notions of system description and bifurcation be used in order to formally develop a theory of surprises? Consider the interconnection between the concept of complexity from Question 3 with the problem of measuring the possibility of surprising behavior of the system Σ.

Suppose you were faced with the spruce budworm system of the text, in which the equilibrium levels of budworm density are governed by the cusp geometry of Fig. 4.5. If you were on the lower sheet of this geometry but near the fold line leading to a jump to the upper sheet, would you consider the magnitude of the jump from lower to upper sheet as an adequate measure of the surprise value of the situation? Why? If your objective as the manager of the forest is to avoid this kind of unpleasant surprise, how would you arrange harvesting, tree planting, spraying of insecticide, etc., to avoid the surprise entirely? Can you imagine situations in which it might be better to allow many small jumps (surprises) in budworm population as opposed to trying to avoid any kind of density discontinuity ? (See Chapter Eight for more details on the concept of surprise.)

5. The primary goal of the managers of most natural resource systems is to minimize the variability of the economic resource represented by the system, while at the same time maximizing the economic yield by harvesting. For example, fishery managers want to take action to generate maximum sustainable yields, as do forestry managers, mink farmers and agricultural planners. Basically, this philosophy involves trying to manipulate controllable parameters in the system so that the stability boundary of the desired equilibrium is very large and the speed of return to that equilibrium is very fast, the ideal situation being a controlled system having a single equilibrium with an infinite stability domain. Elementary considerations from dynam-

ical system theory demonstrate the unattainability of this ideal. Consider how close you could come to achieving this managerial nirvanna when the system equilibria structure is given by one of the elementary catastrophe geometries.

6. A variety of "myths" have arisen regarding the stability properties of natural resource systems. Holling has classified these myths into four categories according to the way Nature has arranged the long-term behavior of the system:

- *Global Stability*—A benign and infinitely forgiving Nature in which the system is able to absorb any type of perturbation and return to its original operating point.
- *Small Is Beautiful*—The system consists of a number of decentralized islands of stability and displays a high level of heterogeneity in its behavior.
- *Nature, the Practical Joker*—Multiple equilibria exist of both good and bad types, and they impose restrictions on behavioral variability and possible movement of the stability boundaries.
- *Nature Resilient*—A system designed to absorb and *benefit* from change, rather than try to control it.

Give examples of natural processes displaying each of these kinds of stability properties, and discuss why we call these stability properties "myths."

7. Comment upon the following research program aimed at developing system management procedures for coping with surprise in natural resource systems.

- Development of methods for stabilizing the system *outputs* without necessarily stabilizing all state variables. This involves policies that decouple the outputs from the rest of the states.
- Design of "probing" policies that involve variable control actions. Such policies are directly contrary to the objective of stabilizing system outputs.
- Identification of key variables, so that monitoring these variables provides the same information as monitoring the entire system state. In the terminology of Chapter Three, this is a problem of observability.

How would you use the ideas developed in the last two chapters to address the items on this research menu?

8. In Hermann's theory of international conflict, he identifies three major components contributing to the level of intensity of a crisis: *magnitude* of the threat, *decision time* available and degree of *surprise*. These factors

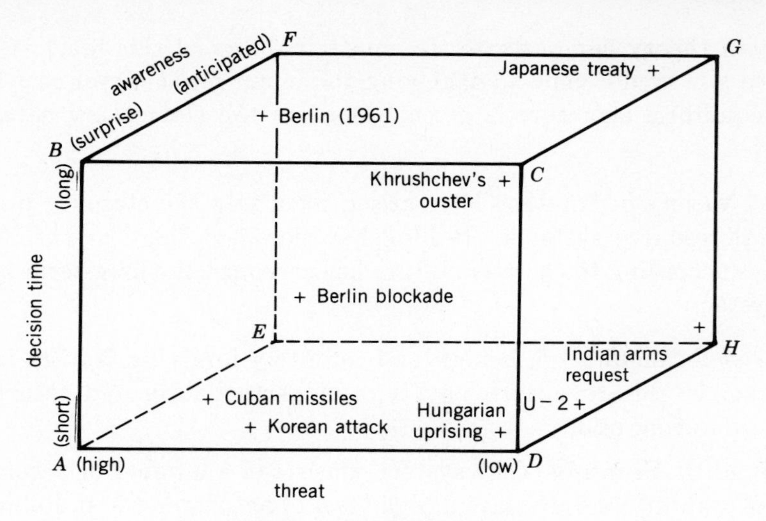

Hermann's Crisis Cube

are displayed in the "crisis cube," together with the positions of several important twentieth-century crises.

Using Hermann's quantities as control variables, how would you use the "metaphysical way" of catastrophe theory to develop a model for international crises? What output variables would you use, and what catastrophe geometry would then be generated? How would you propose to calibrate your model using actual data from real conflicts?

9. Many empirical investigations lead to S-shaped curves describing the manner in which one measured variable changes as a function of another. For example, a worker's salary as a function of time spent on the job, or the growth rate of bacteria on a Petri dish. Mathematically, these relationships are usually described by some variant of the classical logistic equation

$$\frac{dx}{dt} = rx\left(1 - \frac{x}{K}\right),$$

where r represents some growth parameter, and K involves the environmental "carrying capacity." What is the connection between this logistic growth dynamic and the geometry of the fold and cusp catastrophes? How would the higher-order cuspoid catastrophes be used to extend the above logistic dynamics to generate "S-curves" with more bends and levels? What types of physical situations would naturally lead to these "generalized" logistic curves?

10. Thom's original motivation for the development of many of the ideas and techniques of catastrophe theory was to provide a language with which to discuss biological processes of morphogenesis, i.e., the emergence

of physical form. Roughly speaking, the idea was that the distinctive geometries of the various catastrophe manifolds like the cusp, butterfly, and swallowtail would serve as geometric "archetypes" for the type of local physical structure that could unfold from a single fertilized cell. The ways in which these pure archetypes can interact would then provide insight into processes such as cellular differentiation, and would serve to constrain the various phenotypical forms that could come about.

Discuss the pros and cons of basing a theoretical biology upon this idea, and consider the types of issues that would have to be dealt with in order to carry out such a program. Could the same set of concepts be employed to create a program for classifying the form and developmental pattern of other types of life-like objects such as languages, societies or corporations? How would the phenomena of evolution and natural selection be incorporated into any such research program?

11. In the standard set-up from classical von Neumann-Morgenstern game theory, we have two players Al and Bob, who simultaneously and independently take actions a_i and b_j. These actions result in a payoff $V(a_i, b_j)$ to, say, Al (in zero-sum games, this means that Bob's payoff is $-V(a_i, b_j)$). Al tries to maximize his payoff, while Bob attempts to minimize. Von Neumann showed that there exists an optimal strategy for Al and Bob consisting of a selection of various choices according to probability distributions determined by the payoff function V. (We will consider these matters further in Chapter Six.)

Now imagine the situation in which Al is unaware of Bob's existence and only feeds his decision a_i into a "black box" and observes the output $V(a_i, b_j)$. Note here that Bob's choice b_j is incorporated into the interior of the box and is unseen by Al. So from Al's point of view he is playing a game against the box with the goal of maximizing his take over some time sequence of actions, and the original game can now be cast in the form of an input/output pattern of a black box. How would Al interpret the "spirit" of Bob as embodied in the black box?

Now consider the inverse problem: given a black box with a particular input/output pattern, under what circumstances can we associate a zero-sum, two-person game with it? How does this situation relate to elementary catastrophe theory? (*Hint:* When do the internal dynamics of the black box evolve according to the gradient of a potential?)

12. Suppose we have a physical process whose equilibrium state is given by the minimum of the parameterized potential function $V(x, \alpha)$. Let V have a global minimum at the point (x^*, α^*). Then as we vary α, there will be some value $\hat{\alpha}$ at which the minimum at x^* either ceases to be a global minimum or disappears entirely. There are now two possibilities for a change of state of the system:

• *Delay Rule*—The system waits until the local minimum at x^* disappears entirely and then moves to the nearest local minimum associated with the parameter value $\hat{\alpha}$.

• *Maxwell Convention*—As soon as the minimum at x^* ceases to be a global minimum, the system immediately moves to the new global minimum of V corresponding to the parameter value $\hat{\alpha}$.

Give examples in the physical, social and behavioral sciences in which the system would operate according to one of the foregoing rules. In our examples in the text, we have implicitly assumed the Delay Convention. How would the cusp geometry for the budworm example be modified if we had used the Maxwell Convention instead? What difference would this make to the interpretation of the results? Can you think of any other "conventions" that might be used to govern the change of state?

13. In linguistics and biology we can see the following type of hierarchical patterns of morphology:

$$\begin{array}{rcl} \text{Level I—}\{\text{Sentence}\} & \longleftrightarrow & \{\text{Egg}\} \\ \text{Level II—}\{\text{Noun and Verb Phrases,}\ldots\} & \longleftrightarrow & \{\text{Ectoderm, Mesoderm,}\ldots\} \\ \text{Level III—}\{\text{Noun, Verb, Article,}\ldots\} & \longleftrightarrow & \{\text{Bone, Skin,}\ldots\} \\ \vdots & & \vdots \end{array}$$

Discuss the similarities and essential differences between these two morphological diagrams. In particular, comment upon the essential linearity of linguistics versus the intrinsic three-dimensional geometry of biological forms.

14. In applications of singularity and catastrophe theory, the variables appearing in the nondegenerate quadratic form in the Splitting Lemma are often neglected in the problem analysis. Discuss the possible role that these "hidden variables" may play in characterizing the total behavior of the system. As a specific example, consider the case of a simple harmonic oscillator whose position is described by the equation $\ddot{q} = -\omega^2 q$. If all we can observe is the position q, then we see the system behavior as being a linear motion back-and-forth along the q-axis from $+A$ to $-A$, where A is the amplitude of the oscillation. However, if we include the velocity in our observed output, then the motion becomes the transit of a circle with radius A^2 in the $(q, \dot{q})$-plane. This kind of situation recalls Plato's thesis that what we see of Nature is only a projection of the true reality, as with a shadow cast against the wall of a cave. Consider the thesis that the unseen variables in the quadratic form are those that are "projected away" in Plato's Cave.

15. In a well-known (and strongly criticized) attempt to use catastrophe theory in the social sciences, Zeeman suggested the cusp geometry to model the "boom-and-bust" behavior of speculative markets. Taking the control variables to be excess demand for stock by investors and the percentage of speculators in the market, with the observed output being the rate of change of an index of the market's state (e.g., the Dow-Jones average), the cusp geometry depicted below was used to qualitatively model the process of stock market crashes.

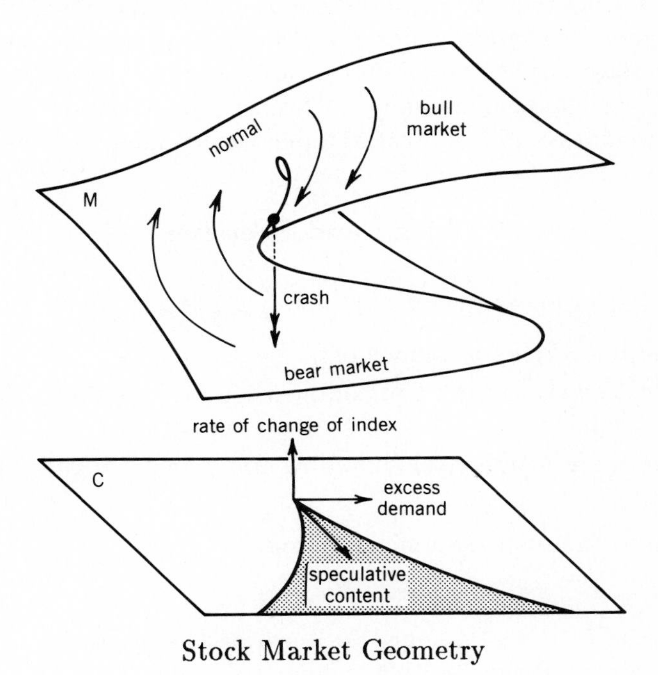

Stock Market Geometry

a) Describe in words the dynamical behavior of a market governed by this geometry.

b) The above model seems to suggest that a purely speculative market (i.e., one with no fundamentalist investors, but only speculators) could never experience a crash. How could you answer this objection?

c) Could a model of the above type be used to describe other types of "crashes" like political revolutions, corporate takeovers and urban decay? How would you define the variables in such situations?

16. Instead of displaying equilibrium behavior that consists of a single *point*, many natural systems have steady-state behavior that is oscillatory (e.g., the human heartbeat, planetary orbits and business cycles). How could you use the ideas of catastrophe theory to account for such cyclic equilibrium behavior?

17. Consider the following experiment: we are given N points in R^n and agree to separate them into two disjoint subsets, X and Y. We now want to find a polynomial p of degree m such that

$$p(x) = \begin{cases} > 0, & \text{for } x \in X, \\ < 0, & \text{for } x \in Y. \end{cases}$$

It can be shown that as n becomes larger than $2m$, the chance of success in finding such a classifier drops dramatically. In other words, if a model classifies more than twice as many points as it has adjustable parameters, it is unusually good in a very strong sense. Discuss the implications of this type of result for assessing the credibility of a catastrophe theory model based upon one of the polynomial canonical forms; i.e.,how could you develop a test for the goodness of fit of a catastrophe theory model to observed data?

Problems and Exercises

1. Consider the function $f(x, y) = x^2y + \frac{1}{3}y^3 + \frac{1}{2}y^2$.

   a) Compute the determinacy of f.

   b) Calculate the corank and codimension of f for the critical point at the origin.

   c) Determine a universal unfolding for f in a neighborhood of the origin.

2. Given the smooth dynamical system

$$\dot{x} = f(x, u),$$

where $f(0, 0) = 0$, it's customary to linearize the system near the origin obtaining the linear approximation system

$$\dot{z} = Fz + Gv,$$

where $F = \frac{\partial f}{\partial x}(0, 0)$ and $G = \frac{\partial f}{\partial u}(0, 0)$. Under what conditions will the trajectory of the linear approximation remain qualitatively the same as that of the original system near the origin? What conditions will cause such a linearization to break down, even locally?

3. Consider the following system of equations sometimes used to describe the behavior of a single species of microorganism growing in a chemostat,

$$\dot{x} = (\mu - D)x,$$

$$\dot{s} = D(s_r - s) - \frac{\mu x}{Y},$$

$$\mu = \frac{\mu_m K_i s}{s^2 + K_i s + K_i K_s}.$$

Here x is the biomass, μ the specific growth rate, s the nutrient concentration, and D the dilution rate, with all other quantities being constants.

Let the system be brought to equilibrium, and then slowly increase the dilution rate D. Interpret the subsequent behavior of x using the "fold" catastrophe. Find the smooth coordinate transformation that takes the system in the physical variables x and s to the canonical variables associated with the fold geometry.

4. The ideal gas law is given by

$$\left(P + \frac{a}{V^2}\right)(V - b) = RT,$$

where P is the pressure, V the volume, T the temperature, R the gas constant, and a, b constants representing the particular gas. If we regard V as the dependent variable, this law can be written as

$$V^3 - \left(b + \frac{RT}{P}\right)V^2 + \left(\frac{a}{P}\right)V - \frac{ab}{P} = 0.$$

This form strongly suggests that the surface it represents is smoothly equivalent to that of the canonical cusp geometry. Thus it should be possible to interpret the behavior of the system in catastrophe theory terms, taking V as the output variable and P and T as controls. However, when we attempt to do this, we find that the predictions do not agree with observations. For example, there is no hysteresis: water generally boils at the same temperature as steam condenses. Nor is there bimodality, since we can predict V uniquely if we know the temperature and pressure. What is the problem here and how can you fix it?

5. Consider the mappings $f_{\pm}$ of the plane defined by

$$\begin{pmatrix} x \\ y \end{pmatrix} \longrightarrow \begin{pmatrix} x^2 \pm y^2 + ax + by \\ xy \end{pmatrix},$$

where a and b are real parameters. Show that the maps f_+ and f_- are not equivalent for any values of the parameters, i.e., there do *not* exist smooth coordinate changes g and h with $g(0) = h(0) = 0$, such that the following diagram commutes:

$$\begin{array}{ccc} R^2 & \xrightarrow{f_+} & R^2 \\ {\scriptstyle g}\downarrow & & \downarrow{\scriptstyle h} \\ R^2 & \xrightarrow[f_-]{} & R^2 \end{array}$$

6. Let $R^{n\times m}$ be the set of all real $n \times m$ matrices, and let M_r be the subset of $R^{n\times m}$ consisting of those matrices of rank r.

a) Show that M_r is a smooth submanifold of $R^{n\times m}$ and that the codimension of M_r equals the product of the coranks $(m-r)$ and $(n-r)$; i.e.,

$$\dim M_r = mn - (m-r)(n-r).$$

b) Specialize the above result to show that square matrices of corank k have codimension k^2 in $R^{n\times n}$.

7. The following model has been suggested to explain drug response in a human when a drug is repeatedly given under the condition that the drug D first binds with the target T according the law of mass-action, and the drug target complex DT then acts as an inducer to depress the binding of further drug administrations:

$$\dot{x} = R(y) - k_1 x(d-y) + k_{-1}y - g(x),$$
$$\dot{y} = k_1 x(d-y) - k_{-1}y,$$

where $k_{\pm 1}$ are mass-action coefficients, x and y are the concentrations of the target complex and target, respectively, $g(x)$ is the degradation rate of x, d is the drug concentration at the biophase (= constant) and $R(y)$ is a sigmoidal function. For definiteness, take

$$R(y) = \alpha_1 \frac{1 + K_1 y^p}{K + K_1 y^p}, \qquad g(x) = k_{10}x,$$

where α_1 = a production constant, k_{10} is the removal rate constant and K_1 represents the measure of tightness of the binding of the target complex to the repressor, while $K > 1$ is a constant. Define the quantities

$$A = \frac{k_1\alpha_1(K_1)^{1/p}d}{k_1\alpha_1 + k_{-1}k_{10}}, \qquad B = \frac{Kk_{10}k_{-1} + k_1\alpha_1}{k_1\alpha_1 + k_{-1}k_{10}}.$$

a) Show that for $A \ll B$ or $A \gg B$, the system has only a single equilibrium with $y > 0$. What do these inequalities imply about the size of the constant d?

b) If A and B are in the range $2\sqrt{B} < A < B/2$, show that there are three equilibria having $y > 0$, and interpret this physically in terms of the amount of drug d that can be administered and still achieve a multiple steady-state concentration in the body.

c) Develop the cusp geometry that governs this situation using A and B as control parameters with y being the behavioral variable.

8. A version of the Lotka-Volterra equations describing the populations of n interacting species is given by

$$\frac{dN_i}{dt} = N_i \left[k_i + \sum_{j=1}^{n} a_{ij} \frac{N_j}{N} \right], \qquad \sum_{i=1}^{n} N_i = N, \quad i = 1, 2, \ldots, n.$$

Here N_i represents the population of the ith species, while k_i and a_{ij} are natural growth and interaction rates, respectively.

a) Show that under the transformation $x_i \doteq N_i/N$, the above system becomes a set of cubic equations for the quantities x_i.

b) Consider the case of $n = 2$ species. Show that if

$$A = k_1 - k_2 + a_{12} - a_{22} < 0, \qquad B = a_{11} + a_{22} - a_{12} - a_{21} > 0,$$

with $|A| < |B|$, the system has two stable equilibria (at $(1, 0)$ and $(0, 1)$) and one unstable equilibrium (at $(-A/B, 1+A/B)$). Determine what initial conditions lead to each of the stable equilibria.

c) Using the initial population level of the first species x_1^0 as a control parameter, show how the shift of equilibrium from one stable point to the other can be modeled via the "fold" catastrophe geometry.

9. Imagine we are given the independent, identically distributed random variables $\{X_i\}_{i=1}^n$, having normal distribution with mean θ and variance τ^2. Assume that τ is known and we want to estimate θ. Our beliefs before the experiment began were that $\theta = \mu$ with probability α. For the *a priori* distribution function of θ, let us take

$$g(\theta|\beta, \mu, \sigma) = \alpha(\sigma) f(\theta|\mu, \sigma^2) + (1 - \alpha(\sigma)) f(\theta|\mu, \sigma^{-2}),$$

where $0 \leq \sigma < 1$, μ and β real, with $\beta > 0$. Here we assume

$$f(\theta|\mu, \sigma^2) = (2\pi)^{-\frac{1}{2}} \sigma^{-1} \exp\{-\tfrac{1}{2}\sigma^{-2}(\theta - \mu)^2\},$$

with

$$\alpha(\sigma) = \frac{\beta\sigma}{1 + \beta\sigma}.$$

Thus, $g(\theta)$ is a mixture of two normal distributions with the same mean μ. Usually we will choose σ small to model the sort of *a priori* distribution described above.

After observing $x = \{X_1, X_2, \ldots, X_n\}$, the *a posteriori* density $g(\theta|x)$ has the form

$$g(\theta|x) = \alpha^* f(\theta|\mu_1^*, V_1^*) + (1 - \alpha^*) f(\theta|\mu_2^*, V_2^*),$$

where

$$\mu_1^* = \frac{\sigma^{-2}\mu + n\tau^{-2}\bar{x}}{\sigma^{-2} + n\tau^{-2}}, \qquad \mu_2^* = \frac{\sigma^2\mu + n\tau^{-2}\bar{x}}{\sigma^2 + n\tau^{-2}},$$

with

$$V_1^* = (\sigma^{-2} + n\tau^{-2})^{-1}, \qquad V_2^* = (\sigma^2 + n\tau^{-2})^{-1}.$$

The quantity α^* is given by the relation

$$\frac{\alpha^*}{1-\alpha^*} = \beta\sigma\left(\frac{\sigma^{-2} + n\tau^{-2}}{\sigma^2 + n\tau^{-2}}\right)^{\frac{1}{2}} \exp\{-\tfrac{1}{2}(\mu - \bar{x})^2[(\sigma^2 + n^{-1}\tau^2)^{-1} - (\sigma^{-2} + n^{-1}\tau^2)^{-1}]\}.$$

This *a posteriori* density will be bimodal provided that $(\mu - \bar{x})^2$ is considerably greater than zero.

Now suppose we want to estimate θ by minimizing the conjgate loss function

$$L(\delta, \theta) = h[1 - \exp\{-\tfrac{1}{2}k^{-1}(\theta - \delta)^2\}].$$

a) Show that the expected loss function $E(\delta)$ satisfies

$$1 - E(\delta) = \hat{\alpha} f(\delta|\mu_1^*, V_1^* + k) + (1 - \hat{\alpha}) f(\delta|\mu_2^*, V_2^* + k),$$

where

$$\hat{\alpha} = \frac{\alpha^*(1 + V_1^* k^{-1})^{-\frac{1}{2}}}{\alpha^*(1 + V_1^* k^{-1})^{-\frac{1}{2}} + (1 - \alpha^*)(1 + V_2^* k^{-1})^{-\frac{1}{2}}},$$

with $f(\delta)$ as defined above.

b) Suppose that either k or n is large so that

$$\frac{V_1^* + k}{V_2^* + k} \approx 1.$$

Show that $E(\delta)$ has a cusp point at

$$(\alpha^*, (\bar{x} - \mu)^2) \approx (\tfrac{1}{2}, 4(V_1^* + k)).$$

c) Using the approximate symmetry of $E(\delta)$, show that the lowest minimum will be the one nearest μ if $\alpha^* > \frac{1}{2}$, and the one nearest $\bar{x}$ if $\alpha^* < \frac{1}{2}$.

d) What happens if k and n are both small?

10. Assume that the dynamics of development in an urban housing area can be described by the quantities

$\dot{N}$ = the rate of growth of housing units in the area at time t,

a = the *excess* number of vacant units in the area relative to the norm,

b = the *relative* accessibility of the area to the regional population base.

Under the assumption that $\dot{N}(t)$ moves so as to maximize the potential

$$V = \pm(\tfrac{1}{4}\dot{N}^4 + \tfrac{1}{2}a\dot{N}^2 + b\dot{N}),$$

show how to characterize the behavior of $\dot{N}$ as a function of a and b using elementary catastrophe theory. How do you know which of the signs, plus or minus, to use with the potential V?

11. In studies of the collapse of ancient civilizations such as the Classic Maya, Mycenaean, Hittite and others, Renfrew has identified the following characteristic features in the "Collapse" phase:

- Collapse of the central administrative organization.
- Disappearance of the traditional elite class.
- Collapse of the centralized economy.
- Settlement shift and population decline.

During the "Aftermath" period, we observe a transition to a lower level of sociopolitical integration and the development of a romantic Dark Age myth.

In addition, these collapses display the temporal properties:

- The collapse may take on the order of 100 years for completion, although in the provinces of an empire the withdrawal of central authority can happen more rapidly.
- Dislocations are more evident early in the collapse period, and are displayed by human conflicts such as wars, destruction, etc.
- Border maintainence declines during the period so that outside pressures can be seen in the historical record.
- The growth of many variables like population, exchange, agricultural activity, etc. follows a truncated S-form.
- There is no obvious single "cause" for the collapse.

a) Let the observed output variable for such collapses be taken to be D, the degree of centrality or control of the governing authority, while the control quantities are I, the accumulated investment in charismatic authority and N, the economic balance for the rural population. Using these quantities, develop a cusp catastrophe model for the decline and fall of the society. Using the geometry developed, trace out the rise and fall of the civilization as a sequence of changes of I and N.

b) Explain why at least two control variables are needed to explain the observed dynamical behavior of these types of collapses.

c) The cusp model describes a bimodal polity involving a rapid transition from a centered to a noncentered society as measured by the degree of central authority. Imagine that you wanted to include a third type of social structure corresponding to, say, a tribe, as opposed to the extremes of a chiefdom or an egalitarian society. The simplest catastrophe geometry that allows for this type of intermediate behavior is the *butterfly*, which involves four control parameters. In addition to the quantities I and M used above, there are now two additional factors, call them D and K, which determine whether or not a tribal structure is possible. What social interpretation can you give to these quantities?

12. Consider the simple Euler arch depicted below.

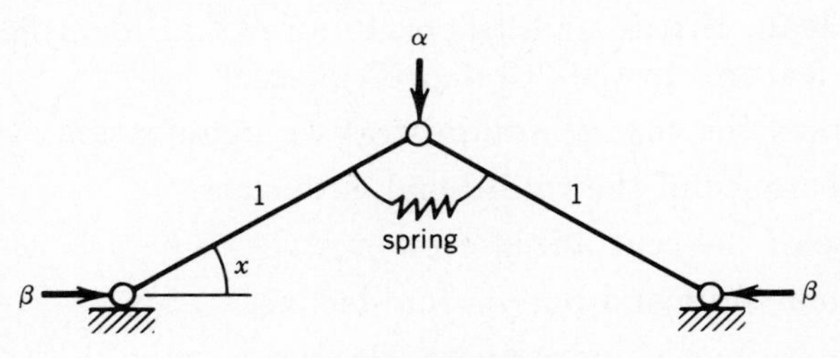

The Euler Arch

a) By consideration of the energy in the spring, the energy gained by the load and the energy lost by compression, show that the total energy in the system is given by

$$V(x, \alpha, \beta) = 2\mu x^2 + \alpha \sin x - 2\beta(1 - \cos x),$$

where μ is the modulus of elasticity for the spring. Using V, calculate the surface of equilibria, the fold lines and the cusp point for the system.

b) Prove that the arch buckles when $\beta = 2\mu$.

c) Show that there exists a coordinate change in x, α and β, such that in the new coordinates V has the form

$$V \sim \frac{1}{6}\mu x^4 + \alpha x - b x^2.$$

Hence, V is the potential for the cusp catastrophe.

13. *Duffing's equation*, given by

$$\ddot{x} + \epsilon k \dot{x} + x + \epsilon \alpha x^2 = \epsilon F \cos \Omega t,$$

is often used to describe nonlinear oscillations in a variety of physical situations. Here ϵ, k and F are positive constants, with ϵ small. The quantities $\Omega = 1 + \epsilon\omega$, ω and α are real parameters. For sufficiently small values of α and ω, the attractors of this system consist of limit cycles with amplitude A and phase ϕ.

a) Use the substitution

$$x = A\cos(\Omega t - \phi),$$

in the original equation to obtain the following estimate (to order ϵ) for the amplitude and phase of the limit cycles:

$$A^2(\tfrac{3}{4}\alpha A^2 - 2\omega)^2 = F^2 - k^2A^2,$$

$$\tan\phi = \frac{4k}{3\alpha A^2 - 8\omega}.$$

b) The first of the above equations gives the amplitude A as a function of the parameters α and ω. Show that the graph of A has two cusp points at

$$(\alpha, \omega) = \pm\left(\frac{\sqrt{3}k}{2}, \frac{32k^3}{9\sqrt{3}F^2}\right).$$

c) Assume $\alpha > \sqrt{3}k/2$ (a *hard spring*). Now slowly increase ω from negative to positive values. Show that A smoothly increases to a maximum at $A = F/k$ at a point $\omega = 3\alpha F^2/8k^2$. What happens now if ω is increased even further?

d) Draw a graph of the function $A(\alpha, \omega)$.

14. What are the critical points and critical values of the mapping $\chi\colon S^2 \to R^2$, taking the sphere to the horizontal plane?

15. Show that the above projection map of the sphere to the plane is stable, but that the map of the circle to the line given by $y = \sin 2x$ is unstable.

16. Consider the *complex* map $w = z^2 + 2\bar{z}$. Show that the set of critical points of this map consists of the entire unit circle $|z| = 1$. What is the geometric structure of the set of critical values?

17. Show that the two functions $f(x) = x^2$ and $g(x) = x^2 - x^4$ can be transformed into each other by means of the coordinate change

$$y(x) = \frac{x}{|x|}\sqrt{\frac{1 - \sqrt{1 - 4x^2}}{2}}.$$

Over what range of x-values is this change valid? Why can't the change work globally?

18. Consider the function $f(x, y) = x^4 + y^4 - 6x^2y^2$.

a) Prove that f is 4–determinate.

b) Show that the codimension of f equals seven.

c) Construct a universal unfolding of f. What terms form a basis for the ideal $m_n/\Delta(f)$?

19. The scalar dynamical system

$$\dot{\theta} = a - \cos\theta, \quad a \text{ real},$$

with state-space $U = \{\theta\colon 0 \leq \theta < 2\pi\}$ has several different types of equilibria, depending upon the value of the parameter a. Classify them.

20. The following equations describe the flow of power at each node in an electric power transmission network:

$$P_i = \alpha_{in}\sin\theta_i + \sum_{j=1}^{n-1}\alpha_{ij}\sin(\theta_i - \theta_j),$$

where P_i is the power injected at the ith node, θ_i is the voltage angle at the ith node and the quantities α_{ij} represent the manner in which the power at each node is transmitted to other nodes in the network. It is assumed that $\alpha_{ij} = \alpha_{ji} > 0$ if $i \neq j$ and $\alpha_{ii} = 0$, $i = 1, 2, \ldots, n$.

a) Show that the trigonometric substitution

$$x_i = \sin(\theta_i - \theta_j), \qquad y_i = \cos(\theta_i - \theta_j),$$

transforms the above equations into the set of nonlinear algebraic equations

$$\sum_{j=2}^{n}\alpha_{1j}x_j = P_1,$$

$$\sum_{\substack{j=2\\ j\neq i}}^{n}\alpha_{ij}(x_jy_i - x_iy_j) - \alpha_{1i}x_i = P_i, \quad i = 2, 3, \ldots, n-1,$$

$$x_i^2 + yi^2 = 1, \quad i = 1, 2, \ldots, n.$$

b) Show that there exists a scalar function V such that the system

$$\begin{pmatrix} \sum\alpha_{1j}\sin(\theta_1 - \theta_j) - P_1 \\ \sum\alpha_{n-1,j}\sin(\theta_{n-1} - \theta_j) - P_{n-1} \end{pmatrix} = \begin{pmatrix} \frac{\partial V}{\partial\theta_1} \\ \frac{\partial V}{\partial\theta_{n-1}} \end{pmatrix},$$

if and only if $P_i = 0$ for all i. That is, the right side is the gradient of a potential field if and only if the power injected at each node is zero.

c) Prove that for $n = 2$ and $\alpha_{ij} = 1$, there are exactly six solutions to the system of part (a), and that of these solutions only one is stable, i.e., the linearized approximation in the neighborhood of this solution has a coefficient matrix whose characteristic values all lie in the left half-plane.

21. The equation describing the temperature at which paper ignites is given by

$$mc^0 \frac{dT}{dt} = q - h_0(T - T_a)^{\frac{4}{3}} - KT^4 + r_2 a_2 m \exp\left(\frac{-e_2}{RT}\right), \qquad T(0) = T_a,$$

where

$T_a =$ the ambient temperature,

$m =$ surface density of the paper,

$c^0 =$ specific heat,

$q =$ the rate at which heat is applied,

$h_0 =$ temperature independent constant,

$K =$ radiation factor,

$r_2 =$ heat of the reaction,

$a_2 =$ preexponential factor,

$e_2 =$ activation energy,

$R =$ universal gas constant.

a) Show that the equation describing the equilibrium temperature T_e is

$$q - h_0(T_e - T_a)^{\frac{4}{3}} - KT_e^4 + r_2 a_2 m \exp\left(\frac{-e_2}{RT_e}\right) = 0.$$

b) Prove that the above relation for T_e is equivalent to the fold catastrophe geometry, and plot the quantity T_e as a function of q, keeping all other quantities constant.

c) The criterion for ignition of the paper at temperature T_c is that

$$\frac{d^2T}{dt} = 0, \qquad \frac{dT}{dt} = 0,$$

at $T = T_c$. Show that this implies that T_c satisfies the transcendental equation

$$a_2 \exp(-e_2/RT_c)\frac{r_2 e_2 m}{RT_c^2} - \frac{4}{3}h(T_c) - 4KT_c^3 = 0,$$

where $h(T_c) = h_0(T_c - T_a)^{\frac{1}{3}}$.

22. Suppose we have the dynamical process

$$\dot{x}_i = f_i(x_1, x_2, \ldots, x_n), \qquad i = 1, 2, \ldots, n.$$

Define the elements

$$u_{ij}(x) \doteq \frac{\partial}{\partial x_j}\left(\frac{dx_i}{dt}\right), \qquad i, j = 1, 2, \ldots, n.$$

The quantity $u_{ij}(x)$ measures the effect of a change in x_j on the *rate of change* of the quantity x_i when the system is in the state x. If $u_{ij}(x) > 0$, we call u_{ij} an *activator* of x_i, whereas if $u_{ij}(x) < 0$, it is an *inhibitor* of x_i. It's clear that if we know the dynamics, then we can compute the *activation-inhibition pattern* $\{u_{ij}(x)\}$. Moreover, we have the relation

$$df_i = \sum_{j=1}^{n} u_{ij} dx_j, \qquad i = 1, 2, \ldots, n.$$

a) Suppose the dynamics are *unknown,* but that the activation-inhibition pattern is given. Show that under these circumstances, it's possible to re-construct the dynamics if and only if the differential forms $\{df_i\}$ are *exact,* i.e.,

$$\frac{\partial}{\partial x_k} u_{ij} = \frac{\partial}{\partial x_j} u_{ik},$$

for all i, j, $k = 1, 2, \ldots, n$.

b) Show that the condition for exactness of a differential form is (highly) nongeneric, i.e., in the space of differential forms, the exact forms do *not* form an open, dense set.

c) The activation-inhibition patterns that generate exact differential forms are the only ones that give rise to systems whose dynamics can be globally described by a set of differential equations. But such patterns are highly nongeneric. Consequently, there are many systems for which we can give an activation-inhibition pattern but for which there does not exist a globally defined dynamical system generating this pattern. Discuss the relevance of this observation for the problem of system complexity.

Notes and References

§1. Early in this century, Hadamard presented the view of a "well-posed" problem in mathematical physics as being one that had the following properties: (1) the solution existed; (2) the solution was unique; (3) the solution was a continuous function of the initial problem data. The discussion of this chapter casts considerable doubt upon the last property, since many of the most important and puzzling questions of modern science do not seem to adhere to it, such as turbulence, cellular differentiation and stock market crashes. Of course, it could be argued that such phenomena are still described by smooth functions of the data in a technical sense, and that the seeming discontinuities are only regions where the derivatives of the function are of large magnitude. But this seems like a weak argument at best, and it appears that it is indeed necessary to account explicitly in a mathematical fashion for the emergence of discontinuous behavior from smooth descriptions. The ideas and methods of this chapter give an indication of just how this program might be carried out.

§3. To replace Hadamard's notion of a well-posed problem, an idea that relies upon a concept of stability for the solution to a *fixed* equation, Thom has emphasized the concept of structural stability which hinges upon stability concepts within a *family* of descriptions. Thom's view is that since we can never make exact observations of any physical process, the only mathematical descriptions that can be accepted are those whose behavioral character is preserved when we perturb the description by small amounts. In other words, if a small change in the model results in large changes in the behavior, then the model cannot be accepted as a description for a persistent natural process. For an account of Thom's ideas, see

Thom, R., *Structural Stability and Morphogenesis,* Benjamin, Reading, MA, 1975.

Technical discussions of the structural stability and genericity concepts are given in

Golubitsky, M., and V. Guillemin, *Stable Mappings and Their Singularities,* Springer, New York, 1973,

Arnold, V. I., S. Gusein-Zade and A. Varchenko, *Singularities of Differentiable Maps,* Vol. 1, Birkhäuser, Boston, 1985.

§4. A good account of Morse's Lemma is found in

Poston, T., and I. Stewart, *Catastrophe Theory and Its Applications,* Pitman, London, 1978.

For a development of Morse's Theorem within the more general context of singularity theory, see

Lu, Y. C., *Singularity Theory and an Introduction to Catastrophe Theory,* Springer, New York, 1976.

§5. The Splitting Lemma is important because it enables us to identify exactly where the essential nonlinearities lie in a given description. For an account of how to use the Splitting Lemma to reduce the dimensionality in nonlinear optimization problems, see the paper

Casti, J., "Singularity Theory for Nonlinear Optimization Problems," *Appl. Math. & Comp.*, 23 (1987), 137–161.

Additional results utilizing singularity theory concepts for optimization problems can be found in

Jongen, H., and G. Zwier, "Structural Analysis in Semi-Infinite Optimization," in *3rd Franco-German Conf. on Optimization,* C. Lemarechal, ed., INRIA, Le Chesnay, France, 1985, pp. 56–67,

Fujiwara, O., "Morse Programs: a Topological Approach to Smooth Constrained Optimization," *Math. Oper. Res.*, 7 (1982), 602–616.

§6. A common situation in physics and engineering is to have a finite number of terms in the Taylor series expansion of a function, and then to try to deduce information about the function from this finite amount of information. Often the assumption is made that the function involved is analytic, presumably on the grounds that this will help in analyzing properties of the function. To see that this is not the case, consider the function $f(x, y) = x^2(e^y - 1)$. The 17-jet of f is given by

$$j^{17}f = x^2y + \frac{1}{2}xy^2 + \cdots + \frac{1}{15!}x^2y^{15}.$$

Consequently, the equation $j^{17}f = 0$ has the x- and y-axes as its roots. But,

$$j^{18}f = j^{17}f + \frac{1}{18!}y^{18},$$

so that there are no solutions with $y > 0$ for the equation $j^{18}f = 0$. Hence, $j^{17}f$ is not sufficient to determine the character of $j^{18}f$, not to mention the character of f itself. Thus, for no finite k does having the information: (1) f is analytic around 0, and (2) $j^kf = j^k(x^2(e^y - 1))$ imply that the roots of f are even approximately those of $x^2(e^y - 1)$ near the origin. In fact, analyticity of f has no bearing on the question! This example forcefully brings out the role of determinacy as the relevant notion when we want to know how far out in the Taylor series we have to go in order to capture all the local structure in f. For a fuller discussion of this point, as well as more elaboration on the foregoing example, see

Poston, T., and I. Stewart, *Taylor Expansions and Catastrophes,* Pitman, London, 1976.

The computational procedure discussed in the text follows that given in

Deakin, M., "An Elementary Approach to Catastrophe Theory," *Bull. Math. Biol.,* 40 (1978), 429–450,

Stewart, I., "Applications of Catastrophe Theory to the Physical Sciences," *Physica D,* 20 (1981), 245–305.

§7. From the standpoint of applications, the idea of a universal unfolding is probably the most important single result from elementary catastrophe theory. The unfolding terms show us the types of perturbations that cannot be neutralized by a smooth coordinate transformation, and give information about the way in which the properties of the function will change when it is perturbed in various ways near a degenerate singularity. Special emphasis on this point of view is found in the book by Poston and Stewart cited under §4 above. See also many of the papers in the collection

Zeeman, E. C., *Catastrophe Theory: Selected Papers, 1972–1977,* Addison-Wesley, Reading, MA, 1977.

§8. A complete proof of the Classification Theorem may be found in the Zeeman book cited above. It appears that Thom was responsible for seeing that such a result must be true, as well as for recognizing the many pieces that would have to be put together in order to actually prove the theorem. He then convinced a variety of mathematicians including Arnold, Mather, Malgrange and Boardman to put together the necessary ingredients for the final result. It makes for fascinating reading today to contemplate the dialogue between René Thom and Christopher Zeeman on the future prospects of catastrophe theory for applications in the physical, life and social sciences. This dialogue is reprinted in the Zeeman book noted above.

§9. For a derivation of the power system dynamics, see

Modern Concepts of Power System Dynamics, IEEE Special Publication #70MG2-PWR, IEEE, New York, 1970.

§10. A detailed geometrical study of the bifurcation diagrams for all the elementary catastrophes is given in

Woodcock, A. E. R., and T. Poston, *A Geometrical Study of the Elementary Catastrophes,* Springer Lecture Notes in Mathematics, Vol. 373, Springer, Berlin, 1974.

§11. The fishery management example follows that in

Jones, D., and C. Walters, "Catastrophe Theory and Fisheries Regulation," *J. Fisheries Res. Board Canada,* 33 (1976), 2829–2833.

Other work along similar lines is reported in

Peterman, R., "A Simple Mechanism that Causes Collapsing Stability Regions in Exploited Salmonid Populations," *J. Fisheries Res. Board Canada,* 34, No. 8, 1977.

§12. The dynamics describing the budworm outbreaks for a single site were first put forth in

Ludwig, D., C. Holling and D. Jones, "Qualitative Analysis of Insect Outbreak Systems: The Spruce Budworm and Forest," *J. Animal Ecology,* 47 (1978), 315–332.

The catastrophe theory analysis presented here showing the impossiblity of eliminating budworm outbreaks by manipulation of system parameters is given in detail in

Casti, J., "Catastrophes, Control and the Inevitability of Spruce Budworm Outbreaks," *Ecol. Modelling,* 14 (1982), 293–300.

§13. An excellent introductory account of the development of singularity theory as an outgrowth of Morse's work is the Lu book cited under §4 above. For more technical accounts of Whitney's Theorem and its subsequent extension to general smooth maps, see the Golubitsky and Guillemin book referred to in §3 above, as well as

Arnold, V. I., *Singularity Theory,* Cambridge U. Press, Cambridge, 1981,

Gibson, J., *Singular Points of Smooth Mappings,* Pitman, London, 1979,

Martinet, J., *Singularities of Smooth Functions and Maps,* Cambridge U. Press, Cambridge, 1982.

The urban spatial structure problem follows the treatment in

Casti, J., "System Similarities and the Existence of Natural Laws," in *Differential Topology, Geometry and Related Fields,* G. Rassias, ed., Teubner, Leipzig, 1985, pp. 51–74.

§14. The very major differences between classification of singularities of smooth functions and the classification of smooth vector fields is brought out in the review of Thom's book by Guckenheimer. See

Guckenheimer, J., "Review of R. Thom, *Stabilité structurelle et morphogénèse,*" *Bull. Amer. Math. Soc.,* 79 (1973), 878–890.

Good accounts of various aspects of the problem of classifying the singularities of vector fields from a variety of perspectives can be found in

Arnold, V. I., *Geometrical Methods in the Theory of Ordinary Differential Equations,* Springer, New York, 1983,

Guckenheimer, J., and P. Holmes, *Nonlinear Oscillations, Dynamical Systems, and Bifurcations of Vector Fields,* Springer, New York, 1983.

A vitally important tool in the classification of smooth functions is, as we have seen, the Splitting Lemma. The corresponding result for dynamical systems is the Center Manifold Theorem, a good account of which is given in the foregoing volumes. (See also Chapter Five, Problem 5.)

The lake pollution example is taken from

Casti, J., J. Kempf and L. Duckstein, "Modeling Phytoplankton Dynamics Using Catastrophe Theory," *Water Resources Res.,* 15 (1979), 1189–1194.

DQ #2. The matter of a "natural law" vs. an "empirical relationship" is one fraught with many perplexing epistemological as well as semantic difficulties, some of which we have already considered in §3 of Chapter One. Additional perspectives from a more pragmatic rather than philosophical point of view are found in

Casti, J., "Systemism, System Theory and Social System Modeling," *Regional Sci. and Urban Econ.,* 11 (1981), 405–424,

Kalman, R., "Comments on Scientific Aspects of Modelling," in *Towards a Plan of Action for Mankind,* M. Marois, ed., North-Holland, Amsterdam, 1974, pp. 493–505.

DQ #5–7. The issues raised by this cluster of problems have been treated in some detail from the perspective of natural resource management by the ecological systems group at the University of British Columbia. A sampling of their work is given by

Holling, C. S., "Resilience and Stability of Ecological Systems," *Annual Rev. of Ecology and Systematics,* 4 (1973), 1–23,

Holling, C. S., "Science for Public Policy: Highlights of Adaptive Environmental Assessment and Management," Report R–23, Institute for Animal Resource Ecology, U. of British Columbia, Vancouver, March 1981,

Walters, C., G. Spangler, W. Christie, P. Manion and J. Kitchell, "A Synthesis of Knowns, Unknowns, and Policy Recommendations from the Sea Lamprey Int'l. Symposium," *Canadian J. Fish. Aquat. Sci.,* Vol. 37, No. 11, (1980), 2202–2208,

Hilborn, R., "Some Failures and Successes in Applying Systems Analysis to Ecological Management Problems," Report R–18, Institute for Animal Resource Ecology, U. of British Columbia, Vancouver, October 1978.

DQ #8. For more details on the uses of system-theoretic ideas in modeling international conflict situations, see

McClelland, C., "System Theory and Human Conflict," in *The Nature of Human Conflict,* E. B. McNeil, ed., Prentice-Hall, Englewood Cliffs, NJ, 1965,

Holt, R., B. Job and L. Markus, "Catastrophe Theory and the Study of War," *J. Conflict Resol.,* 22 (1978), 171–208.

DQ #12. The issue of what convention to use in applied catastrophe theory situations is a delicate one, ultimately coming down to a consideration of the various time-scales at work in the problem. For a more detailed consideration of these matters, see

Gilmore, R., "Catastrophe Time Scales and Conventions," *Physical Rev. A,* 20 (1979), 2510–2515.

DQ #15. The stock market example is taken from the work reported in

Zeeman, E. C., "On the Unstable Behavior of Stock Exchanges," *J. Math. Economics,* 1 (1974), 39–49.

PE #11. The various factors involved in the collapse of civilizations is considered further in

Renfrew, C., "Systems Collapse as Social Transformation: Catastrophe and Anastrophe in Early State Societies," in *Transformations: Mathematical Approaches to Cultural Change,* C. Renfrew and K. Cooke, eds., Academic Press, New York, 1979.

A detailed account of the collapse of the Mayan civilization, complete with computer programs simulating the decline, is found in the volume

Lowe, J. W. G., *The Dynamics of Collapse: A Systems Simulation of the Classic Maya Collapse,* U. of New Mexico Press., Albuquerque, 1985.

PE #12–13. Further discussions of these problems are found in the Zeeman book cited under §7 above.

PE #22. An activation-inhibition network is an important example of a situation for which we can have a dynamical description that cannot be expressed by means of a set of differential equations. In general, the activation-inhibition pattern $\{u_{ij}(x)\}$ does not lead to an exact differential form; hence, there is no differential equation system corresponding to the pattern. The best that can be done is to *approximate* locally the pattern near the state x by means of such a differential system.

We have seen that if there is a differential description, the activation-inhibition pattern is given as

$$u_{ij}(x) = \frac{\partial}{\partial x_j}\left(\frac{dx_i}{dt}\right).$$

There is no reason to stop here. We can consider the quantities

$$v_{ijk}(x) = \frac{\partial}{\partial x_k} u_{ij}(x),$$

which express the effect of changes in x_k on the activation-inhibition pattern. If $v_{ijk}(x) > 0$, then an increase in x_k tends to accelerate the activation of the rate of production of x_i by x_j, and we call x_k an *agonist* of x_i. If $v_{ijk}(x) < 0$, then x_k is termed an *antagonist.* Note that we can carry on this process by defining the quantities

$$w_{ijk\ell}(x) = \frac{\partial}{\partial x_\ell} v_{ijk}(x),$$

and so forth. If there is a differential equation description of the dynamics, each of these levels of description is obtained from the preceding one by means of a simple differentiation of that description. However, if only the patterns are given and there is no global differential description, then the patterns are independent and cannot be obtained from a single "master" description. This is the situation that Rosen claims distinguishes "complex" systems from those that are "simple." A fuller account of this notion of system complexity and its relationship to activation-inhibition patterns can be found in

Rosen, R., "Some Comments on Activation and Inhibition," *Bull. Math. Biophysics,* 41 (1979), 427–445,

Rosen, R., "On Information and Complexity," in *Complexity, Language and Life: Mathematical Approaches,* J. Casti and A. Karlqvist, eds., Springer, Heidelberg, 1986.

CHAPTER FIVE

ORDER IN CHAOS: TURBULENCE AND PATTERN IN THE FLOW OF FLUIDS, POPULATIONS AND MONEY

1. *Deterministic Mechanisms and Stochastic Behavior*

Suppose we consider the temporal growth pattern of an insect population. Let's assume the insects grow at a rate proportional to their current population, and that they die by mutual competition for scarce resources. Assume that the birth rate is given by the proportionality constant r, while the carrying capacity of the environment is specified by a constant K. Further, assume that the interaction is by direct mutual competition. Under these conditions, the population growth dynamics can be described by the quadratic difference equation

$$y_{t+1} = ry_t - \frac{y_t^2}{K}, \qquad y_0 = y^0.$$

If we assume that the environment can support only a finite population of insects, then we can scale the population to a new variable x so that x lies in the unit interval. Incorporating the constant K into the scaling, we can then rewrite the dynamics as

$$x_{t+1} = \alpha x_t(1 - x_t) \doteq F(x_t, \alpha), \qquad x_0 = x^0, \qquad 0 \leq x_t \leq 1.$$

Now suppose that we fix α in the range $0 < \alpha < 1$. In this range, every initial population eventually dies out ($x_t \to 0$). On the other hand, if $\alpha > 4$, it's easy to see that all initial populations except the fixed points $x = 0$ and $x = 1$ diverge to $-\infty$. Thus, non-trivial dynamical behavior is possible only for $1 \leq \alpha \leq 4$. Let's consider what happens in the neighborhood of an equilibrium point of the above system, i.e., a point x^*, for which $x^* = F(x^*, \alpha)$. Such a point x^* is given by $x^* = (1 - \frac{1}{\alpha})$, and we want to know whether or not such a point is stable, i.e., do all points sufficiently close to x^* approach x^* as $t \to \infty$? The answer depends upon the slope of $F(x, \alpha)$ at $x = x^*$. If this slope is less than one in magnitude, then x^* is attracting; if not, then x^* is repelling. An easy calculation yields the slope as

$$\lambda^{(1)}(x^*) \doteq \left.\frac{dF}{dx}\right|_{x=x^*} = 2 - \alpha,$$

so that the point x^* is stable for $1 < \alpha < 3$. What happens when we increase α beyond the value $\alpha = 3$? It's clear that the point x^* is no longer attracting, but rather is repelling. Now what is the ultimate fate of a point near x^*?

To address this question, we first note that the previously stable fixed point x^* can be regarded as a period-1 cycle that becomes unstable at $\alpha = 3$. So it's natural to consider whether or not there may be a cycle of period-2 which *is* stable for $\alpha \geq 3$. To study this question, we examine the population at successive intervals that are two generations apart, i.e., we are interested in the function which relates x_{t+2} to x_t. This is given by the second iteration of the map $F(x, \alpha)$. Thus,

$$x_{t+2} = F(F(x_t, \alpha), \alpha) \doteq F^{(2)}(x_t, \alpha).$$

Populations that recur every *second* generation are now fixed points of the map $F^{(2)}(x, \alpha)$, and can be found by solving the equation

$$F^{(2)}\left(x^*_{(2)}, \alpha\right) = x^*_{(2)},$$

where we have used an obvious notation for the period-2 fixed point(s). By simple geometric arguments or by direct algebraic calulations, it can be seen that there are two solutions to the above equation, both representing stable fixed points if the magnitude of

$$\left.\frac{dF^{(2)}}{dx}\right|_{x=x^*_{(2)}} \doteq \lambda^{(2)} < 1.$$

But a direct calculation shows that $\lambda^{(2)} = {\lambda^{(1)}}^2$ when $x^* = x^*_{(2)} = F(x^*_{(2)}, \alpha)$, so that two new stable fixed points of period-2 are born out of the old fixed point x^* when x^* becomes unstable. This process is perfectly general; as we continue to increase α, the period-2 fixed points become unstable and give rise to two new stable fixed points of period-4; a further increase of α gives stable fixed points of period-8, and so on. This sequence of births and deaths of stable fixed points of periods 2^k continues for all $k = 1, 2, \ldots$; however, the "window" of parameter values α for which any one cycle is stable progressively diminishes, so that the entire process converges to a critical parameter value $\alpha^* \approx 3.57\ldots$.

As we go beyond α^*, there are an infinite number of fixed points with different periodicities, as well as an infinite number of different periodic cycles. In addition, there are also an uncountable number of initial populations x^0 that give rise to totally aperiodic, but bounded, trajectories. Beyond the value $\alpha \approx 3.8284\ldots$, there are cycles with every integer period, as well as an uncountable number of aperiodic trajectories. This is the so-called *chaotic* regime for the system. This terminology suggests that the dynamical trajectories in this regime are indistinguishable from a stochastic process, and many analytical and computational results confirm this intuition. We shall deal extensively with this point below, so for now let's only observe that even though values of α beyond α^* give rise to aperiodic trajectories and "deterministic randomness," it is still the case that almost all initial points x^0 are attracted to a unique cycle. Thus, the set of initial populations that belong to any of the other infinite number of cycles, or to an aperiodic trajectory, though uncountable in number, still form a set of measure zero on the unit interval.

The justification for describing the motion of our system as stochastic comes from the fact that any given stable cycle is likely to occupy a vanishingly small window of parameter values. Also, the long time required for the transients associated with the initial conditions to damp out means that, in practice, the trajectory is very likely to "wander" away from this cycle and "randomly" attach itself to one or another of the other cycles or aperiodic orbits. Thus, a stochastic description seems appropriate despite the totally deterministic nature of the underlying dynamical equations. In passing, it's interesting to note that the above iteration of the function $F(x, \alpha)$ was often used in the early days of computing as a random number-generating device using values of α near $\alpha = 4$, and in fact, this scheme is still useful for pocket calculator computations.

The foregoing example, simple as it is, exhibits all of the types of long-term behavior that can be displayed by a dynamical system: fixed points, cycles and aperiodic motions, stable and unstable. Our goal in this chapter is to characterize the circumstances under which each of these qualitatively different types of behavior emerge, and to give some indication as to how such dynamical systems can be used to model a spectrum of situations arising in the natural sciences, economics and ecology. But first we need a more thorough grounding in the basic elements of dynamical system theory.

2. *Dynamical Systems, Flows and Attractors*

Intuitively speaking, we think of a dynamical system as being the description of the time behavior of a point moving about on some sort of surface according to a rule that describes how one point is to follow another. Thus, we think of a curve drawn on the surface whose initial point corresponds to the state of the system at time $t = 0$, and whose end point represents the system state at time $t = \infty$. The concept of a *dynamical system* is introduced in order to make precise the notions of "curve," "surface" and "prescription," inherent in this intuitive vision of a dynamical process.

Technically, we can formalize the notion of a surface by the idea of a *manifold.* In rough terms, an n-dimensional manifold M is a space in which

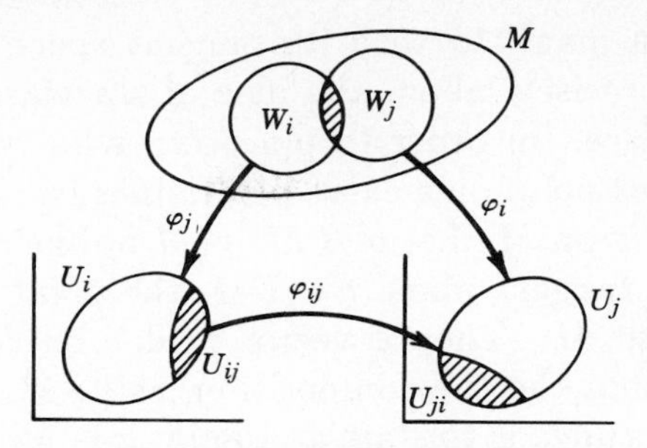

Figure 5.1 A Manifold M

it is possible to set up a coordinate system near each point such that locally the space looks like a subset of the euclidean space R^n. Of course, we may need to use a different coordinate system around each point in M. So, in order to ensure that the same point of M is described consistently using two different local coordinate frames, we need to have maps that enable us to translate back and forth between these different local descriptions. Depending upon the analytic properties of these "dictionary" maps, we speak of M as being a *differentiable, analytic, topological,* etc., manifold. The general picture is shown in Fig. 5.1, where the sets U_i represent subsets of R^n, the maps φ_i are the coordinate maps and the dictionary maps are given by φ_{ij}. Typical examples of smooth manifolds are R^n, an n-torus and an n-sphere. In what follows, we shall concern ourselves only with smooth (i.e., C^∞) manifolds.

Consider a point $x \in M$. The *tangent vector* to M at x consists of the equivalence class of all curves leaving x, with two such curves being equivalent if their images in R^n under any local coordinate system centered at x can be transformed into each other, i.e., are equivalent in R^n. The set of all tangent vectors at x forms a vector space, termed the *tangent space* TM_x to M at x. The set $TM = \bigcup_{x \in M} TM_x$ forms the *tangent bundle* of M (see Fig. 5.2 for an illustration of these objects).

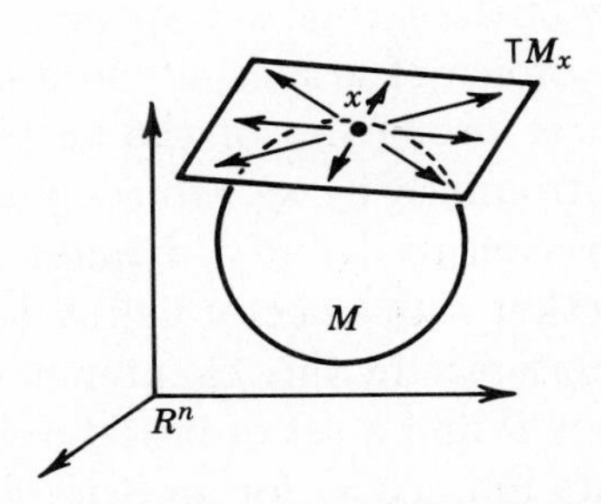

Figure 5.2 The Tangent Vectors and Tangent Space to M at x

The notion of a manifold with its tangent space and tangent bundle allows us to speak precisely about the idea of a surface and the concept of a curve on this surface. In order to pin down what we mean by a "rule" specifying how a given point on a curve determines its successor, we need the notion of a *vector field* on M. Let $p\colon TM \to M$ be the mapping (projection) that assigns to each tangent plane $\tau \in TM$, the point $x \in M$ at which the plane τ is tangent to M. Then a vector field v on M is just a mapping $v\colon M \to TM$ such that the the composition $p \circ v\colon M \to M$ is the identity map on M. Thus, the idea is to start at a point $x \in M$ and apply the map v to get the tangent plane $TM_x \in TM$. The projection p then takes us back to the original point x. Consequently, the role of the vector field v is to *select* a tangent vector in TM_x that then tells the curve which way to "move" away from the point x. If we have a local coordinate system $(x_1, x_2, \ldots, x_n)$ around the point $x \in M$, the vector field v at x can be written in the manner familiar from elementary differential equations as

$$\frac{dx}{dt} = v(x), \qquad x(0) = x.$$

It's rather straightforward to show that given the manifold M and a vector field v on a compact subset of M, there exists a one-parameter family of diffeomorphisms $g_t\colon M \to M$, such that

$$\frac{d}{dt} g_t x = v(g_t x).$$

Furthermore, the curve $g_t x_0$ is the unique solution of the equation

$$\frac{dx}{dt} = v(x), \qquad x(0) = x_0.$$

The mapping $\Phi(t, x_0) = g_t x_0$ is termed the *flow* of the vector field v, and g_t satisfies the semigroup property $g_{t+s} = g_t g_s$, expressing the uniqueness of the solution of the differential equation determined by v.

The main advantage of the formal set-up we have just described is that it is independent of the choice of a special coordinate system at x. Thus, we can study the *intrinsic* properties of the vector field v without being distracted by artifacts introduced by an arbitrary choice of coordinates.

We are now in a position to define a *dynamical system* $\mathcal{D}$ as simply a smooth manifold M, together with a vector field v defined on M. Compactly, $\mathcal{D} = (M, v)$. Our main interest in this chapter is to explore the question: Given a dynamical system $\mathcal{D}$ and a set of initial points U in M, what is the nature of the set of points $\lim_{t\to\infty} x_t$ for $x_0 \in U$? Here we will always take the time-parameter $t \in R$, the reals, or $t \in \mathbf{Z}$, the integers, or possibly the subsets R_+ or $\mathbf{Z}_+$ consisting of nonnegative time. To address this question, we need to consider the *stationary points* of the dynamical system $\mathcal{D}$.

Assume that time is continuous. Then we call $x^* \in M$ a *stationary point* for the dynamical system $\mathcal{D}$ if $v(x^*) = 0$; in discrete time, the corresponding analogue is a *fixed point* of $\mathcal{D}$ defined to be a point $x^* \in M$ such that $v(x^*) = x^*$. Now let $\sigma > 0$. Then we say that a trajectory $\{x_t\}$ of $\mathcal{D}$ starting at the point x_0 is a *periodic orbit* of period σ, if $x_t \neq x_0$ for $0 < t < \sigma$, and $x_\sigma = x_0$. A stationary point x^* is termed *stable* if for every $\epsilon > 0$, there exists a $\delta(\epsilon) > 0$ such that the distance $|x(t) - x^*| < \epsilon$ for all $t > t_0$, whenever $|x(t_0) - x^*| < \delta$. A stationary point is called *asymptotically stable* if it is stable and has a neighborhood U such that if $y \in U$, the distances between the points on the trajectory of $\mathcal{D}$ having initial point y and the stationary point tend uniformly to zero as $t \to \infty$. Analogous definitions apply for the stability of periodic orbits.

Example: Linear Vector Fields

Consider a system of linear differential equations

$$\frac{dx}{dt} = Ax, \qquad x \in R^n.$$

The solution curves of the system are given by $x_t = e^{At}x_0$. If the coordinates in R^n are chosen so that the matrix A is in Jordan form, then each solution curve $\{x_t\} \neq 0$ has one of the following kinds of behaviors:

1) $\|x_t\| \to \infty$ as $t \to \pm\infty$.

2) $\|x_t\| \to 0$ as $t \to +\infty$, $\|x_t\| \to -\infty$ as $t \to -\infty$.

3) $\|x_t\| \to 0$ as $t \to -\infty$, $\|x_t\| \to +\infty$ as $t \to \infty$.

4) $\|x_t\|$ and $\|x_t\|^{-1}$ are bounded.

The last case occurs only if A has a nonrepeated characteristic value on the imaginary axis. In particular, if 0 is a characteristic value of A there is a linear subspace of stationary points. If there is a pair of nonzero, pure imaginary characteristic values of A, then there are periodic orbits, and more than one such pair may lead to almost periodic orbits, depending upon the ratios of these characteristic values.

Since generically a matrix will not have purely imaginary characteristic values, we can expect one of the first three cases to occur. Assume now that A is generic. Those solutions having behavior of type 2 span a linear subspace of R^n. This subspace is spanned by the characteristic vectors of A associated with the characteristic values having negative real parts. This subspace is called the *stable manifold* of the stationary point at the origin. Similarly, the solutions displaying behavior of type 3 span a linear subspace called the *unstable manifold* of the origin. Since we have assumed that A has no purely imaginary characteristic values, the stable and unstable manifolds are complementary subspaces. In such situations, we say that the stationary point at the origin is *hyperbolic,* since all solutions not on the stable or unstable manifold follow trajectories that look like hyperbolas twisted in space.

The situation just described for linear systems carries over to "generic" nonlinear dynamical systems as the following result shows.

STABLE MANIFOLD THEOREM. *Let v be a smooth vector field having a stationary point p on an n-dimensional manifold M. Assume that the flow of v is defined for all t, and that the derivative of v at p has s characteristic values with negative real parts and $n - s$ characteristic values with positive real parts, counted according to their multiplicities. Then there are submanifolds $W^s(p)$ and $W^u(p)$ of M such that:*

i) $p \in W^s(p) \bigcap W^u(p)$.

ii) If $\{g_t\}$ represents the flow of v, then

$$W^s(p) = \{x \colon g_t x \to p \text{ as } t \to \infty\},$$

$$W^u(p) = \{x \colon g_t x \to p \text{ as } t \to -\infty\}.$$

iii) The tangent space to W^s at p is spanned by the characteristic vectors of the derivative of v at p corresponding to the characteristic values having negative real parts. Similarly, the tangent space to W^u at p is spanned by the characteristic vectors corresponding to the characteristic values of the derivative of v at p having positive real parts.

There is a corresponding result for the fixed points of invertible maps, in which the conditions on the derivative of the map at the fixed point p involve the characteristic values being inside or outside the unit circle rather than in either the right or left half-plane. Note that the characteristic values in this case correspond to the exponentials of those in the continuous-time case.

One of the most important features of the stable and unstable manifolds is that they intersect *transversally* at p. This means that their tangent spaces taken together span the tangent space to M at p. Such intersections are important because they persist under perturbations of the submanifolds. This notion forms the basis for the idea of the "structural stability" of the vector field v under perturbations of various sorts.

In many physical situations, the time constants are such that the transient motion of the system is quite rapid and the real questions of interest center about the limiting behavior. Furthermore, even if the transient motion is important for the functioning of the system, it is the behavior near the stationary points that ultimately "organizes" the manner in which the transient motion takes place. Thus, we need to have some means of working with the sets that describe the way the system trajectories behave as $t \to \infty$.

Let the dynamical system $\mathcal{D}$ be given, and let $\{\Phi(x, t): t \in R\}$ be the trajectory through the point $x \in M$. We call x a *wandering point* if there is a neighborhood U of x such that $\Phi(U, t) \bigcap U = \emptyset$ whenever $|t| > t_0$, for some $t_0 > 0$. Points that are not wandering are called *nonwandering* points. Intuitively, a wandering point x has the property that after a short initial time, no point near x returns to a point near x. The set of nonwandering points of the flow Φ is called the *nonwandering set* of Φ, and is denoted by Ω. The basic idea in dynamics is to first describe the nonwandering set Ω, together with its dynamics, and to then describe the way in which trajectories flow from one piece of Ω to another. To do this, we need the idea of an *attractor* of the flow Φ.

A closed, invariant set S for the flow Φ is called *topologically transitive* if there is a trajectory γ inside S such that every point of S is the limit of a sequence of points taken from γ. Thus S is the closure of γ, and γ is dense in S. Then a topologically transitive subset S of Ω is called an *attractor* for the flow Φ if S has a neighborhood U such that Φ carries U into itself as $t \to \infty$, and every point near S asymptotically approaches S as $t \to \infty$. Thus, small changes in initial conditions from those in S don't send the trajectory far away from S, and all such trajectories ultimately return to S. Finally, we call U the *domain of attraction* of S. As a technical point, we note that an attractor is something more than just an *attracting set,* i.e., a set A such that all points nearby to A ultimately go to A as $t \to \infty$. To be an attractor, a set must be an attracting set that also contains a dense orbit. So, for example, an attracting set that contains a stable periodic orbit cannot be an attractor.

Examples

To fix some of the foregoing ideas, consider the following examples:

1) $\dot{x} = x - x^3, \quad \dot{y} = -y, \quad (x, y) \in R^2 (= M).$

The set $\{(x, y): -1 \leq x \leq 1, y = 0\}$ is an attracting set for the system, although most of the points in this set are wandering (proof?). The attractors of the flow are the two points $(\pm 1, 0)$.

2) $\dot{\theta} = 1, \quad \dot{\phi} = \pi, \quad (\theta, \phi) \in T^2$ (= 2-torus = M).

For this system, the irrational linear flow has a dense orbit, and as a result, every point on T^2 is nonwandering. The attractor for this flow is the entire manifold T^2.

There are a few additional concepts from the theory of dynamical systems that we will need in order to analyze the various types of behavior that can occur, but we shall defer their introduction until later. For now, let's move on to a consideration of dynamical processes that present behavior that is halfway between the elementary stationary points and limit cycles and the seemingly random motion associated with chaos, i.e., processes giving rise to *quasi-periodic* trajectories.

3. *Quasi-periodic Trajectories and the KAM Theorem*

Imagine a classical mechanical system whose position and momentum are described in local coordinates by the variables $p = (p_1, p_2, \ldots, p_n)$ and $q = (q_1, q_2, \ldots, q_n)$, respectively. We call such a system *Hamiltonian* if there exists a function $H(p, q)$ such that the dynamical behavior of the system can be written as

$$\dot{q} = \frac{\partial H}{\partial p}, \qquad -\dot{p} = \frac{\partial H}{\partial q}.$$

Heading the long list of reasons why such systems are important are the following:

- *Liouville's Theorem*—Hamiltonian systems preserve the phase volume; i.e., an initial volume element $\{dp_1 \cdots dp_n \cdot dq_1 \cdots dq_n\}$ in the phase space R^{2n} is preserved by the flow of a Hamiltonian system.

- *Conservation of Energy*—The function H is a first integral for the flow of the dynamics, i.e., $dH/dt = 0$ along the trajectories of the system.

The second property leads to the idea of an *integrable* Hamiltonian system as one for which there exist n independent first integrals, i.e., functions $I_1(p, q), \ldots, I_n(p, q)$, such that $dI_i/dt = 0$, $i = 1, 2, \ldots, n$. For such systems, we can obtain an explicit characterization of their dynamical behavior, as will be shown in a moment.

Example: The Harmonic Oscillator

The standard example of a Hamiltonian system is the simple harmonic oscillator, describing the small oscillations of a pendulum. The Hamiltonian for this system is

$$H(p, q) = \frac{\omega}{2}(p^2 + q^2),$$

where ω is the frequency of the oscillation. A first integral for this system is $I = p^2 + q^2$, leading to the well-known conclusion that the trajectories of the system lie on the circles

$$p^2 + q^2 = k^2, \quad k \in R.$$

The constant k^2 represents the energy of the oscillator, and is fixed by the initial displacement $p(0)$ and momentum $q(0)$.

The circular motion of the foregoing linear system in its phase-space (or periodic motion in the state-space) is typical of *nonlinear* integrable Hamiltonian systems, being generalized in higher dimensions to motion on a torus instead of a circle. Since this generalization is important for what follows, let's explicitly indicate how it goes.

Assume we are given a Hamiltonian system as described above. By the integrability assumption, there exists a coordinate transformation into a new set of generalized position and momentum variables x and y, respectively, such that in these new variables the Hamiltonian function H depends only upon the position variables x, i.e., $H = H(x_1, x_2, \ldots, x_n)$. In the new variables, the dynamical equations are

$$\dot{y} = \frac{\partial H}{\partial x} \doteq \Omega(x), \qquad -\dot{x} = \frac{\partial H}{\partial y} = 0.$$

This system can be easily integrated, yielding

$$\begin{aligned} y(t) &= \Omega(x)t + y(0), \\ x(t) &= x(0). \end{aligned}$$

Returning to the old variables, we obtain $2n$ combinations of $\{p_i, q_i\}$ and t:

$$\begin{aligned} y(0) &= y(p(t), q(t), t), \\ x(0) &= x(p(t), q(t), t), \end{aligned}$$

which do not depend upon t. This means that we have solved the equations of motion and obtained $2n$ constants (integrals, invariants) of the motion. This result has been obtained under the assumption that there exists a coordinate change that will make the new Hamiltonian independent of the generalized momenta variables. It can be shown that the existence of n independent integrals of motion will suffice for this assumption to be satisfied.

So what does the foregoing result tell us about the dynamical behavior of an integrable Hamiltonian system? Basically, the motion is similar to that of n coupled oscillators, as can be seen from the above equations. In order to keep the motion confined to a finite segment of R^{2n}, it must be the case that the linearly growing terms in y appear as arguments of periodic functions. We have seen in the preceding example that when $n = 1$, the motion is circular with constant angular velocity; hence, periodic. When $n = 2$, there are two radii determined by the initial values $x_1(0)$ and $x_2(0)$, with angular velocities Ω_1 and Ω_2. Thus, the motion is confined to a two-dimensional torus. However, in addition to periodic motion we now have the possibility for a new kind of behavior. When the ratio Ω_1/Ω_2 is irrational, the motion cannot be periodic and the trajectory winds about the surface of the torus endlessly and densely covers it. Such motion is termed *quasi-periodic.* In general, the motion of an integrable Hamiltonian system with n degrees of freedom is quasi-periodic and takes place on an n-torus.

The two most immediate questions surrounding integrable Hamiltonian systems are the following:

- How "typical" is it for a Hamiltonian system to be integrable?
- If an integrable Hamiltonian system is perturbed "slightly," does the qualitative nature of the motion remain the same? That is, if we use the new Hamiltonian $H = H_0 + P$, where H_0 is Hamiltonian and $P(x, y, t)$ is an analytic perturbation that is of period 2π in the x variables and contains a small parameter, does the system trajectory still remain quasiperiodic?

Both of these questions can be given rather unambiguous answers: it is extremely rare for a Hamiltonian system to be integrable if $n \geq 3$. In fact, it is so rare that a nonintegrable system cannot usually even be *approximated* by a sequence of integrable systems. The second question is answered by the celebrated Kolmogorov-Arnold-Moser (KAM) Theorem, a rough statement of which is as follows:

KOLMOGOROV-ARNOLD-MOSER THEOREM. *Assume that the perturbation P is small, and that the frequencies $\Omega(x)$ of the unperturbed system satisfy the nonresonance condition*

$$\det \frac{\partial(\Omega_1(x), \ldots, \Omega_n(x))}{\partial(x_1, \ldots, x_n)} \neq 0.$$

Then the motion of the system is still confined to an n-torus, except for a set of initial conditions of small measure.

There are several comments in order regarding this basic result:

1) The n-tori, termed *KAM surfaces,* if seen in plane section appear as slightly distorted versions of those for the $P = 0$ case. Nevertheless, the qualitative nature of the motion remains basically the same as for the integrable situation.

2) The set of special initial conditions that lead to trajectories not on an n-torus may wander about freely on the energy surface for the system, i.e., the surface for which H =constant.

3) The boundary of the constant energy surface must be of dimension $2n - 1$; hence, when $n \geq 3$, the n-dimensional KAM surfaces cannot serve as boundaries dividing the energy surface into regions that confine the wandering trajectories (as shown in Fig. 5.3 for the case $n = 2$). Consequently, these trajectories may wander along the whole energy surface and give rise to a mechanism for randomness termed *Arnold diffusion.*

Figure 5.3 Invariant Tori for a Three-Dimensional Energy Manifold

Example: Convex Billiard Tables

As an amusing illustration of the application of the KAM Theorem, consider the closed trajectories of a billiard ball moving on a frictionless table of any convex shape. An example of a closed trajectory that is stable in the linear approximation is the minor axis of an ellipse when the table has an elliptical shape. The KAM Theorem enables us to conclude that a closed trajectory that is near the minor axis of an ellipse is also stable on any billiard table that is a small perturbation of the original ellipse.

There are a number of additional applications of the KAM Theorem to problems in classical mechanics that involve rotations of heavy rigid bodies, planetary motion, and the like. Since these matters are rather far afield from our interests in this book, we leave the reader to consult the Notes and References for further details. We now move on to a consideration of the type of random motion suggested by the process of Arnold diffusion, i.e., randomness generated by a deterministic mechanism.

4. *A Philosophical Digression: Randomness and Determinism in Dynamical System Modeling*

The type of aperiodic behavior displayed by the simple quadratic "logistic type" map discussed in Section 1, as well as the Arnold diffusion type of instability shown by the trajectories of nonintegrable Hamiltonian systems of higher dimension, makes it clear that very simple, totally deterministic mechanisms can (and do) give rise to very complicated behaviors. In fact, from an observational standpoint such behaviors are totally indistinguishable from what one would expect to see if the underlying generating mechanism were what probabilists would describe as a *stochastic* process. Before discussing some of the epistemological aspects of this observation, let's look at two very specific examples of this phenomenon as vehicles upon which to focus our subsequent remarks on the stochastic vs. deterministic dichotomy.

Example: Piecewise-Linear Maps and Coin-Tossing

Consider the piecewise-linear map

$$x_{t+1} = \begin{cases} 2x_t, & 0 \leq x_t \leq \frac{1}{2}, \\ 2 - 2x_t, & \frac{1}{2} \leq x_t \leq 1. \end{cases}$$

It should be noted that this map is topologically (but not diffeomorphically!) equivalent to the quadratic map $x_{t+1} = 4x_t(1 - x_t)$, considered in Section 1 by the continuous change of coordinates $x \to (2 \sin^{-1} \sqrt{x})/\pi$.

Suppose we divide the unit interval into two halves and label the segment $0 \leq x \leq \frac{1}{2}$ by "H," and the segment $\frac{1}{2} \leq x \leq 1$ by "T." Now pick an arbitrary starting point x_0 in the unit interval, and carry out the iteration specified above, labeling each successive iterate by the half of the interval into which it falls. Thus, each such experiment will result in a sequence like HTTHHHTTHTHTH.... It has been shown that after a large number of such iterations, the density function for the variable x describing the fraction of the time that there are more Heads than Tails is given by

$$p(x) = \frac{1}{\pi \sqrt{x(1 - x)}} .$$

This is exactly the probability density function that we would have obtained if each iteration of the above map was regarded as consisting of the flip of a fair coin, and we kept track of the resulting sequence of Heads and Tails. In short, the iterates of the deterministic, piecewise-linear map given above are mathematically and *observationally* indistinguishable from the sample path of a sequence of Bernoulli trials with probability $p = \frac{1}{2}$.

Example: Quadratic Optimization and Stochastic Filtering

Suppose we wish to minimize the quadratic form

$$\int_0^T (x^2 + u^2)\, dt,$$

over all scalar functions $u(t)$ defined on $[0, T]$. Here we assume that x and u are related by means of the linear differential equation

$$\dot{x} = x + u, \qquad x(0) = c.$$

By means of arguments given in Chapter Seven, it's easy to see that the solution of this problem is given by

$$u^*(t) = -p(t)x^*(t),$$

where p is the solution of the Riccati equation

$$\dot{p} = 1 - p^2, \qquad p(T) = 0,$$

and x^* is the solution to the dynamical equation

$$\dot{x}^* = (1 - p(t))x^*, \qquad x^*(0) = c.$$

This is a purely deterministic optimization problem.

On the other hand, in the theory of linear systems it is shown that the solution of the above optimization problem is the precise mathematical dual to a problem of determining the optimal estimate of a signal observed in the presence of Gaussian noise, i.e., a problem whose statement involves an *intrinsic* stochastic component. The solution of the estimation problem leads to the well-known *Kalman filter,* one of the major tools of the modern system theorist. Thus what appears at first glance to be an inherently stochastic problem turns out to be solvable in terms that don't involve any statistical considerations, at all! For more details on this duality, we refer the reader to the Notes and References, as well as to Chapter Seven.

These examples (and there are many others) call into question several axioms of faith in the so-called applied modeling community. Two of the most important such axioms involve the degree to which Nature is *truly* stochastic, and the contention that inherent uncertainty in our knowledge of the "true" system should be handled by incorporating correction factors involving random terms. Since both of these assertions seem to be firmly entrenched in the modeling *Weltanschauung* of the generic applied mathematician and system modeler, it's worth taking a longer look at the epistemological foundations upon which these prejudices are based, and examining the degree to which such claims hold water in view of the kind of "deterministic randomness" we are discussing here.

As to the inherent randomness of Nature, this appears to be as much a question of subjective psychology as it is a matter of physics and mathematics, the recent experiments of Clauser, Aspect, and others notwithstanding. Since the advent of the Copenhagen interpretation of quantum mechanics, most physicists have adhered to the view that, at least at the quantum level, Nature is indeed intrinsically stochastic. Einstein, and more recently Bohm, objected to this interpretation, claiming, in effect, that the necessity of a probabilistic interpretation in the Copenhagen view was a consequence not of Nature but of the restricted notion of what constitutes a measurement. This led to the claim that the supposed inherent randomness could be accounted for by the introduction of hidden variables, i.e., the seeming intrinsic uncertainty of Nature would be eliminated if we had knowledge of these unobservables. In short, the seeming randomness of Nature is due to our methods of observation, and not to any inherent stochastic mechanisms of Nature itself.

This entire circle of thought was given added spice in 1964 with the announcement of Bell's Theorem, asserting that there could be no *local* hidden variable theories. The experiments of Clauser, and especially Aspect, confirmed the predictions of Bell's result leaving us in the position that Nature can be intrinsically deterministic only if the hidden variables are nonlocal in character. Such theories are currently under development by Bohm, Hiley, and others, utilizing the idea of a global quantum field encompassing, in principle, the entire universe.

For us, the main conclusion to be drawn from this body of work is that Nature *could* be deterministic as Einstein felt, but if so, that kind of determinism is far different from that with which we are familiar from everyday life. So, contrary to popular belief and opinion, the way is still open for a totally deterministic reality, but it must be a nonlocal type of deterministic mechanism.

Matters of quantum reality are, for the most part, far removed from the concerns of everyday life for engineers, economists, ecologists, and others of a more macroscopic professional orientation. Such practitioners generally tend to a quite different position on the issue of the inherent randomness of Nature. They are usually quite willing to concede that Nature may indeed be totally deterministic, even at the micro-level, but that no models or measurements we can construct can ever be free of error, and consequently, we must build into our models some additional terms to account for these modeling "hidden variables." The argument then continues by asserting that since we don't *know* what has been left out of the model, the best approach to take is to represent this ignorance by throwing some random variables into the modeling soup. Such an approach is based, of course, on the entirely unjustified thesis that adding a random "fudge factor" will somehow bring the model closer to reality. When stated in such bald fashion, the nonsense underlying such a position is patently obvious; however, there is the germ of a real issue here, i.e., to what degree is it *necessary* to use the machinery of probability theory and stochastic processes in our mathematical descriptions of reality to account for our uncertainty in the underlying mechanisms and/or observed behavior.

The examples given above, as well as what we will see in the remainder of this chapter, show that even with *perfect* measuring instruments it's quite possible for a system to display behavior that looks as if it were the output of a random device or, perhaps, the behavior of a deterministic device whose output is corrupted by noisy measurements. Neither of these situations need be the case. A deterministic mechanism can give rise to random-looking behavior, even when the measurements are exact! Further, as the Quadratic Optimization example showed, it's perfectly possible for a process formulated in terms of stochastic processes to be mathematically identical to a totally deterministic process. These are facts, not opinions, and they cast serious doubt upon the "stochastics are necessary" school of thought in system modeling. In fact, adopting another venerable principle of modeling, Occam's Razor, we might be tempted to conclude that the unnecessary introduction of random terms into a deterministic framework constitutes an inadmissible multiplication of causes beyond what the facts warrant. While we hold no brief against the *utility* of such procedures in carefully controlled and well-understood situations, our main point here is that they are far from being either necesaary or sufficient as a principle of good modeling practice. Even further, Occam's Razor demands that we consider deterministic mechanisms of the above sort before resorting to the *ad hoc* addition of random "correction" terms into a model. So we take the position that the introduction of stochastic components into a model is more a matter of taste and convenience than it is an axiom of necessity, and it will be our goal in the remainder of the chapter to provide examples, in addition to precepts, in support of this contention.

End of digression. Now let's return to mathematical modeling and the problems of deterministic chaos.

5. *Scalar Maps and the "Period-3" Theorem*

The period-doubling and transition to chaotic motion that we saw in Section 1 for the quadratic map $x \to \alpha x(1-x)$ is not a curiosity confined to this map alone, but is characteristic of *almost all* maps of the unit interval possessing a single "hump." Here we want to examine just what kind of one-dimensional maps lead to this period-doubling and chaotic type of motion.

We consider a mapping $f\colon I \to I$, where I is some closed interval containing the origin. Let α be a real parameter, and assume that we generate a sequence of elements from I according to the rule

$$x_{t+1} = f(x_t, \alpha), \quad x_0 \in I.$$

As in Section 1, our interest is in identifying the long-term behavior of the sequence $\{x_t\}$ for different values of α and different initial states x_0.

Let's agree to confine our attention to functions f that are *unimodal*, i.e., have only one maximum on I. Without loss of generality, we can rescale f and x so that

1) the maximum is located at $x = 0$, $f'(0, \alpha) = 0$, and $f(0, \alpha) = 1$.
2) $f'(x, \alpha) > 0$ for $x < 0$ and $f'(x, \alpha) < 0$ for $x > 0$.

In addition, we assume that the the Taylor series of f near the origin looks like

$$f(x, \alpha) = 1 - ax^k + \dots,$$

with $k > 0$ an *even* integer.

After a sufficiently large number of iterations of the map f, the behavior of the sequence $\{x_t\}$ usually settles into either a periodic or aperiodic stationary pattern, where we consider a fixed point to be a periodic orbit of period–1. By definition, all points from a periodic orbit of period r must be fixed points of the rth iterate of the map f, which we denote as $f^{(r)}(x, \alpha) \equiv F(r, x, \alpha)$.

The first question to address concerning any fixed point is whether or not it is stable. If x^* denotes such a fixed point, i.e., a point where $f(x^*, \alpha) = x^*$, then it will be linearly stable if $|f'(x^*, \alpha)| < 1$. Similarly, an orbit of period r is stable if

$$|F'(r, x_i, \alpha)| = \prod_{j=1}^{r} |f'(x_j, \alpha)| < 1, \qquad i = 1, 2, \ldots, r.$$

The foregoing stability condition holds for a finite interval on the α-axis, termed the *periodic window.*

Arguing as in Section 1, for $r = 1$ there will be some periodic window $0 < \alpha < \alpha_1$ for which x^* will be a stable periodic orbit of period r (a fixed point of F). At $\alpha = \alpha_1$, the stability of x^* will break down and a stable cycle of period $r = 2$ will emerge with a periodic window $\alpha_1 \leq \alpha < \alpha_2$. This cycle is, of course, a fixed point for the map $F(2, x, \alpha)$. Again, at $\alpha = \alpha_2$ the stability of this 2-cycle will break down and a new stable cycle of period $r = 4$ will emerge, with its own periodic window $\alpha_2 \leq \alpha < \alpha_3$. This 4-cycle will be a fixed point of the iterate of $F(2, x, \alpha)$, accounting for the reason why the period always doubles in length from the previous cycle. This process continues indefinitely with periods of order $r = 2^k$ emerging at the parameter values $\alpha = \alpha_k$. It can be shown that the sequence $\{\alpha_k\}$ converges rapidly for *any* function f satisfying the above conditions, and that for quadratic maps we have

$$\alpha_k \approx \alpha_\infty - \frac{A}{\delta^k},$$

where A is a constant and $\delta = 4.6692016\ldots$ is a universal constant, termed the *Feigenbaum constant,* which is characteristic of all quadratic maps of the type considered above. Another way of interpreting this universal feature of period-doubling maps is to note that

$$\delta = \lim_{n \to \infty} \frac{\alpha_{n-1} - \alpha_n}{\alpha_n - \alpha_{n+1}}.$$

Thus, even though the numbers $\{\alpha_k\}$ themselves *do* depend upon the particular function f, the relative size of the gaps in the sequence between one periodic window and the next is a universal constant for all one-dimensional quadratic maps. It's clear from this fact, as well as the earlier estimate for α_k, that the sequence of bifurcation values converges very rapidly to a value α_∞ at which entirely new phenomena emerge.

Note that if the map is not quadratic, i.e., if $f = 1 - ax^k + \ldots$, $k = 4, 6, \ldots$, there is a different universal constant for the appropriate class of maps. For example, if $k = 4$ the appropriate $\delta = 7.248\ldots$, while for $k = 6$ the value is $\delta = 9.296\ldots$. What is important is that the number is universal for all maps of the corresponding class.

For $\alpha > \alpha_\infty$, there are an infinite number of periodic windows immersed in the background of an aperiodic regime. A careful examination of the iterates $\{x_t\}$ shows them jumping between 2^k subintervals of I with k decreasing from ∞ to 0 when α increases from α_∞ to α_{max} (the maximum value of the parameter that will keep all iterates within the interval I). Before entering into a more detailed discussion of this aperiodic regime, let's have a look at a few of the more important properties of one-dimensional mappings.

- *Stable Periods*—A unimodal mapping can have at most one stable period for each value of α; in fact, it may have *no* stable period for many values of α. A necessary condition for f to have a stable period is that the *Schwarzian derivative* of f be negative on the interval I, i.e.,

$$\frac{f'''(x)}{f'(x)} - \frac{3}{2}\left(\frac{f''(x)}{f'(x)}\right)^2 < 0,$$

for all $x \in I$. This condition is not sufficient for the stable orbit to exist: even when the Schwarzian derivative is negative, it's possible to get different aperiodic orbits starting from different x_0 and never reach a stable periodic orbit.

- *Scaling Properties*—Consider the function $F(2^{k-1}, x, \hat{\alpha}_k)$, where $\hat{\alpha}_k$ is the value of the parameter α at which

$$\frac{dF(2^k, x, \alpha)}{dx} = 0.$$

Such an α_k is called a *superstable* value of α. There is an invariance relation between the function above and the next iterate of f at the superstable value of α for $f^{(2^k)}$, i.e., the function $F(2^k, x, \hat{\alpha}_{k+1})$. To see this relation, introduce the operator

$$\begin{aligned} Tf(x, \hat{\alpha}_k) &= -\beta f\left(f\left(\frac{-x}{\beta}, \hat{\alpha}_{k+1}\right), \hat{\alpha}_{k+1}\right), \\ &= -\beta F\left(2, \frac{-x}{\beta}, \hat{\alpha}_{k+1}\right). \end{aligned}$$

Here the parameter β is just a scaling factor in the domain and range of the mapping f. It's also easy to see that

$$T^m f(x, \hat{\alpha}_k) = (-\beta)^m F\left(2^m, \frac{x}{(-\beta)^m}, \hat{\alpha}_{m+k}\right), \quad m \geq 1.$$

It was conjectured and later proved that there exists a limiting function

$$g(x) = \lim_{m\to\infty} T^m f(x, \hat{\alpha}_k),$$

and that $g(x)$ is *independent* of the initial function f. Thus, g provides a universal scaling function for the iterates of any unimodal one-dimensional map, and the properties of g enable us to see that the region around any periodic orbit (for a given k) is self-similar to the same region around the corresponding periodic orbit for *any* value of k. The local character of each iterate in the vicinity of its fixed point is universal and is determined by the properties of the function g, with only the scaling factor β being dependent upon the particular map under consideration.

Employing the recursive nature of $F(2^k, x/(-\beta)^k, \hat{\alpha}_{k+m})$, after a little algebra we obtain the functional equation for g as

$$g(x) = -\beta g\left(g\left(\frac{-x}{\beta}\right)\right).$$

The normalization $f(0, \alpha) = 1$ and the superstable condition that the derivative of F vanish at the superstable value of α, lead to the boundary conditions for g as

$$g(0) = 1, \qquad g'(0) = 0.$$

It should be noted that these conditons are *not* sufficient to determine g uniquely, but that a series expansion near the origin can be used to numerically get at the local behavior.

• *Sarkovskii's Theorem*—In the study of periodic orbits of scalar maps, it's of considerable interest to determine when an orbit of a given period implies cycles of other periods. A powerful result in this direction was given in the mid-1960s by Sarkovskii, and later greatly extended by others, to provide actual tests for chaotic behavior.

SARKOVSKII'S THEOREM. *Order the positive integers according to the following scheme:*

$$3 \to 5 \to 7 \to 9 \cdots \to 3\cdot 2 \to 5\cdot 2 \to 7\cdot 2 \to 9\cdot 2 \to \cdots \to 3\cdot 2^2 \to 5\cdot 2^2$$
$$7\cdot 2^2 \to 9\cdot 2^2 \to \cdots \to 3\cdot 2^n \to 5\cdot 2^n \to 7\cdot 2^n \to 9\cdot 2^n \to \ldots\ldots 2^m$$
$$\to \cdots \to 32 \to 16 \to 8 \to 4 \to 2 \to 1,$$

where the symbol "$\rightarrow$" means "precedes." If f is a unimodal map with a point x_0 leading to a cycle of period p, then there must be a point x_0^* leading to a point of period q, for every q such that $p \rightarrow q$.

Note that Sarkovskii's Theorem says nothing about the stability of the various periodic orbits, and is only a statement about various initial values of x for a fixed value of the parameter α.

An extremely important corollary of Sarkovskii's result is that if a map has a cycle of period-3, then it has cycles of *all* integer periods. This observation forms the basis for one of the most famous results in the chaos literature, the celebrated "Period-3 Theorem" of Li and Yorke.

PERIOD-3 THEOREM. *Let $f\colon I \rightarrow I$ be continuous. Assume that there exists a point $\hat{x} \in I$ such that the first three iterates of $\hat{x}$ are given by $f(\hat{x}) = b$, $f^{(2)}(\hat{x}) = c$ and $f^{(3)}(\hat{x}) = d$. Assume further that these iterates satisfy the inequality*

$$d \leq \hat{x} < b < c (\text{ or } d \geq \hat{x} > b > c).$$

Then for every $k = 1, 2, \ldots$, there is a periodic point in I having period k, and there is an uncountable set $U \subset I$ containing no periodic points, which satisfies the following conditions:

i) For every $x, y \in U$ with $x \neq y$,

$$\limsup_{n\rightarrow\infty} \left| f^{(n)}(x) - f^{(n)}(y) \right| > 0,$$

$$\liminf_{n\rightarrow\infty} \left| f^{(n)}(x) - f^{(n)}(y) \right| = 0.$$

ii) For every $x \in U$ and periodic point $x_0 \in I$,

$$\limsup_{n\rightarrow\infty} \left| f^{(n)}(x) - f^{(n)}(x_0) \right| > 0.$$

Note that the second part of the Period-3 Theorem improves upon the Sarkovskii result by characterizing some aspects of the aperiodic points; i.e., that any aperiodic trajectory will always come as close as we want to another aperiodic trajectory but that they will never intersect. Furthermore, the aperiodic orbits will have no point of contact with any of the periodic orbits. We also note that the conditions of the theorem are satisfied by any function f that has a point of period-3, although it's not necessary for f to have such a point for the type of chaos described to emerge.

The Period-3 Theorem provides us with a set of mathematical criteria for a mapping f to display "chaotic" behavior. There are two main components: (1) a countable number of periodic orbits, and (2) an uncountable number of aperiodic orbits. We can add to this the requirement that the limiting behavior be an unstable function of the initial state, i.e., the system is highly sensitive to small changes in the initial state x_0. With such criteria in mind, now let's turn to a consideration of various tests that can be used to identify the presence of such chaos in processes for which we have only numerical observations.

6. *Tests for Chaos*

Imagine that we are given the output $Y = \{x_t\}$ of some scalar dynamical process, where $x_{t+1} = f(x_t, \alpha)$. What kind of mathematical "fingerprints" would we look for to conclude that the system was displaying chaotic behavior? In other words, what types of tests could be applied to the sequence Y to determine whether or not Y is a sample path from a system operating in the chaotic regime? Since a countable number of periodic orbits together with an uncountable number of aperiodic trajectories are a characteristic feature of chaotic phenomena, the most direct attack on the above question is by means of Fourier analysis.

The autocorrelation function of the map f is given by

$$c_j = \lim_{N \to \infty} \frac{1}{N} \sum_{k=1}^{N} x_k x_{k+j} = \langle x_0 x_j \rangle,$$

with the Fourier transform of this function being

$$C(\omega) = c_0 + 2 \sum_{j=1}^{\infty} c_j \cos j\omega.$$

If we let $x(\omega)$ denote the Fourier transform of the sequence Y, then we note the important relation $C(\omega) = |x(\omega)|^2$. Thus, $C(\omega)$ defines the *power spectrum* of the sequence of iterates Y.

We have seen that the transition to chaos for the map f involves a cascade of period-doubling bifurcations. At a value of the parameter α for which there is a stable cycle of period $2^n, C(\omega)$ will consist of a set of "spikes" at the values $\omega = \pi m/2^n$, where m is an integer less than 2^n. As α increases and a bifurcation takes place so that there is a new cycle of period 2^{n+1}, new contributions to $C(\omega)$ will appear as additional spikes at $\omega = \pi(2m - 1)/2^{n+1}$, with m an integer such that $2m - 1 < 2^n$. After $\alpha > \alpha_\infty$, $C(\omega)$ will consist of the same spikes as on the periodic side for α, as well as a broad-band component representing the "noisy," or chaotic, aperiodic trajectories.

A consequence of the above observations is that we can expect to see a rather sharp change in the power spectrum of f as α passes from the regions of periodic to period-doubling to chaotic motion. As illustration of this contention, consider the spectrum shown in Fig. 5.4 which is taken from an actual experiment involving the heating of a liquid layer from below that is lighter than the liquid above and, hence, rises by convective currents as the heat intensity is increased. The top figure shows isolated peaks corresponding to a fundamental frequency and its harmonics; the system is *periodic* in this regime. The middle figure shows several independent frequencies; the motion is *quasi-periodic.* Finally, the last figure shows wide peaks on a background of a continuous spectrum, indicating the transition to *aperiodic* motion.

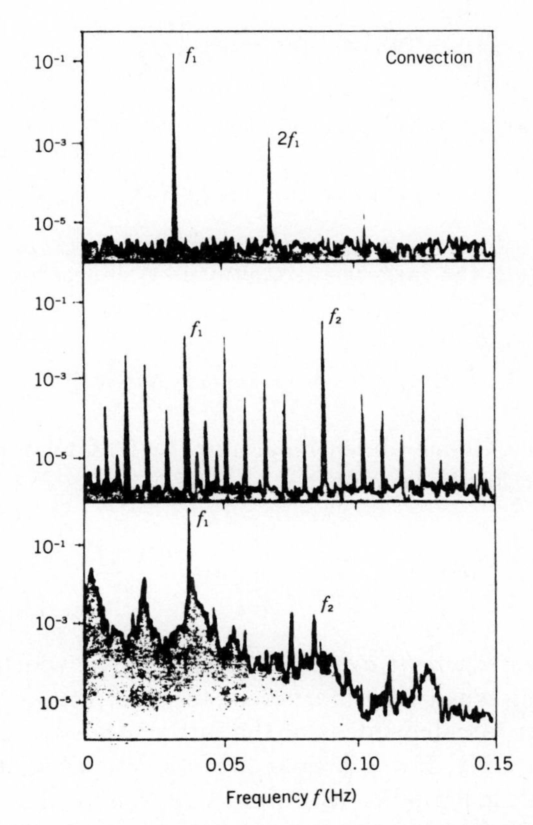

Figure 5.4 Power Spectra in the Transition to Chaos

A principal characteristic of the aperiodic trajectories is that although they remain bounded, each such trajectory diverges exponentially from any other. This observation provides us with another approach to the determination of the onset of chaos for the map f. Consider the initial state x_0 and a nearby initial state $x_0 + \Delta$. These two initial states give rise to trajectories whose difference w_t satisfies the equation

$$w_{t+1} = \mathcal{M}(x_t)w_t, \qquad w_0 = \Delta,$$

where

$$\mathcal{M} = \frac{\partial f}{\partial x},$$

i.e., the Jacobian matrix of f. Note that for one-dimensional maps, $\mathcal{M} = f'(x)$. If we introduce the norm $\|\cdot\|$, then the exponential rate of divergence of the trajectories is given by

$$\sigma(x_0) = \lim_{t\to\infty} \frac{1}{t} \log \frac{\|w_t\|}{\|\Delta\|}.$$

This expression can be simplified to

$$\sigma(x_0) = \lim_{n\to\infty} \log |\hat{\lambda}_n(t)|,$$

where $\hat{\lambda}_n$ represents the largest (in magnitude) characteristic value of the matrix A_n, given by

$$A_n = [\mathcal{M}(x_n) \cdot \mathcal{M}(x_{n-1}) \cdots \mathcal{M}(x_1)]^{1/n}.$$

In the special case of a one-dimensional map f, the foregoing representation simplifies even further to become

$$\sigma(x_0) = \lim_{n\to\infty} \frac{1}{n} \sum_{i=1}^{n} \log \left| \frac{df(x_i)}{dx} \right|.$$

The *Lyapunov exponent* σ represents the rate at which the two trajectories are diverging when they start close together at x_0. Except for a set of measure zero, σ is independent of the initial state x_0, and when $\sigma > 0$, we have a chaotic orbit. If $\sigma < 0$ there is a stable cycle and the orbit, after an initial transient, is periodic. It's interesting to note that near the critical value α_∞ at which chaotic motion begins (i.e., where σ becomes positive), it can be shown that we have the estimate

$$\sigma \approx |\alpha - \alpha_\infty|^\eta,$$

where $\eta = \log 2/\log \delta$, with δ again being the Feigenbaum constant. Thus, knowledge of the sign of the Lyapunov exponent provides us with another test for the onset of chaos.

A final approach to the characterization of chaos is via the notion of the invariant distribution of the mapping f. We say that $P(x)$ is an *invariant distribution* (*invariant measure, probability distribution*) if $P(x) = f(P(x))$, with $\int P(x)\,dx = 1$. Generally speaking, there will be many invariant distributions for a given f; however, we can single out a unique distribution by demanding that the space averages over the initial conditions x_0 equal the time averages over the trajectory $\{x_t\}$.

Basically, what the invariant distribution measures is the likelihood of a given point $x^* \in I$ appearing during the course of iteration of the map f. So, for instance, if the parameter α corresponds to a stable cycle C of period T, then the invariant distribution for f at this value of the parameter will consist of the function

$$P(x) = \begin{cases} \frac{1}{T}, & \text{for } x \in C, \\ 0, & \text{otherwise.} \end{cases}$$

Except for a set of measure zero, every x_0 leads to this distribution under iteration of the map f. If α is a parameter leading to aperiodic motion, again almost all initial conditions lead to a unique equilibrium distribution. However, in this case the distribution may be discontinuous in x, and is typically nonzero over a continuum of x values. But how can we construct the distribution P?

The construction of $P(x)$ makes use of the fact that for maps f having a single hump, there are always two inverse points for a given value of x. See Fig. 5.5 in which the inverse points x_1 and x_2 to the point x are displayed by reflecting the point x in the line $f(x) = x$. Consequently, the number of trajectory points falling into the interval dx equals the number falling into the two intervals dx_1 and dx_2. Thus, we have

$$P(x)\,dx = P(x_1)\,dx_1 + P(x_2)\,dx_2.$$

Now by noting that

$$\frac{dx}{dx_i} = \left|\frac{df}{dx}\right|_{x_i}, \qquad i = 1, 2,$$

we obtain the functional equation for $P(x)$ as

$$P(x) = \frac{P(x_1)}{|df/dx|_{x_1}} + \frac{P(x_2)}{|df/dx|_{x_2}}.$$

This equation can be solved by successive approximations to yield the invariant distribution P.

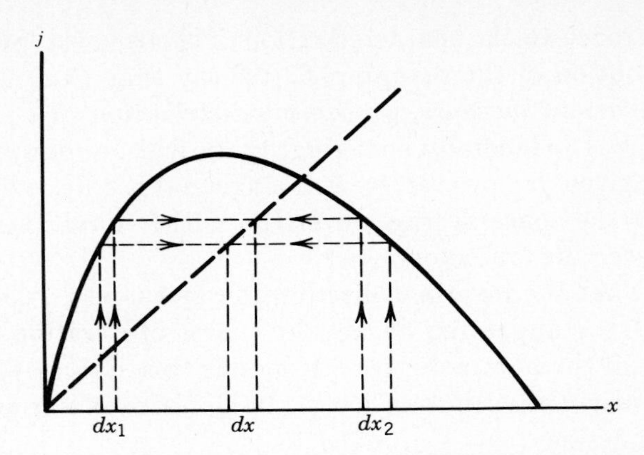

Figure 5.5 Construction of the Distribution $P(x)$

Example: The "Tent" Map

We consider the piecewise-linear map

$$f(x) = \begin{cases} 2x, & \text{for } 0 \leq x \leq \frac{1}{2}, \\ 2 - 2x, & \text{for } \frac{1}{2} \leq x \leq 1. \end{cases}$$

It's easy to see that for this map the functional equation for $P(x)$ is given by

$$P(x) = \frac{1}{2}\left[P(\frac{x}{2}) + P(1 - \frac{x}{2})\right],$$

which has the obvious solution $P(x) = 1$. This computation validates our earlier claim that the "tent" map has iterates that are indistinguishable from those obtained by the flipping of a fair coin, since the distribution of states is identical with that obtained by dividing the x-axis into two halves and labeling each with H or T according to the outcome of the coin toss. We can conclude chaotic motion from this tent map since the distribution P is continuous.

It's also possible to obtain the Lyapunov exponent for f directly from the invariant distribution P as

$$\sigma = \int P(x) \log \left| \frac{df}{dx} \right| dx.$$

For the tent map, we have

$$\sigma = \int_0^1 \log 2 \, dx = \log 2 > 0,$$

again showing that the motion is chaotic.

With the above tests in hand for pinpointing chaos in one-dimensional maps, a natural question that arises is the degree to which these ideas and concepts are peculiar to one-dimensional maps. Or, put another way, can we extend the notion of chaos to higher-dimensional settings?

7. *Snap-Back Repellors and High-Dimensional Chaos*

The main ingredient in the scenario of transition to chaos for scalar maps is the shift in stability of a fixed point of the map; a stable point becomes unstable and two new stable fixed points emerge for the iterated map, and so forth, eventually resulting in chaotic motion. Our concern here is with the degree to which such a scenario carrys over to the setting of a multidimensional map $f\colon R^n \to R^n$, $n \geq 2$.

For scalar maps, the central characterizing result is the Period-3 Theorem which tells us under what circumstances aperiodic motion will occur. The example

$$\begin{aligned} x_{t+1} &= (ax_t + by_t)(1 - ax_t - by_t), \\ y_{t+1} &= x_t, \end{aligned}$$

which has a stable cycle of period-3 when $a = 1.9$ and $b = 2.1$, and no chaotic motion for these parameter values, shows that the Period-3 Theorem does not carry over to higher-dimensional settings. Our question is whether there is a similar type of result that does apply in this more general situation. Insight into this question can be gained by examining the proof of the Period-3 result.

A key ingredient in the Period-3 Theorem's proof is the fact that if I is a compact subset of the real line, and f is a continuous map on R such that $I \subset f(I)$, then f has a fixed point in I. This result no longer holds in R^n; so, to provide a test for chaos in higher dimensions, we need another notion that somehow implies a period-3 cycle when $n = 1$, but also implies aperiodic orbits when $n > 1$. Such a concept is provided by the idea of a "snap-back" repellor.

Let $f\colon R^n \to R^n$ have a fixed point z. Further, assume that f is continuously differentiable in some neighborhood B containing z. We call z a *repelling* fixed point if all characteristic values of the Jacobian matrix $\partial f/\partial x$, lie outside the unit circle for all $x \in B$. The repelling fixed point is called a *snap-back repellor* if there exists a point $x_0 \in B$, $x_0 \neq z$, and an integer M such that $f^M(x_0) = z$ with

$$\det\left(\frac{\partial f(x_0)}{\partial x}\right) \neq 0.$$

It can be shown that when $n = 1$, the existence of a snap-back repellor and the existence of an orbit of period-3 are equivalent. Consequently, we could conjecture that for $n > 1$ snap-back repellors might imply the existence of aperiodic orbits. Such a conjecture was proved by Marotto in the following result.

SNAP-BACK REPELLOR THEOREM. *Snap-back repellors imply chaos in R^n.*

Speaking more precisely, the conclusion of the Snap-Back Repellor Theorem is that the existence of a snap-back repellor implies the same conclusions in R^n as the Period-3 Theorem gives for one-dimensional maps (upon replacing absolute values by the euclidean norm in R^n). That is, periodic orbits of all integer periods, and an uncountable number of exponentially diverging aperiodic orbits. In short, chaos.

Example: The Henon Attractor

We consider the two-dimensional system

$$x_{t+1} = 1 + y_t - ax_t^2,$$
$$y_{t+1} = bx_t,$$

where $a, b \in R$. This system has the fixed points

$$x^* = (2a)^{-1}\left[-(1-b) \pm \sqrt{(1-b)^2 + 4a}\right],$$
$$y^* = bx^*.$$

These points will be real only if we impose the condition $a > -(1-b)^2/4$. In addition, since we are primarily interested in *unstable* fixed points, we make the additional assumption that $a > a_* = 3(1-b)^2/4$, which ensures the instability of both fixed points. For purposes of analysis, we take a value for $b = 0.3$. For values of a in the range $a_* < a < \hat{a} \approx 1.06$, there is a regime of period-doubling. For $\hat{a} < a < a^* \approx 1.55$, x^* is a repelling fixed point that, since $\det(\partial f/\partial x) = -b \neq 0$, suggests x^* could be a snap-back repellor. A wide variety of numerical and analytic work supports this conjecture. For values of $a > a^*$, almost all trajectories tend to infinity.

Extensive analysis on the Henon mapping has shown a number of important features that distinguish its behavior from that seen in one-dimensional maps. Among these properties are the following:

1) For a fixed value of a, the long-term behavior depends upon the initial point (x_0, y_0), i.e., the (x, y)-plane is divided into basins, and we may be led to different periodic or aperiodic orbits depending upon which basin we start within. This is in contrast to the one-dimensional mappings which can have only at most one stable periodic orbit.

2) Some of the attractors of the Henon map can be self-similar, i.e., have *fractal* dimensions (we shall return to this property later in the chapter).

3) Henon's map displays *homoclinic* points, i.e., points of intersection between stable and unstable manifolds. It is exactly the homoclinic points that correspond to the snap-back repellors and that give rise to the chaotic trajectories.

With the foregoing mathematical weapons for the study of dynamical behavior, now let us turn our attention to the way in which chaos can naturally emerge in a variety of settings in the social and life sciences.

8. *Economic Chaos*

The neoclassical theory of capital accumulation provides an explanation of investment cycles that lies exclusively in the interaction of the propensity to save and the productivity of capital when sufficient nonlinearities and a production lag are present. This theory can be used to establish the existence of irregular economic oscillations which need not converge to a cycle of any regular periodicity. Moreover, because they are unstable, errors of parameter estimation or errors in the initial conditions, however minute, will accumulate rapidly into substantial forecasting errors. Such irregular fluctuations can emerge after a period of apparently balanced growth so that the future behavior of a model solution cannot be anticipated from its past.

Although it cannot be proved that any real economies are chaotic in the foregoing sense, the example we give here shows that irregular fluctuations of a highly unstable nature constitute one characteristic mode of behavior in dynamic economic models, and that they may emerge from within conventional economic theories. It's also of interest to note that the past behavior of a nonlinear system may be no guide whatsoever for inferring even *qualitative* patterns of change in its future, since the type of model we are discussing here may evolve through apparently different regimes even though no *structural* changes have occurred. Now let's give a concrete illustration of some of these points.

Assume that the production function of the economy is given in the homogeneous Cobb-Douglas form $f(k) = Bk^{\beta}$, where $B, \beta > 0$. Here k is the capital/labor ratio of the economy. The capital accumulation dynamics are

$$k_{t+1} = \frac{f(k_t) - h(k_t)}{(1+\lambda)},$$

where $h(\cdot)$ is the consumption wealth function and λ is the population growth rate. Per capita consumption depends upon wealth, interest rates and income, but we can use the production function to eliminate income from consideration and equate marginal productivity of capital $f'(k)$ with interest to arrive at the function h. These arguments give

$$h(k) = (1-s)f(k),$$

where s is the marginal propensity to save. Putting all these remarks together, we arrive at the dynamics for capital accumulation as

$$k_{t+1} = \frac{sBk^{\beta}}{1+\lambda}.$$

For $\beta > 0$, investment cycles cannot occur and instead growth converges to an equilibrium capital/labor ratio $k^* = [sB/(1+\lambda)]^{1/(1-\beta)}$.

To illustrate how unstable oscillations and chaos could occur, suppose we introduce a productivity inhibiting effect into the dynamics by changing the production function to

$$f(k) = Bk^{\beta}(m-k)^{\gamma}, \quad \gamma > 0.$$

As $k \to m$, output falls rapidly so we can think of the term $(m-k)$ as representing the harmful effects upon output of an excessive concentration of capital (e.g., too "fat and lazy" to work). Keeping a constant savings rate s, the new capital accumulation dynamics are

$$k_{t+1} = Ak_t^{\beta}(m-k_t)^{\gamma},$$

where we have written $A = sB/(1+\lambda)$.

For small values of A and k_0, growth will be monotonic, converging to a stable equilibrium. For A in the range

$$\frac{\beta m}{\beta+\gamma} < A\left(\frac{\beta}{\beta+\gamma}\right)^{\beta}\left(\frac{\gamma}{\beta+\gamma}\right)^{\gamma} m^{\beta+\gamma} \le m,$$

the system displays period-doubling bifurcations. Let A^* be the value of A for which equality is attained on the right side of the above expression. Then for $A > A^*$, irregular investment and growth cycles occur and chaotic behavior sets in. In fact, it can be shown that there exists an entire interval of values of the productivity multiplier B for which chaotic behavior will occur.

A somewhat sobering aspect of the above result is that it provides a basis for skepticism about *any* modeling effort that relies upon parameter estimation unless it can be demonstrated beforehand that the true parameter values do not lie in the chaotic region. If this is not the case, then there is little hope that observations on the past behavior of the system will provide a basis for identifying the parameter values, and a model based upon such a spurious identification will almost certainly be quite useless.

9. *Population Fluctuations*

To illustrate the way in which chaotic oscillations can arise in simple models of population growth, consider a population of individuals of two age classes. If x_t and y_t represent the levels of the two groups at generation t, then the dynamics governing the population growth are

$$x_{t+1} = b_1(N)x_t + b_2(N)y_t,$$
$$y_{t+1} = sx_t,$$

where $N_t = x_t + y_t$. The quantities $b_1(N)$ and $b_2(N)$ are birthrate functions, and s is the survival rate from the younger population x into the next generation y.

For the sake of straightforward analysis, let's assume that the birthrate functions are the same for both populations and are given by

$$b_1(N) = b_2(N) = b_0 \exp(-\alpha N),$$

where b_0 and α are positive constants. Furthermore, assume that all cohorts from the younger generation survive into the older group, so that $s = 1$. With these assumptions, the dynamics become

$$\begin{aligned} x_{t+1} &= b_0(x_t + y_t)e^{-\alpha(x_t+y_t)}, \\ y_{t+1} &= x_t. \end{aligned}$$

Let's fix the "crowding factor" α at the level $\alpha = 0.1$ and study the behavior of the populations as a function of the joint birthrate b_0.

The behavior of this two-dimensional system is fundamentally different from that seen in the scalar case. First of all, for $b_0 < 8.95$, there is a unique, globally attracting fixed point at $x^* = y^* = [\log(2b_0)]/(2\alpha)$. When $b_0 \approx 8.95$, a pair of 3-point cycles appear surrounding the equilibrium. One of these cycles is stable; the other unstable. However, the equilibrium point remains locally attracting until $b_0 = e^3/2$. Thus, in the range $8.95 < b_0 < e^3/2$, a stable 3-point cycle coexists with the stable equilibrium point. Note the stark contrast of this situation with that in the one-dimensional case. There a 3-point cycle would automatically imply chaos by the Period-3 Theorem. As $b_0 \to e^3/2$, the unstable inner cycle decreases in amplitude until at $b_0 = e^3/2$ it coalesces with the stable equilibrium point leaving an unstable equilibrium.

Further increases in b_0 merely increase the amplitude of the 3-point cycle, which is now globally attracting until $b_0 \approx 14.5$. At this point, a period-doubling bifurcation occurs in which each of the apex points of the 3-point cycle splits into 2 points, creating a stable 6-point cycle. As b_0 is increased further, the 6-point cycle gives way to further period-doubling bifurcations creating $12, 24, 48, \ldots$-point cycles. The parameter intervals corresponding to these $3 \cdot 2^k$-point cycles monotonically decrease just as in the one-dimensional case. A new kind of bifurcation occurs near $b_0 \approx 17$. At this point, the orbit of the system appears more and more chaotic. However, if b_0 is further increased to $b_0 \approx 24$, the chaos is replaced by a stable 4-point cycle. This 4-point cycle then generates a sequence of $4 \cdot 2^k$-point cycles, which eventually terminate in another chaotic regime.

In Fig. 5.6 we show the pattern of alternating bands of chaos and $n \cdot 2^k$-point cycles. Note that the size of the periodic bands progressively decreases

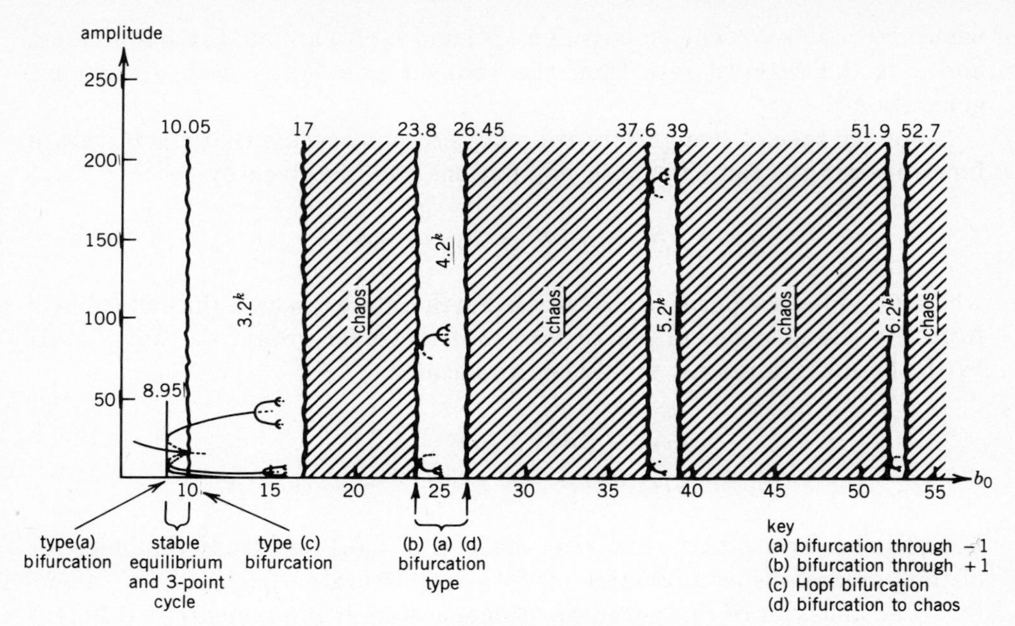

Figure 5.6 Periodic and Chaotic Bands of Population Growth

in width, so that for sufficiently high birthrates chaotic dynamics is virtually assured. There are two remarks worthy of note about the dynamical behavior displayed by this elementary model. First, the emergence of "pattern" out of chaos as shown by the continual reappearance of the periodic cycles for ever-increasing birthrates. In the one-dimensional case, we had cycles of *all* periods together with aperiodic motion, once we passed into the chaotic regime. Here the situation is entirely different: we have *either* periodic cycles or aperiodic orbits, but not both for a given level b_0. A second macroscopic feature is the appearance of periodic "outbreaks": since each successive bifurcation creates a new fixed point near the origin, the orbit tends to spend more and more time at low population levels, punctuated by occasional brief excursions away from the origin culminating in rapid "crashes."

The foregoing example illustrates as forcefully as possible the major differences that emerge between the type of behavior that comes about for scalar processes and that which can be seen in even the simplest sort of multi-dimensional setting. We shall see more evidence of this dichotomy as we go along.

In closing this example, we note that the situation will be even more complicated if the two cohorts do not have identical birthrates. In this case there exists the possibility of quasi-periodic motion, and the emergence of chaos is far more complex when the chaotic orbits come out of the quasi-periodic orbits than when they emerge from periodic motion.

10. *Continuous Chaos and the Lorenz Attractor*

Up to now, our attention has been directed exclusively to discrete-time systems whose dynamical behavior is characterized by the iteration of an n-dimensional map. But most of the processes of classical and modern physics are described by differential equations (flows), not difference equations (maps). The issue we want to explore in this section is to what degree the phenomena of period-doubling bifurcations and aperiodic trajectories go over to continuous-time systems governed by flows.

The first point to note in connection with the attractors of continuous-time systems is the classical result of Poincaré and Bendixson.

POINCARÉ-BENDIXSON THEOREM. *A nonempty, compact attractor of an autonomous planar flow must be either an equilibrium point or a closed orbit.*

Thus, no autonomous, planar dynamical process can display the aperiodic trajectories characterizing chaotic motion. Such two-dimensional systems can have only point equilibria and/or limit cycles (periodic orbits) as their long-term behavior. Consequently, in looking for chaos in continuous-time systems, we must focus attention on systems of dimension $n \geq 3$.

The first such system to display nonperiodic flow was discovered by Lorenz in a paper published in 1963. Briefly, the physical background of the problem involved the heating of a layer of fluid of uniform depth H, with the temperature difference between the upper and lower layer being kept at a constant level ΔT. Lorenz is a meteorologist, and this problem represents a simplified version of atmospheric heating by the Sun. Such a system has a steady-state solution that varies linearly with depth, and if this solution is unstable, convection currents should develop. If we restrict the geometry so that all motions in the fluid are parallel to the (x, z)-plane with no variations in the y-direction, the equations of motion may be written as

$$\frac{\partial}{\partial t}\nabla^2\psi = -\frac{\partial(\psi, \nabla^2\psi)}{\partial(x, z)} + \nu\nabla^4\psi + g\alpha\frac{\partial\theta}{\partial x},$$

$$\frac{\partial}{\partial t}\theta = -\frac{\partial(\psi, \theta)}{\partial(x, z)} + \frac{\Delta T}{H}\frac{\partial\psi}{\partial x} + \kappa\nabla^2\theta.$$

Here ψ is a stream function for the two-dimensional motion, θ is the difference in the temperature from the linear gradient in the case of no convection, and the constants g, α, ν and κ denote the acceleration of gravity, the coefficient of thermal expansion, the kinematic viscosity and the thermal conductivity, respectively. Assume that the upper and lower boundaries are free, so that both ψ and $\nabla^2\psi$ vanish at both boundaries.

It can be shown that convection will occur if the *Rayleigh number*

$$R_a = g\alpha H^3 \Delta T \nu^{-1} \kappa^{-1},$$

exceeds the critical value

$$R_c = \pi^4 a^{-2}(1 + a^2)^3,$$

where a is a parameter expressing the period of the convective oscillation. The smallest value of R_c occurs when $a^2 = \frac{1}{2}$, and is given by $R_c = 27\pi^4/4$.

Now imagine that we expand the above dynamics into a double Fourier series in x and z, with coefficients involving the time t alone. Let us introduce the terms

$$a(1 + a^2)^{-1}\kappa^{-1}\psi = X\sqrt{2}\sin(\pi a H^{-1}x)\sin(\pi H^{-1}z),$$
$$\pi R_c^{-1}R_a \Delta T^{-1}\theta = Y\sqrt{2}\cos(\pi a H^{-1}x)\sin(\pi H^{-1}z) - Z\sin(2\pi H^{-1}z),$$

where X, Y and Z are functions of t alone. Substituting these expressions into the original equations for ψ and θ and equating coefficients of like terms, we obtain the following system for the functions X, Y and Z:

$$\dot{X} = -\sigma X + \sigma Y,$$
$$\dot{Y} = -XZ + rX - Y,$$
$$\dot{Z} = XY - bZ,$$

where $\sigma = \kappa^{-1}\nu$ is the Prandtl number, $r = R_c^{-1}R_a$ and $b = 4(1 + a^2)^{-1}$. These are the famous *Lorenz equations.* The variable X is proportional to the intensity of the convective motion, Y is proportional to the temperature difference between the ascending and descending currents, and Z is proportional to the distortion of the vertical temperature profile from linearity. Similar signs of X and Y denotes that warm fluid is rising and cold fluid is descending, whereas a positive value of Z indicates that the strongest gradients of temperature occur near the boundaries. We will examine the behavior of these equations for various ranges of values of the parameters σ, r and b.

Since it's of greatest physical interest to consider variations in the Rayleigh number, we fix the parameters σ and b and consider only variation in r. For definiteness, let $\sigma = 10$, $b = \frac{8}{3}$. In this range, for $r \approx 25$ stable convection rolls lose their stability and are replaced by another kind of large amplitude motion. This was the kind of chaos that Lorenz was out to study. Let's follow his analysis by setting $r = 28$ and choose an initial condition near the saddle point at $p = (0, 0, 0)$. Lorenz found that the solutions rapidly approached the branched surface S shown in Fig. 5.7. The boundary of S is part of the unstable manifold $W^u(p)$ of the saddle point p. Figure 5.7 shows the first 50 loops of one "side" of the unstable manifold of p, with the surface being shaded and the branch indicated.

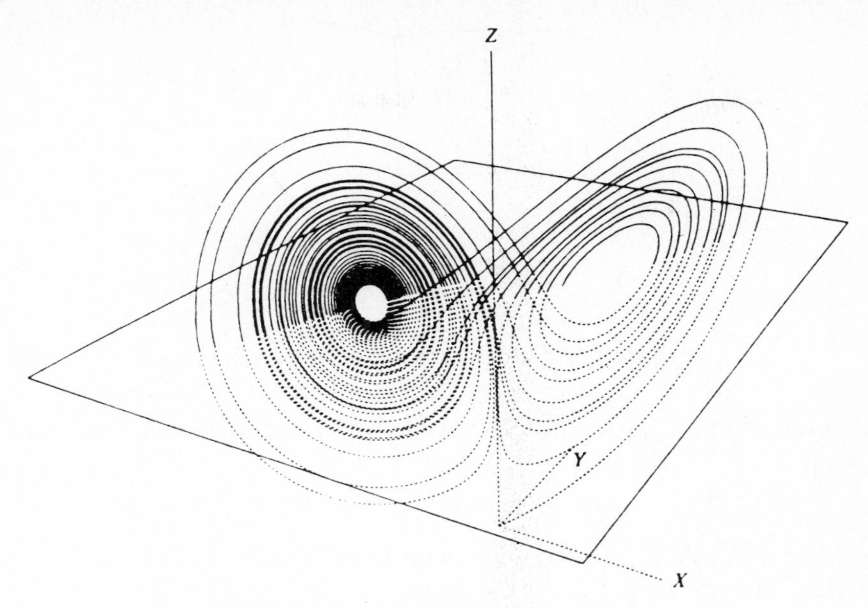

Figure 5.7 The Lorenz Attractor

The presence of the branch in the Lorenz attractor shows that the attractor has infinitely many sheets. Furthermore, solutions do not intersect within the attractor but rather move from sheet to sheet as they circulate over the branch. This kind of motion is exactly analogous to the type of aperiodic behavior displayed by one-dimensional maps, and leads to the conjecture that the Lorenz attractor is indeed an example of continuous chaos. Since the geometric and analytic study of this particular system is by now well-advanced, we refer the reader to the Notes and References for details, passing on to the more general question of *strange attractors* and their identification in actual practice.

11. *Poincaré Maps and Strange Attractors*

Consider a set of differential equations

$$\dot{x} = V(x), \qquad x \in R^n.$$

The phase space for this system is n-dimensional, with coordinates x_i, $i = 1, 2, \ldots, n$. The *Poincaré section* for the system is determined by selecting a surface Σ_R in the phase space such that the surface intersects the trajectory of the system transversally. Roughly speaking, this means that the tangent to the system trajectory does not lie in the surface Σ_R at the point of intersection (see Fig. 5.8). The *Poincaré map* is found by choosing a point $x_k \in \Sigma_R$ and integrating the dynamics forward in time to find the

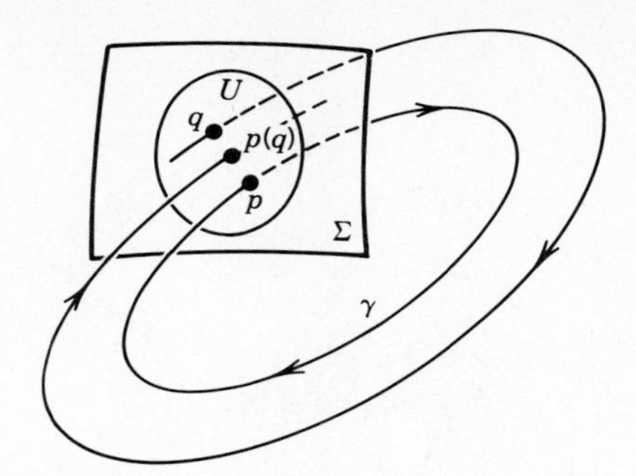

Figure 5.8 The Poincaré Section and Map

next intersection x_{k+1} of the orbit with Σ_R. In this way we can construct the map

$$x_{k+1} = f(x_k), \qquad x_k \in R^{n-1}.$$

If V is smooth and Σ_R is everywhere transverse to V, then it can be shown that the Poincaré map f is also smooth.

Geometrically, it's easy to see that a periodic orbit of the flow (a point x such that $V(x) = 0$) is a fixed point of the Poincaré map, whereas an aperiodic orbit of the flow corresponds to the emergence of discrete chaos for the Poincaré map. Thus, use of the Poincaré map enables us to transfer much of the analysis for the limit sets of flows to the simpler (and lower-dimensional) setting of maps.

Since the phase-space volume must contract for dissipative systems, the stable, steady-state motion for an n-dimensional system must lie on a "surface" of dimension less than n. Loosely speaking, we call such a surface an *attractor* for the flow. For A to be such an attractor for our system, we demand the following properties:

1) A is invariant under the flow of the system.

2) There is an open neighborhood U around A such that all points $x_t \in U \to A$ as $t \to \infty$.

3) No part of A is transient, i.e., only the steady-state motions of the system are in A.

4) A cannot be decomposed into disjoint invariant components.

For scalar dynamics, an attractor A is called a *strange attractor* if in addition to the above properties, we also have periodic orbits of all periods as well as an uncountable number of aperiodic orbits in A. More generally, an attractor is strange if it contains a transversal homoclinic orbit. We refer the reader to the literature cited in the Notes and References for the painstaking (and painful!) details associated with these higher-dimensional situations. Our goal is to understand a little of the structure of the attractor set A when the attractor is strange.

At first glance, it may appear that to study the properties of A we must look at the phase portrait of the system in n-dimensional space. However, due to an observation of Ruelle, supported by deep embedding theorems, it turns out that the multidimensional phase portrait can be constructed from observations of a *single* variable as follows: for almost every coordinate function $x_i(t)$, and almost every time delay T, the m-dimensional portrait constructed from the vectors $\{x_i(t_k), x_i(t_k+T), \dots, x_i(t_k+(m-1)T)\}$, $k = 1, 2, \dots, \infty$, will give an embedding of the original manifold if $m \geq 2n+1$; i.e., if we take a sufficiently large number of observations of $x_i(t)$, we will obtain a manifold that will have many of the same properties of the original phase portrait. An illustration of this result is shown in Fig. 5.9. Here the first part of the figure shows measurements taken on a single variable in the regime of a periodic orbit, while the second half of the figure shows the same variable for parameters in the chaotic regime. Note that in the second part of the figure, the third axis, $x(t_k + 17.6)$, is normal to the page.

It's also possible to use the Poincaré section to study the attractors displayed in Fig. 5.9. The orbits for the strange attractor lie essentially along a sheet. Thus, intersections of this sheetlike attractor with a plane lie, to a good approximation, along a curve and not on a higher-dimensional

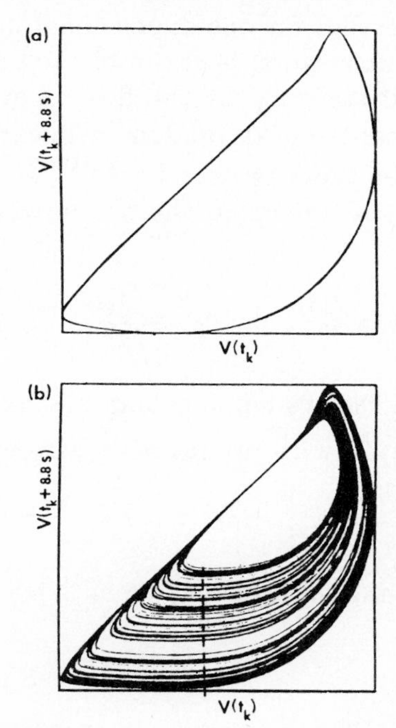

Figure 5.9 Embedded Phase Portraits for Periodic and Strange Attractors

set. The Poincaré map for this attractor is shown in Fig. 5.10, from which we can see the "one-hump" structure familiar from our earlier discussion. Analysis of this map shows that the largest Lyapunov exponent is positive, confirming that the set A is indeed a strange attractor.

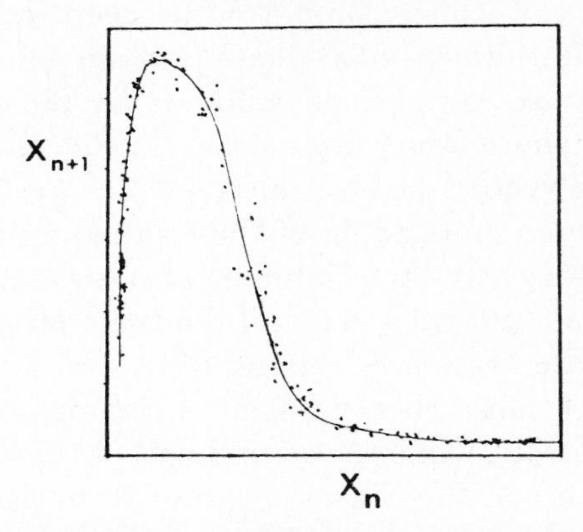

Figure 5.10 The Poincaré Map for a Strange Attractor

Using any reasonable definition of dimension, the dimension D of simple attractors can easily be seen to be integers: $D = 0$ for fixed points; $D = 1$ for limit cycles; $D = n$ for an n-torus. However, for strange attractors D often turns out to have a noninteger dimension. The idea of the Lyapunov exponents, introduced earlier for one-dimensional maps, offers us a scheme for the classification of attractors, both simple and strange.

Consider the initial state x_0 for the flow. Let's build a local coordinate system at x_0 by choosing at random n linearly independent vectors emanating from x_0. Let these vectors be $\{e_0^{(i)}, i = 1, 2, \ldots, n\}$. Now we orthogonalize this set and integrate the set forward using the linearized equations of motion

$$\dot{e}_0^{(i)} = \sum_{i=1}^{n} M_{ij}(x_0)e_0^{(j)}, \qquad e_0^{(i)} = x_0^{(i)}, \qquad i = 1, 2, \ldots, n.$$

Here $M(x_0) = \frac{\partial f(x_0)}{\partial x}$ is the system Jacobian matrix. This procedure yields a new set of vectors $\{e_1^{(i)}\}$ with origins at x_1. Now consider the following set of geometrical ratios:

$$C_1 = \text{ length ratios } |e_1^{(i)}|/|e_0^{(i)}|,$$

$$C_2 = \text{ parallelogram area ratios } |e_1^{(i)} \wedge e_1^{(i)}|/|e_0^{(i)} \wedge e_0^{(i)}|,$$

$$\vdots$$

$$C_n = n\text{-dimensional hyperparallelopiped volume ratio } \frac{\bigwedge_{i=1}^{n} e_1^{(i)}}{\bigwedge_{i=1}^{n} e_0^{(i)}}.$$

Here "$\wedge$" represents the exterior product symbol that generalizes the idea of a vector product to higher dimensions. In the set C_k there are exactly $\binom{n}{k}$ real numbers representing the various geometrical ratios. If we continue this process along each step of the trajectory and take the long-term averages of each of these sets $\{C_k\}$, we obtain a global collection of characteristics of the attractor A. It can be shown that if the vectors $\{e_0^{(i)}\}$ are chosen randomly, then the average of the numbers C_1, i.e., the length ratios, converges to the largest Lyapunov exponent σ_1 for the system. Similarly, the average of the area ratios (the numbers C_2) converges to the sum of the two largest Lyapunov exponents, $\sigma_1 + \sigma_2$, and so on. Finally, the last quantity must converge to the contraction rate of a volume element along the trajectory,

$$\sum_{i=1}^{n} \sigma_i = -\operatorname{div} V.$$

The Lyapunov exponents as calculated above can be ordered as

$$\sigma_1 \geq \sigma_2 \geq \cdots \geq \sigma_n.$$

It has been suggested that the sign pattern of the above ordering can be used as a scheme for the qualitative classification of attractors for dissipative flows. For example, if $n = 3$ we can have the sign pattern $(\sigma_1, \sigma_2, \sigma_3) = (-,-,-)$, indicative of a fixed point with dimension $D = 0$. Similarly, the signature $(0, -, -)$ would correspond to a limit cycle with $D = 1$, whereas $(+, 0, -)$ would characterize a chaotic attractor. Since the signature changes with changes in the control parameters, this kind of classification scheme can be extremely useful in identifying when parametric changes lead to qualitative transitions in the type of the attractor.

Now let's turn some of the foregoing ideas to use in consideration of what some have called *the* unsolved problem of classical physics: turbulence.

12. *Chaos and the Problem of Turbulence*

The great triumphs of nineteenth-century theoretical physics such as Maxwell's theory of electromagnetism and Boltzmann's theory of statistical mechanics, are widely touted as illustrations of the seemingly onward and upward, monotonic progress of science, in general, and physics in particular. What is usually left unsaid is an account of the equally great *failures* of science, failures that most scientists fervently wish would simply disappear, much like the now mythical ether. Perhaps the greatest failure of this sort in classical physics is the inability to give any sort of coherent account of the puzzling phenomenon of "turbulence." The laminar flow of fluids was already well-accounted for by the classical Navier-Stokes equation for the stream function, but the commonly observed transition from smooth laminar flow to periodic eddies and then into total turbulence was completely beyond the bounds of the conceptual framework and techniques of nineteenth-century physics. In fact, it has not been until the discovery of aperiodic orbits, strange attractors, and the like, that even twentieth-century physicists have been able to get a handle on what is beginning to emerge as an actual *theory* of turbulent flow. In this section we give a skeletal outline of this newly emerging paradigm for turbulence, primarily as an introduction to the actual physical interpretations that can be attached to the purely mathematical phenomenon of chaos.

The primary difficulty in giving a mathematical account of turbulence is the lack of any single scale of length appropriate to the phenomenon. Intuitively, and by observation, turbulent flow involves nested eddies of all scales, ranging from the macroscopic down to the molecular, and any mathematical description of the process must take these different scales into account. This situation is rather similar to the problem of phase transitions, where length scales ranging from the correlation length, which approaches infinity at the transition temperature, down to the atomic scale all play an essential role in the overall transition process.

Before entering into an account of turbulence using the ideas of chaos, we emphasize the point that the kind of machinery developed above is relevant only to the problem of the *onset* of turbulence, and has nothing to do with fully developed turbulence of the type studied in mechanical and civil engineering. Furthermore, mathematical chaos is concerned primarily with random behavior in time, while physical turbulence involves stochastic spatial behavior as well. In addition, the mathematical models built upon chaotic dynamics generally involve only planar geometry rather than the full three-dimensional situation of real life. Consequently, until gaps such as these are filled, the theories we present here can only be regarded as an initial approximation to true physical turbulence. Nevertheless, they serve as the kind of first step needed to understand the mechanism by which smooth flow turns turbulent.

The classical Navier-Stokes equations describing the stream velocity and pressure of fluid flow are given in dimensionless units as

$$\frac{\partial v}{\partial t} + v \cdot \nabla v = -\nabla p + \frac{1}{R}\nabla^2 v,$$

with the incompressibility condition

$$\nabla \cdot v = 0.$$

The Poisson equation for the pressure is

$$\nabla^2 p = -\nabla \cdot (v \cdot \nabla v).$$

Both of these equations are subject to whatever boundary conditions are appropriate for the problem under study.

In the foregoing equations, the only dimensionless physical number that appears explicitly is R, the *Reynolds number* of the fluid. It has been known for many years that when $R \gg 1$, turbulence sets in with the typical thresholds being between 1000 and 2000. At these levels of R, the velocity field v becomes highly disordered and irreproducible, so that from a practical point of view it becomes essentially random. This conclusion is an observational one, and has never been established theoretically with anything like the kind of rigor and generality that it has experimentally. The constant R measures the ratio of the two nonlinear terms in the velocity field equation to the linear term, and in practical situations like the atmosphere, the oceans, and so on, it is always a few thousand or greater. So if we are dealing with real fluids, we are dealing with turbulence.

In view of the preceding remarks, the mathematical characterization of the onset of turbulence essentially reduces to a description of how the constant velocity flow fields of laminar flow in the region of small R progress to turbulent flow as the Reynolds number is increased. Over the past several years, a number of mathematical "scenarios" have been presented as to how this transition could take place. Here we shall give a brief account of the major competing positions on this crucial problem.

- *The Landau-Hopf Scenario*—This scenario views the emergence of turbulence as being attributable to the appearance of an increasing number of quasi-periodic motions resulting from successive bifurcations in the system.

When the Reynolds number is small, the fluid motion is laminar and stationary, corresponding to an attractor consisting of a stable *fixed point* in the phase-space. As R is increased, the fixed point loses its stability and begins to repel all nearby trajectories. Since a small change in R cannot cause global disruption to the flow, the fixed point may become repelling locally, but it will remain attracting for regions located far enough away. Thus, the local repulsion and global attraction causes the formation of a *limit cycle*, which is a periodic motion of the flow, i.e., a sort of "whirlpool." This process of generation of a limit cycle from a stable fixed point is exactly the *Hopf bifurcation* discussed in Chapter Four.

Further increase in R yields a point at which the limit cycle loses its stability and becomes repelling. At this point there would appear an attracting 2-torus surrounding the unstable limit cycle. The motion becomes quasi-periodic if the two frequencies on the torus are incommensurable, i.e., rationally independent. Landau and Hopf then let this process continue indefinitely and identify the final state with an infinite number of incommensurable frequencies as being fully turbulent.

The Landau-Hopf scenario suffers from a number of serious difficulties of both a theoretical and an experimental nature. First of all, the scenario requires the successive appearance of new incommensurable frequencies in the power spectrum. In laboratory experiments, turbulent spectra do develop a few independent frequencies, but then soon turn into broad noisy bands. Furthermore, the Landau-Hopf picture ignores the important physical phenomenon of phase-locking. In fact, in nonlinear systems new incommensurable frequencies cannot indefinitely continue to appear without interacting with each other. Nearby frequencies then tend to get locked, thereby diminishing the number of independent frequencies. Finally, the Landau-Hopf scenario contains no provision for sensitive dependence on the initial conditions. For all of these reasons, the Landau-Hopf road to turbulence is unlikely to be followed as a mechanism for the onset of turbulent flow.

• *The Ruelle-Takens Scenario*—In 1971 Ruelle and Takens showed that the Landau-Hopf picture is unlikely to occur in Nature, and provided an alternate mechanism that can be compactly summarized as fixed point $\rightarrow$ limit cycle $\rightarrow$ 2-torus $\rightarrow$ strange attractor; i.e., quasi-periodic motion on a 2-torus can lose its stability and give birth to turbulence directly. In short, the scenario is based upon the mathematical result that if a system undergoes two Hopf bifurcations as the Reynolds number is increased from a stationary initial state, then it is likely (generic) that the system possesses a strange attractor after the second bifurcation.

Experimental evidence involving the fluid velocity field between rotating cylinders and convective flow of the Rayleigh-Bénard type can be interpreted as being consistent with the Ruelle-Takens scheme.

• *The Feigenbaum Scenario*—This picture of turbulence is based upon using the Poincaré map P_R associated with the flow of the Navier-Stokes equation, and invoking the period-doubling $\rightarrow$ aperiodic orbit picture we have already seen for maps. Under some smoothness hypotheses on the map P_R, it can be shown that the bifurcation pattern associated with the Feigenbaum scenario involves a stable fixed point giving way to a stable periodic orbit as R is increased. However, in this scenario the bifurcation is of the *pitchfork* type rather than the Hopf type, as in the previous scenarios. Thus as R increases, the linear part of P_R has a real root that moves out of the unit circle at -1 rather than a pair of complex conjugate roots that leaves the unit circle away from the real axis.

Under experimental conditions, the Feigenbaum scenario claims that if subharmonic bifurcations are observed at values R_1 and R_2 of the Reynolds number, then one can expect another bifurcation near

$$R_3 = R_2 - \frac{R_1 - R_2}{\delta} ,$$

where $\delta = 4.66920\ldots$ is the universal Feigenbaum number. These predictions, as well as several others involving the shape of the power spectrum, have been well tested on a variety of physical problems, including heat transport by convection in liquid helium.

• *The Pomeau-Manneville Scenario*—This view of the onset of turbulence involves the phenomenon of transition through *intermittency.* Thus the laminar phase is not lost entirely but is rather "interlaced" with intermittent bursts of chaotic behavior. Mathematically, the nature of this scenario involves a *saddle-node* bifurcation, in which there is a collision of a stable and an unstable fixed point which then both disappear (into complex fixed points). The phenomenon underlying this scenario can be illustrated by the map

$$x_{t+1} = f(x_t) = 1 - \mu x_t^2, \qquad x_0 \in [-1, 1], \quad \mu \in [0, 2].$$

The third iterate $f^{(3)}$ can be shown to have a saddle-node when $\mu = 1.75$. For $\mu > 1.75$, $f^{(3)}$ has a stable periodic orbit of period-3, and an unstable one nearby. These orbits collide at $\mu = 1.75$, and both then have characteristic value 1. When μ is slightly less than 1.75, it can be shown that if $\mu - 1.75$ is $O(\epsilon)$, then a typical orbit will need $O(\epsilon^{-\frac{1}{2}})$ iterations to cross a fixed, small interval around $x \approx 0$. As long as the orbit is in this small interval, an observer will have the impression of seeing a periodic orbit of period–3; as soon as one leaves this small interval, iterations of the map will look chaotic.

We can summarize the Pomeau-Manneville scenario as follows: assume the Poincaré map of the system has a saddle-node at a critical value of the Reynolds number R_c. Then as R is varied, one will see intermittently turbulent behavior of random duration, with laminar phases of mean duration $\sim (|R - R_c|^{-\frac{1}{2}})$ in between.

A major difficulty with this scenario is that it has no clear-cut precursors, since the unstable fixed point that is going to collide with the stable one may not be visible. There are some theoretical ways for getting out of this difficulty, but the degree to which they are physically plausible remains in doubt. As far as experimental evidence is concerned, although intermittent transition to turbulence can be seen in a number of experiments, only a small subset of these experiments seem to fit the scenario portrayed above. It should be noted that for this scenario, observation of the power spectrum is of little use in its support and one must instead look at the time behavior of the system. In Fig. 5.11 we display a summary of the three main scenarios (excluding the Landau-Hopf model).

With the preceding mathematically distinct pictures in mind, now let's take a look at a natural system outside the realm of physics and engineering where phenomena suggestive of turbulence seem to occur.

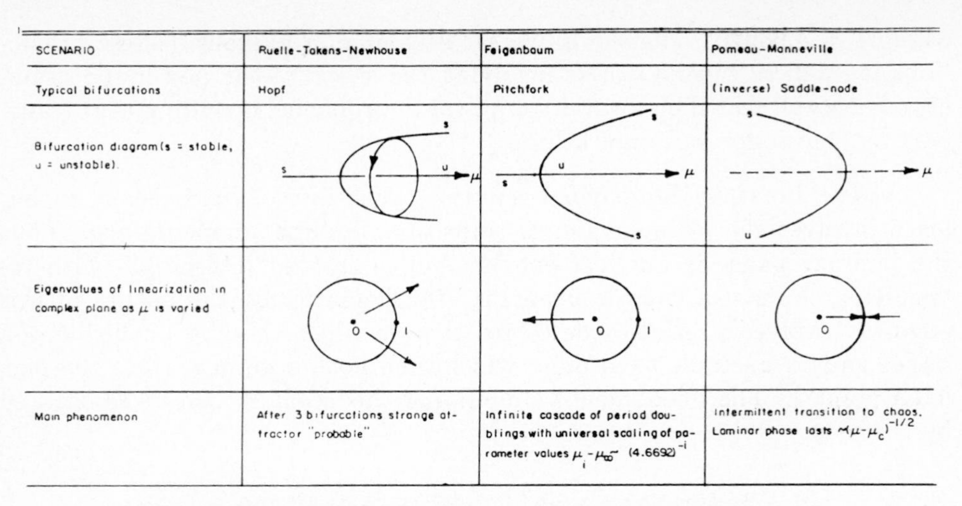

Figure 5.11 Scenarios for the Onset of Turbulence

13. *Turbulent Behavior in the Spread of Disease*

In classical models for the spread of disease, it is assumed that infection is caused by contact sufficiently close that the infectious agent is spread from the infective individual to one who is susceptible. Assume that, on the average, each individual will have the same number of contacts with every other individual over a given period of time. Further, assume that the population is constant and that the unit of time is taken to be the infectious time for an individual. In S–I–S-type models, at the end of the infectious period the individual does *not* gain immunity from the disease but rather goes back into the pool of susceptibles with no immunity. Diseases like the common cold and malaria follow this pattern.

Under the foregoing assumptions, if we let $I(t)$ be the number of infectives at time t, with $S(t)$ the number of susceptibles, then the dynamics of the spread of disease can be given as

$$I(t+1) = S(t)[1 - e^{-\alpha I(t)}],$$

$$S(t+1) = I(t) + S(t)e^{-\alpha I(t)},$$

where $e^{-\alpha}$ is the probability of no effective contact between two individuals in one time period. In addition, let the total population be given by $N = I(t) + S(t)$. If we normalize the variables by the population as

$$x_1(t) = \frac{I(t)}{N}, \qquad x_2(t) = \frac{S(t)}{N}, \qquad A = \alpha N,$$

we obtain the normalized dynamics as

$$x_1(t+1) = x_2(t)[1 - e^{-Ax_1(t)}],$$

$$x_2(t+1) = x_1(t) + x_2(t)e^{-Ax_1(t)},$$

with $x_1(t) + x_2(t) = 1$ for all t. In view of this constraint, we can write $x_1 \doteq x$, and study the single equation

$$x_{t+1} = (1 - x_t)(1 - e^{Ax_t}).$$

For the above one-dimensional model, it's easy to show that there is a unique fixed point x^* satisfying

$$x^* = (1 - x^*)(1 - e^{-Ax^*}),$$

and that this point is globally stable for all $A \geq 0$. So if there is to be turbulent behavior in the spread of disease, the above model must be expanded.

The most straightforward way to extend the foregoing model is to assume that an infective individual spends one time-unit in the infected state, then two time-units in a removed, or isolated, state and then returns to the susceptible state. If we let $R(t)$ represent the number of individuals in the removed state at time t, arguments of the type given above yield the normalized equations

$$\begin{aligned}
x_1(t+1) &= [1 - x_1(t) - x_2(t) - x_3(t)][1 - e^{-Ax_1(t)}],\\
x_2(t+1) &= x_1(t),\\
x_3(t+1) &= x_2(t).
\end{aligned}$$

For $A > 1$, this system has a unique fixed point at $x_1 = x_2 = x_3 = x^*$ satisfying

$$x^* = (1 - 3x^*)(1 - e^{-Ax^*}).$$

The stability of this fixed point is lost at

$$A = \frac{4e - 3}{e - 1} \approx 4.58,$$

at which point the system enters into a stable periodic cycle. At $A = 10$, the solution appears to be asymptotic to a stable 5-cycle, at $A = 30$ there appears to be convergence to a 10-cycle and at $A = 40$ the limiting behavior looks like a 20-cycle. This numerical evidence strongly suggests that for large values of the contact rate A, the period-doubling regime will terminate in the kind of aperiodic motion characteristic of a strange attractor. It's tempting to speculate that should such behavior be analytically confirmed, the chaotic state would be interpretable as a random outbreak of disease, seemingly uncorrelated with the number of initial infectives. Furthermore, such a pattern of outbreak would bear strong spatial and temporal resemblance to the onset of turbulent fluid flow, in the sense that there would be transitions between infectives and susceptibles at all scales of time and space. Operationally, such a situation would imply that the usual procedures of quarantine of the infected could no longer be expected to stem the disease, and other methods would have to be taken to reduce the contact rate A to a level below the turbulence threshold.

Discussion Questions

1. In a by now famous series of "debates" over the physical *interpretation* of the state vector in quantum mechanics, Bohr argued the so-called Copenhagen position that Nature is intrinsically random, and that it is impossible, *in principle,* simultaneously to measure with complete accuracy conjugate variables like position/momentum or time/energy. Einstein, on the other hand, took the point of view expressed in his famous aphorism, "God does not play dice with the Universe," claiming that the Copenhagen interpretation was incomplete, and the Uncertainty Principle underlying the measurement problem was an artifact of the measurement process rather than an intrinsic property of Nature. Do the results we have seen on deterministic chaos appear to shed any light on this debate?

2. In Chapter Three we discussed the notion of a *completely observable* system as one whose measured output enabled us to determine the initial state uniquely. By extending the linear arguments given there to polynomial systems, it is straightforward to see that the "chaotic" logistic equation $x_{t+1} = 4x_t(1 - x_t)$ is completely observable in the system-theoretic sense, i.e., by observing the output, we can uniquely identify the initial state. How do you reconcile this result with the fact that this equation displays random-looking behavior?

3. In his kinetic theory of gases, Boltzmann argued that molecular motion should be considered random, and that every molecule explored the entire region of phase-space that was accessible to it, subject to energy constraints. This point of view is termed the *ergodic hypothesis,* and has served as the foundation for the theory of statistical mechanics. What does the KAM Theorem say about this hypothesis? In particular, does the KAM Theorem provide a means for validating or refuting Boltzmann's hypothesis?

4. In the period-doubling to chaos process for one-dimensional maps, the range of parameter values for a cycle of length 2^k to give way to a cycle of length 2^{k+1} decreases monotonically and rapidly. Thus, small changes in the parameter can result in the emergence of a cycle of much longer length, or even a cycle of infinite length (an aperiodic orbit). Discuss the implications of this fact for practical identification of chaos from *real* data. In particular, do you see any practical difference between an aperiodic orbit and one that is a long stable cycle?

5. The logistic map of Question 2 has been seen to be equivalent to a sequence of flips of a fair coin (Bernoulli trials with $p = \frac{1}{2}$). This fact leads to the description of the behavior of the system in terms of the Bernoulli probability distribution. And conversely, given a Bernoulli distribution, we

can always use the logistic map as a mechanism that realizes it. What about other probability distributions? That is, given an arbitrary probability distribution $P(x)$, can we always find a deterministic mechanism whose dynamics unfold according to that distribution? What bearing does this problem have upon the issues presented in Question 1? (*Hint:* Consider computer random number generators.)

6. Intuitively, it seems that a fixed point is a "simpler" kind of attractor than a periodic orbit (limit cycle) which, in turn, seems simpler than an aperiodic orbit. Can you think of any good scheme that would formalize this intuition into a procedure for defining the *complexity* of an attractor? In this regard, discuss the Kolmogorov-Chaitin concept of complexity (see Chapter Two, Discussion Question 4).

7. Consider a measuring instrument with a uniform scale of resolution ϵ, so that the measurement of a quantity yields one of $1/\epsilon$ numbers. If we have an n-dimensional dynamical system and assign such an instrument for the measurement of each of the n variables, then the phase-space of the system is partitioned into ϵ^{-n} boxes. Let $N(\epsilon)$ be the number of boxes that cover the attractor of the system. Let $P_i(\epsilon)$ be the "probablility density" of occurrence of the attractor in the ith box, i.e., the fraction of the time that the trajectory stays in the ith box. The average information contained in a single measurement is then

$$I(\epsilon) = -\sum_{i=1}^{N(\epsilon)} P_i(\epsilon) \log_2 P_i(\epsilon).$$

Define the *information dimension* D_I of the attractor to be

$$D_I = \lim_{\epsilon \to 0} \left[\frac{I(\epsilon)}{|\log_2 \epsilon|} \right].$$

The information dimension achieves its maximal value when the probability of all boxes is equal, in which case $I(\epsilon) = \log_2 N(\epsilon)$.

Discuss the desirability of trying to reduce the observational resolution ϵ to zero. In particular, since the dynamic storage capacity of the attractor is characterized by the number D_I, is there any *a priori* reason to believe that the maximal storage capacity of the attractor occurs when $\epsilon = 0$? Consider this question within the context of survivability of biological organisms.

8. Reliable information processing requires the existence of a good code or language, i.e., a set of rules that *generate* information at a given hierarchical level, and then *compress* it for use at a higher cognitive level. To accomplish this, a language should strike an optimum ratio between

variety (stochasticity) and the ability to detect and correct errors (memory). Consider the degree to which the type of chaotic dynamical systems of this chapter could be used to model this dual objective. In particular, discuss the notion that entropy (randomness) increases when the volume in the state-space occupied by the system flow *expands,* and is *compressed* (thereby revealing information) when the volume contracts toward the attractor.

9. The C^0-Density Theorem asserts that if a vector field is structurally stable, the only attractors of the field are fixed points and closed orbits. Further, the Stability Dogma claims that good mathematical modeling requires that we use structurally stable models as a hedge against the inevitable uncertainty inherent in our knowledge of the true dynamics and the imprecision in our data; roughly speaking, we can't put much faith in predictions resulting from a structurally unstable model. Consider these facts and arguments in the light of the claim that the structurally unstable models of chaotic flow serve as a good mathematical metaphor for modeling turbulent fluid flow, the outbreak of epidemics and the growth of certain types of populations.

10. A topic of popular interest in some circles is the *Kondratiev wave,* which describes what proponents claim is a periodic cycle in world economic affairs. Assuming such a wave is not just an artifact of the economists' imagination, discuss the relative merits of the following mathematical "explanations" of the wave:

- The wave is a stable limit cycle arising from the Hopf bifurcation mechanism.
- The wave is one of the period-doubling orbits emerging during bifurcation of a logistic-type map.
- The wave is a quasi-periodic orbit on an n-torus.

How would you go about using real data and economic facts to test the foregoing hypotheses?

Assuming that the dynamical process generating the wave contains a parameter μ representing the rate of technological development, what kind of behavior would you expect for small, medium and large values of μ?

11. The human heart provides an example of a physical process whose proper functioning demands a strongly attracting periodic orbit. One of the characteristic signs of heart failure is the loss of this stability, together with the ensuing aperiodic phenomenon of *ventricular fibrillation.* Consider how you might use chaotic dynamics to model such a process. What quantity would you use as the bifurcation parameter?

12. The characteristic feature of dissipative systems is that the phase volume contracts as $t \to \infty$, ultimately leading to the attractor. However, we have also seen that such systems may have strange attractors implying

orbits that *diverge* exponentially fast. How can you reconcile the contracting volume property with the exponentially divergent aperiodic orbits? Try to relate this observation with the property of *equifinality* in biological systems, whereby many organisms tend to develop the same final forms independently of their vastly different initial genotypes.

13. In physical experiments, one expects that the data will be well approximated by a deterministic model. Differences between the predictions of the "best" deterministic model and the data are generally attributed to stochastic effects. When the appropriate deterministic models are chaotic, how does one go about separating the deterministic and the stochastic components of the data? Do you see this kind of issue as possibly constituting a general form of the Uncertainty Principle of quantum mechanics, whereby scientists are *forced* to accept a fundamental level of uncertainty, or "fuzziness," in their conception of the world?

Problems and Exercises

1. Consider the dynamical system

$$\dot{x} = x^2,$$
$$\dot{\theta} = 1,$$

which has the solution (flow)

$$\phi_t(x_0, \theta_0) = \frac{tx_0}{1 - tx_0} + \theta_0.$$

a) Using the section

$$\Sigma = \{(x, \theta) \colon \theta = 0\},$$

show that the Poincaré map for this flow is given by

$$P(x_0) = \frac{x_0}{1 - 2\pi x_0}, \qquad x_0 \in (-\infty, 2\pi).$$

(This shows that the Poincaré map may not always be globally defined.)

b) For the system

$$\dot{r} = r(1 - r^2),$$
$$\dot{\theta} = 1,$$

use the section $\Sigma = \{(r, \theta) \colon r > 0,\ \theta = 0\}$, to show that the Poincaré map is given by

$$P(r_0) = \left(1 + \left(\frac{1}{r_0^2} - 1\right) e^{-4\pi}\right)^{-\frac{1}{2}}.$$

Show that the closed orbit is stable by computing the derivative dP/dr_0 at $r_0 = 1$ to be equal to $e^{-4\pi} < 1$.

2. Consider the system

$$\dot{x} = x,$$
$$\dot{y} = -y + x^2,$$

which has a single equilibrium at the origin.

a) By eliminating the time variable t, show that the system can be written as a linear equation and integrated directly to yield

$$y(x) = \frac{x^2}{3} + \frac{c}{x},$$

where c is the constant of integration.

b) Show that the stable and unstable manifolds for the origin are given by

$$W^s(0) = \{(x, y): x = 0\},$$
$$W^u(0) = \left\{(x, y): y = \frac{x^2}{3}\right\}.$$

c) Draw a graph of these two manifolds, and compare it with the corresponding graph of the same system linearized about the origin.

3. Consider the system

$$\dot{x} = -\zeta x - \lambda y + xy,$$
$$\dot{y} = \lambda x - \zeta y + \frac{1}{2}(x^2 - y^2).$$

a) Show that for $\zeta = 0$, the system is integrable with Hamiltonian

$$H(x, y) = -\frac{\lambda}{2}(x^2 + y^2) + \frac{1}{2}\left(xy^2 - \frac{x^3}{3}\right).$$

This system has three saddle points at

$$p_1 = (\lambda, \sqrt{3}\lambda), \quad p_2 = (\lambda, -\sqrt{3}\lambda), \quad p_3 = (-2\lambda, 0),$$

as well as a center at the origin. Draw the phase portrait of the system showing that the saddle points are connected. What happens to these connections if $\zeta > 0$?

b) Prove that when $\lambda = 0$, the above system is a gradient dynamical system with potential function

$$V(x, y) = \frac{\zeta}{2}(x^2 + y^2) + \frac{1}{2}\left(\frac{y^3}{3} - x^2 y\right).$$

Show that in this case the origin is a sink (for $\zeta > 0$), and that there are saddle points at

$$q_1 = (\sqrt{3}\zeta, \zeta), \quad q_2 = (-\sqrt{3}\zeta, \zeta), \quad q_3 = (0, -2\zeta).$$

What does the phase portrait look like?

4. The map

$$x_{t+1} = \begin{cases} +1 + \beta x_t, & x_t \in [-1, 0), \\ -1 + \beta x_t, & x_t \in (0, 1], \end{cases}$$

has the points $x = \pm \frac{1}{\beta}$ as the first two preimages of zero, i.e., points that map to zero. Show that the kth preimages of zero are given by the points

$$x = \sum_{j=1}^{k} \left(\pm \frac{1}{\beta^j} \right).$$

Hence, conclude that those points constituting the stable manifold of zero are dense in the interval $[-1, 1]$.

5. Suppose we are given the system

$$\begin{aligned} \dot{x} &= Ax + f(x, y), \\ \dot{y} &= By + g(x, y), \qquad f, g \in C^\infty, \end{aligned} \tag{$\dagger$}$$

where $x \in R^n$, $y \in R^m$ with A and B constant matrices. Suppose further that B is a stability matrix (i.e., $\operatorname{Re} \lambda_i(B) < 0$, $i = 1, 2, \ldots, m$), and A has all of its roots on the imaginary axis ($\operatorname{Re} \lambda_j(A) = 0$, $j = 1, 2, \ldots, n$). In addition, let $f(0) = g(0) = f'(0) = g'(0) = 0$, where $'$ denotes the Jacobian matrix.

If $y = h(x)$ is an invariant manifold for the system and h is smooth, we call h a *center manifold* for the system if $h(0) = h'(0) = 0$. Note that if $f = g = 0$, all solutions tend exponentially fast to solutions of $\dot{x} = Ax$. That is, the equation on the center manifold determines the asymptotic behavior of the entire system, up to exponentially decaying terms. The Center Manifold Theorem enables us to extend this argument to the case when f and/or g are not zero.

CENTER MANIFOLD THEOREM.

i) There exists a center manifold $y = h(x)$ for the system ($\dagger$) if $|x|$ is sufficiently small. The behavior of ($\dagger$) on the center manifold is governed by the equation

$$\dot{u} = Au + f(u, h(u)). \tag{$*$}$$

ii) The zero solution of ($\dagger$) has exactly the same stability properties as the zero solution of ($*$). Furthermore, if the zero solution of ($*$) is stable, and if $x(0)$, $y(0)$ are sufficiently small, there exists a solution $u(t)$ of ($\dagger$) such that as $t \to \infty$

$$\begin{aligned} x(t) &= u(t) + O(e^{-\gamma t}), \\ y(t) &= h(u(t)) + O(e^{-\gamma t}), \qquad \gamma > 0. \end{aligned}$$

iii) If ϕ: $R^n \to R^m$ is a smooth map with $\phi(0) = \phi'(0) = 0$, and we define the operation

$$[M\phi](x) \doteq \phi'(x)\{Ax + f(x, \phi(x))\} - B\phi(x) - g(x, \phi(x)),$$

then if $[M\phi](x) = O(|x|^q)$, $q > 1$, *as* $|x| \to 0$, *we have* $|h(x) - \phi(x)| = O(|x|^q)$ *as* $|x| \to 0$.

Note that part (iii) enables us to approximate the center manifold h by the function ϕ up to terms $O(|x|^q)$.

a) Show that the system

$$\dot{x} = xy + ax^3 + by^2x,$$
$$\dot{y} = -y + cx^2 + dx^2y,$$

has a center manifold $h(x) = cx^2 + cdx^4 + O(|x|^6)$. Hence, conclude that the equation governing the stability of the original system is

$$\dot{u} = (cd + bc^2)u^5 + O(|u|^7).$$

So if $a + c = 0$, the origin is stable for the original system if $cd + bc^2 < 0$ and unstable if $cd + bc^2 > 0$. What happens if $cd + bc^2 = 0$?

b) Show how to extend the Center Manifold Theorem to the case of maps, and to the situation in which the linearized part of (†) may have characteristic roots in the right half-plane.

6. Consider the general two-dimensional quadratic map

$$\begin{pmatrix} u_{n+1} \\ v_{n+1} \end{pmatrix} = \begin{pmatrix} \Lambda_{11} & \Lambda_{12} \\ \Lambda_{21} & \Lambda_{22} \end{pmatrix} \begin{pmatrix} u_n \\ v_n \end{pmatrix} + \begin{pmatrix} \Gamma_{11} & \Gamma_{12} & \Gamma_{13} \\ \Gamma_{21} & \Gamma_{22} & \Gamma_{23} \end{pmatrix} \begin{pmatrix} u_n^2 \\ u_n v_n \\ v_n^2 \end{pmatrix}.$$

a) Show that this system can always be transformed to the standard form

$$x_{n+1} + Bx_{n-1} = 2Cx_n + 2x_n^2,$$

where B is the determinant of the Jacobian of the original system and $C = (\lambda_1 + \lambda_2)/2$, with λ_1 and λ_2 being the characteristic roots of the linear part of the original system.

b) Show that there is a stable fixed point at the origin for

$$|C| < \frac{1 + B}{2},$$

which becomes unstable when

$$C < -\frac{1 + B}{2}.$$

At this point a stable 2-cycle appears.

c) Prove that the above system undergoes period-doubling bifurcations as C and B decrease until chaotic motion sets in at the levels $B_\infty = 0$, $C_\infty = (1 - \sqrt{17})/4 \approx -0.781$.

d) Would this same type of behavior take place for a Hamiltonian map, i.e., one for which $B \equiv 1$?

7. Show that the noninvertible map

$$\begin{aligned} x_{n+1} &= 2x_n \mod 1, \\ y_{n+1} &= \alpha y_n + \cos 4\pi x_n, \end{aligned}$$

has Lyapunov exponents $\sigma_1 = \log 2$, $\sigma_2 = \log \alpha$. Hence, conclude that the system displays chaotic motion.

8. A simple model of host–parasite dynamics is given by

$$\begin{aligned} x_{t+1} &= x_t e^{[r(1-x_t)-\gamma y_t]}, \\ y_{t+1} &= x_t(1 - e^{-\gamma y_t}), \end{aligned}$$

where x represents the host population and y is the parasite. Here the parameters r and γ represent the host population's net rate of increase and the parasites' ability to locate and reproduce in the host's larvae, respectively.

a) Show that the (r, γ)-plane can be partitioned into four regions in which the system behavior can be one of the following stable modes: (i) a fixed point; (ii) a periodic orbit; (iii) a quasi-periodic orbit, or (iv) chaos.

b) By regarding the parameters r and γ as "strategies" selected by the host and parasite, respectively, it can be shown that the above system leads to a competitive equilibrium (r^*, γ^*) that is a *Nash equilibrium* for the system, i.e., a point from which neither "player" can deviate without decreasing his utility—in this case, the average population. It turns out that this Nash equilibrium can very easily lie in the cyclic or even chaotic region of the (r, γ)-plane. How would you interpret the meaning of such a strategy?

9. Probably the simplest continuous-time system displaying a strange attractor is

$$\begin{aligned} \dot{x} &= -(y + z), \\ \dot{y} &= x + \frac{1}{5}y, \\ \dot{z} &= \frac{1}{5} + z(x - \mu), \end{aligned}$$

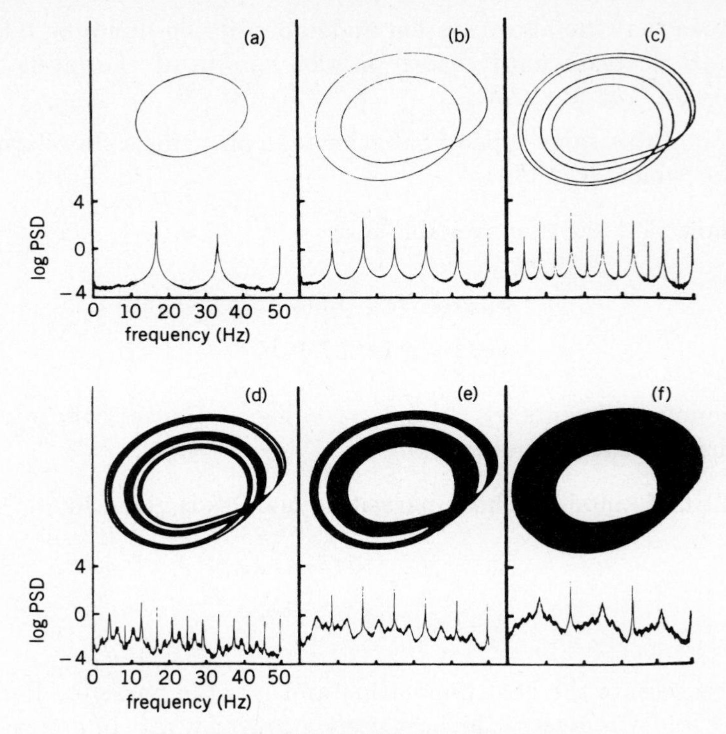

Behavior of the Rössler Attractor

where μ is a real parameter. This system is called *Rössler's attractor.*

a) The behavior of this system in the (x, y)-plane, as well as the power spectral density, is shown above for various values of the single parameter μ. At what value of μ (approximately) do you think chaotic motion sets in?

b) The Rössler attractor is sometimes called a *baker's transformation.* Can you see why?

10. Consider the following integral equation used to model the spread of infectious disease:

$$x(t) = \frac{b}{L_1} \int_{t-L_1-L_2}^{t-L_1} x(s)[1 - x(s)]\, ds,$$

where b, L_1 and L_2 are parameters.

a) Show that for $L_2 = 0$ this equation reduces to the difference equation

$$x(t) = bx(t - L_1)(1 - x(t - L_1)),$$

which displays chaotic motion for $b > 3.57$.

b) The above result suggests that for L_2 "small," but not zero, the original equation may display chaotic motion as well. How would you go about testing this hypothesis? (*Hint:* Consider the linear approximation to the original equation in the neighborhood of the fixed point $x^* = (b-1)/b$ with $L_2 = 0$.)

11. The following mapping takes the circle S^1 to itself

$$\theta_{n+1} = \theta_n + \Omega - \frac{\mu}{2\pi} \sin 2\pi\theta_n \mod 1, \qquad \mu, \Omega \geq 0.$$

a) Prove that for $\mu = 0$ the only orbits are periodic when Ω is rational, and quasi-periodic when Ω is irrational.

b) Show that in the (Ω, μ)-plane, when $\mu > 0$ there are "tongues" that emerge from each rational point on the $\mu = 0$ axis, and that the values of μ and Ω within each tongue correspond to periodic orbits for the map. Further, show that these tongues get wider and wider as μ increases, and the points outside the tongues correspond to quasi-periodic orbits.

c) The quasi-periodic regions do not approach each other closely, even when $\mu \to 1$ from below. Thus there are gaps where the transition from quasi-periodic motion to chaos, as indicated in the Ruelle-Takens scenario, can take place. Show that this can occur *only* if both parameters μ and Ω are varied together. In other words, the transition from quasi-periodic motion to chaos is a "codimension two" bifurcation.

12. Consider the *linear* control system

$$x_{t+1} = \alpha x_t + u_t,$$

with α real.

a) Show that if there is no control ($u_t \equiv 0$), the solutions to this system do not oscillate.

b) Let the control law be given in feedback form as

$$u_t = \beta |x_t| - 1,$$

with β real. Prove that if the parameters satisfy the conditions

$$\beta > 0, \qquad \beta^2 - \alpha^2 \geq \alpha + \beta + 1,$$

the controlled system has at least N_{n-2} periodic trajectories of period n, where N_k = the kth Fibonacci number.

c) Show that if strict inequality holds in part (b), there are an uncountable number of aperiodic trajectories such that if X and Y are two such trajectories, then

$$|X_t - Y_t| \geq \frac{2(\beta^2 - \alpha^2 - \alpha - \beta - 1)}{(\alpha - \beta)^2(\alpha + \beta)^2} > 0,$$

for sufficiently large t.

13. The generalized baker's transformation of the square is given by

$$x_{n+1} = \begin{cases} a x_n, & \text{if } y_n < \alpha, \\ \frac{1}{2} + b x_n, & \text{if } y_n > \alpha, \end{cases}$$

and

$$y_{n+1} = \begin{cases} \frac{1}{\alpha} y_n, & \text{if } y_n < \alpha, \\ \frac{1}{1-\alpha}(y_n - \alpha), & \text{if } y_n > \alpha, \end{cases}$$

where $0 \leq x_n,\ y_n \leq 1$. Assume the parameters satisfy $a,\ b,\ \alpha \leq \frac{1}{2}$ and $b \geq a$.

a) Show that the Jacobian matrix of this system is diagonal and depends only upon y.

b) Calculate the Lyapunov exponents σ_1 and σ_2 for this system as

$$\log \sigma_1 = \alpha \log \frac{1}{\alpha} + \beta \log \frac{1}{\beta},$$

and

$$\log \sigma_2 = \alpha \log a + \beta \log b.$$

c) Define the *Lyapunov dimension* d_L of the system's attractor to be

$$d_L = 1 + \frac{\log \sigma_1}{\log(1/\sigma_2)} \,.$$

Compute the Lyapunov dimension of the baker's transformation to be $d_L = 2$ when $\alpha = \frac{1}{2}$, $a = b = k > \frac{1}{2}$.

14. Consider the one-dimensional quadratic map

$$x_{t+1} = C - x_t^2,$$

with C a real parameter.

a) Show that for $|C| > \frac{1}{4}$ the orbits all tend to $\pm\infty$.

b) Prove that at $C = -\frac{1}{4}$ a period-doubling regime sets in followed by a chaotic region, with the domain of attraction for the chaotic orbits given by $|x| \leq x_*$, where $x_* = -\frac{1}{2} + (\frac{1}{4} + C)^{1/2}$. Show that the period-doubling and chaotic behavior holds for all C in the range $-\frac{1}{4} \leq C < 2$.

c) What happens when $C \geq 2$?

15. In the quadratic family of Problem 14, assume the parameter C is chosen so that there is exactly one periodic orbit of every period $n > 0$. The *kneading sequence* of the map f is then given by the binary sequence defined by $b_i = 0$ or 1 as $f^i(0)$ is less than or greater than zero.

a) For a particular value of C in the periodic range above, prove that the corresponding kneading sequence for f can be calculated recursively as

$$b_i = \begin{cases} b_{i-2^n}, & \text{if } 2^n + 1 \leq i < 2^{n+1}, \\ 1 - b_{2^n}, & \text{if } i = 2^{n+1}. \end{cases}$$

The section of the sequence from $i = 2^n + 1$ to 2^{n+1} is generated by repeating the first 2^n terms and then changing the last.

b) Prove that the kneading sequence of f^2 is obtained by taking the subsequence of $\{b_i\}$ of terms having even indices, yielding a sequence that is obtained from $\{b_i\}$ by changing every term. Show that this fact implies that $f(x)$ and $-f^2(x)$ have the same kneading sequences.

c) Show that if two quadratic maps f and h have the same kneading sequence, there exists a homeomorphism g such that $f \circ g = g \circ h$; i.e., f and h are topologically conjugate. Prove that the map g is the universal Feigenbaum function discussed in the text.

16. Instead of the usual quadratic map with a single hump, consider the cubic map with *two* humps

$$x_{t+1} = ax_t^3 + (1 - a)x_t,$$

where a is a real parameter.

a) For what values of a do the two humps lie within the interval $[-1, 1]$? Where are the attractors for this system if $a < 0$?

b) Show that the dynamical behavior of this cubic system is given by

$$\begin{aligned} 0 < a < 2 \quad & \text{(a stable point)}, \\ 2 < a < 3 \quad & \text{(a stable cycle of period 2)}, \\ 3 < a < 1 + \sqrt{5} \quad & \text{(two distinct cycles, each with period 2)}, \\ 1 + \sqrt{5} < a < 4 \quad & \text{(distinct periodic attractors, as well as aperiodic orbits)}. \end{aligned}$$

17. Consider the two-dimensional map

$$\begin{aligned} \theta_{n+1} &= 2\theta_n \mod 2\pi, \\ z_{n+1} &= \lambda z_n + \cos\theta_n, \end{aligned}$$

where $1 < \lambda < 2$, $0 \leq \theta < 2\pi$.

a) Show that the Jacobian matrix of the map has characteristic values λ and 2, indicating that there can be no attractors with finite z.

b) Since almost all initial conditions lead to $z_\infty = \pm\infty$, there must be a function $f(\theta)$ such that initial points (θ_0, z_0) having $z_0 > f(\theta_0)$ lead to $z_\infty = +\infty$, while initial points (θ_0, z_0) with $z_0 < f(\theta_0)$ lead to $z_\infty = -\infty$. Show that such a function $f(\theta)$ is given by

$$f(\theta) = -\sum_{k=1}^{\infty} \lambda^{-(k+1)} \cos(2^k \theta).$$

c) Prove that $f(\theta)$ is nondifferentiable, and has infinite length with a *fractal* dimension $d = 2 - \log \lambda / \log 2$.

18. Let $\mathcal{S} = \{X_i\}$, $i = 1, 2, \ldots, N$ be a finite collection of points lying on the attractor of a dynamical system whose phase space is of dimension n, i.e., $X_i(t) \in R^n$. Define the *correlation integral* $C(\ell)$ of the points to be

$$C(\ell) = \lim_{N\to\infty} \frac{1}{N^2} \times \{ \text{ number of pairs } (i, j) \text{ such that the distance } |X_i - X_j| < \ell \}.$$

a) Prove that $C(\ell) \sim \ell^\nu$, where ν is a positive, real number.

b) Show that if the "signals" $\mathcal{S}$ arise from random noise, then $C(\ell) \sim \ell^n$; but if $\mathcal{S}$ comes from determinisitic chaos, we will have $\nu < n$. Thus, the correlation exponent ν provides a means for distinguishing between deterministic and stochastic randomness.

c) Let the system Lyapunov exponents be given by $\sigma_1 \geq \sigma_2 \geq \ldots$, with j the largest integer such that

$$\sum_{i=1}^{j} \sigma_i \geq 0.$$

Prove that the exponent ν satisfies the inequality

$$\nu \leq j + \frac{1}{|\sigma_{j+1}|} \sum_{i\leq j} \sigma_i.$$

d) Show that for the Lorenz system with the parameter values of the text, we have $\nu = 2.05 \pm 0.01$, while for the logistic map with $\alpha = 3.57$, the correlation exponent is bounded by $0.492 < \nu < 0.502$. Hence, it is possible to conclude that these maps represent true deterministic chaos and not stochastic randomness.

19. In simple models of two competing technologies, if a represents the annual adoption rate of one of the technologies and f_t is the fractional market share enjoyed by that technology in period t, then the dynamics of adoption of this technology are described by

$$f_{t+1} = \begin{cases} (a+1)f_t - af_t^2, & \text{if } -1 < a \leq 1, \\ (a+1)f_t - \left(\frac{a+1}{2}\right)^2 f_t^2, & \text{if } 1 < a \leq 3. \end{cases}$$

a) Show that for $a > 2.57$ the above system displays chaotic behavior.

b) Since a represents an adoption **rate**, how can you interpret values of a greater than one?

c) Do you think that the possibility of chaotic motion for these dynamics in any way invalidates the use of such a model for describing the dynamics of technological substitution?

Notes and References

§1. A good introductory account of the period-doubling phenomenon for quadratic maps and its relevance to insect population studies is given in

May, R., and G. Oster, "Bifurcations and Dynamic Complexity in Simple Ecological Models," *Amer. Naturalist,* 110 (1976), 573–599.

It's worthwhile to note that quadratic maps can always display unpredictable behavior, and it's exactly this class of maps that forms the basis for the well-known Lotka-Volterra-type predator–prey relations that lie at the heart of a large number of ecological analyses. Hence, the reason why the systems ecology community has been one of the primary consumers of the theoretical results in chaos theory. For a detailed account of the predator–prey relations, see

Peschel, M., and W. Mende, *The Predator–Prey Model,* Springer, Vienna, 1986.

§2. An undergraduate-level account of dynamical systems from a modern point of view is provided in

Arnold, V. I., *Ordinary Differential Equations,* MIT Press, Cambridge, MA, 1973,

Hirsch, M., and S. Smale, *Differential Equations, Dynamical Systems and Linear Algebra,* Academic Press, New York, 1974.

A somewhat more advanced, but still very readable, treatment is

Irwin, M. C., *Smooth Dynamical Systems,* Academic Press, New York, 1980.

§3. A good elementary discussion of the KAM Theorem is found in Appendix 8 of the book

Arnold, V. I., *Mathematical Methods of Classical Mechanics,* Springer, New York, 1978.

For a more detailed account, see the work

Arnold, V. I., "Small Denominators—II: Proof of a Theorem of A. N. Kolmogorov on the Preservation of Conditionally Periodic Motions under a Small Perturbation of the Hamiltonian," *Russ. Math. Surveys,* 18 (1963), 9–36.

A very interesting treatment of the entire problem of "small divisors," and its historical origins in celestial mechanics in connection with the stability of the solar system, is found in

Sternberg, S., *Celestial Mechanics,* Vols. 1 & 2, Benjamin, New York, 1969.

§4. A more detailed discussion of the "tent-map" and its relationship to quadratic functions is found in

Lichtenberg, A., and M. Lieberman, *Regular and Stochastic Motion,* Springer, New York, 1983.

The duality between a quadratic optimization problem and stochastic filtering is treated extensively in the linear system theory literature. Typical references are

Casti, J., *Linear Dynamical Systems,* Academic Press, Orlando, 1987,

Anderson, B. D. O., and J. Moore, *Linear Optimal Control,* Prentice-Hall, Englewood Cliffs, NJ, 1971.

The problem of hidden variables as a way out of the probabilistic interpretation of quantum phenomena has had a rather rocky history, coming in and out of favor with the work of Einstein, Bohm, Bell, and others. Very readable popular accounts of this problem, as well as many others associated with the epistemological content of the quantum measurement problem, may be found in

Herbert, N., *Quantum Reality,* Doubleday, New York, 1985,

d'Espagnat, B., *In Search of Reality,* Springer, New York, 1983,

Davies, P. C. W., and J. R. Brown, eds., *The Ghost in the Atom,* Cambridge U. Press, Cambridge, 1986.

§5. The best elementary accounts of chaotic phenomena are the introductory sections of the following volumes devoted primarily to reprints of the bench mark papers in the field:

Hao, Bai-Lin, *Chaos,* World Scientific, Singapore, 1984,

Cvitanović, P., *Universality in Chaos,* Adam Hilger, Ltd., Bristol, UK, 1984.

For additional information on the Feigenbaum number and the question of self-similarity, see

Feigenbaum, M., "Universal Behavior in Nonlinear Systems," *Los Alamos Science,* 1 (1980), 4–27 (Reprinted in the Cvitanović book cited above),

Feigenbaum, M., "The Universal Metric Properties of Nonlinear Transformations," *J. Stat. Physics,* 21 (1979), 669–706.

A fuller account of the famous Period-3 Theorem and its proof is found in

Li, T., and J. Yorke, "Period-3 Implies Chaos," *Amer. Math. Monthly,* 82 (1975), 985–992.

§6. A much more detailed account of the tests for chaos is given in the work by Lichtenberg and Lieberman cited under §4 above. See also

Grossmann, S., and S. Thomae, "Invariant Distributions and Stationary Correlation Functions of One-Dimensional Discrete Processes," *Z. Naturf.,* 32 (1977), 1353–1363,

Crutchfield, J. J. D. Farmer, N. H. Packard, R. Shaw, R. J. Donnelly, and G. Jones, "Power Spectral Analysis of a Dynamical System," *Phys. Lett.,* 76A (1980), 1–4,

Grassberger, P., and I. Procaccia, "Characterization of Strange Attractors," *Phys. Rev. Lett.,* 50 (1983), 346–349,

Wolf, A., "Quantifying Chaos with Lyapunov Exponents," in *Chaos,* A. Holden, ed., Princeton U. Press, Princeton, 1986, pp. 273–290.

§7. The notion of a snap-back repellor as the natural analogue in higher dimensions for the scalar period-3 attractor is treated in detail in

Kaplan, J., and F. Marotto, "Chaotic Behavior in Dynamical Systems," in *Nonlinear Systems and Applications,* V. Lakshmikantham, ed., Academic Press, New York, 1977, pp. 199–210.

The Henon attractor is by far the most well-studied example of a higher-dimensional chaotic process. The original reference is

Hénon, M., "A Two-Dimensional Mapping with a Strange Attractor," *Comm. Math. Physics,* 50 (1976), 69–77.

Other work on higher-dimensional strange attractors is given by

Collet, P., J. P. Eckmann and H. Koch, "Period Doubling Bifurcations for Families of Maps on R^n," *J. Stat. Physics,* 25 (1980), 1–14,

Zisook, A., "Universal Effects of Dissipation in Two-Dimensional Mappings," *Phys. Rev. A,* 24 (1981), 1640–1642.

§8. The example of economic chaos follows that given in

Day, R., "The Emergence of Chaos from Classical Economic Growth," *Quart. J. Econ.,* May 1983, 201–213.

If you think the emergence of chaos is due to the somewhat out-moded assumptions built in to the classical theory of economic processes, consider the arguments given in

Day, R., "Emergence of Chaos from Neoclassical Growth," *Geographical Analysis,* 13 (1981), 315–327,

Benhabib, J., and R. Day, "Rational Choice and Erratic Behavior," *Rev. Econ. Stud.,* 48 (1981), 459–471,

Ford, J., "Ergodicity for Economists," in *New Quantitative Techniques for Economic Analysis,* G. Szegö, ed., Academic Press, New York, 1981, pp. 79–96.

§9. A more complete account of this example is given in

Oster, G., "The Dynamics of Nonlinear Models with Age Structure," in *Studies in Mathematical Biology, Part II,* S. Levin, ed., Math. Assn. America, Washington, DC, 1978, pp. 411–438.

§10. An extensive treatment of the Lorenz attractor and its extensions and generalizations is given in

Sparrow, C., *The Lorenz Equations: Bifurcations, Chaos, and Strange Attractors,* Springer, New York, 1982.

For Lorenz's original work, see

Lorenz, E., "Deterministic Non-Periodic Flows," *J. Atmos. Sci.,* 20 (1963), 130–141.

It's interesting to note that chaotic behavior, in the strict mathematical sense, has not yet been proved for the Lorenz attractor. All that is available in this direction is extensive computer experiments giving all the indications of chaotic motion. This situation is in contrast to that of discrete-time dynamics (maps) for which hard theorems confirming true chaotic behavior are known (e.g., the Period-3 Theorem).

§11. A textbook discussion of the Poincaré map is provided in the volumes cited under §2 above. See also the graduate text

Guckenheimer, J., and P. Holmes, *Nonlinear Oscillations, Dynamical Systems, and Bifurcations of Vector Fields,* Springer, New York, 1983.

A popular exposition of the concept of a strange attractor, together with some historical remarks on the genesis of the idea, is provided by one of the founders of the field in

Ruelle, D., "Strange Attractors," *Math. Intelligencer,* 2 (1980), 126–137.

The idea of using embedding theorems to reduce the problem of finding a strange attractor in R^n to that of the properties of a time-series is considered in

Packard, N., J. P. Crutchfield, J. D. Farmer, and R. Shaw, "Geometry from a Time Series," *Phys. Rev. Lett.,* 45 (1980), 712,

Swinney, H., "Observations of Order and Chaos in Nonlinear Systems," *Physica D,* 7D (1983), 3–15.

§12. The scenarios to turbulence, as well as a more extended version of Fig. 5.11, are discussed in far greater detail in the survey

Eckmann, J. P., "Roads to Turbulence in Dissipative Dynamical Systems," *Rev. Mod. Physics,* 53 (1981), 643–654,

Procaccia, I., "Universal Properties of Dynamical Complex Systems: The Organization of Chaos," *Nature,* 333 (1988), 618–623.

Other treatments of the turbulence problem from the viewpoint of modern dynamical system theory are

Helleman, R., "Self-Generated Chaotic Behavior in Nonlinear Mechanics," in *Fundamental Problems in Statistical Mechanics,* E. Cohen, ed., North-Holland, Amsterdam, Vol. 5, 1980, pp. 165–233,

Ruelle, D., and F. Takens, "On the Nature of Turbulence," *Comm. Math. Physics,* 20 (1971), 167–192.

Some of the classical papers on the turbulence question are

Landau, L., "On the Problem of Turbulence," *Akad. Nauk Doklady,* 44 (1944), 339,

Hopf, E., "A Mathematical Example Displaying Features of Turbulence," *Comm. Pure & Appl. Math.,* 1 (1948), 303–309.

A thought-provoking account of chaotic behavior in weather patterns, suggesting the existence of a low-dimensional strange attractor, is presented in

Tsonis, A., and J. Elsner, "The Weather Attractor Over Very Short Timescales," *Nature,* 333 (1988), 545–547.

The model for "turbulent" behavior in the spread of disease is taken from

Cooke, K. L., D. Calef, and E. Level, "Stability or Chaos in Discrete Epidemic Models," in *Nonlinear Systems and Applications,* V. Lakshmikantham, ed., Academic Press, New York, 1977, pp. 73–93.

DQ #1. For a deeper account of the quantum measurement problem, see the volumes cited under §4 above as well as

Jammer, M., *The Philosophy of Quantum Mechanics,* Wiley, New York, 1974,

Wheeler, J. A., and W. Zurek, eds., *Quantum Theory and Measurement,* Princeton U. Press, Princeton, 1983.

DQ #7–8. These matters pertaining to the preservation of information and its role as a survival mechanism for biological organisms are taken up in much greater detail in

Nicolis, J., *Dynamics of Hierarchical Systems,* Springer, Berlin, 1986.

DQ #10. The question of long waves in the ebb and flow of global economic indicators was first put forth by Kondratiev in the classic paper

Kondratiev, N., "The Long Waves in Economic Life," *Rev. Econ. Statistics,* 17 (1935), 105–115.

A discussion of the existence and relevance of such postulated waves is given from several points of view in

Freeman, C., ed., *Long Waves in the World Economy,* Butterworth, London, 1983.

Additional work in this area includes

Sterman, J., "A Simple Model of the Economic Long Wave," *IIASA Collaborative Paper CP-85-21,* IIASA, Laxenburg, Austria, April 1985,

Senge, P., "The Economic Long Wave: A Survey of Evidence," *MIT Systems Dynamics Group Working Paper D-3262-1,* MIT, Cambridge, MA, 1982,

Mass, N., *Economic Cycles: An Analysis of Underlying Causes,* MIT Press, Cambridge, MA, 1975.

The general concept of equifinality as a property of biological systems is discussed by Driesch, who pointed out that the characteristic feature of equifinality, the independence of the final state from the initial conditions, was completely at variance with the traditional equilibrium view of classical physics. This argument, far from being a convincing proof of "vitalism," serves mainly to underscore the inadequacy of traditional physics for dealing with biological processes. For a fuller discussion of these matters, see

von Bertalanffy, L., *Problems of Life,* Harper & Row, New York, 1960.

PE #5. For a full treatment of the Center Manifold Theorem, see the Guckenheimer and Holmes book cited under §11 as well as

Carr, J., *Applications of Centre Manifold Theory,* Springer, New York, 1981.

PE #9. Rössler's attractor is treated in more detail in

Rössler, O., "An Equation for Continuous Chaos," *Phy. Lett.,* 57A (1976), 397.

PE #12. The introduction of a control element into the dynamics opens up the possibility for many additional types of dynamical behavior to be *induced* (perhaps inadvertently) by the controller. This exercise shows only the tip of the iceberg, and was taken from the paper

Baillieul, J., R. Brockett and R. Washburn, "Chaotic Motion in Nonlinear Feedback Systems," *IEEE Tran. Circuits & Sys.,* CAS-27 (1980), 990–997.

PE #15. Kneading sequences play an important role in the study of the behavior of many types of dynamical processes. For an introductory account, see

Guckenheimer, J., "Bifurcations of Dynamical Systems," in *Dynamical Systems,* J. Guckenheimer, J. Moser, and S. Newhouse, eds., Birkhäuser, Boston, 1980.

CHAPTER SIX

Strategies for Survival: Competition, Games and the Theory of Evolution

1. *Evolutionary Epistemology*

In the Darwinian battle for survival, Fortune's formula may be compactly expressed as

$$\text{variation} + \text{selection} = \text{adaptation},$$

encapsulating in three everyday words over a century's worth of scholarly and public debate about the nature of change in living organisms, and the degree to which that process, whatever it may be, is homologous to the flow of human affairs. The intensity of the debate as to the precise interpretation of the terms in the foregoing formula has often been elevated (or lowered?) to levels of hysteria and metaphysics quite far beyond the bounds of mere science, occasionally even spilling over from the professional journals and popular press into the courts and pulpits. We have two principal goals in this chapter: to avoid (like the plague) entering into any discussion of the metaphysical or theological aspects of Darwinian theories and, what's more important, to provide an account of current system-theoretic and mathematical approaches to the formalization of the basic ingredients of evolutionary theory as contained in the "evolutionary equation" above.

In the preceding chapters, we have been primarily occupied with matters of dynamics and an exposition of the manner in which modern mathematics attempts to formalize processes of change. Now we shift our emphasis a bit and consider less matters of analytic technique, and more the employment of our tools to address that most quintessential of dynamical processes: (r)evolutionary change. The mathematical investigation of the nature of evolutionary processes offers an almost unlimited playpen for the favorite toys of the system scientist: *bifurcation theory, catastrophes, complexity, randomness, stability, self-organization* and *self-reference* all make their appearance, as well as many other tricks and subterfuges of the modern applied mathematician. All of these weapons are brought to bear upon the following handful of basic questions:

- What is the relationship between a change at the (micro) level of the organism's genotype, and the resultant change at the (macro) level of its physical and/or behavioral phenotype?

• What does it mean to say that a particular phenotypic property is "selected for" in the competition for survival, i.e., what is the relationship between selection criteria and the notion of adaptive fitness?

• What are the causal paths linking environmental, genetic and phenotypic change?

• At what point do we draw the dividing line between evolutionary and revolutionary change?

• To what degree, if any, are the concepts and results of *biological* evolution relevant to the changes observed in social and behavioral systems?

Despite the virtually limitless variations possible upon the above themes, we can compactly summarize the problem of evolution as: "What changes what, how and why?" When stated in such bald fashion, it becomes clear that a procedure for the formalization of evolutionary phenomena is tantamount to a procedure for the formalization of *any* dynamical process. In this sense the study of evolution serves as a *universal* metaphor for any changing natural process; in short, for *all* processes. With these thoughts in mind, let's take a more detailed look at the major components of the Darwinian world view in its modern incarnation.

2. *The Neo-Darwinian Paradigm*

Darwinian theory is based upon a population that:

1) Varies in heritable traits influencing the reproduction and survival (fitness) of its members.

2) Has offspring that resemble their parents more than they resemble randomly chosen members of the population.

3) Produces more offspring, on average, than needed to replace members who are removed from the population (death).

We can refer to these properties of the population as *variation, inheritance* and *reproduction.*

On intuitive grounds it seems evident that populations producing more offspring that survive until the age of reproduction will be favored in an environment having limited resources, since other groups will not be able to establish themselves in competition with a population producing as many offspring as possible that survive to reproductive age. This thesis was at the heart of what Darwin called *natural selection.* According to Darwinian theory, all aspects of the biological structure and behavior of living organisms are molded by this process of natural selection. The current *Neo-Darwinist* paradigm for species change adds to the classical Darwinian picture a theory of inheritance based on a genetic mechanism, as well as a theory of how genes spread in a population (population genetics).

In the Neo-Darwinian set-up, it's important to note the controversial point that selection may occur at several levels of organization: thus, the unit of selection may be the gene, the chromosome, the individual, the population, the society, etc. There is some evidence to support the claim that selection at levels higher than the classical Darwinian level of the individual are far less efficient, but it is still of considerable research interest to understand how these various levels of selection are integrated within a biological community.

Probably the most mysterious component involved in the entire Darwinian framework is the concept of *fitness.* To understand the nature of this idea a little better, suppose we have a population of asexually-reproducing individuals. Assume that each individual gives rise to α offspring per unit time, and that these offspring mature within one time-unit. Let q_o represent the probability that each such offspring survives one time-unit, and q_p the corresponding survival probability for an adult. Then if N_t represents the expected population level at time t, we clearly have

$$N_{t+1} = (\alpha q_o + q_p) N_t,$$

where the quantity $\lambda \doteq \alpha q_o + q_p$ is the natural definition of *individual fitness.* Thus, λ is the net rate of population growth and, hence, represents a quantitative formalization of the third component of Darwin's world as listed above. In the case of asexual reproduction, natural selection acts so as to select those strategies (determined by α, q_o and q_p) that maximize λ. In common parlance, we use the term "fittest" to characterize those individual strategies that are so selected. Similar arguments involving generalized notions of fitness can be given for sexually-reproducing populations, as well as for inhomogeneous populations.

Darwin really never had a proper understanding of the processes of variation and inheritance, primarily because he had no knowledge of the concept of a *gene,* nor any notion of how new, inheritable traits arise and spread through a population. It was left to the German biologist A. Weissman to formulate the basis for our current understanding of these two central ingredients in the Darwinian picture with his theory of "germ plasm." Basically, Weissman's argument was that any fertilized egg at an early stage gives rise to two independent populations of cells within the organism: the *germ line* constituting the sex cells, and the *soma line* constituting the body. Weissman then hypothesized that genetic changes occurring only in the germ line are independent of genetic changes taking place in the soma line, at least insofar as acquired characteristics are not passed along to subsequent generations.

Fundamentally, the distinction between the germ line and the soma line corresponds to the distinction between what are now termed the *genotype* and the *phenotype*. An organism's genotype consists of its total assemblage of genes (i.e., it's *genome*); the phenotype is comprised of its total assemblage of traits. The phenotype is determined by the individual's environment as well as its genotype, so the fitness of an individual is a phenotypic property. Consequently, natural selection operates on the phenotype. However, through this action the average genotypic properties of the population are changed over generations in response to selective pressures generated by the environment. Thus there is an interplay between the intrinsic features of the individual (its genotype), and the features of the environment that ultimately determine the survival of the species. The reader will note that Weissman's idea, first presented in 1866, corresponds directly to what we have already seen described earlier as the Central Dogma of Molecular Biology: "DNA $\rightarrow$ RNA $\rightarrow$ Protein," with the germ line corresponding to the DNA/RNA, while the phenotype pairs with the proteins. So genetic information, once it has passed into the phenotype, cannot get back out again. Lamarckian inheritance, in which acquired traits can be passed on to subsequent generations, is directly opposite to the Weissmanian view and directly contravenes the Central Dogma. There is strong observational evidence to support the Weissman thesis, at least for biological inheritance; however, Lamarck's vision is exactly what we see when it comes to learning and cultural inheritance, and must be taken into account when considering the matter of evolutionary processes in the social realm. But more of this later.

It's important to emphasize the point that the overall morphology M of an organism is determined *directly* by the gene products (proteins) and the environment E, i.e.,

$$M = f(p_1, \dots, p_n; E).$$

But each protein is determined by some combination of the genes and the environment as $p_i = h(g_1, \dots, g_m; E)$, where both f and h are usually highly nonlinear functions. Because of this nonlinearity, small changes in the genotype and/or environment and/or proteins *may* result in large or small changes in the phenotype. However, if evolution through natural selection brings about adaptation, we would expect that organisms will be close to their adaptive peak ($\lambda \approx \lambda_{\max}$) so that forms that deviate greatly from those currently present will, in general, be characterized by lower fitness. For this reason, evolutionary biologists have not traditionally been too concerned with such macromutations. However, this is now becoming an area of active interest as empirical evidence begins to suggest that such "hopeful monsters" may play a more important role than previously thought in evolutionary trends. With the above bit of Darwiniana under our belts, let's move on to a consideration of how we might formulate in more precise terms some of the questions surrounding the processes of evolutionary change.

3. *Genetic vs. Phenotypic Change*

Organisms can change at either the level of genes, in which case our interest focuses upon how such changes (mutations) affect phenotypic traits like size, shape or behavior, or they can change at the behavioral level, leading to a consideration of how such changes diffuse throughout the population. These are qualitatively different types of evolutionary change and, consequently, require quite different types of modeling frameworks for their treatment. To fix ideas for future reference, here we consider a typical example of each type of situation.

- *Mendelian Genetic Inheritance*—The modern theory of inheritance is based upon principles set down by Mendel governing the transmission of genes from parents to offspring. A particular gene may occur in different forms termed *alleles,* and the resultant genetic pattern of the child is determined by the pairing of various alleles contributed by the parents to the fertilized egg. Let's consider the simplest case of a gene with two alleles A and B.

The three possible pairs AA, AB and BB determine the possible genotypes of the organism, with AB being identical to BA. Since the sex cells of the parents have only a single copy of the gene on each chromosome, the genes of the child are fixed by the pairing of a chromosome from each parent. So if the parents are *homozygous,* i.e., either both of type AA or BB, then the child must also be homozygous; however, if the parents are *heterozygous,* i.e., AB, then the child's genotype is not fixed and can be any of AA, BB or AB with probabilities $\frac{1}{4}$, $\frac{1}{4}$ and $\frac{1}{2}$, respectively. Finally, there is the case when one parent is AA while the other is BB. In this situation, the child is *always* heterozygous. The allele A is said to be *dominant* if the two genotypes AA and AB have indistinguishable effects on the phenotype. In this case, the allele B is termed *recessive* if the genotype BB is observably different from AA and AB.

Now suppose we have a large population with the alleles A and B present in proportions p and $q = 1 - p$, respectively. Further, assume there is random mating in the population and that the genotypes AA, AB and BB are equally viable. What will be the proportion of the population bearing the three different genotypes in the first and subsequent generations? To address this question, note that an individual of the first generation will be of type AA only if both parents contribute the allele A to its genotype. Since this allele occurs with likelihood p, the probability of the genotype AA equals p^2. Similarly, the genotypes BB and AB occur in the first generation with proportions q^2 and $2pq$, respectively (since the AB and BA genotypes are indistinguishable). Consequently, the proportion of allele A in the first generation is

$$p_1 = \text{Prob(AA)} + \tfrac{1}{2}\,\text{Prob(AB)} = p^2 + \tfrac{1}{2}(2pq) = p(p+q) = p,$$

while the proportion of allele B is $q_1 = 1 - p_1 = q$. Thus, under the random mating and equal viability hypothesis, the proportions of the two alleles remains invariant from generation to generation.

Now let's change the situation by allowing a kind of non-Darwinian evolutionary effect so that one allele may displace the other, leading to a genetic drift in the population. Suppose we have a population of N reproducing individuals. This means that our gene with alleles A and B has 2N representatives in each generation of this population. Imagine that allele A occurs α times in generation m, whereas allele B occurs $2N - \alpha$ times. What is the likelihood that A occurs μ times in generation $m + 1$ for $\mu = 0, 1, \ldots, 2N$? Again assuming random mating and equal viability of the three genotypes, this question involves the execution of 2N Bernoulli trials with probability of "success" (appearance of allele A) being $p = \alpha/2N$. Thus, the probability of exactly μ appearances of the allele A in generation $m + 1$ is

$$p_\mu = \binom{2N}{\mu} \left(\frac{\alpha}{2N}\right)^\mu \left(1 - \frac{\alpha}{2N}\right)^{2N-\mu}.$$

As a corollary, we find the probability that allele A will disappear in the next generation is given by

$$p_0 = \left(1 - \frac{\alpha}{2N}\right)^{2N},$$

while the chance of disappearance of allele B is

$$p_{2N} = \left(\frac{\alpha}{2N}\right)^{2N}.$$

These results come about from standard random walk theory in probability, and show the overwhelming likelihood that the population will fix on one genotype or the other. Note also here the crucial importance of the population size N. If N is small, as it often is in real-life situations, then the fixation on one or the other genotype is not assured, and we face a much more difficult situation for an acccount of which the reader is directed to the literature.

Thus far we have assumed that neither allele confers a selective advantage, thereby increasing its chances of appearing in the next generation. To model this situation, we must modify the probability of occurrence of alleles A and B in a particular trial. So instead of the earlier probability of occurrence $p = \frac{\alpha}{2N}$ for allele A, suppose we take the probability of A as $\left(\frac{\alpha}{2N}\right)^\psi$, where $\psi \geq 0$. Thus if $\psi < 1$ A has an advantage, but if $\psi > 1$ allele B has a differential advantage. The parameter ψ represents a "fitness" index.

The inclusion of the fitness index giving rise to a selective advantage for one of the alleles shows that the process of genetic transmission has the character of a "game," in which the rules and the outcome are fixed by Nature without any consideration of the phenotypes (individuals) at all! We will return to these genetic matters in a later section in our discussion of replicator dynamics, but as motivation for the next section let's first follow up this idea of an evolutionary game, though now within the context of changes of *behavior* at the phenotypic level.

• *Hawks and Doves*—In a famous conflict resolution game, Maynard Smith considered a population consisting of animals competing for a resource (food, territory, mates, etc.). Each animal can choose one of two strategies: HAWK, an aggressive and potentially dangerous line of action, or DOVE, an unaggressive and safe strategy. Given these strategies, HAWK will do well against DOVE, but badly against another HAWK because of the risk of injury. The results of any contest between two animals will be measured in some units of Darwinian fitness (e.g., expected number of offspring) and payoffs of the following sort can be expected:

	HAWK	DOVE
HAWK	-2	2
DOVE	0	1

Here, by convention, we assume that the payoffs are to the animal using the strategy on the left if the opponent employs the strategy at the top. The actual numbers appearing in the payoff matrix are not too important; only their relative differences count for the arguments that follow.

Suppose a large population of animals plays this game repeatedly, pairing off randomly. Further, assume that a given animal always uses either the strategy HAWK or DOVE when engaged in a contest, and that the offspring of that animal inherit this strategic choice. Finally, let the number of offspring produced equal the payoff in the matrix plus, perhaps, some constant value. This situation then describes a model of evolution by natural selection involving asexual reproduction. Our interest is in the evolution of such a population, i.e., as time unfolds, what are the steady-state proportions of HAWKS and DOVES in the population?

To understand the evolution of the population, consider a strategy that is "uninvadable" in the sense that if a large population adopts this strategy, then any mutation causing individuals to adopt some other strategy will be eliminated from the population by natural selection. Such a strategy is termed an *evolutionary stable strategy* (ESS). Considering the payoff matrix above, it's clear that HAWK is not ESS since DOVE does better against HAWK than HAWK does against itself. A similar argument shows that DOVE is also not an ESS. Thus there is no "pure" ESS for this game. But now suppose that for a given encounter an animal adopts the strategy

HAWK with probability p, and DOVE with probability $q = 1 - p$. Such a strategy is termed "mixed," and we will show later that a mixed strategy with $p = \frac{1}{3}$ is ESS for this particular game. One way of interpreting this result is to say that if mixed strategies were not allowed, then over the course of time the population would evolve to a distribution consisting of $\frac{1}{3}$ HAWKS and $\frac{2}{3}$ DOVES.

There are a number of ways to extend the above game to bring it into closer contact with real-life situations: sexual reproduction, more strategies, asymmetric contests, and the like, some of which will be taken up in a later section. For now the important point to note is that the evolutionary trend of the population is defined in terms of the phenotypic *behaviors*, implicitly assuming that these behaviors are indicators of underlying genetic variations. Thus, in contrast to the Mendelian inheritance situation described earlier, the HAWK-DOVE game makes no explicit reference to genes, only to phenotypic behavioral patterns. What is common to the two situations is the notion of competition between behavioral patterns or gene pools being formally characterized in terms of a game. This is an important development in the modern mathematical treatment of evolutionary change, forming the basis for some of the most exciting contemporary applications of game theory, so let's take a brief *intermezzo* here to review the basics of the theory of games of strategy. With these concepts in hand, we will then return to a more detailed look at evolutionary dynamics.

4. *Game Theory—The Basics*

The distilled essence of the classical theory of games is captured in the following example involving Sherlock Holmes and his archenemy, Professor Moriarty, first treated by von Neumann and Morgenstern. Holmes wants to take a train from London to Dover, at which point he can take the boat to the Continent and make a safe escape from Moriarty. He boards the train and then sees Moriarty on the platform. Holmes can safely assume that Moriarty has managed to get on the train as well, and is now faced with one of two choices: continue with the train to Dover, or get off at Canterbury which is the only stop. Moriarty has the same two choices. It can be assumed that if they both meet, either in Canterbury or Dover, Holmes' chances of survival are zero. But if Holmes gets off at Dover and Moriarty gets off at Canterbury, his chances of survival are then 100%, since he can then make good his escape to the Continent; on the other hand, if Holmes departs at Canterbury and Moriarty stays on till Dover, then Holmes' survival odds are only 50% since the chase could then continue. Both Holmes and Moriarty are ultrarational men and consider their options accordingly, so where should they get off the train?

To analyze this situation, consider the payoff matrix

		Moriarty	
		Canterbury	Dover
Holmes	Canterbury	0	50
	Dover	100	0

Here it's reasonable to assume that Holmes wants to choose his strategy to maximize his return, while Moriarty's interests are diametrically opposed and he wants to take his action to make the payoff to Holmes as small as possible. This is an example of a *zero-sum* game in which the payoff to one player equals the loss to the other, so that the sum of the total payoffs to both players is zero. Since Holmes wants to maximize while Moriarty wants to minimize, it's clear that both will be happy if there is an entry in the payoff matrix that simultaneously is the largest element in its row (to satisfy Holmes), and the smallest element in its column (to make Moriarty happy). Such a point corresponds to what is termed a *saddle-point,* and represents an obvious solution to the game. It's easy to see that the Holmes-Moriarty dilemma has no saddle-point, implying that there is no clear-cut choice for either of them. That is, a choice that a rational player will *always* make in the absence of information about his opponent's action, other than that his opponent is rational too. Thus, the game has no solution in what are termed pure strategies for either Holmes or Moriarty. To decide what to do, they are going to have to consider *mixing* their actions with the relative weight attached to getting off in Dover or Canterbury determined by the relative payoffs for the two alternatives.

Imagine that Holmes encounters Moriarty many times in this kind of situation, and that in each encounter there is a probability p that Holmes will get off in Canterbury and a probability $1 - p$ that he will disembark in Dover. The expected payoff to Holmes for using such a strategy is then

$$\mathcal{E}(\text{Holmes}) = p \cdot 0 + (1 - p) \cdot 100 = (1 - p) \cdot 100$$

if Moriarty gets off in Canterbury, and it is

$$\mathcal{E}(\text{Holmes}) = p \cdot 50 + (1 - p) \cdot 0 = p \cdot 50$$

if Moriarty leaves the train in Dover. Of course, Moriarty has his own set of probabilities for leaving the train. So let's assume that he decides to get off in Canterbury with likelihood r, and in Dover with probability $1 - r$. Thus Holmes' overall expected payoff is

$$\mathcal{E}^*(\text{Holmes}) = 100r(1 - p) + 50p(1 - r) = 100r + 50p - 150pr.$$

Holmes wants to pick p to maximize this quantity. Arguing in a similar fashion, we can compute Moriarty's overall expected return as

$$\mathcal{E}^*(\text{Moriarty}) = 100r(1-p) + 50p(1-r) = 100r + 50p - 150pr,$$

exactly the same as Holmes' (since this is a zero-sum game). Moriarty wants to minimize this quantity while, as we saw above, Holmes wants to maximize it. Equating the above expressions for Holmes and Moriarty, and doing the resulting algebra, we see that the players will achieve their objectives if $p = \frac{2}{3}$ and $r = \frac{1}{3}$; i.e., two-thirds of the time Holmes should alight in Canterbury, and one-third of the time he should leave the train in Dover. Moriarty should get off in Canterbury one-third of the time, and leave in Dover two-thirds of the time. With these strategies, the expected payoff to Holmes is

$$\frac{1}{3}(0) + \frac{2}{3}(50) = 33\frac{1}{3}.$$

Thus, Holmes can guarantee himself a one-third chance of staying alive (on the average) no matter what Moriarty chooses to do, and Moriarty can ensure himself at least a one-third chance of doing Holmes in, regardless of where Holmes decides to leave the train. It's a simple matter to verify that weighing their actions according to these chances gives both Holmes and Moriarty at least as great a return (on the average) as by playing any pure strategy of always getting off at either Canterbury or Dover. These combinations of p and r represent *mixed strategies* for Holmes and Moriarty. It's amusing to note that the most likely outcome of a single play of this game will have Holmes getting off in Canterbury with Moriarty continuing on the train to Dover—exactly what happened in the actual story by Conan Doyle. Not only is this the most likely outcome, but it also has the virtue of enabling the chase to continue!

The above game shows that the expected maximal return to Holmes equals the expected minimal return to Moriarty under the assumption that both players are rational and play optimally. The existence of mixed strategies guaranteeing such an outcome for all zero-sum, two-person games regardless of the number of pure strategies available to each player, forms the core of the celebrated Minimax Theorem, first proved by von Neumann in 1928.

If we let

$$A = (a_{ij}), \qquad i = 1, 2, \ldots n; \; j = 1, 2, \ldots, m,$$

be the payoff matrix for a zero-sum, two-person game, with $p = (p_1, \ldots, p_n)$ being the probability vector representing the likelihood of Player I choosing action i, while $q = (q_1, q_2, \ldots, q_m)$ is the corresponding probability vector for Player II, then we can state von Neumann's result more formally:

THE MINIMAX THEOREM. *For a two-person, zero-sum game with payoff matrix A, there exist probability vectors* p^* *and* q^* *such that*

$$V(p^*, q^*) \doteq \max_p \min_q V(p, q) = \min_q \max_p V(p, q),$$

where $V(p, q) \doteq (p, Aq)$*(the payoff to Player I using strategy* $p = -$*the payoff to Player II using strategy* q*).*

The quantity $V(p^*, q^*)$ is termed the *value* of the game.

Unfortunately, the Minimax Theorem provides no information on how to actually *find* the optimal mixed strategies. It can be shown that the computation of the optimal strategy vectors p^* and q^* can always be reduced to a standard linear programming problem whose solution can be obtained by the techniques based upon the Simplex Method (see Problem 1) or one of its many variants, although more efficient, specialized algorithms are usually employed . There are a number of ways in which we can attempt to extend the foregoing, admittedly artificial, situation to encompass more realistic conflicts. The most obvious is to allow more than two players, leading to the thorny difficulties of "n-person game theory." For now, we prefer to steer away from these deep waters and introduce an equally important generalization to the case when the interests of the two players are not diametrically opposed. This is the *mixed-motive* or, as it is sometimes termed, *nonzero-sum* game.

In mixed-motive games, the interests of the two players are neither in direct opposition nor in direct cooperation, leading to the nonzero-sum outcome of any play of the game. For this reason, the entries in the payoff matrix consist of two numbers each, the first representing the payoff to Player I, the second to Player II for the given pair of actions. Suppose, for simplicity, that each player has two actions at his disposal which, for reasons that will become apparent later, we label "C" and "D." Further, let's assume that the game is *symmetric* so that the outcome is unchanged if we reverse the roles of the two players. Under these hypotheses, the general form of the payoff matrix will be

		Player II	
		C	D
Player I	C	(R, R)	(S, T)
	D	(T, S)	(P, P)

where P, R, S and T are real numbers. Since the two-person, mixed-motive games play an extremely important role in the modeling of evolutionary processes, we examine their structure in more detail.

5. *A Typology of Mixed-Motive Games*

There are 24 different ways to order the four numbers P, R, S and T, each of which gives rise to a different mixed-motive game. But in view of the symmetry hypothesis, only 12 of these games are qualitatively different. And of these, there are eight that possess optimal equilibrium strategies, i.e., games with an equilibrium point representing the existence of an optimal pure strategy for each player. Such games are conceptually uninteresting, so we focus attention on the remaining four two-person, mixed-motive games without equilibrium points. Each of these games has been extensively studied in the literature and can be represented by a prototypical situation capturing the concepts peculiar to that particular type of game. These qualitatively different games are given by the following classification scheme.

- *Leader*—($T > S > R > P$).
- *The Battle of the Sexes*—($S > T > R > P$).
- *Chicken*—($T > R > S > P$).
- *The Prisoner's Dilemma*—($T > R > P > S$).

To see the basic differences in character and strategy for each of these games, let's look at the prototypical example from each category.

- *Leader*—Consider the case of two drivers attempting to enter a busy stream of traffic from opposite ends of an intersection. When the cross traffic clears, each of the drivers must decide whether to concede the right-of-way to the other (C), or drive into the gap (D). If both concede, they will both be delayed, whereas if both drive out together, there may be a collision. However, if one drives out while the other waits, the "leader" will be able to carry on with his trip, while the "follower" may still be able to squeeze into the gap left behind the leader before it closes again. A typical payoff matrix for this Leader game would be

		Driver I	
		C	D
Driver II	C	(2,2)	(3,4)
	D	(4,3)	(1,1)

There is clearly no dominant strategy in the game of Leader; according to the minimax principle, to avoid the worst possible outcome each driver should choose strategy C, ensuring that neither will receive a payoff less than 2 units. However, the minimax strategies are not in equilibrium as each driver would have reason to regret his choice when he finds out the other's choice. This simple observation shows that the minimax principle cannot be used as a means for prescribing rational actions in mixed-motive games.

In fact, there are *two* equilibrium strategies in Leader: the strategies (C,D) and (D,C) appearing at the off-diagonal corners of the payoff matrix. If Driver I chooses D, the second driver can do no better than choosing C, and vice versa. In other words, neither can do better by deviating from such an equilibrium outcome. However, in contrast to zero-sum games in which such equilibrium points are always equivalent, in Leader the first driver prefers the (D,C) equilibrium, while Driver II prefers (C,D). There is no mathematical way of resolving this difference of opinion. But in real-world situations of this type, the impasse is often resolved by the fact that one of the equilibrium points is more *visible* to the players than the other. Thus, in the simple traffic example above, cultural and/or psychological factors may enter to break the deadlock, and various tacit rules like "first come, first served," or signaling schemes like blinking of lights are used. This kind of signaling, incidentally, is in sharp contrast to zero-sum games in which such signals would definitely not be to a driver's advantage.

• *Battle of the Sexes*—In this game a married couple has to choose between two options for his/her evening entertainment. The husband prefers one type of entertainment (e.g., a movie), whereas the wife prefers another, say, going out for a pizza. The problem is that they would both rather go out together than alone. If they each go for their first choice (call it action D), they end up going out alone and each receives a payoff of 2 units. If both make sacrifices and each goes to the activity that they don't like (denoted action C), they each suffer and receive only 1 unit of reward. But if one sacrifices while the other gets their first choice, then they still go out together, but the "hero" who sacrifices receives 3 units of return while the other party gets 4. The payoff matrix for this game is shown below.

		Wife	
		C	D
Husband	C	(1,1)	(3,4)
	D	(4,3)	(2,2)

There are a number of features in common between Battle of the Sexes and Leader: (1) neither the Husband nor the Wife has a dominant strategy, (2) the minimax strategies intersect in the nonequilibrium result (D,D), (3) both strategies (C,D) and (D,C) are in equilibrium. However, in contrast to Leader, the player who deviates unilaterally from the minimax strategy rewards the other player more than himself in Battle of the Sexes. This is just the opposite of what happens in Leader where the deviator rewards himself more than his opponent. But just as in Leader, in Battle of the Sexes a player can gain by communicating with the other player in order to obtain some sort of commitment to strategy C. So, for instance, the husband could announce that he was irrevocably committed to his first choice of entertainment, in which case this will work to his advantage if his wife then acts in her own best interests. The only difficulty is in convincing her that he's serious. The main point here is that some kind of commitment is needed in order for both parties to achieve the best possible outcome in the Battle of the Sexes game.

• *Chicken*—A well-known game, whose principles date back at least as far as the Homeric era, involves two motorists driving toward each other on a collision course. Each has the option of being a "chicken," and swerving to avoid the collision (C), or of continuing on the deadly course (D). If both drivers are chicken, they each survive and receive a payoff of 3 units. But if one "chickens out," while the other drives straight on, the chicken loses face (but not his life) and the "macho man" wins a prestige victory. In this case the chicken receives 2 units, whereas the opponent receives 4. Finally, if they both carry on to a fatal collision, they each receive Death's reward of 1 unit. The payoff matrix is

		Driver I	
		C	D
Driver II	C	(3,3)	(2,4)
	D	(4,2)	(1,1)

Again, in Chicken there is no dominant strategy and the minimax strategies intersect in the outcome that both drivers "chicken out." Also as before, an exploiter who deviates from a minimax strategy can gain an advantage for himself, but this time he *invariably* affects the other player adversely by such a deviation. So not only does the deviator harm the other player, he also puts himself and the other player in a position where they may have a disastrous outcome. Chicken also has the peculiar feature that it is impossible to avoid playing the game with someone who is insistent, since to refuse to play is effectively the same as playing and losing. In addition, the player who succeeds in making his commitment to the dangerous D option and appears convincing will *always* win at the expense of the other player, assuming the other player is rational. Thus, a player who has a deserved reputation for recklessness enjoys a decided advantage in Chicken over one who is merely rational. Perhaps this accounts for the aversion of the typical academic to this kind of irrational winning strategy, as most academics seem to pride themselves on being both risk averse and ultrarational, an unhappy losing combination when engaged in the game of Chicken. This becomes especially true if Chicken is played a number of times because the player who gains an early advantage usually maintains or even increases that advantage later on. For once he has successfully exploited the other player, he gains confidence in his ability to get away with the risky strategy in the future while making his opponent all the more fearful of deviating from the cautious minimax alternative.

- *The Prisoner's Dilemma*—The last basic type of nonzero-sum game, and by far the most interesting, is the famous game of two prisoners who are accused of a crime. Each of them has the option of concealing information (C) from the police or disclosing it (D). If they each conceal it (i.e., they cooperate), they will both be acquitted with a payoff of 3 units to each. If one conceals and the other discloses the information, the one who "squealed" receives a reward of 4 units, while the payoff to the "martyr" will be only 1 unit, reflecting his role in the obstruction of justice. Finally, if they both disclose the information, they will both be convicted thereby obtaining a payoff of only 2 units. The appropriate payoff matrix for the Prisoner's Dilemma game is

		Prisoner I	
		C	D
Prisoner II	C	(3,3)	(1,4)
	D	(4,1)	(2,2)

Incidentally, our use of the symbols C and D to represent the possible actions by the players in all of these games is motivated by the usual interpretation of the actions in the Prisoner's Dilemma game. Here C represents "cooperating" with your pal and not confessing, whereas D signifies "defecting" to the police and giving the information needed for a conviction.

The Prisoner's Dilemma is a real paradox. The minimax strategies intersect in the choice of mutual defection, which is also the only equilibrium point in the game. So neither prisoner has any reason to regret a minimax choice if the other also plays minimax. The minimax strategies are also dominant for both prisoners since each receives a larger payoff by defecting than by cooperating against either strategy played by the other. Thus, it appears to be in the best interests of each prisoner to defect to the police *regardless* of what the other prisoner chooses to do. But if both prisoners choose this individually maximizing action of defecting, the outcome (2,2) is less than they could have each received (3,3) if they had chosen their dominated strategy of remaining silent.

The essence of the paradox in the Prisoner's Dilemma lies in the conflict between the ideas of individual and collective rationality. According to individual rationality, it's clearly better for a prisoner to defect and give information to the police. Yet if both try to be "martyrs" and remain silent, it's paradoxically better for both of them. What's needed to guarantee a better outcome for both of them is some sort of selection principle that is based upon collective interests. Perhaps the oldest, and most well-known, such principle is the Golden Rule of Confucius: "Do unto others as you would have them do unto you." However, in this regard note that the Golden Rule can be disastrous in another kind of game as seen by the Battle of the Sexes. Here if both the husband and wife adopt the Golden Rule, the outcome is the worst possible with each of them going out alone to an entertainment that they don't like.

The foregoing archetypal two-person, mixed-motive games have many features in common, especially Leader, Battle of the Sexes and Chicken. For instance, each of these games share the properties:

1) There exists a "natural" outcome if each party plays minimax, i.e., chooses option C.

2) The outcome (C,C) is a nonequilibrium point that is vulnerable in that both players are tempted to deviate from it.

3) Each of the games possesses two asymmetric equilibrium points neither of which is stable, since the players are not in agreement as to which of the two points is preferable.

4) None of the games possesses a dominant strategy for either player.

5) The worst possible outcome for both players results if both choose their non-minimax strategies, i.e., option D.

By way of contrast, the Prisoner's Dilemma game possesses none of these features.

From the standpoint of evolutionary models, the Prisoner's Dilemma is by far the most interesting of the above games and the one that we shall focus considerable attention upon later. For now, let's return to the issue of dynamics and spend some time looking at how we can infuse the essentially static character of two-person games with some notion of dynamical change.

6. *Replicators, Hypercycles and Selection*

We have seen that evolutionary models come in two basic types: genetic and phenotypic. Despite the radically different physical interpretation that we attach to these two classes, it turns out to be possible to give a common mathematical framework suitable for investigating many of the important properties of both types of models. This formalism goes under the rubric of *replicator dynamics,* and involves postulating an abstract unit of selection, termed a *replicator.* Roughly speaking, the properties characterizing a replicator consist of being able to give rise to an unlimited number (in principle, at least) of copies of itself, and the occurrence of many variants whose properties can influence the number of copies of each variant that will be produced. Thus, genes can be replicators as can chemical molecules and behavioral patterns. We shall consider these examples later, but for now let's just assume that we have at hand an abstract replicator that occurs in one of n different forms or variants, which we label $i = 1, 2, \ldots, n$, and that the frequency of the variants is given by the vector $x = (x_1, x_2, \ldots, x_n)$.

To develop the general form of the replicator dynamics, assume that we have a function $f(x)$ expressing the interaction of the different variants of the replicator. Further, let's normalize things so that the flow describing the change of the variants takes place on the unit simplex S^n in R^n. This entails scaling x so that

$$0 \leq x_i \leq 1, \qquad \sum_{i=1}^{n} x_i = 1.$$

If we assume that the rate of increase of the frequency of variant i is directly proportional to the current level of x_i, as well as the interaction of variant i with the other variants, we are led to the dynamics for x_i as

$$\dot{x}_i = x_i \left[f_i(x) - \Phi(x) \right], \qquad i = 1, 2, \ldots, n. \tag{$\dagger$}$$

Here the term $\Phi(x)$ is given by the average fitness of the population

$$\Phi(x) = \sum_{j=1}^{n} x_j f_j(x),$$

and is introduced to ensure that the system trajectory remains within S^n. It's clear that the nature of the change in the replicator is determined solely by the interaction function $f(x)$.

To make contact with our earlier examples, assume that we select the interaction function $f(x)$ to involve only first-order, i.e., linear interactions. Then we have

$$f_i(x) = (Ax)_i = \sum_{j=1}^{n} a_{ij} x_j, \qquad i = 1, 2, \ldots, n.$$

Here $A = (a_{ij})$ is a matrix whose terms specify the nature and degree of interaction between variants i and j. Under this linearity assumption, the replicator dynamics ($\dagger$) describe the effect of selection on allele frequencies in a gene pool, as well as the distribution of behavioral phenotypes in a species population. The difference is not in the mathematics, but in the interpretation of what we mean by the replicator variants. In the first case,

x_i represents the frequency of allele A_i in the gene pool, whereas in the second case x_i is the frequency of behavioral phenotype B_i in the species population. In both cases, the elements $\{a_{ij}\}$ represent the manner in which the variants interact and "select" the variants in the next generation. To see how this set-up works in a familiar setting, let's revisit the animal conflict problem considered earlier.

Let $B_1, B_2, \ldots, B_n$ be the behavioral phenotypes within a population, with $x_1, x_2, \ldots, x_n$ their respective frequencies. Further, let a_{ij} be the expected payoff to an animal using behavior B_i in competition with one using behavior B_j, $i, j = 1, 2, \ldots, n$. Assuming random encounters as before, it's easy to see that the quantity $(Ax)_i$ is the average payoff to a contestant using behavior B_i when the population is in the state x, with the overall mean payoff for the entire population then being

$$\Phi(x) = (x, Ax) = \sum_{i=1}^{n} x_i (Ax)_i.$$

Again assuming asexual reproduction, the relative rate of increase of phenotype B_i is given by

$$\frac{\dot{x}_i}{x_i} = (Ax)_i - (x, Ax) = (Ax)_i - \Phi(x),$$

which are the original dynamics (†) for the case of linear interactions. A similar argument can be used to derive *exactly* the same equations for the case of genetic selection. Rather than going into this boring and repetitious case, let's look at another quite different physical setting in which the replicator dynamics describe chemical processes of interest—the case of *prebiotic evolution.*

Let $C_1, C_2, \ldots, C_n$ be self-replicating polynucleotides (RNA or DNA) swimming about in a primordial "soup." The concentration of nucleotide C_i is given by x_i, $i = 1, 2, \ldots, n$, and we assume there is a dilution flow $\Phi(x)$ that keeps the total concentration constant, normalized to be equal to 1. In this setting, independent replication of the polymers leads to a situation in which all but one of the molecular species vanishes, with loss of the corresponding encoded information. To prevent this information loss, Eigen and Schuster considered *networks* of catalytically interacting polynucleotides. Of special interest are those closed feedback loops of interactions in which each molecular species is catalyzed by its predecessor. These loops are termed *hypercycles,* and evolve according to the dynamics

$$\dot{x}_i = x_i[x_{i-1}H_i(x) - \Phi(x)], \qquad i = 1, 2, \ldots, n,$$

where the indices are taken modulo n and the functions $H_i(x)$ are strictly positive on S^n. The simplest case of a hypercycle is when $H_i(x) \equiv k_i > 0$, which is obtained from the linear replicator dynamics by letting A be the permutation matrix

$$A = \begin{pmatrix} 0 & 0 & \dots & 0 & k_1 \\ k_2 & 0 & \dots & 0 & 0 \\ 0 & k_3 & \dots & 0 & \\ \vdots & \vdots & & \vdots & \vdots \\ 0 & 0 & \dots & k_n & 0 \end{pmatrix}.$$

The above discussion indicates the ubiquitous nature of the replicator equations (†) for describing a wide variety of processes involving the selection of replicating variants from a given population of genes, animals or molecules. With these physical situations in mind, we turn now to a more detailed look at the mathematical properties of the replicator dynamics.

7. *Equilibrium Properties of Replicator Systems*

There are two basic questions of interest surrounding the dynamical behavior of replicator systems: (1) given an initial distribution of replicators, what is the equilibrium distribution as $t \to \infty$, and (2) in the limit, do any species (behavior patterns, molecular types) die out? Obviously, the second question is a special case of the first, so we begin by attacking the standard question of the equilibrium behavior of a dynamical system, a problem that we have already faced in a variety of contexts in the preceding chapters. Here we make use of general results in the theory of differential equations in the specific setting of the replicator system

$$\dot{x}_i = x_i[f_i(x) - \Phi(x)], \qquad i = 1, 2, \dots, n.$$

We begin by noting that the manifold on which the replicator dynamics unfold is the n-simplex S^n, defined by the following subset of R^n:

$$S^n = \left\{ x \in R^n : 0 \le x_i \le 1, \sum_{i=1}^{n} x_i = 1 \right\}.$$

Since $\sum \dot{x}_i = 0$ on S^n, each face of the simplex (which consists of subsimplices characterized by $x_i = 0$ for all i in a non-empty subset of $\{1, 2, \dots, n\}$) forms an invariant set for the replicator dynamics. In particular, the corners e_i are equilibria. For the moment, let's ignore these "trivial" equilibria and concentrate attention upon the nontrivial equilibria located in the interior of the simplex S^n.

The equilibria in the interior of S^n are given by the positive solutions of the equations

$$f_1(x) = f_2(x) = \cdots = f_n(x), \qquad x_1 + x_2 + \cdots + x_n = 1.$$

Any solution x^* of the above system must be such that $f_i(x^*) = \Phi$. In the special case of a first-order replicator system, the interior equilibria satisfy the equations

$$\sum a_{1j}x_j^* = \sum a_{2j}x_j^* = \cdots = \sum a_{nj}x_j^*.$$

Such solutions form an affine space, and it has been shown that generically there is at most a single interior solution for first-order replicator systems. That is, there is an open, dense subset of matrices $A \in R^{n \times n}$ such that the corresponding first-order replicator system has at most one equilibrium in the interior of S^n and in the interior of each face. A specific test for the absence of such an interior equilibrium is provided by considering the adjoint matrix of A, $\operatorname{adj} A$, and the unit vector $u = (1, 1, \cdots, 1)$. If the vector $(\operatorname{adj} A)u$ has both positive and negative entries, then there is no interior equilibrium and no interior periodic orbits for the first-order replicator system.

We should note also that the robustness of the interior equilibrium, when it exists, also rules out the appearance of elementary catastrophes. If such bifurcations could occur in the dynamic, there would be a perturbation of the system that would lead to more than one interior equilibrium. So, for first-order replicator systems, there is at most a single candidate for an equilibrium distribution that preserves *all* the species or behaviors in the system, leading to the speculation that Nature's way is to let some perish that others may survive. We shall explore this point in more detail later on.

Suppose that the general replicator equations have an equilbrium point $p \in S^n$ (not necessarily an interior point). Then an easy calculation shows that the Jacobian matrix of the system has $\Phi(p)$ as one of its characteristic values, with a corresponding characteristic vector that does not belong to the tangent space to S^n at p. Thus, this characteristic value is irrelevant as far as determination of the stability properties of the equilbrium at p and can be ignored. So, for instance, the relevant characteristic values for the first-order system at a corner e_i are the $n-1$ values $a_{ij} - a_{ii}$, $j \neq i$.

Example: The Hypercycle

For the hypercycle system discussed in the last section, the matrix A is a permutation matrix with entries $\{k_i\}$, and it's straightforward to show there always exists an equilibrium point $p \in \operatorname{int}\ S^n$. The coordinates of this point are given by

$$p_i = \frac{k_{i+1}^{-1}}{\sum k_j^{-1}}, \qquad i = 1, 2, \ldots, n \quad \text{mod } n.$$

In this case, the characteristic values are (up to a positive multiplier) the n roots of unity with the irrelevant root being 1 itself. Thus, for $n \leq 3$ the point

p is locally asymptotically stable, but for $n \geq 5$ the point is always unstable. The case when $n = 4$ is more delicate since we have two characteristic values lying on the imaginary axis. But by use of the Lyapunov function $\prod x_i$, it can be shown that in this case p is actually globally asymptotically stable. It's also interesting to note that when $n \geq 5$, numerical results suggest the existence of a stable periodic attractor, although there is as yet no rigorous proof of this behavior for the hypercycle.

By the above discussion, if the replicator system has an interior equilibrium we can employ standard techniques of stability analysis to determine the local, and sometimes the global, stability characteristics of the equilibrium point. If there is no interior equilibrium, then it has been shown there is a constant vector $c \in R^n$ with $\sum_i c_i = 0$ such that the function $V(x) = \prod x_i^{c_i}$ increases along the orbits of the replicator system. Hence, by Lyapunov's Theorem, it follows that each orbit of the system has its ω-limit contained in the boundary of S^n. Thus, if there is no interior equilibrium point, there are no periodic, recurrent or even nonwandering points in int S^n. However, this does **not** mean that $\lim_{t\to\infty} x_i(t) = 0$ for some i. There exist examples of systems whose ω-limit of every interior orbit is a cycle consisting of the corners and edges of the simplex S^n. Such a limiting trajectory would imply only a temporary die-off of a given species or behavior, with the species re-appearing infinitely often. Since this question of persistence of species is of vital importance in many biological settings, let's consider conditions on the replicator dynamics that ensure the survivability of all species initially present in the system.

8. *Perseverance of Species in Replicator Systems*

As emphasized above, the issue of persistence centers about the question of when a species i with initial fraction $x_i(0) > 0$ *perseveres* for all $t > 0$. More formally, we say that the system is *perseverant* if there exists a $\delta > 0$ such that $\lim_{t\to\infty} \inf x_i(t) \geq \delta$ for all i whenever $x_i(0) > 0$. Systems having this property are clearly of great practical importance since, first of all, small fluctuations in the population of any species cannot result in a species being wiped out, and second, if the system starts on the boundary with one or more components absent, mutations that introduce these components will spread, resulting in a system that is safely cushioned away from extinctions.

Remarks

1) Perseverance is *not* a structurally stable property.

2) Nonperseverance doesn't necessarily imply that some component of the system is driven to extinction. For example, there exist systems with attractors on the boundary, as well as in the interior of S^n. It can also happen that each orbit of the system remains bounded away from the boundary,

but that the bound depends upon the particular orbit. For perseverance, the bound must be uniform for all orbits.

Since there appear to be no useful conditions that are both necessary and sufficient for perseverance, we must make do with the following separate conditions:

• *Sufficiency*—A replicator system is perseverant if there exists a function $P: S^n \to R$ such that: (1) $P(x) > 0$ for $x \in \text{int}\, S^n$ and $P(x) = 0$ for $x \in \partial S^n$; (2) $\dot{P} = P\Psi$, where Ψ is a continuous function such that there exists a $T > 0$ making the integral

$$\frac{1}{T}\int_0^T \Psi[x(t)]\, dt > 0,$$

for all $x(t) \in \partial S^n$. In other words, near the boundary of S^n the function P is increasing on the average, so that P acts as a kind of average Lyapunov function.

An example of the use of this result arises for the hypercycle system in which the function $P(x) = x_1 x_2 \cdots x_n$ satisfies the condition for an average Lyapunov function. Thus, we can conclude that the hypercycle equations are perseverant, even though we have already seen that the interior equilibrium is, in general, unstable for $n \geq 5$. Now let's look at some simple necessary conditions.

• *Necessity*—The Brouwer Fixed Point Theorem implies that a necessary conditon for perseverance is the existence of an equilibrium in the interior of S^n. In the case of first-order replicator systems, this equilibrium is necessarily unique, and for perseverance it can be shown that it's necessary that the trace of the system Jacobian matrix be strictly negative at this interior equilibrium point.

For the special (but practically important) case of first-order systems with matrices of the form

$$A = \begin{pmatrix} 0 & - & - & \cdots & + \\ + & 0 & - & \cdots & - \\ - & + & 0 & \cdots & - \\ \vdots & \vdots & & & \vdots \\ - & - & \cdots & + & 0 \end{pmatrix},$$

where "+" indicates a positive entry and "−" denotes an entry that is either negative or zero, the following result of Hofbauer and Amann characterizes the perseverance of the corresponding replicator system.

Perseverance Theorem. *For matrices A of the special form above, the following conditions are equivalent for first-order replicator systems:*

i) The system is perseverant.

ii) There is a unique interior equilibrium p such that $\Phi(p) > 0$.

iii) There is a vector $z \in R^n$, with $z_i > 0$ for all i, such that all components of zA are positive.

iv) The matrix C whose components are $c_{ij} = a_{i+1,j}$ (with the indices taken mod n) is such that all of its principal minors are positive.

Note that the matrix C described in part (iv) of the theorem is obtained from A by moving the first row of A to the bottom. Such matrices C having strictly positive diagonal elements, with all other elements nonpositive, play an important role in mathematical economics, where $-C$ is termed a *Metzler* matrix.

As an easy application of the above theorem, we note that the hypercycle equation with interaction matrix

$$A = \begin{pmatrix} 0 & 0 & \cdots & 0 & k_1 \\ k_2 & 0 & \cdots & 0 & 0 \\ 0 & k_3 & \cdots & 0 & 0 \\ \vdots & \vdots & \cdots & \vdots & \vdots \\ 0 & 0 & \cdots & k_n & 0 \end{pmatrix}$$

leads to a matrix C that is diagonal with positive entries; hence, perseverant by part (iv) of the Perseverance Theorem.

There exist a variety of additional conditions for the persistency of replicator systems, some of which are explored in the Problems and Exercises. But for now we take a different tack and look at another important question surrounding replicator systems, the matter of classification. In Chapter Four, we saw that it was possible to classify all smooth functions of codimension no greater than five, but that a similar classification for smooth vector fields presented far more subtle difficulties. Here we look at the same question, but for the far more restricted case of vector fields specified by replicator systems.

9. *Classification of Replicator Systems*

The basic idea in classifying first-order replicator systems is to find conditions under which two matrices A and B generate the same flow on S^n. For instance, if a constant is added to every element of A, the flow remains unaltered. Thus, given any matrix, we can reduce its diagonal to zero without altering the flow by subtracting an appropriate constant from each column. Let K_n be the set of matrices in $R^{n \times n}$ all of whose columns are multiples of $u = (1, 1, \cdots, 1)'$, and let Z_n be the set of matrices whose diagonals are zero. It's clear that any matrix $M \in R^{n \times n}$ can be written as the direct sum

$$M = Z \oplus K, \qquad Z \in Z_n, \qquad K \in K_n.$$

The following resultof Zeeman's provides a simple test for two matrices to yield the same flow on S^n.

EQUIVALENCE LEMMA. *The matrices A and B yield the same flow for the first-order replicator system if and only if $A - B \in K_n$.*

As a consequence of the above lemma, we see that every equivalence class is of the form $E \oplus K_n$, where E is an equivalence class in Z_n. Thus, it suffices to classify matrices $A \in R^{n \times n}$ in the smaller subset Z_n.

For the case $n = 2$, the above remarks lead to the consideration of matrices of the form

$$A = \begin{pmatrix} 0 & a \\ b & 0 \end{pmatrix}.$$

There are four distinct cases to consider, depending upon the zero/nonzero pattern of the two elements a and b.

1) $a, b > 0$. In this case, the point $(\frac{a}{a+b}, \frac{b}{a+b})$ is an interior equilibrium that is an attractor of the flow.

2) $a, b < 0$. This case is the reverse of the previous case except that the interior equilibrium is now a repellor, hence, unstable.

3) $a \geq 0 \geq b$, but not both zero. In this situation, the corner point $(1, 0)$ is an attractor. This case is also equivalent to the case $a \leq 0 \leq b$, but not both zero, except that now the point $(0, 1)$ is an attractor.

4) $a = b = 0$. Here all points are equilibria.

We call the matrix A *stable* if there is a neighborhood of A such that every matrix in the neighborhood generates a flow that is equivalent to that of A. The above analysis shows that there are three stable classes for two-dimensional replicator dynamics given by

$$a, b > 0, \qquad a, b < 0, \qquad a > 0 > b \text{ or } a < 0 < b.$$

It should be noted that if we allow time reversal, the first and second cases are the same, leading to only 2 stable classes for $n = 2$. Similar arguments lead to the strongly supported conjecture that for $n = 3$ there are 19 stable classes.

For higher-dimensional replicator systems there appears to be no hope for a simple classification, since the number of qualitatively distinct possbilities grows geometrically with the system dimension n. The only general result that can be given is that stability implies that all equilibria of the system are hyperbolic; i.e., if p is such an equilibrium, the system Jacobian matrix at p has no characteristic values on the imaginary axis.

With the foregoing excursion into the dynamical behavior of replicator systems completed, let's now return to the connection between such dynamical systems and the game-theoretic concepts underlying advantageous survival strategies.

10. *Game Theory and Replicator Systems*

In our earlier discussions using game-theoretic concepts for the determination of competitively advantageous strategies, there was no explicit notion of a dynamic. However, the first-order replicator systems considered above provide us with a natural means to attach a dynamical process to the competitive situation considered before in the special case when both players have the same strategy set and the payoffs are symmetric; i.e., the payoff matrix A is the same for both competitors. By the arguments already given, the rate of growth of those individuals playing strategy i is proportional to the advantage of that strategy. But the advantage of i is measured against the average payoff against *all* strategies which, if we let x_i be the fraction of the population playing i, leads to the replicator dynamics

$$\dot{x}_i = x_i[(Ax)_i - (x, Ax)], \qquad i = 1, 2, \ldots, n.$$

We have considered earlier the concept of an evolutionary stable strategy (ESS) in an informal way. Now we define the concept formally as a strategy p possessing the following equilibrium and stability properties: (1) p is at least as good a reply against itself as any other strategy x, and (2) if x is a best reply against p, then p is a better reply against x than x itself. Mathematically, we can state these two conditions compactly as

$(p, Ap) \geq (x, Ap)$ for all $x \in S^n$ (Equilibrium);

if $(p, Ap) = (x, Ap)$ for $x \neq p$, then $(p, Ax) > (x, Ax)$ (Stability).

The most natural question that arises at this point is to ask what the relationship is, if any, between an asymptotically stable equilibrium of the replicator system associated with the payoff matrix A and the ESS p. The following result goes most of the way toward answering this question.

ESS THEOREM. *The following statements are equivalent:*

i) p is an ESS.

ii) For all $q \in S^n$, $q \neq p$, we have

$$(p, A[(1-\epsilon)p + \epsilon q]) > (q, A[(1-\epsilon)p + \epsilon q]),$$

for all $\epsilon > 0$ sufficiently small.

iii) For all $x \neq p$ in some neighborhood of p, we have

$$(p, Ax) > (x, Ax).$$

iv) The function $V = \prod x_i^{p_i}$ is a Lyapunov function at p for the replicator dynamics.

Remarks

1) The last condition of the ESS Theorem shows that an ESS must be an asymptotically stable attractor of the replicator dynamics. However, the converse is not true; there exist asymptotically stable attractors that are not ESS.

2) The second statement of the theorem can be interpreted as saying that if the state of the population is p and a subpopulation (mutation) in state q arises, such a subpopulation will become extinct since the p population does better than q against the mixture $(1-\epsilon)p+\epsilon q$ as long as the q fluctuation in the original population remains sufficiently small. In other words, a population using the strategy p is uninvadable.

We have now explored a number of theoretical aspects of both static games, replicator dynamics and evolutionary processes, as well as several of the interconnections between these areas. In the next few sections we put much of this machinery to work by examining in some detail two evolutionary processes involving plants that suggest an interpretation of Nature's strategy for survival in terms of ESS. Our first set of examples of the uses of ESS in an ecological context involves the behavioral patterns observed in the attempt by plants to maximize their ability to propagate their seeds into future generations.

11. *Desert Plant Root Systems and ESS*

A good illustration of how ESS arise in Nature is provided by the way desert plants organize their root structures to acquire enough water to ensure their survival in an arid environment. Such plants can develop two types of root systems: a *lateral* system that spreads out horizontally utilizing water sources near the surface, and a *tap root* system that uses deeper sources of water. Most desert plants specialize in one or the other of these two systems; however, some shrubs have the capacity to develop either a lateral or a tap root system. Generally, surface water is limited so there is considerable competition between neighboring plants for this source, but underground water is usually quite plentiful so that there is little competition between plants using tap root systems. However, underground water is only available in spots, so that the development of a tap root system entails a certain amount of risk associated with failing to find an underground water supply.

Now consider the situation of two neighboring plants that each have the option of developing either a lateral or a tap root system. Assume that the evolutionary utility, in the sense of Darwinian fitness, is directly proportional to the rate of water intake. Let S be the utility of a lateral root system in the absence of competition, and let U be the utility of a tap root system. Finally, suppose that if both plants develop a lateral root system they divide equally the utility associated with a lateral root system in the absence of competition. Under the foregoing hypotheses, the payoff matrix in the evolutionary game between these two plants is given by

		Plant I	
		L	T
Plant II	L	(S/2, S/2)	(S,U)
	T	(U,S)	(U,U)

Here L indicates the strategy: develop a lateral root system, and T represents the tap root option. Let us also emphasize that the payoff U is an *expected* payoff obtained by multiplying the payoff associated with the presence of underground water by the probability of actually finding water at the site where the tap root system is developed.

The equilibria of the above game depend upon the relative magnitudes of the quantities $S/2$ and U. The case $S/2 > U$ leads to a single equilibrium associated with both plants developing lateral root systems. In fact, in this case L is a dominant strategy for each plant, being best against both strategies of the other plant. The more interesting case comes about when we have $S/2 < U \leq S$.

If $S/2 < U$, the game has two equilibria: (T,L) and (L,T); i.e., whatever one plant does, the other does the opposite. However, since the game is symmetric, neither the pure strategy "always build lateral" nor the pure strategy "always build tap root" can be ESS, as neither of these strategies is the best reply to itself. Thus the ESS is a mixed strategy. Let x be the fraction of the time the plant "plays" L, with $(1 - x)$ being the fraction of the time the plant plays T. Then for $(x, 1 - x)$ to be an ESS, we must have

$$\left(\frac{S}{2}\right) x + (1 - x)S = U,$$

which leads to

$$x = 2\left(1 - \frac{U}{S}\right).$$

Thus, the strategy play L a fraction $2(1 - U/S)$ and T a fraction $2U/S - 1$ of the time is uninvadable.

It's of some interest to derive the above result from first principles using the replicator dynamics discussed in earlier sections. If we let x_1 be the fraction of the plants playing the strategy L, with x_2 being the fraction playing T, the interaction matrix is

$$A = \begin{pmatrix} S/2 & S \\ U & U \end{pmatrix},$$

leading to the first-order replicator system

$$\dot{x}_1 = x_1 \left[S\left(\frac{x_1}{2} + x_2\right) - \frac{Sx_1^2}{2} - Sx_1x_2 - Ux_2 \right],$$

$$\dot{x}_2 = x_2 \left[U - \frac{Sx_1^2}{2} - Sx_1x_2 - Ux_2 \right].$$

Using the fact that $x_1 + x_2 = 1$, we can eliminate x_2 from the foregoing system arriving at the final system for $x \doteq x_1$ as

$$\dot{x} = x \left[(S - U) - x(S - U) + \frac{S\left(x^2 - x\right)}{2} \right].$$

It's easy to verify that the point $x^* = 2(1 - U/S)$ is an equilibrium point of the above system. Furthermore, it's a straightforward matter to see that x^* is a globally stable attractor, thereby satisfying both the equilibrium and stability conditons for an ESS.

The assumption that a plant has only a single competing neighbor is somewhat unrealistic, so let's consider a situation in which a given plant is equidistant from a set of neighbors. This means that each plant in the region is at the center of a hexagon whose vertices are occupied by similar plants. Thus, each shrub has six neighbors, and competition for surface water is with those of the six that choose the lateral root strategy. Competition for underground water is global, with the amount of underground water available assumed to be $U(1 - uq)$, where U is the amount that would be available if there were no competitors for underground water, q is the proportion of tap root players in the population and u is a parameter denoting the relationship between the number of competitors and water availability in the underground reservoir.

In the above many-player game, we cannot analyze the strategies in payoff matrix form. However, if the game consists of a large number of players, we can replace the payoff matrix with the payoff function $W(I, J)$, which represents the expected change in fitness received by a plant playing strategy J in a population of I players. Using this idea, it can be shown that the strategy $(x, 1 - x)$, with

$$x = \frac{(S/U) + u - 1}{u + (S/U)/2},$$

is ESS. As expected, this strategy reduces to the earlier case when $u = 0$.

12. *ESS vs. Optimality*

It's often asserted in ecological analyses that natural selection has arranged matters so that the population is driven to a state in which the resources are optimally exploited. Here we want to indicate briefly how this argument can break down in situations in which the selection mechanism is frequency-dependent, as in the foregoing desert plant example.

The goal of a shrub population, in the sense of species optimality, is to adopt a strategy J that permits the maximum water uptake per individual. If p represents the fraction of the population that adopts the lateral root strategy, the total water uptake arising from considering the four possible interactions between tap and lateral root players is

$$f(p) = U - pU + pS - \tfrac{1}{2}p^2 S.$$

The shrubs should choose p in an attempt to maximize this function. It's easily seen that the maximum for this function is attained at the value

$$p_{\max} = 1 - \frac{U}{S}\,.$$

Thus, maximum water uptake per individual occurs if the probability of adopting a lateral root strategy is $1 - U/S$. On the other hand, the ESS probability of adopting a lateral root system is 1 if $U < S/2$ and $2(1 - U/S)$ if $U > S/2$. So ESS analysis predicts that *twice* as many plants will adopt a tap root strategy than would be suggested by optimality considerations.

The two criteria, ESS and optimality, also differ considerably in their predictions about the actual amount of water uptake achieved by the two strategies. In the case when $U < S/2$, the ESS strategy predicts a water uptake of $S/2$; on the other hand, using the optimality strategy we obtain a water uptake of $(S^2 + U^2)/2S$. Thus, the water uptake per individual as predicted by the optimality criterion solution is consistently higher than that from the ESS solution, and can be as much as 25% greater depending upon the particular value of U/S.

The main point to note about the optimality vs. ESS results is that there can be a significant difference between the predictions based upon one or the other criterion. Which criterion is "correct" cannot really be settled by armchair analysis alone but is dictated by the considerations of the problem and, most importantly, by the experimental data. So, for instance, in the desert shrub example above, the difference between the two criteria in predicted numbers of shrubs using a lateral root system is so large that it should be a relatively straightforward matter to construct a laboratory experiment to test whether Nature is using an ESS or an optimality selection criterion. Now let's turn to another type of plant growth problem illustrating some of the same features seen above but in a context where there can be a *continuum* of strategies.

13. *Flowering Times for Plants under Competition*

Consider an annual flowering plant having a growing season of length T. During the growing season, the plant can devote energy to either vegetative growth or to seed production. Under the hypotheses of a single plant in the absence of competition, it has been shown that the optimal strategy for maximizing seed production is to grow only vegetatively up to a time $u < T$ and to then shift over and devote the remainder of the time $T - u$ solely to seed production. Under plausible assumptions, this leads to an optimal crossover value of

$$u^* = T - \frac{1}{RL},$$

where R is the net photosynthetic production per unit leaf mass, and L is the ratio of leaf mass to remaining vegetative mass. The total seed production in this case will be

$$H(u) = (T - u)RA_0 \exp(RLu),$$

where A_0 is the initial leaf biomass of the plant.

Here we are interested in the more general case where there are many such plants in competition. Let S_i be the seed production of the ith plant in the absence of competitors, with u_i being the time when the ith plant switches from vegetative to reproductive growth. In this situation, we have

$$S_i = (T - u_i)f(u_i),$$

where $f(\cdot)$ is a function that measures the plant's ability to produce seeds given that flowering starts at time u_i. Since $(T - u_i)$ represents the time remaining in the growing season to produce seeds, if $f(0)$ is small and $f(u_i)$ increases rapidly with u_i, the maximum value for S_i will occur between 0 and T. Further, it's reasonable to suppose that $f'(u_i) > 0$, since $f(u_i)$ is a measure of the plant's size and photosynthetic capability. Note that we assume the plant population to be homogeneous; i.e., the function f is the same for all plants in the population.

Now let's assume there are two phenotypes in the population: Type 0 with a flowering time U_0, and Type 1 with a flowering time U_1. Let's postulate that the generating function for the average fitness of the two phenotypes is given by

$$G = \frac{(T - u_i)f(u_i)}{1 + W(U_0, U_1, u_i, P_0, P_1, n)},$$

where P_0 and P_1 are the initial populations of the two phenotypes, and n is the total number of plants (players) in the population. The function W

represents the effect of competition and is zero when $n = 1$. Furthermore, it should be possible for one phenotype to diminish the competitive effect of the other by flowering earlier or to intensify the competitive effect by postponing its flowering time. These considerations lead to the conditions

$$\frac{\partial W(s)}{\partial u_i} < 0, \qquad \frac{\partial W(s)}{\partial s} > 0.$$

In the first expression, the partial derivative is evaluated at $s = U_0 = u_i$, but in the second expression s is substituted for U_0, U_1 and u_i before taking the partial derivative. In what follows, it's convenient to introduce the function $E(s)$ as

$$E(s) = (T - s)\frac{df(s)}{du_i} - f(s).$$

We now examine flowering times using the average fitness function G under the conditions of maximizing seed production under no competition, and under competition when we desire an ESS. Finally, we compare both results with the solution obtained for a community of plants experiencing competition, where *community* seed production is maximized.

In the case of no competition, $W = 0$ and if we set $\partial G/\partial s = 0$, we obtain the condition $E(s) = 0$ or, equivalently,

$$\frac{df(s)}{ds} = \frac{f(s)}{T - s}.$$

This condition for optimal seed production is similar to a well-known result in economics associated with the cost of production: by producing more of a product, the average cost of production goes down, but the marginal cost of production goes up. Cost is minimized at the point where marginal and average costs are equal. For plants, seed production is maximized when the marginal rate of increase in seed production equals the average rate of seed production taken with respect to the time remaining in the growing season. This relationship is expressed graphically in Fig. 6.1 where the average and marginal rates of production are displayed, with their crossover point being the optimal time of flowering s^*.

Now let's assume competition ($n > 1$) and calculate the ESS strategy. The necessary condition for an ESS strategy is that

$$\frac{\partial G}{\partial u_i} = 0.$$

Computing this quantity gives

$$\frac{\partial G}{\partial u_i} = \frac{(1 + W)[-f + (T - u_i)(\partial f/\partial u_i)] - (T - u_i)(\partial W/\partial u_i)f}{(1 + W)^2}.$$

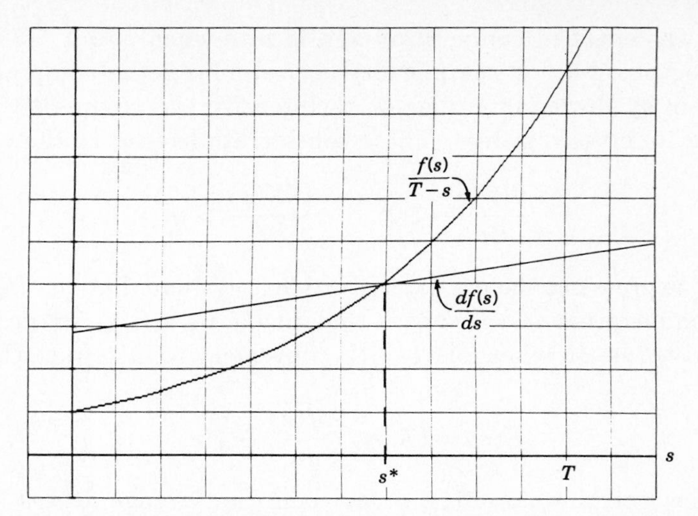

Figure 6.1 The Average and Marginal Rates of Seed Production

Substituting s for U_0, U_1 and u_i in the above terms, and setting the resulting expression to zero, yields

$$E(s) = \left[\frac{(T-s)f(s)}{1+W(s)}\right] \frac{\partial W(s)}{\partial u_i} = G(s)\frac{\partial W(s)}{\partial u_i}\ .$$

From the assumptions on W, this means $E(s) < 0$ at an ESS solution. And from the definition of $E(s)$ this requires that

$$\frac{df(s)}{ds} < \frac{f(s)}{T-s}\ .$$

Comparing this result with that obtained under no competition, we find that the above inequality is satisfied only if the ESS flowering time is *greater* than s^*. That is, the plants should begin to flower later when faced with competition. Just as in our desert shrub example, it can be seen that the number of seeds produced per plant under competition will always be less than the number produced in a noncompetitive environment.

In order to obtain the optimal *community* seed production level, we must have

$$\frac{\partial G(s)}{\partial s} = 0.$$

We compute this expression using the function G as above, obtaining

$$E(s) = \frac{\partial W(s)}{\partial s}\left[\frac{(T-s)f(s)}{1+W(s)}\right].$$

From the assumptions on W, this requires

$$\frac{df(s)}{ds} > \frac{f(s)}{(T-s)} .$$

This will be satisfied only if the flowering time is *earlier* than s^*, the optimal flowering time in the absence of competition. This solution will yield more seeds per plant than the ESS solution, which is clearly agriculturally desirable. The catch is that this cooperative solution is an unstable solution and, consequently, cannot be naturally maintained.

The foregoing plant examples have shown some of the simpler interconnections between the concepts of ESS, optimality, and the properties of replicator dynamics within the rather simple behavioral context of plant growth behaviors. At this point it's tempting to make the rash speculation that perhaps *human* behavioral patterns are also determined in somewhat the same fashion as for plants and animals, leading to the core of the idea underlying the field now termed *sociobiology.* We now examine the pros and cons of this hypothesis as a stepping stone to consideration of the role of evolutionary concepts within the broader context of cultural and social change.

14. *Sociobiology*

During the past decade or so, there has emerged an amalgamation of the fields of genetics, evolutionary biology, ethology and anthropology termed "sociobiology," which can be broadly construed as the study of the biological basis of behavior in humans and primates. Basically, what this amounts to is the investigation of the degree to which changes at the genetic level *determine* behavioral patterns at the level of the phenotype. In more prosaic terms, are human behavioral patterns "hard-wired in" by the genes? The sociobiologists would claim that the answer is fundamentally yes, and that cultural processes play only a secondary role in the determination of individual propensity to behave in certain fashions. As one might suspect, such a contention strikes a raw nerve in circles devoted to maintaining the position that humans, by virtue of culture, are somehow immune from the principles by which the rest of the animal world operates, and that the sociobiologists by denying this special role of humans only provide ammunition and the weight of science to those who adopt various political and sociological positions of genetic superiority in matters of intelligence, health, industrial productivity, and the like. Other critics have attacked sociobiology on more scientific grounds, stating that there is no recognizable difference between sociobiological theory and traditional Darwinian evolution. In short, sociobiology is nothing more than classical Darwinism dressed up in new clothes. In this brief discussion, we take no position on these controversial emotional and scientific issues, leaving the interested reader to consult the items listed in the Notes and References for a full account of all positions. Our goal here will be merely to give a short introduction to the mathematical relationships involved in bridging the gap between the level of genes and the level of behaviors.

As a discipline, sociobiology is concerned with the implications of natural selection and evolutionary biology on the social behavior of organisms. An animal's behavior can be examined for its effect on the animal's fitness and the resultant changes in the frequency of the genes associated with or "causing" that behavior. From cost/benefit analyses of possible behaviors relative to their effects on fitness and gene frequencies, various models of animal behavior can be derived. The basic behavioral modes that have been examined in this regard tend to fall into four distinct categories:

- *Sex and Mating Systems*—Mating behavior relates to the strategies adopted by males and females in attempting to maximize the likelihood of their genes surviving into the next generation. The majority of these studies focus upon the issue of monogamous vs. polygamous mating strategies.

- *Kin Selection and Inclusive Fitness*—If one accepts the contention that the only important factor in evolution is a change in gene frequency independent of how that change comes about, then it can be argued that a definition of fitness limited to the number of offspring of an individual and his direct descendants is too narrow, since an individual could increase the frequency of his genes in the next generation by either increasing his own fitness or the fitness of his genetic relatives. This idea forms the basis for the theory of kin selection and leads to the concept of inclusive fitness as the sum of the individual and *kin* fitness.

- *Aggression*—Active aggression is a means by which an animal can gain resources and thereby increase its individual fitness. The cost/benefit analysis determines the adaptive value of aggression in any given situation, and studies in this area have been mostly devoted to an analysis of the factors that tend either to increase or decrease aggressive behavior as in our simple Hawk-Dove game discussed earlier in the chapter.

- *Dispersal*—The desire for males to increase their gene frequency often leads to behavior patterns that involve migration, i.e., males leaving their social group and either joining another or carrying on in a solitary fashion. Dispersal studies center upon conditions under which such behavior is optimal, and the form such migratory patterns take for different species.

Since it has been stated that the problem of altruism and kin selection is the central theoretical problem of sociobiology, let's take a closer look at what is involved in the analysis of this class of behaviors.

In the classical Darwinian interpretation, an altruistic act involving the sacrifice of an individual for the sake of the group is difficult to comprehend. Darwinian theory claims that individuals act so as to maximize the chances of their own survival, leading to behaviors that, by definition, can never be altruistic. Thus, the existence of altruistic behaviors stands as an experimental counterexample to the strong Darwinian claim that behavioral adaptations take place so as to maximize individual fitness. As noted above, the idea of kin selection and inclusive fitness provides a framework within which we can make sense of both altruistic acts and evolutionary change by gene frequency modification.

As a simple example of the above kin selection idea, consider the actions of an animal A when he sights an approaching predator. He can either give an alarm call (an altruistic act) and warn three fellow group members, thereby attracting attention to himself leading to predation, or he can remain silent and hide (a selfish act), thereby leading to his safety but to the loss of the others in the group. Should A give the alarm call or not? Let's assume that r represents the level of genetic relatedness between A and the group members. This number represents the proportion of genes that two individuals share and are common by descent. In sexually reproducing species, without considering inbreeding, if the group members are full siblings of A, $r = \frac{1}{2}$. Thus, if A gives the alarm call we have the evolutionary equation

$$(\tfrac{1}{2})A + (\tfrac{1}{2})A + (\tfrac{1}{2})A - A = +(\tfrac{1}{2})A,$$

but if A does the selfish thing and hides, we have

$$A - (\tfrac{1}{2})A - (\tfrac{1}{2})A - (\tfrac{1}{2})A = -(\tfrac{1}{2})A.$$

Clearly, the *inclusive* fitness of A is greater if he gives the call, so he should do so. However, it's easy to see that if the group members were only *cousins,* so that $r = \frac{1}{8}$, then it would be best for A not to give the call. These simple arguments lead to the altruistic inequality $K > 1/r$, expressing the fact that K, the ratio of cost to benefit, must be greater than the reciprocal of the genetic relatedness between the recipient and the beneficiary of the act in order for an altruistic act to be evolutionarily selected. Despite the obvious artificiality of this example, there are numerous situations in nature where kin selection has been found to be a dominant behavioral mode. Now let's shift over to math mode and take another look at the issue of aggression, but this time by explicitly linking together the genetic level with the phenotypic.

In the genetic model considered earlier in the chapter, we assumed that selection operated through viability differences between genotypes, with viability meaning survival probability from zygote to adult. Let's now expand upon this model by also taking into account fertility differences between mating pairs.

Let x_{ii} represent the frequency of the homozygote genotype A_iA_i, and let $2x_{ij}$ be that of the heterozygote $A_iA_j, i \neq j$, so that

$$\sum_{i,j} x_{ij} = 1.$$

Let $x_i = \sum_j x_{ij}$ be the frequency of allele A_i in the gene pool, and let $w(ij, st)$ denote the average fecundity (number of offspring) of a mating between an ij-male and an st-female. Assuming random mating, the frequency of the genotype A_iA_j in the next generation is

$$x'_{ij} = \frac{1}{2\phi} \sum_{s,t} [w(is, jt) + w(jt, is)] x_{is} x_{jt},$$

with ϕ a normalizing constant that corresponds to the mean fecundity of the population.

This relation is far too general to say anything interesting about, so let's further assume that the fertility of a couple can be factored into a male and a female contribution. Thus,

$$w(ij, st) = m(ij) f(st), \qquad 1 \leq i, j, s, t \leq k.$$

So the average fecundity of gene A_i in the male gene pool is

$$M(i) = \sum_j m(ij) x_{ij},$$

whereas the average fecundity of gene A_j in the female pool is

$$F(j) = \sum_k f(jk) x_{jk}.$$

This leads to the differential equation for the genotype A_iA_j as

$$\dot{x}_{ij} = \tfrac{1}{2}[M(i)F(j) + M(j)F(i)] - x_{ij}\phi,$$

with

$$\phi = \sum_i M(i) \sum_j F(j).$$

If we further assume symmetry between the sexes so that $m(ij) = f(ij)$, the above relation simplifies to

$$\dot{x}_{ij} = M(i)M(j) - x_{ij}\phi,$$

with

$$\phi = \left(\sum_i M(i)\right)^2 .$$

Now we return to the sociobiological situation and consider how the above genetic changes affect behaviors. Assume the behavior is determined by a single gene locus with alleles $A_1, \ldots, A_k$. The payoff, which is dependent on the strategy employed and on that of the competitors, must somehow be related to reproductive success. There are many ways to express this, one of the simplest being to assume that the payoff is independent of sex and that the number of offspring of a given couple is proportional to the product of parental payoffs. Such a situation would correspond, for example, to fights that are not sex-specific like competition for food.

Under the foregoing assumption, to each genotype A_iA_j there corresponds one of the strategies $E_1, \ldots, E_n$ or, more generally, a mixed strategy

$$P(ij) = (p_1(ij), \ldots, p_n(ij)) \in S^n.$$

The frequency of strategy E_i in the population is then given by

$$b_i = \sum_{st} p_i(st) x_{st},$$

with the average payoff for strategy E_k being

$$a_k = \sum_{\ell} a_{k\ell} b_\ell,$$

where $\{a_{k\ell}\}$ is given by the payoff matrix. Thus, the fitness for the genotype A_iA_j is

$$m(ij) = \sum_k p_k(ij) a_k = \sum_{k\ell st} a_{k\ell} p_k(ij) p_\ell(st) x_{st}.$$

If we assume that the number of offspring of an A_iA_j male/female is proportional to the fitness $m(ij)$, we are led to exactly the same dynamics for x_{ij} as above—with the significant difference that now the quantity $m(ij)$ is not a constant as before, but is linearly dependent upon x_{ij} as in the above expression. Hence, the relation for the gene frequency change $\dot{x}_{ij}$ is quartic in the genotype frequencies.

The above equations can be used to analyze the Hawk-Dove game considered earlier in the chapter. Such an analysis shows that the genetic constraints considered here may permit the establishment of an evolutionary stable equilibrium in the population, with the corresponding mixture of strategies being exactly that arising from the stable equilibrium of the above dynamical equations. However, it should be noted that there are cases in which the genetic structure *prevents* the establishment of evolutionary stable equilibria. But in most situations the genetic fine structure does not conflict with the purely game-theoretic analysis.

We have already noted that a central question in sociobiology is to explain how cooperative behavior (altruism) can emerge at the individual level as the optimal survival act in a given situation. The preceding arguments suggest that the concepts of kin selection and inclusive fitness provide a basis for such an explanation at the genetic level. But it's also of considerable interest to consider the same question from the purely phenotypic level, and look at how we might see cooperative behavior emerge in a situation in which the *rational* strategy of any individual is to act selfishly in any interaction with another individual.

A stylized version of such a situation is provided by the Prisoner's Dilemma game considered earlier, where we saw that the rational choice for each player was to defect, leading to a nonoptimal *joint* outcome for each. This noncooperative outcome was obtained under the crucial assumption that the game would be played only once. We now examine the means by which cooperation may emerge under the conditions that the players will play the game many times having the opportunity to learn from past interactions about the strategy employed by the other in the population. As will be demonstrated, the forcing of continued interactions dramatically changes the structure of the Prisoner's Dilemma situation, offering the possibility for cooperative actions to emerge naturally from selfish motives. The material of the following section amply justifies the attention we have given to the Prisoner's Dilemma and its pivotal role in bringing game theory into contact with social behavior.

15. *The Evolution of Cooperation*

Consider again the Prisoner's Dilemma with the payoff matrix

		Player I	
		C	D
Player II	C	(R,R)	(S,T)
	D	(T,S)	(P,P)

where the choices of each player are either to Cooperate (C) or Defect (D). The payoffs are assumed to satisfy the ordering

$$T > R > P > S,$$

where T is the "temptation" payoff associated with defecting when the opponent cooperates, R is the "reward" obtained when both players cooperate,

P is the "punishment" attached to a pair of defecting players, and S is the "sucker's" payoff received by a player who cooperates when the other defects. As we have already seen, when the game is played only once the rational strategy for each player is defection, leading to the suboptimal joint punishment payoff P to each player. Now we change the ground rules and assume the same players are to engage in a sequence (finite or infinite) of plays of the Prisoner's Dilemma game, with each player's total payoff being the sum of his payoffs from each play. This is the so-called *iterated Prisoner's Dilemma,* and our interest will be in identifying circumstances under which joint cooperation may emerge as the rational strategy when the players know they will have to face each other many times. In such iterated situations, we make the additional assumption that

$$\frac{T+S}{2} < R,$$

ensuring that if the two players somehow get locked into an "out of phase" alternation of cooperations and defections, each will do worse than if they had both cooperated on each play.

Assume the game will be played a *finite* number of times N, and that N is known to both players *before* the game begins. Then it's easy to dispose of this case since it's clear that the optimal choice at the last stage N is for each player to defect; hence, one then faces an $(N-1)$-stage process where again the optimal choice is defection at play $N-1$, and so on. Consequently, in an iterated Prisoner's Dilemma for which the number of plays is finite and known in advance, the optimal strategies are just as in the "single-shot" game: always defect! So the only possibility for cooperation to emerge is when the number of interactions is potentially infinite or, what is the same thing, when after any play is completed, there is a nonzero probability that the game will continue for another round.

Assume there is a fixed probability w that the game will continue another round after the current round of play is completed. In the biological context, this parameter measures the likelihood of two individuals (or behavioral phenotypes from the same species) competing again after an initial interaction. Factors such as average life span, mobility and individual health can affect w, which we shall always assume is held constant. The parameter w can also be interpreted as a *discount factor,* enabling us to measure the value of future payoffs against return received from the initial round of play. As a simple example, suppose two players each employ the strategy of always defecting (ALL D). Then they each receive a payoff of P per round of play, so that the discounted total payoff received by each player is

$$V(D|D) = P(1 + w + w^2 + w^3 + \dots) = P\left(\frac{1}{1-w}\right).$$

In the iterated Prisoner's Dilemma, a *strategy* is any rule that tells a player what option to choose (C or D) in light of the entire past history of play by both sides. So, for instance, the selfish rule ALL D is a strategy; so is the slightly more complicated rule: cooperate (C) on the first round, thereafter do whatever the opponent did in his last choice. This is the well-known strategy TIT-FOR-TAT which, as we shall see, plays a central role in analysis of the iterated Prisoner's Dilemma. For any value of w, the strategy of unconditional defection (ALL D) is ESS; however, other strategies may also be ESS. In fact, we can show that when w is sufficiently close to 1, there is no single best strategy. Suppose that the other player is using ALL D. Since the other player will never cooperate, the best you can do yourself is to adopt ALL D. Now suppose that the other player is using a strategy of "permanent retaliation," i.e., cooperating until you defect, and then always defecting after that. In this case, your best strategy is to never defect yourself, provided that the temptation to defect on the first round will eventually be more than compensated for by the long-term disadvantage of getting nothing but the punishment payoff P, rather than the reward R, on future plays. This will be true whenever the discount parameter w is great enough. Consequently, whether or not you should cooperate, even on the first move, depends on the strategy being used by your opponent. This implies that if w is sufficiently large, there is no one best strategy.

Now we turn to the evolution of cooperative strategies that are ESS. This problem can be thought of as consisting of three separate questions:

- *Robustness*—What type of strategy can thrive in an environment composed of players who each use a variety of more or less sophisticated strategies?

- *Stability*—Under what conditions can such a strategy, once established, resist invasion by mutant strategies?

- *Initiation*—Even if a strategy is both robust and stable, how can it ever get started in an environment that is predominantly hostile, i.e., noncooperative?

First of all, to test the kind of strategies that can be robust, Axelrod conducted a well-chronicled computer tournament in which a variety of strategies submitted by game theorists, evolutionary biologists, computer scientists, political scientists and others of this ilk were pitted against each other in a round-robin manner. The result of the tournament was that the highest average score was attained by the simplest of all strategies submitted, TIT FOR TAT. An analysis of the results showed that the robustness of TIT FOR TAT was due to three factors: (1) it was never the first to defect, (2) it could be provoked into retaliation by a defection of the other party, and (3) it was forgiving after just a single act of retaliation. As an added measure of its robustness, as the less successful rules were weeded out of the tournament by better ones, TIT FOR TAT continued to do well with the other rules that initially scored near the top and in the long run displaced all other rules. This set of experiments provides strong empirical evidence that TIT FOR TAT is a robust strategy that can thrive in an environment in which there are many players, each playing different types of strategies.

Once a strategy has become established, the question of how well it can resist invasion by mutants enters into play. It's a rather simple matter to show that once TIT FOR TAT is established in a community of players, it's impossible for any mutant strategy to displace it, provided that the probability of future interactions w is sufficiently large. So if there is a great enough likelihood that two players will encounter each other again during the course of the game, and if the dominant strategy in the community is TIT FOR TAT, then any deviation from this strategy will result in a loss in fitness to the mutant. Thus, TIT FOR TAT is a weak ESS strategy. We term it "weak" because TIT FOR TAT is no better a reply to ALL C than is ALL C itself, only just as good. Therefore, the stability requirement for ESS is only "weakly" satisfied. In what follows, we use ESS in this weak sense. In fact, we can give a precise value for how large w must be in order for TIT FOR TAT to be ESS.

COOPERATION THEOREM. *In the iterated Prisoner's Dilemma, TIT FOR TAT is an ESS strategy if and only if*

$$w > \max\left\{\frac{T-R}{T-P}, \frac{T-R}{R-S}\right\}.$$

We have already noted that TIT FOR TAT is not the only ESS for the iterated Prisoner's Dilemma; ALL D is another such ESS strategy. So how could TIT FOR TAT ever get a foothold in a population initially consisting of all defectors? There appear to be at least two different mechanisms that provide a plausible path whereby TIT FOR TAT could get started, even in the face of a fundamentally hostile environment. The first such mechanism is *kin selection,* a concept that we used earlier to help explain how altruism could emerge in Darwin's universe. Not defecting in a Prisoner's Dilemma game is altruism of a kind since the individual is foregoing returns that might have been taken. So in this way cooperation can evolve if two players are sufficiently closely related. In effect, recalculation of the payoff matrix in such a way that an individual has a part interest in the partner's gain (i.e., computing the payoffs in terms of inclusive fitness) can often eliminate the inequalities $T > R$ and $P > S$, in which case cooperation becomes unconditionally favored. Thus, the benefits of cooperation can be obtained by groups of sufficiently closely related players, and once the genes for cooperation exist, selection will promote strategies that involve cooperative behavior.

Another mechanism through which cooperation can come about in an essentially ALL D environment is *clustering.* Suppose that a small group of individuals is using a strategy like TIT FOR TAT and that a certain proportion p of the interactions of members of this cluster is with others from the cluster. Then the average return attained by members of the cluster if they are all playing TIT FOR TAT is

$$\frac{pR}{1-w} + (1-p)\left(S + \frac{wP}{1-w}\right).$$

If the members of the cluster provide a negligible fraction of the interactions for the remainder of the population, then the return obtained by those using ALL D is still $P/(1-w)$. If p and w are large enough, a cluster of TIT FOR TAT individuals can then become initially viable, even in an environment in which the overwhelming majority are using ALL D. We note, in passing, that clustering is often associated with kinship and that the two mechanisms can often reinforce each other. But it's still possible for clustering to be effective even without kinship.

Example: Sea Bass Spawning

A case of cooperation that fits into the above framework involves the spawning behavior of sea bass. These fish have both male and female sexual organs, form pairs, and can be said to take turns at being the high investment (laying eggs) partner and the low investment (providing sperm to fertilize the eggs) partner. Up to ten spawnings occur in a day, with only a few eggs provided each time. Pairs of fish tend to break up if sex roles are not divided evenly. It appears that such cooperative behavior got started at a time when the sea bass was scarce as a species, leading to a high level of inbreeding. Such inbreeding implies a relatedness in the pairs and this would have initially promoted cooperation without the need for further relatedness.

Both the experimental evidence, as well as the results of the computer tournament pitting strategies against each other, suggest that in order to do well in an iterated Prisoner's Dilemma, a strategy should possess several features that have been summarized by Axelrod in the following form:

- *Nice*—A strategy should never be the first to defect.
- *Retaliatory*—A successful strategy always punishes defections by the opponent.
- *Forgiving*—A good strategy is not vindictive: punishment is meted out to fit the crime, and as soon as the other party begins to cooperate again, the punishment ceases.
- *Optimistic*—A winning strategy will be maximally discriminating in the sense that it will be willing to cooperate even if the other party has never cooperated yet.

The ideas we have touched upon in this chapter only begin to scratch the surface of the manifold complexities and subtle complications involved in using evolutionary ideas in social, psychological and cultural environments only loosely related to biological foundations. Since this is an area of considerable intellectual as well as practical importance, a good part of the next chapter will be devoted to the exploration of this theme. In addition, the reader is urged to consider carefully the Discussion Questions and Problems and Exercises of both this chapter and the next in which many additional aspects of evolutionary theory are treated in more detail.

Discussion Questions

1. The notion of adaptation to the environment plays a key role in the Darwinian picture of evolution, in the sense that the selection criterion is based upon the ability of a phenotype to survive in a given environment. If we *define* evolution to be a change in genetic frequency and employ the foregoing environentally based fitness criterion, are we justified is assuming that evolution will stop in a *constant* environment?

2. In the book *The Battle for Human Nature,* B. Schwartz draws the striking parallels between the arguments of sociobiology involving the "selfish gene" and its concern with its own proliferation—the so-called rational economic agent with his concern for utility maximization—and the picture of pleasure maximization by reinforcement as advocated by behavioral psychologists as the basis for human behavior patterns. It's often argued that these three quite disjoint fields, evolutionary biology, economics, and behavioral psychology, provide a *scientific* basis for the claim that self-interest is a human trait originating in *natural,* rather than *moral* law, and the above sciences are used by many people as a vehicle to explain and justify their actions and the actions of others. Consider the pros and cons of this argument.

3. The *Red Queen Hypothesis* states that any evolutionary change in a species is experienced by coexisting species as a deterioration in their environment. Hence, a species must evolve as fast as it can in order to continue its existence, and if it doesn't evolve as fast as it can it will become extinct. (The terminology comes from the Red Queen in *Alice in Wonderland,* who remarked that "it takes all the running you can do, to keep in the same place.") Does this hypothesis seem plausible to you? Imagine that evolution is of the Red Queen type, continuing forever in a physically stable environment. Does it then follow that increasingly more complex forms (defined in some appropriate way) tend to evolve?

4. The following payoff matrix represents a simplified version of the situation governing the Cuban Missile Crisis of 1962.

		USSR	
		Withdraw	Maintain
USA	Blockade	(3,3) (Compromise)	(2,4) (USSR victory)
	Air strike	(4,2) (US victory)	(1,1) (Nuclear war)

a) Does this represent a mixed-motive game? If so, what type is it, i.e., Chicken, Prisoner's Dilemma, etc?

b) What were the strategies actually employed to settle the crisis?

c) Construct additional "games" of this sort to represent other types of political conflict. For example, the everlasting Middle East situation, the crisis in Lebanon, the Turkish-Cypriot conflict, or the Nicaraguan war.

d) Referring to the discussions of Chapter Four, try modeling these crises using catastrophe theory arguments. Then compare your models and conclusions with those obtained using more conventional game-theoretic approaches.

5. Adam Smith claimed that the individual pursuit of self-interest in economic transactions promotes the welfare of society more effectively than by a directed effort to do so. Smith's view was that the "invisible hand" of the economy would set things in order and that no concerted collaborative effort would be needed. Consider this view in relation to the situation in collective bargaining in which it is in the interest of an individual union to achieve a wage settlement that is in excess of the inflation rate, *irrespective* of whether other unions exercises restraint in their own wage demands. However, if all unions adopt this selfish policy the prices of goods and services go up, and everyone is worse off than if they had all exercised restraint. Thus, the invisible hand becomes the invisible claw that can tear the entire fabric of the economy apart. Discuss this situation as a case of an *n-player Prisoner's Dilemma.* Can you think of other situations in which the n-person Prisoner's Dilemma accurately reflects a social conflict situation?

6. Classical equilibrium-centered economics regards the operation of a firm as a sequence of choices taken to maximize some global objective function characterizing the firm's overall "profitability." This is again Adam Smith's world stepped up from the individual level to the level of the firm. Recently, spearheaded by the work of Nelson, Winter, and others, an alternative *evolutionary* view of the firm has been proposed, which emphasizes

the firm's operation as the execution of a set of "routines" that comprise the genotype of the firm. In this theory, firms make incremental changes (mutations) from time to time in their operating routines, with successful mutations being "selected" according to various criteria of profitability or ability of the firm to compete in its market niche. Compare the similarities and differences between evolution in a biological sense and evolution in the economic terms considered in this view of industrial operation. What are the relative strengths and weaknesses between firms operating according to Adam Smith's prescription and those carrying out their activities in an evolutionary mode? Are there any points of contact between the two theories? How would you go about formalizing such an evolutionary theory of economic processes using replicator dynamics?

7. In Spengler's view of historical change, as well as the cyclic pattern of the rise and fall of civilizations described by Toynbee, evolutionary concepts occupy a central position. Recently, Colinvaux has outlined a similar view emphasizing the biologically-based idea that the engine of historical change is the desire by humans to broaden their "eco-niche"—defined as a set of capabilities for extracting resources, surviving hazard and competing with others in a given environment. The primary thesis of this view is that human demography can be explained by the process of regulating family size in accord with both perceived niche and available resources. A corollary of this theory is that society is then divided into "castes" in which those of high caste occupy a broader niche than those of lower castes. As population grows, there are only a few ways in which higher-quality niches can be found: technical innovation, trade, colonies, empire building and oppression. Thus, the historical record is interpreted as being the result of group phenomena arising from the activities of individuals acting independently under the dictates of natural selection.

a) Compare this theory of historical change with the ideas underlying the sociobiological view of human behavior.

b) The ecological view of history seems to downplay the "great man" theory which asserts that the moving force for great historical change is attributable to the doings of a few individuals (Alexander the Great, Napoleon, Hitler, etc.). What role could the great men play in an evolutionary view of history? How would you incorporate the role of such dominant figures into a mathematical model?

c) Is such an ecological view of history *falsifiable;* i.e., can you think of any experiments that could be performed using the historical record that would confirm or deny the evolutionary picture of historical change? (*Remark:* It was the lack of such experimentally testable hypotheses that led the philosopher of science Karl Popper, to condemn the historical views of

Marx, Spengler, and others as being nonscientific. See Chapter Nine for a fuller account of these matters.)

8. The term "culture" is generally assumed to include those aspects of thought, speech, behavior and artifacts that can be learned and transmitted. Does the Darwinian equation

$$\text{adaptation} = \text{mutation} + \text{selection},$$

appear to be relevant to the problem of cultural change? How would you define the analogues of "adaptation," "mutation" and "selection" in the cultural context? Do you see any significant differences between the way phenotypes change in a biological sense, and the way that cultural patterns in society shift during the course of time? Consider the change in language as a specific example in which new words emerge and enter the language of a particular social group, while old words become archaic and eventually disappear.

9. The British biologist Richard Dawkins has argued eloquently the position that the basic unit upon which natural selection operates is not the phenotype, but rather the gene itself. This "selfish gene" theory assumes that the body is just a vehicle that the genes commandeer as survival machines, and the conventional interpretation of the phenotypic fitness as the measure of adaptability is only a means to the greater end of the genes' propagation into the next generation.

a) Consider the merits of the selfish gene theory as compared with the traditional Darwinian view of the organism as the basic unit of selection.

b) How can you explain the emergence of altruism in a selfish gene context?

c) Dawkins also introduced the notion of "memes" as units of cultural transmission, or units of *imitation,* that would propagate themselves through a "meme pool" in much the same ways that genes propagate themselves through a gene pool, spreading ideas, catchwords, cooking receipes, techniques for making houses, and the like, thereby acting as agents of cultural evolution. Compare this idea of a cultural meme with that of a biological gene, in both structure and mode of transmission. How would you go about trying to model such a cultural transmission process using replicator dynamics?

10. An idea closely related to Dawkins' memes is the concept of a *sociogene,* as postulated by Swanson. His view is that both a biological gene and a sociogene are carriers of information.

a) If we think of the information content of a gene as being coded into its DNA, in what manner could the cultural information of a sociogene be carried?

b) Information has many attributes: source, replication, transmission, expression, mutation, etc. Compare these attributes for both "biogenes" and sociogenes. Are there any significant differences?

c) In a viral infection of a living cell, the invading virus parasitizes the genetic machinery of the host cell, inducing it to make copies of itself rather than copies of the cell itself. Does the spread of a "fad" like the hula hoop, designer jeans or Bruce Springsteen songs correspond to a kind of cultural infection of this sort? If so, how would the virus spread, and what would it take to get rid of the "bug?"

11. In earlier chapters, we've considered a number of alternative formulations for expressing the *complexity* of a dynamical process. It's often claimed that the very nature of evolution requires that the replicators evolve from simpler to more complex forms or face extinction. How could you formulate a program for testing this hypothesis using the replicator equations of the text? Do you think it makes any difference which concept of complexity you use, depending upon whether you're characterizing evolutionary processes at the biological, cultural or historical level? Do you think your formal theory will discern any measurable difference between the complexity of a human and an ape? Or a whale? Or a bird?

12. Evolution is an historical process, a unique sequence of events. Theories about evolution come in two types:

A) General theories that say something about the mechanisms that underly the whole process.

B) Specific theories that account for particular events.

It's often held that theories of Type B are untestable, hence unscientific, since it is impossible to run the historical process again with some one factor changed to see whether the result is different. Such theories are unfalsifiable, in a Popperian sense. Do you think a strict adherence to the criterion of falsifiability is valid as a means to assess the merits of a specific evolutionary theory (or model)? Can you think of any simple models of evolutionary behavior that are heuristically valuable but yet untestable? Is testability only relevant when we want to use the models to say something about a real-world situation?

13. Consider the following statements about altruism:

A. Within each group of individuals, altruists are at a reproductive disadvantage compared with nonaltruists.

B. In the ensemble of groups (within which the group selection process takes place), whether altruists have a higher average fitness than nonaltruists is a contingent empirical matter not settled by statement A.

These two statements appear to be contradictory. How can you reconcile them? (*Hint:* Consider the difference between two events being *causally* related as opposed to being *correlated.*)

14. Darwinian evolution predicts a continuous change of phenotypic form as a result of genetic mutation and natural selection; the fossil record, on the other hand, shows strong evidence of discontinuities in the pathway from ancient forms to those observed today. The "punctuated equilibrium" theory of evolutionary change postulates that the pace of evolution moves in fits and spurts, a far cry from the steady, if boring, pace of the Darwinian picture. If the "punctuationist" vision is correct, what factors could you identify that would account for the long periods of stasis, followed by the rapid emergence of new species? Do you think that mass extinctions, like that which killed off the dinosaurs 65 million years ago, form a consistent part of the punctuationist view? Do you think that the catastrophe theory concepts discussed in Chapter Four could be used to formally account for the main features of the punctuated equilibrium view of evolutionary change?

15. The selection mechanisms in most evolutionary models are based upon optimization ideas of one sort or another. Mathematically, all optimization problems involve the same set of basic components:

A) A state set consisting of the variables of the problem.

B) A strategy set describing the admissible courses of action that can be taken.

C) A performance function that enables us to measure the "goodness" (or fitness) of any particular strategy.

D) A constraint set specifying various physical and mathematical limitations on the strategies that can be pursued.

In evolutionary models, there are serious difficulties in specifying completely each of the above components. Discuss these difficulties within the specific context of some of the examples given in the text. Under what circumstances do you think that optimization models would be useful in analyzing evolutionary change?

16. Conventional Darwinian evolution is a fundamentally *reductionistic* theory, the principal thesis being that atomistic changes at the genetic level somehow percolate upward to the phenotypic form, at which point the inexorable pressures of natural selection can work to separate the winners from the losers in the gene pool. Recently, Sheldrake has postulated a more *holistic* procedure for the emergence of physical forms by the medium of what the geneticist Conrad Waddington termed a *morphogenetic field.* Basically, the idea is that there exist fields of potential for biological form just as there are fields for other physical quantities such as gravity and electromagnetism.

These fields shape the forms that both living and nonliving objects assume, with the initial form emerging more or less by chance (Sheldrake's theory is a little vague on this point). Thereafter, once a particular form has gained a foothold, those forms that have been successful in the past have a greater chance of being chosen in the future for organisms of the same species. Such forms are termed "morphogenetic fields," and bear the same role to the material components of biological organisms that, say, the electromagnetic field bears to an electron in physics.

a) Can you think of any experimental tests that could be performed to test Sheldrake's theory?

b) In physics we have equations like Maxwell's field equations, the Schrödinger equation and Einstein's equations to describe the various fields of importance. All of these equations implicitly assume that the fields they describe arise out of material entities like electrons and other sorts of stellar materials. Assuming Sheldrake's theory is correct, it would follow that there must exist analogous field equations describing how biological matter would physically organize itself into various shapes. What mathematical form do you think such equations would take?

c) If morphogenetic fields do indeed exist, how would one go about measuring them?

d) Do you think there could be morphogenetic fields for such nonphysical structures as information patterns (languages, fashions, tunes, etc.)?

17. The French ethnologist Claude Lévi-Strauss has distinguished between two basic forms of human societies, the "clockworks" and the "steam engines." The clockwork societies live practically historyless in a sociocultural equilibrium without evolution of structures. The steam-engine societies, in contrast, undergo vivid evolution such as in most modern industrial countries. According to Levi-Strauss, the distinction between them is due to writing. Explain how you think writing could have come about as an evolutionarily adaptive trait. Within the framework of your explanation, describe why all societies haven't adopted written language. Does conventional evolutionary theory give a credible account of this phenomena?

18. Many species have approximately equal numbers of males and females, though they could produce just as many offspring in each generation if there were fewer males, each fertilizing several females. Why are there so many males?

19. In the fascinating book *The Evolution of Cooperation,* Axelrod gives the following precepts for improving cooperation in iterated Prisoner's Dilemma situations:

- *Enlarge the Shadow of the Future*—Make sure the future is sufficiently important relative to the present.

• *Change the Payoffs*—Rearrange the payoffs so as to decrease the incentives for double-crossing your partner.

• *Teach People to Care about Each Other*—Cooperation can begin, even in a hostile environment, if a sufficient degree of altruism is present.

• *Teach Reciprocity*—Insist that defectors be punished and cooperators be rewarded.

• *Improve Recognition Abilities*—Teach techniques for recognizing players that you have interacted with in the past, so that it will be possible to employ forms of reciprocity.

a) Discuss how you would apply these principles to the Arms Race situation involving the United States and the USSR. Do you think this sort of analysis would be appropriate for the Cuban Missile Crisis discussed earlier in Discussion Question 4?

b) In America's early days the members of the Congress were known to be quite unscrupulous and deceitful. Yet, over the years, cooperation developed and proved stable. Explain how this could have come about using the principles of "cooperation" theory.

Problems and Exercises

1. Consider the two-person game G with payoff matrix $A = (a_{ij})$, where G is assumed to be *zero-sum,* i.e., if Player A uses strategy i and B uses strategy j, the payoff to A is a_{ij}, while the payoff to B is $-a_{ij}$, $i = 1, 2, \ldots, n$; $j = 1, 2, \ldots, m$.

a) Let A's mixed strategy be given by $x = (x_1, x_2, \ldots, x_n)$, i.e., x_i is the probablility that A plays pure strategy i, $\sum_i x_i = 1$. Show that if all $a_{ij} \geq 0$, x can be determined as the solution to the *linear programming* problem

$$\min \sum_{i=1}^{n} x_i,$$

subject to the constraints

$$\sum_{i=1}^{n} a_{ij} x_i \geq 1, \qquad x_j \geq 0, \qquad j = 1, 2, \ldots, m.$$

b) If some $a_{ij} < 0$, show how to modify the payoff matrix to obtain the same reduction to an LP problem.

c) What LP formulation could you use to determine B's optimal mixed strategy?

d) Can this solution procedure work for determining the ESS for mixed-motive games? That is, can we also reduce the determination of the solution to a nonzero-sum game to a special linear programming problem?

2. In the text, we have considered only continuous-time replicator dynamics.

a) Show that if replicators change in *discrete-time*, the corresponding dynamical equations are

$$x_i' = x_i \left[\frac{f_i(x)}{\Phi}\right],$$

where the normalization term Φ is given by

$$\Phi = \sum_{i=1}^{n} x_i f_i(x),$$

as in the continuous-time case.

b) In discrete-time, the genetic model discussed in the text is

$$x_{ij}' = \frac{1}{2\phi} \sum_{s,t} [w(is, jt) + w(jt, is)] x_{is} x_{jt},$$

for the frequency of genotype $A_i A_j$ in the next generation. Assuming multiplicative fecundity and symmetry between the sexes as in the text, we have

$$w(ij, st) = m(ij)m(st).$$

Letting

$$M(i) = \sum_j m(ij) x_{ij},$$

we obtain the dynamics

$$x_{ij}' = \frac{1}{\phi} M(i)M(j),$$

with

$$\phi = \left(\sum_i M(i)\right)^2.$$

If $x_i' \doteq \sum_j x_{ij}'$, show that $x_{ij}' = x_i' x_j'$. Hence, show that after one generation the mean fecundity ϕ increases. Is this result also true for the continuous-time dynamics?

c) How would you use the above result to quantitatively investigate the Red Queen Hypothesis considered in Discussion Question 3?

3. The classical Lotka-Volterra equation describing the predator-prey interactions among n species is given by

$$\dot{y}_i = y_i \left(c_i + \sum_j b_{ij} y_j \right), \qquad i = 1, 2, \ldots, n.$$

Such equations "live" on R^n_+, and generally do not satisfy replicator equations. However, show that by setting $y_n \equiv 1$, *relative* densities, defined by the barycentric transformation

$$x_i = \frac{y_i}{\sum_{j=1}^n y_j}, \qquad i = 1, 2, \ldots, n-1,$$

do satisfy the replicator dynamics

$$\dot{x}_i = x_i[(Ax)_i - \Phi], \qquad x \in S^n \setminus \{x : x_{n-1} = 0\}, \qquad i = 1, 2, \ldots, n-1,$$

with $a_{ij} = b_{ij} - c_i$.

4. Let $p \in \text{int}\, S^n$, the unit simplex in R^n. The *Shahshahani metric* on S^n is defined by the inner product

$$< x, y >_p = \sum_i \frac{1}{p_i} x_i y_i.$$

a) Show that the inner product defined above satisfies the conditions for a metric, i.e., $< x, y >_p \geq 0$, with equality only for $x = y$, $< x, y >_p = < y, x >_p$, and the triangle inequality $< x, y >_p \leq < x, z >_p + < z, y >_p$.

b) Let $V\colon R^n \to R$ be a smooth function. Define the *Shahshahani gradient* of V to be

$$< \text{grad } V(p), y >_p \doteq DV(p)y,$$

for all $y \in T_pS^n$, the tangent space to S^n at p, where $DV(p)$ is the derivative of V at p. Using the fact that $y \in T_pS^n$ if and only if $y \in R^n$ satisfies $\sum_i \dot{y}_i = 0$, show that the replicator equation

$$\dot{x}_i = x_i[f_i(x) - \Phi], \qquad i = 1, 2, \ldots, n,$$

is a Shahshahani gradient of V if and only if f is equivalent to grad V, in the sense that there exists a function $c\colon S^n \to R$ such that

$$f_i(x) - (\text{grad } V)_i(x) = c(x)$$

for all $x \in S^n$ and all $1 \leq i \leq n$.

c) Show that when $V = \frac{1}{2}\sum_{i,j} a_{ij}x_i x_j$ with $a_{ij} = a_{ji}$, the Shahshahani gradient leads to the replicator equations

$$\dot{x}_i = x_i[(Ax)_i - \Phi], \qquad i = 1, 2, \ldots, n,$$

whereas if $V = \sum a_i x_i$, the Shahshahani gradient leads to the equations for haploid organisms

$$\dot{x}_i = x_i(a_i - \Phi), \qquad i = 1, 2, \ldots, n,$$

where x_i is the frequency of chromosome G_i and a_i is its fitness. That is, under the Shahshahani metric, the above replicator systems are *gradient* dynamical systems.

d) Can the bifurcation behavior of the equilibria of the above gradient systems be analyzed using the tools of elementary catastrophe theory developed in Chapter Four?

e) Consider the case when V is a homogeneous function of degree s, i.e., $V(sx) = sV(x)$. Show that in this case $\Phi(x) = sV(x)$ and the average fitness Φ grows at the fastest possible rate with the orbits of the dynamics being orthogonal (in the sense of the Shahshahani inner product) to the constant level sets of Φ.

f) Prove that the first-order replicator equations

$$\dot{x}_i = x_i[(Ax)_i - \Phi], \qquad i = 1, 2, \ldots, n,$$

are obtainable from the Shahshahani gradient of a scalar function V if and only if

$$a_{ij} + a_{jk} + a_{ki} = a_{ji} + a_{ik} + a_{kj},$$

for all indices i, j and k. Show that this is equivalent to the requirement that there exist constants c_i such that $a_{ij} = a_{ji} + c_i - c_j$, $i, j = 1, 2, \ldots, n$.

5. Consider a replicator system of the form

$$\dot{x}_i = x_i[g_i(x_i) - \Phi], \qquad i = 1, 2, \ldots, n,$$

where, without loss of generality, we assume that

$$g_1(0) \geq g_2(0) \geq \cdots \geq g_n(0) > 0.$$

a) Show that the above system is a Shahshahani gradient system.

b) Prove that there exists a number $K > 0$ and a point $p \in S^n$ such that

$$g_1(p_1) = \cdots = g_m(p_m) = K,$$
$$p_1 > 0, \ldots, p_m > 0, \; p_r = 0, \qquad r > m,$$

where m is the largest integer such that $g_m(0) > K$. Show that this implies the system has a unique, global attractor.

c) Prove that the coordinates of the point $p = (p_1, p_2, \ldots, p_n)$ satisfy the relations

$$\lim_{t \to \infty} x_i(t) = p_i, \qquad i = 1, 2, \ldots, n.$$

6. Consider a collection of N species comprising a given ecosystem. Let $\widehat{W}_i$ be the maximal possible fitness of species i in the current environment if the species had all possible favorable alleles, and let $\overline{W}_i$ be the current mean fitness of species i, $i = 1, 2, \ldots, N$. Define the *evolutionary lag load* of species i as

$$L_i = \frac{\widehat{W}_i - \overline{W}_i}{\widehat{W}_i}, \qquad i = 1, 2, \ldots, N.$$

Let β_{ij} represent the increase in the lag load of species i due to a unit change in the lag load of species j.

a) If the average lag load is given by $\overline{L} = (1/N)\sum_i L_i$, show that

$$\frac{d\overline{L}}{dt} = \frac{1}{N}\left[\sum_j (L_j \sum_i \beta_{ij}) - \sum_j L_j\right].$$

Show that this equation has a stationary equilibrium point only if $\sum_i \beta_{ij} \equiv 1$ for all j; otherwise $\overline{L}$ will either decrease (*convergent* evolution) or increase (*divergent* evolution), depending upon whether $\sum_i \beta_{ij} < 1$ (resp. > 1) for most j. This result seems to cast doubt upon the plausibility of the Red Queen Hypothesis, since it implies that there is only a special set of values β_{ij} for which the average lag load remains constant, i.e., for which there is no overall evolutionary change.

b) Show that the above conclusion is based upon the erroneous assumption that the coefficients β_{ij} are constants, *independent* of the number of species N.

c) Consider the number of species N to be a variable, depending upon immigration, speciation, extinction, etc. Let the following model describe the change in average lag load and in number of species:

$$\frac{d\overline{L}}{dt} = (a + b\overline{L} + cN)\overline{L},$$
$$\frac{dN}{dt} = h + (d - e)\overline{L} + (f - g)N,$$

where we assume on physical grounds that $b < 0$, $h > 0$ and $(f - g) < 0$ with no assumptions about the signs of a, c and $(d - e)$. Show that by suitable selection of the coefficients, there can be either a steady Red Queen type of continued evolution, or a stationary state without any evolutionary change. The first corresponds to the gradualistic pattern of Darwinian theory, whereas the second represents a punctualistic pattern of evolution.

7. Consider the childhood game of "Rock–Scissors–Paper" (R-S-P) with the payoff matrix

$$\begin{array}{c} \\ R \\ S \\ P \end{array} \begin{array}{c} \begin{array}{ccc} R & S & P \end{array} \\ \begin{pmatrix} -\epsilon & 1 & -1 \\ -1 & -\epsilon & 1 \\ 1 & -1 & -\epsilon \end{pmatrix} \end{array}$$

Show that if $\epsilon > 0$ (i.e., there is a small positive payment to the bank for a draw), the mixed strategy $(\frac{1}{3}, \frac{1}{3}, \frac{1}{3})$ is ESS, but if $\epsilon < 0$ there is no ESS. What about the usual case when $\epsilon = 0$?

8. We can "soup up" the Hawk-Dove game of the text by adding a third strategy, Retaliator, which behaves like a Dove against Dove but if its opponent escalates, it then also turns into a Hawk. Suppose the payoff matrix for such a Hawk-Dove-Retaliator game is

$$\begin{array}{c} \\ H \\ D \\ R \end{array} \begin{array}{c} \begin{array}{ccc} H & D & R \end{array} \\ \begin{pmatrix} -1 & 2 & -1 \\ 0 & 1 & 0.9 \\ -1 & 1.1 & 1 \end{pmatrix} \end{array}$$

a) Show that $(0, 0, 1)$, i.e., always play Retaliator, is an ESS.

b) If Retaliator is a strategy that is a "compromise" between Hawk and Dove, is the strategy $(\frac{1}{2}, \frac{1}{2}, 0)$, i.e., "Half Hawk, Half Dove," ESS?

c) Show that, in general, if any entry on the diagonal of a payoff matrix is greater than any entry in the same column, then the corresponding pure strategy is ESS (as for R in this example).

9. In *asymmetric* contests, the two participants have different roles to play that may affect both their actions and the payoffs. For example, in the simple Hawk-Dove game each contest may be between the owner of a property and an intruder, with each participant knowing beforehand which role he or she will play. In such situations, Selten has proved there there can be no mixed strategy that is ESS.

a) Can you prove this result? (*Remark:* It *is* possible for a mixed strategy I to be *neutrally* stable, in the sense that I is as good as a pure strategy.)

b) Show that Selten's Theorem does not hold if the roles of the two participants are not known in advance.

10. Let I be a mixed ESS with support $A, B, C, \ldots$; i.e., A, B, etc., are the strategies played with nonzero probabilities in I. Let $E(X, I)$ be the expected return from playing pure strategy X against the mixed ESS strategy I.

a) Prove the Bishop-Canning Theorem that

$$E(A,I) = E(B,I) = \cdots = E(I,I).$$

b) Use the Bishop-Canning Theorem to show that the payoff matrix

$$\begin{array}{cc} & \begin{array}{cc} I & J \end{array} \\ \begin{array}{c} I \\ J \end{array} & \begin{pmatrix} a & b \\ c & d \end{pmatrix} \end{array}$$

admits an ESS if $a < c$ and $d < b$. In such a case, prove that the ESS is to play strategy I with probability

$$p = \frac{(b-d)}{(b+c-a-d)} \, .$$

c) Apply the foregoing result to the Hawk-Dove game of the text.

11. Consider a situation in which food or some other resource is patchily distributed, and exploitation of a patch gives diminishing returns. Assume that when one patch is abandoned, appreciable time is needed to find or travel to the next one. Further, assume that the foragers search the various patches randomly. Prove the Marginal Value Theorem which states that the rate of benefit is maximized by exploiting each patch until the rate of benefit falls to the maximum *mean* rate that can be sustained over a long period.

As a concrete illustration of the Marginal Value Theorem, consider the feeding pattern of ladybird larvae. Such larvae feed on aphids, eating the soft tissue and leaving behind the exoskeleton. Assume that a larvae spends time t feeding on each aphid, and that the mass of food extracted in this time is $m(t)$. Let T be the mean time required to find a new aphid after leaving a partially eaten one. The mean intake rate of food is then

$$Q = \frac{m(t)}{T+t} \, .$$

The Marginal Value Theorem then states that the optimal value of t is that which maximizes Q.

12. In the text we've considered only *symmetric* contests in which each of the participants was assumed to have the same strength and the same strategies and payoffs. In actual contests this assumption is seldom valid, and one has to deal with *asymmetric* conflicts, conflicts between owner and intruder, male and female, ...

Let $E_1, E_2, \ldots, E_n$ be the strategies available to Player I and $F_1, \ldots, F_m$ those available to Player II. If an E_i-strategist meets an F_j-strategist, his payoff is a_{ij}, whereas that of his opponent is given by b_{ji}, $1 \leq i \leq n$, $1 \leq j \leq m$.

a) Show that the replicator dynamics for asymmetric contests are given by

$$\dot{x}_i = x_i[(Ay)_i - \phi], \qquad i = 1, 2, \ldots, n,$$
$$\dot{y}_j = y_j[(Bx)_j - \psi], \qquad j = 1, 2, \ldots, m.$$

On what space does this dynamic unfold?

b) An evolutionarily stable equilibrium is now given by a pair of points (p, q), $p \in S^n, q \in S^m$. What are the conditions for (p, q) to be evolutionary stable?

c) Prove that a mixed equilibrium (p, q) can never be a sink; i.e., the characteristic values of the linearized system at (p, q) cannot all have negative real parts. Thus, show that there can never exist a mixed ESS, thereby proving Selten's Theorem stated earlier in Problem 9.

d) What are the corresponding discrete-time replicator dynamics for asymmetric contests?

13. In developing the ESS, it was assumed that each individual in a population played a *pure* strategy, with the ESS determining only the steady-state fraction of the population that would adopt one or another of the pure strategies. Now assume the more general case of *individuals* playing mixed strategies. Suppose that an individual plays strategy i with probability p_i, $i = 1, 2, \ldots, n$. Thus, the individual is represented by a point $p \in S^n$, and the population is represented by a *distribution function* f on S^n such that

$$\int df = \int_{S^n} f(p)\, dp = 1.$$

a) Show that the *mean* of f (the probable strategy of an opponent) is given as

$$x = \int p f(p)\, dp,$$

and that the *covariance matrix* of the population f is given as

$$F = \int (p - x)(p - x)'\, df(p).$$

b) Assume, as before, that the growth rate of a strategy equals its differential advantage. Show that this means

$$\frac{df(p)}{dt} = f(p)[(p, Ax) - (x, Ax)],$$

where x is the mean defined above, and A is the payoff matrix of the game.

c) Prove that the mean evolves according to the dynamics

$$\dot{x} = FAx.$$

d) Use the above results to prove that if a game has an ESS at a vertex or an edge of the simplex S^n, then it is an attractor both for populations playing pure strategies and for the means of populations playing mixed strategies.

14. Contests between pairs of animals who are competing for a prize (a mate, territory, etc.) in which each engages in display behavior as a stimulus inhibiting an attempt to take possession of the prize by the other, and in which eventually one of the combatants departs leaving the prize to the one who held out the longest, are termed War of Attrition games.

Assume the cost of display is proportional to duration and that the value of the prize is $V > 0$. Further, assume the competitors choose strategies x and y, respectively, where x denotes "wait an amount of time x," and y is defined similarly. Let the payoffs of the game be given by

$$E[x, y] = \begin{cases} V - y, & \text{if } x > y \\ \frac{1}{2}V - x, & \text{if } x = y \\ -x, & \text{if } x < y \end{cases}$$

a) Show that neither player has an optimal pure strategy by showing that no strategy x is best against itself.

b) Define a mixed strategy as the frequency density $p(x)$ of a random variable X. That is, the mixed strategy is to wait an amount of time x, where x is determined as the realization of a random variable X having density function $p(x)$. Prove that the exponential density function

$$p(x) = \frac{\exp(-x/V)}{V}, \qquad x \geq 0,$$

is an ESS against any pure strategy for the War of Attrition by establishing the following relations:

$$E[x, p(y)] = 0 \text{ for all strategies } x,$$

and

$$E[p(x), y] > E[y, y] \text{ for all } y.$$

The first condition shows that all strategies x (and their mixtures) are equally good against $p(y)$, and the second shows that $p(x)$ is a better reply against y than y itself, for all strategies y.

c) The results of part (b) show that $p(x)$ is ESS against any *pure* strategy y. Show that the same result also holds against any *mixed* strategy y as well.

15. Consider the following model for the mutation and selection process. Let there be one gene locus with alleles $A_1, \ldots, A_n$ and let $x_1, \ldots, x_n$ be their relative frequencies in the gene pool of the population. Due to natural selection, only a fraction $w_{ij}x_ix_j$ of the gametes (genotype A_iA_j) will survive to procreative age, where $w_{ij} = w_{ji} \geq 0$ are the *fitness parameters* of the population. Let ϵ_{ij} be the *mutation rate* from A_j to A_i, $(i \neq j)$, where $\epsilon_{ij} \geq 0$, $\sum_{i=1}^{n} \epsilon_{ij} = 1$, for all $j = 1, 2, \ldots, n$.

a) Show that the frequency x_i' of genotype A_i in the next generation is given by

$$x_i' = \sum_{j=1}^{n} \frac{\epsilon_{ij} x_j (Wx)_j}{W(x)},$$

where $W(x) = (x, Wx)$.

b) Show that the continuous-time version of the above result is

$$\dot{x}_i = W(x)^{-1} \sum_{j,k} \epsilon_{ij} w_{jk} x_j x_k - x_i, \qquad i = 1, 2, \ldots, n.$$

c) Suppose the model without mutation admits a stable interior equilibrium. Under the conditions of parts (a) and (b), prove that for mutation rates of the form $\epsilon_{ij} = \epsilon_i$, $i \neq j$, the differential system has a globally stable interior equilibrium, while the difference equation has an equilibrium that is at least locally stable.

d) Show that the differential system of part (b) forms a gradient system using the Shahshahani gradient for the function

$$V(x) = \frac{1-\epsilon}{2} \log W(x) + \sum_{i=1}^{n} \epsilon_i \log x_i,$$

where $\epsilon = \sum_{j=1}^{n} \epsilon_j$. Thus, conclude that the change in gene frequency occurs in such a fashion so as to maximize the increase in mean fitness.

16. By a *learning rule,* we mean a rule specifying which of a set of possible actions A, B, ... an animal will perform on any occasion given its previous experiences. A "rule for ESSs" is a rule such that if all the members of a population adopt it, they will in time come to adopt the ESS for the game in question.

a) Show that the rule

$$\text{probability of A} = \frac{\text{total payoff received so far for doing A}}{\text{total payoff received so far for all actions}},$$

is an ESS learning rule.

b) Suppose different individuals in a population adopt different learning rules and reproduce their kind. Further, suppose the number of offspring are proportional to the payoff an animal accumulates by playing games against others in the population. If after a time a given learning rule evolves for the population, we call it an "evolutionary stable learning rule," or ES. Note that an ES and a rule for ESSs are two quite different concepts: one is the learning rule we expect to evolve; the other is a rule that takes the population to the ESS of a game. Nevertheless, show that an ES rule is necessarily a rule for ESSs.

17. Consider a population playing the iterated Prisoner's Dilemma with the strategy TIT FOR TAT.

a) Prove that if the mutant strategies ALL D and ALTERNATE (i.e., play DCDCDC...) cannot invade (displace) TIT FOR TAT, then *no* mutant strategy can invade the population.

b) Prove that neither ALL D nor ALTERNATE can invade if the probability of further interaction w is sufficiently large.

c) Show that if a nice strategy cannot be invaded by a single individual, then it cannot be invaded by any cluster of individuals either.

18. In the iterated Prisoner's Dilemma, we say a strategy A *invades* a strategy B if A gets a higher score with B than B gets with itself. The strategy B is called *collectively stable* if B cannot be invaded.

a) Prove the following theorem:

CHARACTERIZATION THEOREM. *A strategy B is collectively stable if and only if B defects on move n whenever the opponent's score so far is too large.*

b) Show that for a nice strategy to be collectively stable, it must defect upon the very first defection of the other player.

c) We say that a strategy A *territorially invades* B if every location in the territory will eventually convert to A. B is called *territorially stable* if no strategy can territorially invade it. Prove that if a strategy is collectively stable, then it is also territorially stable. Thus, protection from invasion is at least as easy in a territorial system as in a freely mixing system. Hence, mutual cooperation can be sustained in a territorial system at least as easily as in a freely mixing system.

Notes and References

§1. The philosophical underpinnings of evolutionary epistemology have far broader currency than just to support the response of biological organisms to changes in their environment, as we have endeavored to point out during the course of this chapter. For introductory accounts of evolution as a conceptual theme in the philosophy of science, see the volumes

Wuketits, F., ed., *Concepts and Approaches in Evolutionary Epistemology,* Reidel, Dordrecht, 1984,

Popper, K. R., *Objective Knowledge: An Evolutionary Approach,* Clarendon Press, Oxford, 1972,

Jantsch, E., *Design for Evolution: Self-Organization and Planning in the Life of Human Systems,* Braziller, New York, 1975.

For those interested in the always entertaining battle raging over the scientific content of the book of Genesis, we recommend

Montagu, A., ed., *Science and Creationism,* Oxford U. Press, Oxford, 1984,

Godfrey, L., ed., *Scientists Confront Creationism,* Norton, New York, 1983.

§2. An easily digestible, nonmathematical overview of the entire Darwinian and neo-Darwinian theories can be found in

Arthur, W., *Theories of Life,* Penguin, London, 1987.

More technical expositions are given by

Mayr, E., *The Growth of Biological Thought,* Harvard U. Press, Cambridge, MA, 1982,

Smith, J. Maynard, *The Theory of Evolution,* Penguin, London, 1975.

A critical account of the entire issue of fitness and selection is found in

Sober, E., *The Nature of Selection: Evolutionary Theory in Philosophical Focus,* MIT Press, Cambridge, MA, 1984.

For an account of Weissman's theory of the "germ plasm," see

Webster, G., and B. C. Goodwin, "The Origin of Species: A Structuralist Approach," *J. Soc. Biol. Struct.*, 5 (1982), 15–47,

Stenseth, N., "Darwinian Evolution in Ecosystems: A Survey of Some Ideas and Difficulties Together with Some Possible Solutions," in *Complexity, Language and Life: Mathematical Approaches,* J. Casti and A. Karlqvist, eds., Springer, Heidelberg, 1986.

§3. The theory of genetics was put on a solid mathematical foundation by the work of the great statistician Ronald A. Fisher. His classic work in the area is

Fisher, R. A., *The Genetical Theory of Natural Selection,* Clarendon Press, Oxford, 1930.

More recent accounts are

Futuyama, D., *Evolutionary Biology,* Sinauer, Sunderland, MA, 1979,

Roughgarden, J., *Theory of Population Genetics and Evolutionary Ecology: An Introduction,* Macmillan, New York, 1979.

The classic paper that sparked off the explosion of interest in utilizing game-theoretic ideas in the context of evolutionary ecology is

Smith, J. Maynard, and G. Price, "The Logic of Animal Conflict," *Nature,* 246 (1973), 15–18.

A much more complete account, together with peer commentary, is given in the review article

Smith, J. Maynard, "Game Theory and the Evolution of Behavior," *Behav. & Brain Sci.,* 7 (1984), 94–101.

§4. Seldom can the origin of a field of intellectual activity be traced specifically to a single book or paper; game theory is one of the few exceptions. The classic work from which the subject emerged almost full-grown is

von Neumann, J., and O. Morgenstern, *The Theory of Games and Economic Behavior,* Princeton U. Press, Princeton, 1944.

Since the appearance of the above "encyclopedia" of games of strategy, numerous introductory accounts have appeared. Some of our favorites are

Williams, J., *The Compleat Strategist,* McGraw-Hill, New York, 1954,

Dresher, M., *The Mathematics of Games of Strategy,* Prentice-Hall, Englewood Cliffs, NJ, 1961,

Hamburger, H., *Games as Models of Social Phenomena,* Freeman, San Francisco, 1979,

Jones, A. J., *Game Theory: Mathematical Models of Conflict,* Ellis Horwood, Chichester, UK, 1980.

§5. The material of this section follows the treatment given in the following book, which is notable for its detailed account of experimental research on strategic interaction, together with a wealth of examples taken from political science, philosophy, biology and economics:

Colman, A., *Game Theory and Experimental Games,* Pergamon, London, 1982.

The Prisoner's Dilemma seems to have first been explicitly formulated by Merrill Flood at the RAND Corporation and named by Albert Tucker in the early 1950s. It has been the subject of an extensive literature which by now numbers well over 1000 papers and books. Good introductory accounts of the intricacies of the game are

Rapoport, A., and A. Chammah, *Prisoner's Dilemma: A Study in Conflict and Cooperation,* U. of Michigan Press, Ann Arbor, 1965,

Rapoport, A., *Mathematical Models in the Social and Behavioral Sciences,* Wiley, New York, 1983,

Brams, S., *Game Theory and Politics,* Free Press, New York, 1975.

§6–7. The best compact reference for the mathematical details of replicator systems is the review

Sigmund, K., "A Survey of Replicator Equations," in *Complexity, Language and Life: Mathematical Approaches,* J. Casti and A. Karlqvist, eds., Springer, Heidelberg, 1986, pp. 88–104.

The example of the hypercycle is treated in great mathematical and chemical detail in

Eigen, M., and P. Schuster, *The Hypercycle: A Principle of Natural Self-Organization,* Springer, Berlin, 1979.

§8. In the ecological community, the phenomenon we have termed perseverance often goes under the rubric persistence. Unfortunately, this latter term is also used to mean something quite different in circles outside population ecology, and there is as yet no real consensus on the use of the overworked word persistence. So we have preferred to employ the related word perseverance which captures equally well the idea of this section. Important references on this topic are

Hutson, V., and C. Vickers, "A Criterion for Permanent Coexistence of Species with an Application to a Two-Prey/One-Predator System," *Math. Biosci.,* 63 (1983), 253–269,

Amann, E., *Permanence for Catalytic Networks,* Dissertation, Dept. of Mathematics, U. of Vienna, Vienna, 1984,

Hofbauer, J., "A Difference Equation Model for the Hypercycle," *SIAM J. Appl. Math.,* 44 (1984), 762–772.

§9. The mathematical interconnections between two-player games and replicator dynamics, including a very detailed consideration of the classification scheme considered here, may be found in

Zeeman, E. C., "Population Dynamics from Game Theory," in *Global Theory of Dynamical Systems, Proceedings,* Springer Lecture Notes in Mathematics, Vol. 819, New York, 1980, pp. 471–497.

Another paper containing further developments along the same lines is

Zeeman, E. C., "Dynamics of the Evolution of Animal Conflicts," *J. Theor. Biol.,* 89 (1981), 249–270.

§10. For further details on the role of ESS in replicator dynamics, see

Hofbauer, J., P. Schuster and K. Sigmund, "A Note on Evolutionary Stable Strategies and Game Dynamics," *J. Theor. Biol.,* 81 (1979), 609–612,

Schuster, P., and K. Sigmund, "Towards a Dynamics of Social Behavior: Strategic and Genetic Models for the Evolution of Animal Conflicts," *J. Soc. & Biol. Struct.,* 8 (1985), 255–277.

§11–12. The desert plant example, together with its extension to the case of many players as well as a consideration of the optimality vs. stability issue, is given in

Riechert, S., and P. Hammerstein, "Game Theory in the Ecological Context," *Ann. Rev. Ecol. Syst.,* 14 (1983), 377–409.

For an account of field experiments supporting the theoretical properties of ESS in an ecological setting, see

Riechert, S., "Spider Fights as a Test of Evolutionary Game Theory," *American Scientist,* 74 (1986), 604–610.

§13. For additional information on the matter of plants and game-theoretic flowering strategies, see

Vincent, T., and J. Brown, "Stability in an Evolutionary Game," *Theor. Pop. Biol.,* 26 (1984), 408–427.

§14. The classic work that started all the ruckus over the degree to which social behavioral patterns are biologically determined is

Wilson, E., *Sociobiology,* Harvard U. Press, Cambridge, MA, 1975.

Other introductory accounts of the basic ideas underlying the sociobiology thesis are

Barash, D., *Sociobiology and Human Behavior,* Elsevier, New York, 1977,

Alexander, R., *Darwinism and Human Affairs,* U. of Washington Press, Seattle, 1979,

Baer, D., and D. McEachron, "A Review of Selected Sociobiological Principles: Applications to Hominid Evolution," *J. Soc. & Biol. Struct.,* 5 (1982), 69–90.

The very idea that human behavior could be determined by biological considerations seems to strike a raw nerve with those devoted to the bizarre notion that humans are somehow masters of their own destiny. For blow-by-blow accounts of this highly political and emotionally charged debate, see

Kitcher, P., *Vaulting Ambition,* MIT Press, Cambridge, MA, 1985,

Caplan, A., ed., *The Sociobiology Debate,* Harper & Row, New York, 1978,

Lewontin, R., S. Rose and L. Kamin, *Not in Our Genes,* Pantheon, New York, 1984,

Ruse, M., *Sociobiology: Sense or Nonsense?,* Reidel, Dordrecht, 1979.

For a fuller account of the genetic to phenotype model of animal behavior, see the Schuster and Sigmund article cited under §10 above, as well as

Bomze, I., P. Schuster and K. Sigmund, "The Role of Mendelian Genetics in Strategic Models of Animal Behavior," *J. Theor. Biol.,* 101 (1983), 19–38.

§15. The complete story of the computer tournament in which various strategies for the Prisoner's Dilemma game were pitted against each other in a round-robin competition is given in the fascinating book

Axelrod, R., *The Evolution of Cooperation,* Basic Books, New York, 1984.

Further discussion of the computer tournament, along with extensive commentary on the possible social implications of the results, can be found in

Hofstadter, D., "The Prisoner's Dilemma Computer Tournament and the Evolution of Cooperation," in *Metamagical Themas,* Basic Books, New York, 1985, pp. 715–734.

Some of the implications of the Prisoner's Dilemma for biological behavior are outlined in

Axelrod, R. and W. Hamilton, "The Evolution of Cooperation," *Science,* 211 (1981), 1390–1396.

DQ #2. For detailed arguments on the parallels between economics, evolutionary biology and behavioral psychology, see

Schwartz, B., *The Battle for Human Nature,* Norton, New York, 1986,

von Schilcher, F., and N. Tennant, *Philosophy, Evolution and Human Nature,* Routledge and Kegan Paul, London, 1984.

DQ #3. The Red Queen Hypothesis was introduced into the ecological literature by Van Valen in

Van Valen, L., "A New Evolutionary Law," *Evol. Theory,* 1 (1973), 1–30.

For a review of this somewhat controversial hypothesis, see

Stenseth, N., "Darwinian Evolution in Ecosystems—The Red Queen View," in *Evolution,* P. Greenwood, *et al.,* eds., Cambridge U. Press, Cambridge, 1985.

DQ #4. A detailed consideration of the Cuban Missile Crisis from a modeling and systems-analytic perspective is provided by the classic work

Allison, G., *Essence of Decision: Explaining the Cuban Missile Crisis,* Little Brown, Boston, 1971.

DQ #5. For a full discussion of the n-Player Prisoner's Dilemma, see the book by Colman cited under §5 above.

DQ #6. The idea of harnessing evolutionary principles to economic phenomena is not a new one, as is recounted in the Schwartz book cited under DQ #2. However, there has recently been considerable interest in utilizing the same approach to study the economics of the firm rather than at the level of the individual. A pioneering work in this area that builds upon the earlier work of Schumpeter is

Nelson, R. R., and S. G. Winter, *An Evolutionary Theory of Economic Change,* Harvard U. Press, Cambridge, MA, 1982.

In a somewhat different direction, see also

Boulding, K., *Evolutionary Economics,* Sage, Beverly Hills, CA, 1981.

DQ #7. The classic works of Spengler and Toynbee on the dynamics of historical change are

Spengler, O., *The Decline of the West,* Knopf, New York, 1926,

Toynbee, A., *A Study of History,* Oxford U. Press, Oxford, 1972.

An interesting use of linear system theory to support some of Toynbee's ideas is

Lepschy, A., and S. Milo, "Historical Events Dynamics and 'A Study of History' by Arnold Toynbee," *Scientia,* 11 (1976), 39–50.

The work of Colinvaux emphasizes the role of an eco-niche in determining the flow of historical events on a broad scale. For a detailed account, see

Colinvaux, P., *The Fates of Nations: A Biological Theory of History,* Simon & Schuster, New York, 1980.

DQ #9. Dawkin's theory of the selfish gene is entertainingly presented in his well-known books

Dawkins, R., *The Selfish Gene,* Oxford U. Press, Oxford, 1976,

Dawkins, R., *The Extended Phenotype,* Freeman, San Francisco, 1982.

Further discussion of the idea of a cultural meme is given in the article

Hofstadter, D., "On Viral Sentences and Self-Replicating Structures," *Scientific American,* January 1983 (reprinted in *Metamagical Themas,* Basic Books, New York, 1985).

DQ #10. Swanson's ideas are given in detail in

Swanson, C., *Ever-Expanding Horizons,* U. of Massachusetts Press, Amherst, MA, 1983.

DQ #14. The most prominent proponents of the punctuated equilibrium view of evolutionary change are the well-known paleontologists Stephen J. Gould and Nils Eldredge. Their views were first presented in

Eldredge, N., and S. Gould, "Punctuated Equilibria: An Alternative to Phyletic Gradualism," in *Models in Paleobiology,* T. Schopf, ed., Freeman, San Francisco, 1972.

DQ #16. The Sheldrake theory of morphogenetic fields is viewed by mainline biologists in much the same light that Stalinists viewed Trotsky. Nevertheless, a number of predictions of the theory have been borne out by experiments carried out in recent years, and the idea now seems more alive than ever before. For an account of the underlying ideas, together with reprints of reviews of the first edition of the pioneering book and a description of the experimental evidence, see

Sheldrake, R., *A New Science of Life, A New Edition,* Anthony Blond, London, 1985,

Sheldrake, R., *The Presence of the Past,* Times Books, New York, 1988.

DQ #19. See the Axelrod book cited under §15 for a fuller account of how the results of the Prisoner's Dilemma tournament suggest methods for improving cooperation among individually selfish competing parties.

PE #7–10. For these as well as many more examples, including a discussion of the Bishop-Canning Theorem, see

Smith, J. Maynard, *Evolution and the Theory of Games,* Cambridge U. Press, Cambridge, 1982.

PE #15. A fuller account of this problem can be found in

Hofbauer, J., "The Selection-Mutation Equation," *IIASA Collaborative Paper, CP-85-2,* IIASA, Laxenburg, Austria, January 1985.

PE #17. These results are considered in greater detail in the Axelrod book cited under §15.

CHAPTER SEVEN

Taming Nature and Man: Control, Anticipation and Adaptation in Social and Biological Processes

1. *Classical Optimal Control Theory*

At various points in our narrative, we've touched upon the matter of influencing the behavior of a dynamical process by regulating its input. In Chapter Three we saw how the input to a system could be considered as a stimulus pattern generating a behavioral response, and in subsequent chapters we dealt with variation in constitutive parameters that could, under appropriate circumstances, cause the system to display radically different modes of behavior. However, in most of these cases it was at least tacitly assumed that the variation in the input was brought about more by the vagaries of Nature than by the designs of man. This is a pretty passive version of "control," and hardly agrees with what we usually mean when we think of a process as being "controlled." Consequently, in this chapter we take a more activist point of view and look at how we can regulate the behavior of a dynamical process in order to make it behave according to the wishes of its controller, rather than the whims of Nature. Such a perspective carries along with it the notion that the benefits of control don't come for free, and that we have to weigh those benefits against the cost of exerting the controlling actions. This trade-off leads directly to the idea of an *optimal control process,* in which optimality is measured according to some economic criterion that measures the relative cost of control against the benefits of desired system behavior.

Classical optimal control theory arose out of the so-called simplest problem of the calculus of variations. The prototypical control process is that of finding the control input $u(t)$, $0 \leq t \leq T$, such that the integral

$$J(u) = \int_0^T g(x,u)\,dt,$$

is minimized, where the system state $x(t)$ is connected to the control input through the differential equation

$$\dot{x} = f(x,u), \qquad x(0) = x_0.$$

There are many variations upon the basic theme of this problem, depending upon the assumptions made regarding the space Ω of admissible control inputs, the constraints imposed upon the system state, the degree of smoothness in the functions f and g, whether T is finite or infinite, other types of nonintegral criteria and so forth. But quite irrespective of these considerations, there have emerged two complementary approaches to the determination of the optimal control law $u^*(t)$: the *open-loop* approach, which emphasizes use of the Pontryagin Maximum (or Minimum) Principle, and the *closed-loop* approach, which uses Bellman's Principle of Optimality and dynamic programming. Let's briefly look at each of these procedures.

2. *The Maximum Principle and Open-Loop Control*

For the sake of exposition and to avoid annoying and distracting technical caveats, let's assume that the functions g and f, as well as the space of admissible inputs and the constraint space, possess whatever analytic properties we need in order to have our subsequent operations make sense. To find the function $u^*(t)$ that minimizes the integral J under the differential equation side constraint, the Pontryagin approach is to introduce a Lagrange multiplier function $\lambda(t)$ and to use this function to define a Hamiltonian $H(x, u, \lambda, t)$ for the system. The optimal control $u^*(t)$ is then determined as the function that minimizes H. Let's take a look at how this method goes.

Let the control input $u(t) \in R^m$ with the state $x(t) \in R^n$. Using the *co-state* vector $\lambda(t) \in R^n$, we form the Hamiltonian function H as

$$H(x, u, \lambda, t) = g(x, u, t) + \lambda' f(x, u, t),$$

where "$'$" denotes the transpose operation. At each moment t, H is a scalar-valued function of x, u and λ. According to the Pontryagin Principle, the control $u^*(t)$ that minimizes the criterion J is exactly that element $u \in \Omega$ that minimizes H pointwise in t. Thus,

$$u^*(t) = \arg\min_{u \in \Omega} H(x, u, \lambda, t).$$

Of course, the function H contains the unknown co-state function λ. So in order to minimize H and find u^*, we must develop a relation between u and λ. Following an approach very similar to that employed in the calculus of variations to obtain the classical Euler-Lagrange equation, it can be shown that the functions x, u and λ are related through the following two-point boundary value problem:

$$\dot{x} = \frac{\partial H}{\partial \lambda}(x, u, \lambda, t), \qquad x(0) = x_0,$$

$$-\dot{\lambda} = \frac{\partial H}{\partial x}(x, u, \lambda, t), \qquad \lambda(T) = 0, \qquad 0 \leq t \leq T.$$

Assuming there are no constraints on the allowable variation in the control u, a necessary condition for the optimizing input u^* is that u^* be the solution to the equation

$$\frac{\partial H}{\partial u}(x, u, \lambda, t) = 0,$$

where the above boundary-value problem is used to express both x and λ as functions of u.

Example

To illustrate the foregoing approach, consider the case when the criterion function g is quadratic and the state dynamics f are linear. We then have

$$g(x,u,t) = \tfrac{1}{2}[(x, Qx) + (u, Ru)],$$
$$f(x,u,t) = Fx + Gu,$$

where we impose the conditions $Q \geq 0$, $R > 0$ to ensure that the problem has a well-defined solution. In this case, the Hamiltonian function is

$$H(x,u,\lambda,t) = \tfrac{1}{2}[(x, Qx) + (u, Ru)] + (\lambda, Fx + Gu),$$

and the corresponding two-point boundary-value problem becomes

$$\dot{x} = Fx + Gu, \qquad x(0) = x_0,$$
$$-\dot{\lambda} = Qx + F'\lambda, \qquad \lambda(T) = 0.$$

The condition for the minimizing control is

$$Ru^* + G'\lambda = 0,$$

or

$$u^*(t) = -R^{-1}G'\lambda(t).$$

Thus we see that in order to compute the optimizing control u^*, it's necessary to solve the boundary-value problem for the co-state function λ.

As the above discussion shows, use of the Maximum Principle always yields the optimal control law $u^*(t)$ as a function of the time t. Intuitively speaking, we look at a clock to determine what time it is. We then exert the controlling action $u^*(t)$ called for at that moment. This is what is termed *open-loop* control; determination of what action to take at any moment is determined solely by what the current time is and not by what the system is actually doing, i.e., by what the current system state $x(t)$ may be. Note also that from a computational point of view, determination of $u^*(t)$ may not always be an entirely straightforward matter since it involves solution of a two-point boundary-value problem for the co-state λ. This can be a nontrivial problem, especially if the interval length T is large. On the other hand, the Maximum Principle approach is very flexible in regard to the kind of constraints that can be imposed, especially constraints on the control law u. And the determination of $u^*(t)$ involves only the solution of *ordinary* differential equations, although generally of a nonlinear nature. Now let's look at the alternative approach based upon dynamic programming considerations.

3. *Feedback Control and Dynamic Programming*

The essence of the dynamic programming approach to optimal control is encapsulated in Bellman's Principle of Optimality. Assume that $u^*(t)$ is the optimal control function and that $x^*(t)$ is the associated optimal trajectory. Let $v = u^*(0)$ denote the optimal action to take at the initial moment, with the initial state being given as $x(0) = x^*(0) = c$. Then the Principle of Optimality states that the part of the optimal trajectory starting at time Δ from the state $c + f(c, v, 0)\Delta$ is also the optimal trajectory for the problem that begins not at time $t = 0$ but at time $t = \Delta$, though now in the state $c + f(c, v, 0)\Delta$ rather than the state c. In other words, any part of an optimal trajectory is also optimal. This idea is depicted in Fig. 7.1.

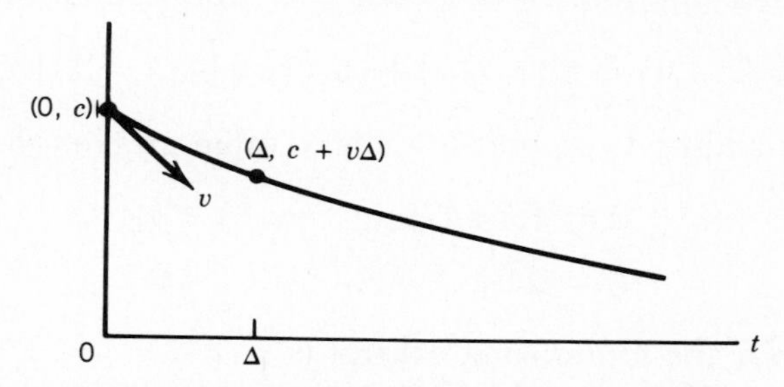

Figure 7.1 The Principle of Optimality

Let's now use the Principle of Optimality to develop an equation for the control that minimizes

$$J = \int_t^T g(x, u)\, dt,$$

where

$$\dot{x} = f(x, u), \qquad x(t) = c.$$

The minimal value of J, as well as the associated optimal control law, is determined by two quantities: the initial state c, and the length of time of the process which is characterized by $T - t$, since T is assumed to be fixed. So we introduce the function

$I(c, t) =$ the value of J obtained for a process of duration $T - t$ that starts in state c when an optimal control is used.

Then regardless of what the initial control action $v = u(0)$ may be, the Principle of Optimality asserts that we must have the inequality

$$I(c, t) \leq \{g(c, v)\Delta + I(c + f(c, v)\Delta, t + \Delta)\} + o(\Delta). \qquad (\dagger)$$

Here we have made use of the additivity property of integrals

$$\int_t^T = \int_t^{t+\Delta} + \int_{t+\Delta}^T,$$

as well as the Mean-Value Theorem for integrals. We clearly want to make the right side of (†) as small as possible, so we choose v to minimize, leading to the recurrence relation for I as

$$I(c,t) = \min_v \{g(c,v)\Delta + I(c + f(c,v)\Delta, t + \Delta)\} + o(\Delta). \qquad (*)$$

For a process of length $T - t = 0$, we have the trivial starting condition

$$I(c,T) = 0.$$

The relation $(*)$ linking the solution to a problem starting at time t with one starting at time $t + \Delta$, together with the initial condition above, enables us to calculate the *optimal-value function* $I(c,t)$ as well as the optimal decision function $v(c,t)$ for each possible initial state c and every interval length $T - t$. This procedure then solves not only the original problem that we started with for some *fixed* values of c and t, but also a host of additional problems involving various initial states and various interval lengths. Let's look for a moment at what's involved in actually computing the functions $I(c,t)$ and $v(c,t)$.

Suppose we agree to quantize each component of the state into K discrete levels. Further, suppose that each component of the control is also quantized into L levels. Then if the state is n-dimensional, there are K^n distinct state values that must be considered at each time moment, and if the control is m-dimensional we must test each of the L^m control levels against each of the states in order to compute the function $I(c,t)$ for a single value of t. In addition, to calculate $I(c,t)$ it's necessary to have in memory the function $I(\cdot, t + \Delta)$ for *every* possible state value. This means that we need at least K^n locations of high-speed memory, not counting what may be needed for other purposes. Let's say $K = 100$. Then if n is small, say, $n = 1$ or 2, the memory needed is modest, on the order of a few thousand; but real problems have n more in the neighborhood of $n = 10, 20$ or up to even a few hundred. In such cases no brute force use of $(*)$ can be contemplated, even with the biggest and fastest of modern computers, and more sophisticated techniques are required in order to cut this computational requirement down to size. Some of these methods are discussed in the material cited in the Notes and References so we won't go into them here. Note also that though not as severe a restriction as the foregoing memory difficulty, the computational burden also increases markedly as we increase the number of control components and/or refine their discretization. However, the computational requirements only increase linearly in this case rather than displaying the geometric increase seen in state dimensionality.

If we pass to the limit and let $\Delta \to 0$ in (*), we obtain the famous Bellman-Hamilton-Jacobi equation for the function I as

$$-\frac{\partial I}{\partial t} = \min_v \left\{ g(c,v) + \left(f(c,v), \frac{\partial I}{\partial c} \right) \right\}, \qquad t < T,$$
$$I(c,T) = 0.$$

Example

Let's use the above equation to show how dynamic programming can be used to solve the simple linear-quadratic problem considered earlier. Here we have

$$g(x,u) = \tfrac{1}{2}[(x,\, Qx) + (u,\, Ru)],$$
$$f(x,u) = Fx + Gu.$$

The Bellman-Hamilton-Jacobi equation in this case becomes

$$-\frac{\partial I}{\partial t} = \min_v \left\{ \frac{1}{2}[(c,\, Qc) + (v,\, Rv)] + \left(Fc + Gv, \frac{\partial I}{\partial c} \right) \right\}, \qquad T > 0,$$
$$I(c,T) = 0.$$

It's a fairly simple matter to verify that in this case the function $I(c,t)$ is a quadratic form in c; hence, there exists a matrix function $P(t)$ such that

$$I(c,t) = (c,\, P(t)c),$$

which, after some tedious although straightforward calculation, yields

$$v_{\min}(c,t) = -R^{-1}G'P(t)c, \qquad (\ddagger)$$

where the function $P(t)$ satisfies the matrix Riccati equation

$$-\dot{P}(t) = Q + PF + F'P - PGR^{-1}G'P, \qquad P(T) = 0.$$

The optimal state trajectory is the solution of the *closed-loop* dynamics

$$\dot{x}^*(t) = [F - GR^{-1}G'P(t)]x^*(t), \qquad x^*(t) = c.$$

The most important point to note about the above development is that ($\ddagger$) expresses the optimal control as a function of the system state c, rather than as a funtion of the time. This means that we must *measure* what the system is actually doing and then apply the control that is appropriate for that state instead of just looking at a clock and determining the control as a function of the time alone. Basically, the state measurement is "fed back" from the system output to its input, giving rise to the concept of *feedback*, or *closed-loop*, control as shown in Fig. 7.2.

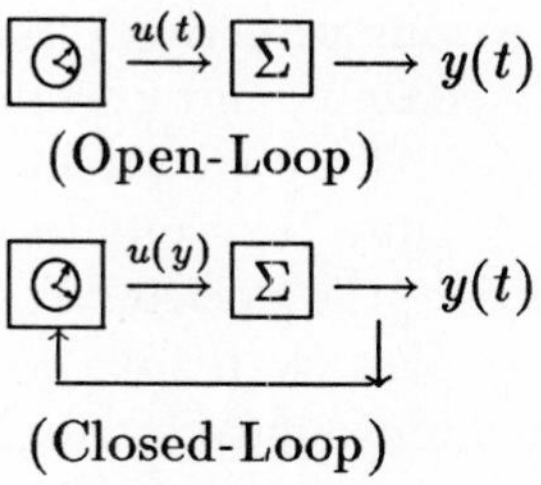

$$\boxed{\circlearrowleft} \xrightarrow{u(t)} \boxed{\Sigma} \longrightarrow y(t)$$

(Open-Loop)

$$\boxed{\circlearrowleft} \xrightarrow{u(y)} \boxed{\Sigma} \longrightarrow y(t)$$

(Closed-Loop)

Figure 7.2 Open versus Closed-Loop Control

We have already noted the "curse of dimensionality" that greatly complicates the computational solution of high-dimensional control problems by dynamic programming methods. It should also be observed that the basic equation for the optimal-value function I is a partial differential equation rather than the ordinary differential equation obtained using the Maximum Principle. However, instead of being a boundary-value problem, the dynamic programming equation for I is an *initial-value* problem which is far easier to deal with computationally than a boundary-value problem. Thus, we can trade off ordinary differential equations and boundary-value problems for partial differential equations and initial-value problems. In general, which method to use depends upon a whole host of circumstances that we have no space to enter into here. The interested reader is invited to consult the many excellent references on this circle of questions cited in the chapter Notes and References.

4. *Anticipatory Systems*

In either open-loop or feedback mode, the foregoing classical control processes are inherently *reactive* in character; i.e., the control action is always taken in response to the past and present behavior of the system itself. Here we want to consider another class of control processes in which the controlling action depends not only upon what the system is (and has been) doing, but also upon what we *think* the system will be doing in the future. We term such processes *anticipatory.*

When we speak of anticipatory processes, we appear to edge dangerously close to ignoring one of the pillars upon which modern theoretical science is based, namely, that effects should not precede their causes. However, as we shall see below, there is no cause for alarm as the type of process we have in mind is perfectly consistent with traditional notions of causality since we don't claim that the system behavior is determined by *actual* future events, but rather by our current *beliefs* about the nature of those events. As a trivial example, suppose we are standing on a street corner waiting to cross to the other side of the street. We look up the street and see a car coming at some speed in our direction. Using the personal computer between our ears, we quickly examine our current position, the position and velocity of the car, and the width of the street and *predict* that if we step off the curb into the street, it may turn out to be our last living act. As a result of this *prediction,* we take the decision (controlling action) to remain where we are and let the car pass by before crossing. In light of this simple example, we see that such anticipatory processes abound in modern life and it's somewhat puzzling as to why they have not been more extensively studied. Much of the remainder of this chapter is devoted towards rectifying this deplorable state of affairs.

Consideration of both the general idea of an anticipatory process, as well as the simple example above, shows that the crucial ingredient separating the type of classical process discussed earlier from an anticipatory process is the notion of *self-reference.* The anticipatory process contains an internal model of itself, and this model is used in a time-scale faster than real-time to influence the actions taken at any given moment. Furthermore, as the process unfolds and more information is gained about the operating environment, this internal model is continually updated according to one scheme or another, thereby providing a means for both adaptation as well as evolution. Before entering into a detailed consideration of how such anticipatory systems arise in living systems, let's first take a longer look at the basic components of an anticipatory process.

For the type of anticipatory system we have in mind, three components are needed: (1) the system Σ itself, (2) a controller C, and (3) a model M of Σ. These three objects are connected together as shown in Fig. 7.3.

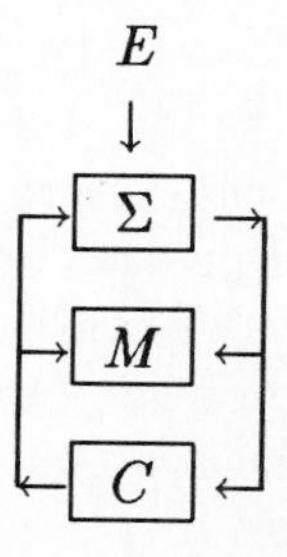

Figure 7.3 The Components of an Anticipatory System

In the diagram, the model M sees the behavior of Σ as well as the results of past control actions from C. As a result of this information, M *predicts* what the future states of Σ will be and transmits this information to the controller C, who integrates the prediction with his measurements of Σ and the environment E in order to generate the *current* controlling action. This decision is then input to Σ and the cycle repeats itself.

It's of fundamental importance to note two aspects of this set-up: (1) the model M must operate on a time-scale faster than that of Σ, and (2) the model must utilize the information about past behaviors of Σ and past actions of C to improve its predictions about what Σ will do in the future; i.e., M must possess some sort of learning and adaptation mechanism. How much faster the time-scale of M needs to be over that of Σ is determined by the speed of information processing in C, as well as by the speed of signal transmission from C to Σ and from Σ to C and M. Basically, what's needed is that M be able to predict far enough into the future of Σ that the predictions can effectively be used to influence the actual behavior of Σ at that future time.

At this point, a traditional control engineer would be likely to comment that the type of anticipatory scheme outlined above seems not to differ much in character from the standard adaptive control procedures already well-scouted in the control engineering literature. It's worthwhile to pause for a moment and address this point.

The standard adaptive control problem is structured along the following lines. We are given a dynamical process

$$\dot{x} = f(x, u, a),$$

where $a \in R^k$ is some vector of unknown parameters. Furthermore, we have the observations

$$y = h(x, u),$$

and our job is both to control the system and learn about the values of a by choice of the control law u. In general, the control is a function not only of the state x, but also our current best estimate of the parameters a, i.e., $u = u(x, \hat{a})$, where we have written $\hat{a}$ to represent the estimated value of the parameters. As the process unfolds, we obtain new observations y and we process them according to one learning scheme or another to update our estimate of a. The important point to note here is that the actual controlling action u is determined only as a function of the current (and, perhaps, past) state and our current best estimate of the parameters $\hat{a}$.

By way of contrast, in an anticipatory mode besides the above dynamics, observations, and learning scheme for a, we have a model M that predicts future states of the system based upon all this past information. Thus, the control that is actually applied at the current moment is a function not only of the current and past states and the best estimate of the parameters, but also of the predicted future states. Thus,

$$u = u(x(t), \hat{a}, \hat{x}(t + T_i)), \qquad T_i > 0, \qquad i = 1, 2, \ldots, s,$$

where we have written $\hat{x}(\cdot)$ to represent the predicted future states. So we see here that there is an additional factor that enters into the computation of the control law to use at each time, namely, the predicted future states. One might object that since the predicted future states $\hat{x}(t+T)$ are created by the model M from past states and controls, the anticipatory scheme is basically just a special type of classical feedback control process in which the control law is a somewhat more complicated function of the state than usual. This observation is true, but it's far from clear that it constitutes a fatal flaw. Quite the contrary, in fact.

First of all, the fact that the predicted states are functions of the past enables us to preserve the causal relationship between the system state and its inputs: the next state depends only upon the current and past states and controls. Second, if it's an objection that the predicted states used in generating the control law u are functions of the past states and controls, then the same objection applies to the standard adaptive control problem involving the estimated parameters $\hat{a}$, which themselves are determined solely from past states and inputs. Needless to say, there are good reasons to retain the distinction between the states and estimated parameters in the classical set-up, and we shall attempt to show in the remainder of the chapter that the same kind of distinction is crucial when we come to consider anticipatory processes. There really is something new that's been added when we pass to anticipatory systems, the model M, and it's something that dramatically changes the character of the kind of behavior we can expect and the ways in which we can interact with the underlying system Σ. Rather than belabor this point in the abstract, let's look at a specific example.

Example: A Chemical Reaction Network

To illustrate some of the ideas associated with anticipatory processes, we consider an example due to Rosen of a biosynthetic reaction network. If we let A_i represent a chemical substrate at stage i, and E_i the enzyme that catalyzes it, we can graphically depict the network as in Fig. 7.4.

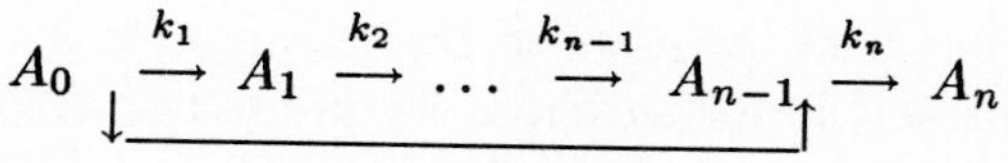

$$A_0 \xrightarrow{k_1} A_1 \xrightarrow{k_2} \cdots \xrightarrow{k_{n-1}} A_{n-1} \xrightarrow{k_n} A_n$$

Figure 7.4 A Chemical Reaction Network

In the diagram, the quantity k_i is the reaction rate for the ith reaction, $i = 1, 2, \ldots, n$. Further, we have assumed there is a forward activation step in the network so that the concentration of substrate A_0 serves to activate the production of A_n. That is, the concentration $A_0(t)$ *predicts* the concentration of A_n at some *future* time $t + T$. So we choose $k_n = k_n(A_0)$ to embody this forward activation step, leaving all other reaction rates k_i constant, $i = 1, 2, \ldots, n-1$.

Under the foregoing hypotheses, the rate equations for the system are

$$\frac{dA_i}{dt} = k_i A_{i-1} - k_{i+1} A_i,$$
$$\frac{dA_n}{dt} = k_n(A_0) A_{n-1}, \qquad i = 1, 2, \ldots, n-1.$$

Let's assume that the purpose of the forward activation step is to stabilize the level of substrate A_{n-1} in the face of ambient fluctuations in the initial substance A_0. We can represent this requirement mathematically by the condition

$$\frac{dA_{n-1}}{dt} = 0,$$

independently of A_0. Thus, from the rate equations this condition means that we must have

$$k_{n-1} A_{n-2} = k_n(A_0) A_{n-1}.$$

We shall achieve this condition by choosing the functional form of $k_n(A_0)$, which will then embody the predictive model implicit in the forward loop of the network.

From the linear dynamics, it's easy to see that

$$A_{n-2}(t) = \int_0^t K_1(t-s) A_0(s)\, ds,$$
$$A_{n-1}(t) = \int_0^t K_2(t-s) A_0(s)\, ds,$$

where K_1 and K_2 are functions determined entirely by the reaction rates k_i, $i = 1, 2, \ldots, n-1$. These expressions show explicitly that the value of A_0 at a given moment determines the values of A_{n-1} and A_{n-2} at a *later* time.

The control condition on $k_n(A_0)$ now becomes

$$k_n(A_0) = k_{n-1} \frac{\int_0^t K_1(t-s) A_0(s)\, ds}{\int_0^t K_2(t-s) A_0(s)\, ds}, \tag{$*$}$$

i.e., the reaction rate k_n at any time t is determined by the value of A_0 at a prior instant $t - T$ or, equivalently, the value of $A_0(t)$ determines k_n at a *future* time $t + T$. Consequently, we see the manner in which the initial substance A_0 serves to *adapt* the pathway so as to stabilize the condition that

$$\frac{dA_{n-1}}{dt} = 0.$$

Finally, we note that the homeostasis maintained in the pathway is obtained entirely through the modeling relation between A_0 and A_{n-1} by virtue of the relation (*) linking the prediction of the model to the actual rate k_n. That is, the homeostasis is preserved entirely through adaptation generated on the basis of a *predicted* value of A_{n-1}. In particular, there is no feedback in the system and no mechanism available for the system to "see" the value of the quantity that is controlled.

In the above scheme, it's important to note that the anticipatory linkage between A_0 and A_{n-1} will be adaptive only as long as the relation (*) holds, i.e., only as long as the linkage "wired-in" by the forward activation step and the actual linkage dictated by the chemical kinetics remain the same. If there should be a departure, then the rate change given by (*) will become *maladaptive* to a degree measured by the magnitude of the departure. Since such deviations can be expected to occur for *all* real processes, we conclude that a forward activation step of the above type can retain its adaptive character only for a characteristic that is time-dependent upon the nature of the larger system within which it is embedded and upon the character of its interactions with that larger system. This phenomenon, which is termed *temporal spanning,* has no analogue in nonanticipatory systems. In the next section we demonstrate how the notion of temporal spanning can give rise to another kind of feature that is also absent from classical reactive systems: global system failure.

5. *Global Failures and Perfect Subsystems*

In today's world we are all too depressingly aware of the phenomenon of global system failure without any apparent local cause. National economies limp along like war-tattered refugees as politicians, economists, and others attribute the problems to inflation, labor unions, interest rates, foreign competition and a plethora of other local "causes," calling for action to address the global problem by addressing one or the other of these individual components of the overall system. Other examples of this inability to attribute system failure to local causes abound in both the social and biological spheres, so it's of interest to show how this is exactly the type of behavior that can be *expected* from any system of an anticipatory type, i.e., a system that contains a predictive feed-forward loop and uses the predictions of the loop as a basis for system control.

Imagine we have a system Σ composed of a collection of subsystems S_i, $i = 1, 2, \ldots, N$. Assume that the behavior of each subsystem is given by the input/output relation

$$\phi_i(u_i(t), y_i(t+h)) = 0, \qquad u_i \in R^m, \quad y_i \in R^p, \qquad i = 1, 2, \ldots, N,$$

where $u_i(t)$ and $y_i(t+h)$ are the inputs and outputs of subsystem S_i at times t and $t+h$, respectively. Thus, the subsystems receive inputs that may be from either the external environment or the outputs of other subsystems, and produce outputs at some later instant according to the prescription ϕ_i. We assume that as far as the subsystem S_i is concerned, everything is operating as it should be whenever the input/output pair $(u_i, y_i) \in \Omega_i$, where Ω_i is some given subset of $R^m \times R^p$. Furthermore, the overall system Σ has its own input/output relation

$$\Phi(v, w) = 0, \qquad v \in R^n, \quad w \in R^q,$$

where both the input and output v and w are from/to the external environment of Σ. Let's agree that Σ is operating satisfactorily whenever the input/output pair $(v, w) \in \Omega$, where Ω is some appropriate subset of $R^n \times R^q$.

We now examine the logical possibilities between the following propositions:

A. Each subsystem S_i is operating properly, $\quad i = 1, 2, \ldots, N$.

B. The system Σ is operating as it should.

There are four cases to consider:

- $A \Rightarrow B$—In this case, proper operation of the subsystems implies proper operation of Σ. Note that this leaves open the possibility that the system Σ could function properly even if some of the subsystems were operating incorrectly. Using the contrapositive of the above implication, we see that failure of Σ implies failure of at least one subsystem. Hence, every global failure of Σ arises from the failure of some local subsystem.

- $B \Rightarrow A$—Here the correct functioning of the overall system implies that each subsystem is functioning correctly. The contrapositive implication is now that if a component subsystem functions incorrectly, then so does the global system. However, note that this conclusion leaves open the possibility that there may exist modes of failure of the total system that do not stem from the failure of any individual subsystem. It is this possibility that we will explore in more detail in a moment.

- A and B are logically equivalent—This case implies that every global failure gives rise to the failure of some subsystem, and conversely. There can be no global failures in this situation that do not arise from the failure of some subsystem.

- A and B are logically independent—In this situation there is no causal connection between the conditions under which Σ functions correctly and those conditions under which the subsystems operate properly. Thus, the set Ω cannot be decomposed into properties of the local component sets Ω_i, and hence, we cannot define Ω without reference to the total system.

Now let's show how any system possessing a feedforward loop of the type described earlier leads to the second type of logical implication discussed above.

Consider the anticipatory system displayed in Fig. 7.5.

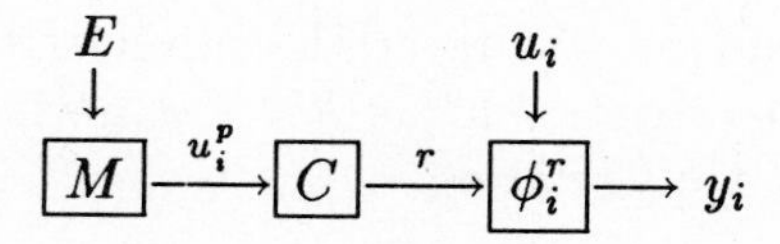

Figure 7.5 An Anticipatory Loop for Subsystem S_i

Here we have the model M of subsystem S_i processing the environmental input E (as well as other information) in order to produce the quantity u_i^p, the *predicted* value of the input to S_i at some future time $t+h$. This predicted value is transmitted to the local controller C who then uses it to compute the value of a parameter vector r that is employed in the input/output law ϕ_i^r which governs how subsystem S_i will process its *actual* input u_i at $t+h$. Now let's see how this set-up can lead to a global system failure.

First of all, we have assumed that the correct behavior of the overall system Σ means that the system output lies in the acceptable region Ω. Assume now that all components of Σ aside from those involved in the above loop for S_i are functioning properly. Under these circumstances, the acceptable functioning of Σ becomes directly related to the adaptation of the component ϕ_i^r in the feedforward loop above. As long as the predicted input u_i^p agrees with the actual input u_i, we necessarily have the state of Σ in Ω. In other words, proper behavior of the global system Σ is now directly related to the fidelity of the model M in our feedforward loop.

Since the predictive model is, in general, closed to interactions that the subsystem S_i is open to, there will necessarily be an increasing deviation between the predictions of the model and the actual inputs experienced by the subsystem. As a result of these deviations between $u_i(t+h)$ and $u_i^p(t+h)$, the preset value of the parameter vector r will render the rule ϕ_i^r increasingly less competent to deal with the actual inputs. In general, there will be some characteristic time T at which the deviation becomes so great that ϕ_i^r is actually *incompetent,* and because of the direct relation between the competence of subsystem S_i and the maintainence of the state of the overall system Σ in Ω, the overall system will fail at this time T. We can identify this time T as the intrinsic life span of the system Σ. Finally, we can now see that the global failure of Σ is not the result of a local failure since, by hypothesis, every nonmodeled component of Σ (including those involved in the feedforward loop) is functioning correctly *according to any local criterion.*

The heart of the argument just presented lies in the observation that there is no local criterion that will enable the controller of subsystem S_i to recognize that anything has gone wrong. As far as C is concerned, he receives the prediction u_i^p and processes it *correctly* into the parameter r. Similarly, the operator of subsystem S_i receives the instructions r from the controller and *correctly* executes the rule ϕ_i^r specifying how the actual input to i is to be transformed into the output. The problem is back upstream at the level of the model M, a problem that *cannot be seen* by either C or the operator of the subsystem. The loss of competence of ϕ_i^r to process the input to S_i can only be expressed in terms of the behavior of the entire system Σ, and not with reference to the feedforward loop considered in isolation. Thus, we have demonstrated that any system with a feedforward loop possesses a global mode of failure that is not identifiable with any kind of local failure. Now let's see how this general scheme relates to the biochemical example discussed earlier.

Example: A Biosynthetic Pathway

We consider the chemical reaction network displayed earlier in Fig. 7.4. In this simple case the substrate A_0 acted as a *predictor* for future values of concentration of the substrate A_n. Thus our model M is just $M = \{A_0\}$. The subsystem S_i in this case is just the substrate A_n, so we take $i = n$ and $S_n = \{A_n\}$. Finally, the parameter r that's generated by the "controller" of S_n is just the reaction rate k_n involved in the dynamical processing of substrate A_{n-1}. So we take $r = k_n$ with the dynamics ϕ_n^r being just the relation linking A_n with A_{n-1}. The crucial point of this situation is that the adaptation quantity k_n is determined solely by the *predicted* value of the input.

Having now seen some of the major differences between feedback and feedforward controllers, it's perhaps useful to pause briefly to summarize what we have learned so far.

6. *A Taxonomy of Adaptive Systems*

In the preceding pages we have seen a variety of properties of both feedback and feedforward control systems. One of the most important distinctions that we can draw between these two classes of systems is the type of information that they employ in carrying out their regulatory function. Since the utilization of information lies at the heart of any sort of adaptive control procedure, it's evident that any significant differences in the manner in which information is used to control the system will generally result in radically different types of control and, hence, system behavior. Here we briefly outline a crude taxonomy of adaptive control mechanisms to show the fundamental differences between those based upon feedback as opposed to those that employ feedforward loops of the type discussed above.

Roughly speaking, there are four types of qualitatively distinct adaptive mechanisms, depending upon whether the adaptive loop is feedback or feedforward with or without memory. Let's take a brief look at each of these types of systems and their characteristic behaviors.

- *Class I: Feedback Regulators, No Memory*—The identifying feature of feedback controllers is that they operate solely from measurements of the controlled system; there is no sensing of the environment. Typically the system has some desired "set-point," and the controller monitors the deviation of the system from this set-point, exerting control appropriate to the deviation of the actual system behavior from its desired level. It should be emphasized that this is error-actuated control, and that the controller does nothing if the current state of the system is in the acceptable region. The simplest types of feedback homeostats have no memory, with the present activity of the controller being independent of its past behavior.

- *Class II: Feedback Regulators with Memory*—In this class we put all feedback controllers that can make a memory trace of their past actions and employ this memory to change various aspects of the system, such as the set-point or parameters like the time constants in the control loop. Any such system can be thought of as a *learning* system, but just as with feedback regulators without memory, such systems are still special-purpose devices. That is, once fabricated and operating, their behavior is fixed; the special-purpose hardware needed for their operation is difficult to modify for other purposes.

- *Class III: Feedforward Regulators, No Memory*—Whereas feedback regulators are characterized by using only the state of the controlled system for their operation, feedforward controllers have a quite different set of properties. First of all, the controller can sense aspects of its environment that can then be correlated with future states of the controlled system. This correlation between the environment at time t and the state at some time $t + h$ is "hard-wired" into the controller. Further, such a controller can modify the dynamics of the controlled system in accordance with the values of both the current state of the environment and the *predicted* future state of the system. Thus, feedforward controllers contain a *model* of the external world embodied in the correlation between the controlled system state and the state of the environment.

It is important to note that the feedforward controller operation is no longer primarily determined by an error signal as in feedback control systems. Now the operation is off of regularities in the environment rather than off of the randomness which generates the error signal used by feedback controllers. This difference is significant, because in a feedback system the behavior of the system has already deteriorated before the controller can begin to act. Since any feedback loop has an intrinsic time constant, it can happen that performance has irreversibly deteriorated before the controller can take action. In addition, in environments that fluctuate too rapidly, the controller may itself track only the fluctuations rather than stabilize the system. Feedforward controllers avoid these difficulties entirely.

The model of the external world built into a feedforward controller introduces the concept of "software" into adaptive mechanisms, since by changing the model we can modify the entire system without completely redesigning it. However, the behavior of such a controller is still primarily hardware-dependent.

- *Class IV: Feedforward Regulators with Memory*—Feedforward controllers with memory can employ their past behavior to modify their internal models. Thus, in some sense these systems are capable of re-programming themselves. Perhaps ironically, such types of adaptive systems must employ features of the Class I feedback regulators in their operation, since they must employ an error signal in order to modify their internal model. So they suffer from the same limitations as all feedback systems. Nevertheless, there is an enormous gain in adaptive power with systems that can modify their models of the world on the basis of past experience.

Keeping the above properties of controllers in mind, we can begin to address a host of questions involving both biological and social organization pertaining to the theme of adaptation. In particular, we can consider the degree to which the adaptive capability of individual units interacts with the adaptive capacity of the overall system of which they are a part. In the last section we saw that if the control mechanism were of either Class III or IV such local adaptivity could nevertheless lead to global maladaptation. On the other hand, it's through defects in the adaptability of local units that higher-level associations with new properties can be generated and maintained, the corollary being that too much capacity in the units precludes or destroys such global associations.

Much of our interest in studying adaptive processes in a social context is derived from the feeling that many of the social organizations of modern life are malfunctioning, and that we must take steps to make them more "adaptive," in some sense. The preceding discussion suggests that at least some of the maladaptation arises from the fact that our systems are *too adaptive.* They have, in effect, too comprehensive a view of what constitutes their own survival, too much access to environmental variables through technology and the media, too many models correlating behavior and the environment from a multitude of special interest groups—in short, too much adaptation at too low a level generated by too much special-purpose hardware. Utilizing the more technical aspects of the above feedback and feedforward systems, together with the lessons from biological organization, we have at least the raw ingredients needed to begin to attack these sorts of questions. But for now let's set these cosmic themes aside and lower our level of rhetoric to consider an example of an anticipatory process in which the role of the predictive model can be more explicitly displayed—the case of a living cell.

7. *Metabolism-Repair Systems*

The quintessential example of a system containing an internal model of itself used to modify its behavior is a living cell. Functionally, we can regard the cell as possessing two components: a *metabolic* part representing the basic chemical activity characterizing the particular cellular function, and a *maintainence* or *repair* component that ensures the continued viability of the cell in the face of external disturbances of many sorts. Biologically, these two quite disparate functional activities of the cell are what we generally attribute to the *cytoplasm* and *nucleus* of the cell, respectively. Thus, we may think of the cell as a small chemical factory that receives inputs from its external environment and processes these inputs into outputs prescribed by the instructions coded into the cellular DNA. Further, the DNA also contains the necessary instructions for both how the cell should operate and how it should be constructed. This genetic material provides a complete internal model of all aspects of the cell. Our goal is to abstract the essence of this situation into a mathematical representation that will enable us to see each of these functional activities and their mutual interrelations.

Let's assume that Ω represents the set of environmental inputs that the cell receives, while Γ is the set of possible outputs that a cell could produce. These sets are determined primarily by the kind of operating environment that the cell can exist within, as well as by the laws of chemistry and physics constraining the nature and type of chemical reactions that can take place in a given environment. Further, let $H(\Omega, \Gamma)$ represent the set of physically realizable metabolisms that the cell could display. Again, $H(\Omega, \Gamma)$ is determined by the particular operating environment, the laws of Nature, and so on. With these notations, we can represent the metabolic activity of the cell as a map

$$\begin{aligned} f\colon \Omega &\to \Gamma, \\ \omega &\mapsto \gamma. \end{aligned}$$

At this stage, we immediately recognize cellular metabolism as an illustration of the type of input/output relation discussed earlier in another context in Chapter Three. Consequently, if the metabolic map f were linear we would have at our disposal all the machinery developed there to bring to bear upon our study of cellular metabolic behavior. But there is much more to a living cell than metabolism alone, so we must now examine how to add the maintainence/repair function to the metabolic machinery.

We begin by noting that the functional role of the repair component is to restore the metabolism f when there is some fluctuation in either the ambient environment ω or a disturbance to the metabolic machinery f itself. It's crucial to note the way in which the cell's metabolic component acts: the external input ω, which is actually a *time-series* of inputs, is processed into the metabolic output γ, also a time-series of output products. At this point the cellular metabolism f disappears, and it is the role of the repair function to utilize some of the cellular output to produce a new metabolic map that will be used to process the next input sequence, call it $\hat{\omega}$. We can imagine that during the course of evolutionary time, the cell has evolved in a manner such that its design, or basal, metabolism has been wired-up to produce the output γ when the input is ω. Thus it must be the case that if $\hat{\omega} = \omega$, then the cellular repair mechanism should produce the metabolism f; i.e., if nothing has changed from the "design specs," then the repair operation doesn't have to do any real "repair" but only produce more of the same, so to speak. Putting all these remarks together, we see that the repair function can be abstractly represented by the map

$$P_f \colon \Gamma \to H(\Omega, \Gamma),$$
$$\gamma \mapsto f,$$

where we have written the subscript f to indicate the "boundary condition"

$$P_f(f(\omega)) = f,$$

imposed on the repair machinery P_f. The role of the repair operation is to counteract, or at least stabilize, the cell's metabolic behavior in the face of the inevitable fluctuations in the outside world and/or the inherent errors of the metabolic machinery f. In short, P_f is an error-correcting mechanism. But who (or what) corrects the errors in P_f? Who repairs the repairers? It does no good to introduce another level of repair into the situation, as the incipient infinite regress is obvious. But what's the alternative?

Nature's solution to the repair problem is simplicity itself: just throw the old repair mechanism away and build a new one! In other words, *replication.* The basic scheme is very straightforward: the cell's genetic component (essentially the DNA) receives information about the cell's metabolic processing apparatus f and then transforms this information into a copy of the repair map P_f. We can represent this reproductive role of the cell by the map

$$\beta_f \colon H(\Omega, \Gamma) \to H(\Gamma, H(\Omega, \Gamma)),$$
$$f \mapsto P_f.$$

Just as with the repair map P_f, the replication operation is subject to the boundary condition that "if it ain't broken, don't fix it," leading to the condition that if the design metabolism is presented, the map β_f should transform it into the associated repair map P_f. Thus, $\beta_f(f) = P_f$. We can summarize our discussion so far by the diagram shown in Fig. 7.6.

$$\Omega \xrightarrow{f} \Gamma \xrightarrow{P_f} H(\Omega, \Gamma) \xrightarrow{\beta_f} H(\Gamma, H(\Omega, \Gamma))$$

Figure 7.6 An Abstract Metabolism-Repair System

The preceding development provides only the bare bones minimum insofar as describing the operation of a living cell. Let's now consider a number of points about Fig. 7.6 in its relation to both the ideas of anticipation considered earlier, as well as to the relationship between the *abstract* metabolism-repair (M,R)-system sketched in the figure and the behavior of *real* cells.

- *Implicit Control*—Classical control systems involve a direct intervention in the system dynamics by the controller. In the (M,R) set up above, there are no direct control inputs being injected into the system at any stage of the operation; in fact, there is no controller in the sense that that term is used in control engineering. The only signals coming into the cell from the outside are those generated by the environment. So in this sense the control is *implicit,* being generated solely from the model requirements that $P_f(f(\omega)) = f$ and $\beta_f(f) = P_f$, and the feedbacks from the repair component to the metabolism and from the replication component to the repair. This distinction between implicit vs. explicit control makes it clear that the kind of system we are speaking about here differs radically from the sort of adaptive model-matching procedures commonly employed in the control engineering literature.

- *Translation and Transcription*—In our discussion of cellular processes in Chapter Two, we saw that the information coded into the cellular DNA is used in two quite distinct fashions. It's *translated* into instructions for assembling RNA and various types of amino acids and, hence, protein structures, and it's *transcribed* in order to replicate the DNA in the process of cellular reproduction. Since the translation operation involves the production of material for both the construction of amino acids and construction of the repair machinery itself, it's fair to say that this use of the DNA information is tantamount to building the metabolic machinery f. But from our diagram in Fig. 7.6, this is exactly the role of the repair map P_f ; hence, we can make the rough identification of P_f with the operation of *translating* the cell's DNA. It follows that the transcription operation leading to genetic replication is identifiable with the replication map β_f. Putting these observations together, we see that the abstract diagram of Fig. 7.6 can be identified with the actual cellular operation as indicated in Fig. 7.7.

$$\underbrace{\Omega \xrightarrow{f} \Gamma}_{\text{metabolism}} \underbrace{\xrightarrow{P_f}}_{\text{translation}} H(\Omega,\Gamma) \underbrace{\xrightarrow{\beta_f}}_{\text{transcription}} H(\Gamma, H(\Omega,\Gamma))$$

Figure 7.7 Abstract (M,R)-Systems and Real Cellular Functions

• *Genetic Mutations and "Lamarckism"*—The Central Dogma underlying modern molecular biology is that modifications to the cellular genotype cannot arise as a result of transmission from the environment ω through the repair component, but only through direct intervention in the replication map β_f itself. In short, mutations come about either from direct external perturbation of β_f, as by radiation or cosmic rays, or by means of internal perturbations such as copying errors of one sort or another. But there can be no transmission of changes in ω generating changes in β_f. Since we have not yet addressed the issue of *how* the maps P_f and β_f are constructed, it's not clear from the diagrams of Figs. 7.6 and 7.7 whether such types of *Lamarckian* changes can take place within the framework of an (M,R)-system. Since it would be of some value to be able to employ the above abstract framework in other contexts such as economics and business, in which there most assuredly is the possibility for Lamarckian change, it would be desirable for the (M,R)-systems to allow the possibility of such changes, at least under suitably restricted circumstances. We shall address these matters in detail in the next section.

The abstract cellular structure outlined here gives rise to an endless list of questions as to how the mathematical skeleton can be given flesh in order to address matters pertaining to the way in which the genetic and metabolic components of the cell interrelate in the performance of its function. Two questions of the most immediate import are: Under what circumstances can we make the repair and replication components emerge directly from the metabolic machinery alone? And what type of disturbance in the the environmental input ω and/or in the metabolic map f can the repair component counteract?

The first of these questions has obvious relevance to the origin of life issue, since a principal point of debate in such circles centers about what came first, metabolism or replication. Any mathematical light we can shed upon conditions that may favor one school or the other, not to mention providing some insight into the possibilities for various types of exo-biologies, would be of considerable value. The second question bears upon the issue of cellular "resilience" to a spectrum of disturbances, both internal and external. Answers to this kind of question will go a long way toward furthering our understanding of the mechanisms by which living systems adapt to change. But in order to address either of these questions, we need to move the abstract framework described above onto a more concrete level by imposing additional structure on the maps and sets in Fig. 7.7. Although it's fine for gaining an overview of the basic concepts of living systems to remain at the very general levels of abstract sets and relations, when it comes to giving concrete answers to concrete questions, we need more structure in order to say anything even halfway interesting and detailed. Our goal in the next section is to provide a starting point for such a program directed to the study of metabolism-repair systems.

8. *Linear (M,R)-Systems: Repair*

The simplest setting in which to address the matters outlined above is when everything in sight is linear. In this case the input and output spaces Ω and Γ are vector spaces, while the metabolic map f is linear and, to make things as simple as possible, autonomous, i.e., constant. Thus we take our mathematical setting to be exactly that considered earlier in Chapter Three, and we regard the cellular metabolism as an *information-processing* object as that term was used in our previous discussion of behaviorism.

Speaking more precisely, the cellular input and output spaces are assumed to be sequences of vectors taken from R^m and R^p. Thus,

$$\Omega = \{\omega\colon \omega = (u_0, u_1, u_2, \ldots, u_N)\}, \qquad u_i \in R^m,$$
$$\Gamma = \{\gamma\colon \gamma = (y_1, y_2, y_3, \ldots)\}, \qquad y_i \in R^p.$$

Note that to respect causality, we have assumed the cellular outputs begin one time unit after receipt of the inputs. Further, for technical reasons alluded to already in Chapter Three, we assume there are only a *finite* number of inputs applied but that the outputs could, in principle, continue indefinitely. In view of the linearity assumptions made on the metabolic map f, we can express the relationship between the cellular inputs and outputs in the form

$$y_t = \sum_{i=0}^{t-1} A_{t-i} u_i, \qquad t = 1, 2, \ldots, \tag{I/O}$$

where the coefficient matrices $A_k \in R^{p \times m}$. Under mild technical conditions (See Problem 4) on the pair (ω, γ), the elements $\mathcal{B} = \{A_i\}$ are uniquely determined by the input/output sequence. Thus, for all intents and purposes we can identify the metabolic map f with the behavior sequence $\mathcal{B}$, and we shall routinely move back and forth between the two in what follows. Hence,

$$f \longleftrightarrow \mathcal{B}.$$

The relation (I/O) can also be written as

$$y_t = \sum_{i=0}^{t-1} [A_{t-i}^{(1)} | A_{t-i}^{(2)} | \cdots | A_{t-i}^{(m)}] u_i, \qquad t = 1, 2, \ldots,$$

where $A_j^{(r)}$ denotes the rth column of the matrix A_j, and the block row vector $[A_j^{(1)} | A_j^{(2)} | \cdots | A_j^{(m)}]$ consists of those elements comprising the columns of A_j. We will employ this notation throughout the remainder of this section.

In Chapter Three we saw that the input/output pair (ω, γ) can be canonically realized by an internal model of dimension $n < \infty$ if and only if the infinite Hankel array

$$\mathcal{H} = \begin{pmatrix} A_1 & A_2 & A_3 & \cdots \\ A_2 & A_3 & A_4 & \cdots \\ A_3 & A_4 & A_5 & \cdots \\ \vdots & \vdots & \vdots & \end{pmatrix}$$

has rank n. In this case there exist matrices (F, G, H) such that the input/output behavior of the system

$$\begin{aligned} x_{t+1} &= F x_t + G u_t, \qquad x_0 = 0, \qquad x_t \in R^n, \qquad t = 0, 1, \ldots, \\ y_t &= H x_t, \end{aligned} \tag{Σ}$$

agrees with that of the system (I/O); i.e., $A_t = H F^{t-1} G$. Furthermore, the system Σ is completely reachable and completely observable, i.e., canonical. Under the *assumption* that the system (I/O) possesses a finite-dimensional realization Σ, we saw in Chapter Three how to compute the elements of this realization $\Sigma = (F, G, H)$ directly from the Hankel array $\mathcal{H}$. These by now almost classical facts from linear system theory will form the basis for our attack upon the question of how to calculate the repair and replication maps P_f and β_f from the cellular metabolism f.

Our first task is to see how to calculate the cellular repair map P_f from the metabolic ingredients f, Ω, Γ and $H(\Omega, \Gamma)$. For future reference, we denote the design, or *basal,* metabolism by

$$\begin{aligned} f^* : \Omega &\to \Gamma, \\ \omega^* &\mapsto \gamma^*. \end{aligned}$$

That is, the cell is operating at its design level when in the environment ω^* the output produced is γ^*, with the corresponding metabolic map being f^*. From our earlier discussion, we know that the repair map must satisfy the relation

$$P_f(\gamma) = P_f(f(\omega)) = f,$$

for any basal metabolism f and associated output γ. Since we have assumed the repair map is linear, the above relation can be written

$$w_\tau = \sum_{i=0}^{\tau-1} \mathcal{R}_{\tau-i} v_i, \qquad \tau = 1, 2, \ldots,$$

where (v_i, w_i) are the input and output to the repair subsystem, with the elements $\{\mathcal{R}_j\}$ being linear maps determined by γ and f. However, when the metabolism is operating correctly, the repair system must accept the input γ^* and produce the output f^*. Consequently, we must have

$$v_\tau = \mathcal{S}(y^*_{\tau+1}), \qquad w_\tau = A^*_\tau,$$

where $\mathcal{S}$ = the "stack" operator having the action of stacking the columns of an $n \times m$ matrix into a column vector of nm components. Note that here we have used a time parameter different than for the metabolic system as it will usually be the case that the time-scale of operation of the repair system differs from that of the metabolism.

It's an easy exercise to see that the coefficients for the repair map must have the form

$$\mathcal{R}_j = [B_{j1}|B_{j2}|\cdots|B_{jp}], \qquad B_{js} \in R^{p \times m}, \qquad j = 1, 2, \ldots .$$

Consequently, in component form we can write the input/output relation for the repair map as

$$w_\tau = \sum_{i=0}^{\tau-1} [\mathcal{R}^{(1)}_{\tau-i}|\mathcal{R}^{(2)}_{\tau-i}|\cdots|\mathcal{R}^{(p)}_{\tau-i}]\mathcal{S}(v_i),$$

where we have written $\mathcal{R}^{(s)}_j \doteq B_{js}$.

Just as the metabolism f was represented by the sequence $\{A_1, A_2, \ldots\}$, we can now see that the repair system P_f can be represented by the sequence

$$P_f \longleftrightarrow \{\mathcal{R}_1, \mathcal{R}_2, \ldots\}.$$

There are two important points to note about the foregoing development:

1) If we write each element A_i as

$$A_i = [A^{(1)}_i|A^{(2)}_i|\cdots|A^{(m)}_i], \qquad A^{(j)}_i \in R^p,$$

the *complexity* of each component of the metabolic map f is $O(pm)$. By the same reasoning, the complexity of each element $\mathcal{R}_j$ of the repair map P_f is $O(p^2 m)$. Thus, already we see that the often noted complexity increase associated with living systems begins to emerge through purely natural mathematical requirements.

2) An immediate calculation shows that $\dim \Sigma = n < \infty$ implies the elements $\{A_1, A_2, \ldots, A_{2n}\}$ are linearly dependent. This condition also implies that the canonical realization of the repair sequence $\{\mathcal{R}_1, \mathcal{R}_2, \ldots\}$ has dimension $n_P \leq n$. Consequently, we can again employ Ho's Algorithm to produce a system $\Sigma_P = (F_P, G_P, H_P)$ realizing the repair map P_f.

Example: The Natural Numbers

To fix the foregoing ideas, suppose we have a cell whose "correct" environmental input is

$$\omega^* = \{1, 1, 0, 0, \ldots, 0\},$$

and whose proper metabolic action is to produce the set of natural numbers

$$\gamma^* = \{1, 2, 3, 4, \ldots\},$$

i.e., $f^*(\omega^*) = \gamma^*$. In this case it's easy to work out that the metabolic map f^* has the form

$$f^* \approx \{A_1^*, A_2^*, \ldots\} = \{1, 1, 2, 2, 3, 3, \ldots\}.$$

It can be shown that this behavior sequence has a canonical realization of dimension $n^* = 3$, so using Ho's Algorithm as in Chapter Three we obtain the canonical system matrices

$$F^* = \begin{pmatrix} 0 & 1 & 0 \\ 1 & -1 & 1 \\ 1 & -2 & 2 \end{pmatrix}, \qquad G^* = \begin{pmatrix} 1 \\ 1 \\ 2 \end{pmatrix}, \qquad H^* = (1 \quad 0 \quad 0).$$

The dynamics for the metabolic subsystem are then

$$x_{t+1}^* = \begin{pmatrix} 0 & 1 & 0 \\ 1 & -1 & 1 \\ 1 & -2 & 2 \end{pmatrix} x_t^* + \begin{pmatrix} 1 \\ 1 \\ 2 \end{pmatrix} u_t^*, \qquad x_0 = 0, \qquad (\Sigma^*)$$

$$y_t^* = (1 \quad 0 \quad 0)\, x_t^*, \qquad t = 0, 1, 2, \ldots.$$

Turning now to the repair component, we must have $P_{f^*}(\gamma^*) = f^*$ which leads to

$$\mathcal{R}_i^* = \begin{cases} (+1), & i \text{ odd} \\ (-1), & i \text{ even}. \end{cases}$$

Thus, the Hankel array associated with the map P_{f^*} is given by

$$\mathcal{H}_{P_{f^*}} = \begin{pmatrix} 1 & -1 & 1 & -1 & 1 & \cdots \\ -1 & 1 & -1 & 1 & -1 & \cdots \\ 1 & -1 & 1 & -1 & 1 & \cdots \\ \vdots & \vdots & \vdots & \vdots & \vdots & \cdots \end{pmatrix}.$$

Since we know the repair sequence has a canonical realization of dimension $n_P^* \leq n^* = 3$, experimenting a bit with Ho's Algorithm (or computing rank $\mathcal{H}_{P_{f^*}}$) gives $n_P^* = 1$, with the resultant canonical repair realization $\Sigma_P^* = (F_P^*, G_P^*, H_P^*)$, where

$$F_P^* = (-1), \qquad G_P^* = (1), \qquad H_P^* = (1).$$

The repair dynamics are then

$$\begin{aligned} z_{\tau+1}^* &= (-1)z_\tau^* + (1)v_\tau^*, \qquad z_0^* = 0, \\ w_\tau^* &= (1)z_\tau^*, \qquad \tau = 0, 1, 2, \ldots . \end{aligned} \qquad (\Sigma_P^*)$$

From our earlier remarks, we connect this system with the metabolic map f^* via inputs and outputs as $w_\tau^* = A_\tau^*$, $v_\tau^* = y_{\tau+1}^*$.

Remark

At first glance, there appears to be a contradiction here between our earlier claim that the repair system is more "complex" than the metabolism, and the fact that $\dim \Sigma_P^* = 1 < \dim \Sigma^* = 3$, at least if one measures complexity by dimensionality of the system state-space. However, our earlier observation used a somewhat different notion of complexity, one involving the objects of our behavioral description, the elements A_i and $\mathcal{R}_i$. Unless $p = 1$, the objects $\{\mathcal{R}_i\}$ always contain more elements than the objects $\{A_i\}$. Therefore, by this measure of complexity the repair system will always be at least as complex as the metabolism. Roughly speaking, it's more difficult to *describe* the behavior of the repair process, but it may be simpler to *realize* its dynamics. In engineering terms, there may be fewer "integrators" but of a more complicated type.

The preceding development has shown how we can always construct the system repair map P_f directly from any prescribed metabolism f. For the moment let's defer consideration of how to construct the associated replication map β_f and instead turn our attention to the second of our two main classes of questions: under what circumstances will the canonical repair map P_{f^*} restore, or at least stabilize, disturbances in the cellular environment ω and/or the metabolic map f? We consider each of these possibilities in turn.

• *Case I: Fixed Environment ω^* and a Fixed Genetic Map P_{f^*} with Variable Metabolism f*—In this case our concern is with perturbations to the basal metabolism f^* when the environmental inputs and the genetic repair map are fixed. If we assume the metabolic disturbance $f^* \to f$, we are interested in categorizing those maps f such that either

$$P_{f^*}(f) = f \quad \text{or} \quad P_{f^*}(f) = f^*.$$

In the first case, the repair machinery stabilizes the metabolism at the new level f, whereas in the latter case the repair machinery restores the original metabolism f^*.

To study this situation, it's useful to introduce the map

$$\Psi_{f^*,\omega^*} : H(\Omega, \Gamma) \to H(\Omega, \Gamma),$$
$$f \mapsto P_{f^*}(f(\omega^*)).$$

The case in which the repair system stabilizes the system at the new metabolism f corresponds to finding the fixed points of the map Ψ_{ω^*,f^*}, i.e., those metabolisms f such that

$$\Psi_{f^*,\omega^*}(f) = f,$$

but the situation in which the repair system restores the basal metabolism f^* corresponds to finding those perturbations f such that

$$\Psi_{f^*,\omega^*}(f) = f^*.$$

The map Ψ_{f^*,ω^*} is clearly linear, being a composition of the linear maps f and P_{f^*}. Further, by construction we must have

$$\Psi_{f^*,\omega^*}(f^*) = f^*,$$

so that f^* is a trivial fixed point of Ψ_{f^*,ω^*}. In addition, we also have the further trivial fixed point $f = 0$. Since each $f \in H(\Omega, \Gamma)$ has the representation $f \approx \{A_1, A_2, \ldots\}$, we can formally represent the map Ψ_{ω^*,f^*} by the infinite array

$$\Psi_{f^*,\omega^*} = \begin{pmatrix} \Psi_{11}^* & \Psi_{12}^* & \Psi_{13}^* & \cdots \\ \Psi_{21}^* & \Psi_{22}^* & \Psi_{23}^* & \cdots \\ \vdots & \vdots & \vdots & \vdots \end{pmatrix},$$

where each $\Psi_{ij}^* \in R^{p \times p}$. We shall return in a moment to a more detailed consideration of the structure of this matrix. For now it's sufficient just to note that any metabolism $f = \{A_1, A_2, \ldots\}$ will be a fixed point of the map Ψ_{f^*,ω^*} if and only if the vector $(A_1, A_2, \ldots)'$ is a characteristic vector of the linear map Ψ_{f^*,ω^*} with associated characteristic value 1. In view of the foregoing remarks and observations, we can state the answer to the question of what types of metabolic disturbances will be either stabilized or neutralized by the repair map P_{f^*} as follows:

METABOLIC REPAIR THEOREM.

i) The cellular metabolic disturbance $f \approx \{A_1, A_2, \cdots\}$ *will be stabilized by the repair map* P_{f^*} *if and only if the vector* $(A_1, A_2, \ldots)'$ *is a characteristic vector of the map* Ψ_{f^*,ω^*} *with associated characteristic value 1.*

ii) The perturbation f *will be "repaired," i.e., restored to the metabolism* f^* *by the repair system* P_{f^*} *if and only if* $f = f^* + \ker \Psi_{f^*,\omega^*}$. *In other words,*

$$(A_1 - A_1^*, A_2 - A_2^*, \ldots)' \in \ker \Psi_{f^*,\omega^*}.$$

The Metabolic Repair Theorem shows clearly that the two properties of stabilization at the new metabolic level f and restoration of the old metabolism f^* are diametrically opposed. If we want to be able to restore the old metabolism f^* in the face of many types of disturbances, we need to have $\ker \Psi_{f^*,\omega^*}$ "large," since in this case there will be many perturbations f that could satisfy condition (ii) of the theorem. On the other hand, if $\ker \Psi_{f^*,\omega^*}$ is big, then there must necessarily be only a "small" number of characteristic vectors of Ψ_{f^*,ω^*} with associated characteristic value 1, i.e., there are only a relatively small number of disturbances f that can be stabilized by P_{f^*} at this new metabolic level. Thus, a cell has the "choice" of being easily adaptable to a wide variety of metabolic disturbances or being able to restore the original metabolism in the face of a wide variety of metabolic changes, but not both! We examine this point in greater detail below, but for the moment let's return to the map Ψ_{f^*,ω^*}.

The above discussion makes evident the central role of the map Ψ_{f^*,ω^*} in determining those metabolic changes that can be either reversed or stabilized. But in order to say anything more definite about the structure of such types of changes, we need to be able to express the map Ψ_{f^*,ω^*} in terms of its component maps f^* and P_{f^*}.

From the component representation of the input/output behavior of the repair system, we have

$$A_\tau = \sum_{j=0}^{i} \sum_{i=0}^{\tau-1} [\mathcal{R}_{\tau-i}^{*(1)} | \cdots | \mathcal{R}_{\tau-i}^{*(p)}] A_{i-j+1} u_j^*,$$

which is clearly a triangular (in fact, Toeplitz) representation, since A_τ depends only upon the elements $A_1, A_2, \ldots, A_\tau$. Under our standard assumptions discussed above, we can always find a solution to this system of equations in the components of the matrices $\{\mathcal{R}_{\tau-i}^*\}$ and the elements $\{u_j^*\}$, $i = 0, 1, 2, \ldots, \tau - 1$; $j = 1, 2, \ldots, m$. On the other hand, the requirement that f^* be a fixed point for the map Ψ_{f^*,ω^*} means that there must be a solution to the system

$$\begin{pmatrix} \Psi^*_{11} & \Psi^*_{12} & \Psi^*_{13} & \cdots \\ \Psi^*_{21} & \Psi^*_{22} & \cdot & \cdots \\ \vdots & \vdots & \vdots & \cdots \end{pmatrix} \begin{pmatrix} A^*_1 \\ A^*_2 \\ \vdots \end{pmatrix} = \begin{pmatrix} A^*_1 \\ A^*_2 \\ \vdots \end{pmatrix},$$

for some *triangular* choice of the elements Ψ^*_{ij}. In particular, this means that we must have $\Psi^*_{ij} = 0$, $j > i$ and, hence,

$$A^*_\tau = \Psi^*_{\tau 1} A^*_1 + \Psi^*_{\tau 2} A^*_2 + \cdots .$$

But we also have the expression for A^*_τ from above involving the elements $\{\mathcal{R}^*_{\tau-1}\}$ and $\{u^*_j\}$. Setting these two expressions equal, we obtain

$$\Psi^*_{\tau 1} A^*_1 + \Psi^*_{\tau 2} A^*_2 + \cdots = A^*_\tau = \sum_{j=0}^{i} \sum_{i=0}^{\tau-1} [\mathcal{R}^{*(1)}_{\tau-1} | \cdots | \mathcal{R}^{*(p)}_{\tau-1}] A^*_{i-j+1} u^*_j, \qquad (\S)$$

$$\tau = 1, 2, \ldots .$$

The relation (§) enables us to identify the elements of the triangular array $\{\Psi_{ij}\}, j > i$. Unfortunately, the algebra in carrying out the identification of these elements is better done by a computer when p and/or m is larger than 1, but we can give a simple expression for Ψ_{f^*,ω^*} in the single-input/single-output case. When $m = p = 1$ we have

$$\Psi_{f^*,\omega^*} = \begin{pmatrix} \mathcal{R}^*_1 u^*_0 & 0 & 0 & \cdots \\ \mathcal{R}^*_2 u^*_0 + \mathcal{R}^*_1 u^*_1 & \mathcal{R}^*_1 u^*_0 & 0 & \cdots \\ \mathcal{R}^*_3 u^*_0 + \mathcal{R}^*_2 u^*_1 + \mathcal{R}^*_1 u^*_2 & \mathcal{R}^*_2 u^*_0 + \mathcal{R}^*_1 u^*_1 & \mathcal{R}^*_1 u^*_0 & \cdots \\ \vdots & \vdots & \vdots & \vdots \end{pmatrix}.$$

Example: The Natural Numbers (Cont'd.)

Let's use the above result to examine the repair mechanism for our earlier example involving the natural numbers. In that case we had

$$\begin{aligned} \omega^* &= (1,1,0,0,\ldots) = \{u^*_0, u^*_1, u^*_2, \ldots\}, \\ f^* &\approx \{1,1,2,2,3,3,\ldots\} = \{A^*_1, A^*_2, A^*_3, \ldots\}, \\ P_{f^*} &\approx \{1,-1,1,-1,\ldots\} = \{\mathcal{R}^*_1, \mathcal{R}^*_2, \mathcal{R}^*_3, \ldots\}. \end{aligned}$$

Let's suppose the design metabolism f^* is perturbed to the new metabolism

$$f \approx \{1,2,2,2,3,3,4,4,\ldots\} = \{A_1, A_2, A_3, \ldots\},$$

i.e., there is a change in the second element only. The cellular output under f is now

$$\gamma = f(\omega^*) = (1, 3, 4, 4, 5, 6, 7, \dots).$$

Thus we see that the metabolic change results in a change of output from γ^* = the natural numbers, to the closely related sequence γ, which differs from γ^* only in the second and third entries. The question is what effect the cellular repair mechanism P_{f^*} will have upon this new metabolism.

To address this issue, we compute the matrix Ψ_{f^*,ω^*} which, using the result given above for the case $m = p = 1$, yields

$$\Psi_{f^*,\omega^*} = \begin{pmatrix} 1 & 0 & 0 & \cdots \\ 0 & 1 & 0 & \cdots \\ 0 & 0 & 1 & \cdots \\ \vdots & \vdots & \vdots & \vdots \end{pmatrix} = \text{identity}.$$

(In fact, it can be shown that this relation holds for *all* scalar metabolisms. See Problem 11 for a multiple-input, multiple-output example where Ψ_{f^*,ω^*} does not equal the identity.)

So in the scalar case *every* metabolism f is a fixed point of the map Ψ_{f^*,ω^*} with associated characteristic value 1; hence, by the Metabolism Repair Theorem the repair mechanism will *stabilize* the cell at the new metabolism f (see Problem 8). In other words, the cell will immediately "lock on" to the new metabolism f, and thereafter operate with this metabolism rather than the old metabolism f^*. We see that there can be no restoration with this kind of cellular repair process, only a stabilization at whatever kind of metabolic perturbation may present itself. This is the ultimate in adaptability as such a cell can immediately adjust to whatever change may come to its original metabolic machinery.

Now let's turn to the case of disturbances in the environment ω^* leaving the cellular metabolism f^* fixed.

Case II: A Fluctuating Environment ω with Fixed Basal Metabolism f^*—Now we turn to the situation in which there is a change of the external environment $\omega^* \to \omega$, and we want to identify all such changes that can be neutralized by the repair mechanism P_{f^*}. Thus, we seek to characterize all environments ω such that

$$P_{f^*}(f^*(\omega^*)) = P_{f^*}(f^*(\omega))(= f^*)$$

implies

$$f^*(\omega^*) = f^*(\omega).$$

In other words, we want to know when the repair map P_{f^*} is one-to-one.

But it's easy to see from the earlier definition of P_{f^*} that the repair map can be represented in Toeplitz form by the linear operator

$$P_{f^*} = \begin{pmatrix} \mathcal{R}_1^* & 0 & 0 & \cdots \\ \mathcal{R}_2^* & \mathcal{R}_1^* & 0 & \cdots \\ \mathcal{R}_3^* & \mathcal{R}_2^* & \mathcal{R}_1^* & \cdots \\ \vdots & \vdots & \vdots & \cdots \end{pmatrix}, \qquad \mathcal{R}_i^* \in R^{p \times pm}.$$

This representation implies that P_{f^*} is one-to-one if and only if $\ker \mathcal{R}_1^* = \{0\}$. This will be the case if and only if $m = 1$ and rank $\mathcal{R}_1^* = p$. We can summarize this simple observation in the following theorem:

ENVIRONMENTAL CHANGE THEOREM. *If $m = 1$ and rank $\mathcal{R}_1^* = p$, all environments ω such that $f^*(\omega) = f^*(\omega^*)$ are given by $\omega = \omega^* + \ker f^*$.*

On the other hand, if $m > 1$ and/or rank $\mathcal{R}_1^* = r < p$, then any environmental change of the form $\omega = x + \omega^*$ will be repaired by P_{f^*}, where x is any solution of the equation $f^*(x) = \hat{\gamma}$, $\hat{\gamma} \in \ker \mathcal{R}_1^*$.

PROOF: Let $m = 1$ and rank $\mathcal{R}_1^* = p$. Then the operator P_{f^*} is one-to-one and all environments ω such that $P_{f^*}(f^*(\omega)) = P_{f^*}(f^*(\omega^*))$ implies $f^*(\omega) = f^*(\omega^*)$ consist of those ω satisfying $\omega = \omega^* + \ker f^*$.

Now let $m > 1$ and/or rank $\mathcal{R}_1^* = r < p$, i.e., $\ker \mathcal{R}_1^*$ is nonempty. Let $\hat{\gamma} \in \ker R_1^*$, and let x be a solution of the equation $f^*(x) = \hat{\gamma}$. Then any environmental change of the form $\omega^* \to \omega = x + \omega^*$ will be neutralized (repaired) by the genetic repair mechanism P_{f^*} since

$$\begin{aligned} P_{f^*}(f^*(\omega)) &= P_{f^*}(f^*(x) + f^*(\omega^*)) \\ &= P_{f^*}(\hat{\gamma}) + P_{f^*}(f^*(\omega^*)) \\ &= 0 + f^* = f^*. \end{aligned}$$

The Metabolic Repair and Environmental Change Theorems provide a clear, complete and computable answer to the questions surrounding the types of changes in the metabolism and/or environment that the cellular repair machinery is capable of fixing. Now let's direct our attention to the problem of repairing the repairer: the cellular replication process.

9. *Linear (M,R)-Systems: Replication*

The cellular replication map

$$\beta_f \colon H(\Omega, \Gamma) \to H(\Gamma, H(\Omega, \Gamma)),$$

can be formally considered in much the same fashion as discussed above for the repair map P_f. However, since the functional role of the replication map is quite different from that of P_f, a number of important questions arise that are absent in the case of repair, questions involving mutation, adaptation, Lamarckian inheritance, and so forth. We shall consider these matters in due course, but for now let's focus upon the steps involved in the *formal* realization of the map β_f.

Since by assumption β_f is a linear map accepting as inputs metabolisms of the form $f \approx \{A_1, A_2, \ldots\}$, and producing outputs $P_f \approx \{\mathcal{R}_1, \mathcal{R}_2, \ldots\}$, we must have a representation of the action of the replication map as

$$c_\sigma = \sum_{i=0}^{\sigma-1} U_{\sigma-i} e_i, \qquad \sigma = 1, 2, \ldots,$$

for an appropriate set of matrices $\{U_j\}$, where the input $e_i \doteq \mathcal{S}(A_i)$ and the output $c_i = \mathcal{R}_i$. Here again the operator $\mathcal{S}(\cdot)$ is the "stacking" operation discussed in the last section. Arguing just as for the repair map, we conclude that the matrices $\{U_j\}$ must have the form

$$U_j = [C_{j1} | C_{j2} | \cdots | C_{j,mp}], \qquad j = 1, 2, \ldots,$$

where each $C_{jr} \in R^{p \times mp}$. In what follows, we shall write $U_j^{(r)} \doteq C_{jr}$. So just as with f and P_f, we have the representation of β_f as

$$\beta_f \longleftrightarrow \{U_1, U_2, \ldots\}.$$

Note that in the above set-up, since the inputs for the replication system must correspond to the metabolism f while the outputs must be the associated repair map P_f, we have the relations

$$e_\sigma = \mathcal{S}(A_{\sigma+1}), \qquad c_\sigma = R_\sigma.$$

These relationships are expressed in the time-scale σ of the replication system, which is generally different from that of either the metabolic or repair subsystems.

Using the same arguments as for the repair map, it's easy to establish that if f has a finite-dimensional realization, so does β_f and $\dim \beta_f \leq \dim f$.

Example: The Natural Numbers (Cont'd.)

Continuing with the example started in the preceding section, we have

$$f \approx \{1, 1, 2, 2, 3, 3, \ldots\},$$
$$P_f \approx \{1, -1, 1, -1, 1, \ldots\},$$

which, after a bit of algebra, gives rise to the replication map

$$\beta_f \approx \{1, -2, 1, 0, 0, \ldots\}.$$

Thus, only the terms U_1, U_2 and U_3 are nonzero. Note the apparent decrease in the complexity of the sequences f, P_f and β_f as we pass from metabolism to repair and on to replication. We will return to this point below.

Applying Ho's Algorithm for realization of the map β_f, we obtain the canonical system

$$q_{\sigma+1} = \begin{pmatrix} 0 & 0 & 0 \\ 1 & 0 & 0 \\ 0 & 1 & 0 \end{pmatrix} q_\sigma + \begin{pmatrix} 1 \\ 0 \\ 0 \end{pmatrix} e_\sigma, \qquad q_0 = 0, \qquad q_\sigma \in R^3,$$

$$c_\sigma = (1 \quad -2 \quad 1) q_\sigma, \qquad \sigma = 0, 1, 2, \ldots .$$

In regard to replication, there are two immediate questions that present themselves:

1) When can phenotypic changes $f^* \to f$ give rise to permanent changes in the genootype?

2) If external disturbances modify P_f^*, what kinds of changes in f^* can result?

The first of these questions is the issue of Lamarckian inheritance, and the second addresses the problem of mutations. For the sake of brevity, we address here only the Lamarckian problem.

Suppose we begin with a cell with the usual basal metabolism $f^*(\omega^*) = \gamma^*$ and associated repair system P_{f^*}. The Lamarckian problem comes down to the question of when a metabolic (phenotypic) change $f^* \to f$ is ultimately "hard-wired" in to the cellular genotype by means of the new phenotype being processed by the cellular repair and replication operations. Conventional wisdom in biology says that such types of inheritance of "acquired characteristics" can never take place; renegade modern biologists say that (under very special circumstances) maybe they can. Here we discuss the situation within the context of the (M,R)-systems, finding mathematical evidence at least to support the claims of the "neo-Lamarckians."

The most compact way in which to mathematically express what's at issue in the Lamarckian problem is by tracing the individual steps in the following diagram:

$$f^* \longrightarrow f \xrightarrow{\omega^*} \gamma \xrightarrow{P_{f^*}} \hat{f} \xrightarrow{\beta_{f^*}} \hat{P}$$

The initial step in the above chain is the phenotypic change, with the remaining steps being the subsequent effects of this change on both the next phenotype ($\hat{f}$) and the next repair map ($\hat{P}$). The following three conditions must **all** hold if the change $f^* \to f$ is to constitute genuine Lamarckian inheritance:

A. The new phenotype f must be stabilized ($\hat{f} = f$);

B. The new genotype P_f must also be stabilized ($\hat{P} = P_f$);

C. The genetic "mutation" $P_{f^*} \to P_f$ must be a real mutation ($\hat{P}(\gamma) = \hat{f}$, with $\hat{P} \neq P_{f^*}$).

Mathematically, these conditions can be stated as:

$$(A) \iff f \text{ is a fixed point of the map } \Psi_{\omega^*,f^*};$$

$$(B) \iff P_f \text{ is a fixed point of the map } \chi_{\omega^*,f^*};$$

$$(C) \iff \gamma \in \ker\,(P_f - P_{f^*}).$$

Satisfaction of all of the above conditions is a tall order, requiring that the kernels of Ψ_{ω^*,f^*} and χ_{ω^*,f^*} be "small," while at the same time demanding that the kernel of the map $P_f - P_{f^*}$ be "large." It's clear that whether or not a particular phenotypic change can be inherited will depend greatly upon the way that change interfaces with the original cellular design parameters ω^* and f^*. One case in which we can guarantee that such changes cannot occur is when the cell is a scalar system, i.e., when $m = p = 1$. In this situation both conditions (A) and (B) are satisfied since the maps Ψ_{ω^*,f^*} and χ_{ω^*,f^*} both equal the identity. But the last condition (C) can never be satisfied since all three conditions together imply that

$$\beta_{f^*} * f * \gamma = f,$$

which, using the commutativity of the convolution product for scalar sequences, leads to the requirement that

$$\beta_{f^*} * \gamma = \text{ identity.}$$

This condition in turn implies that $\gamma = \gamma^*$, leading to a contradiction of condition (A). Thus, the fine line that the cell needs to walk in order to display Lamarckian inheritance cannot be achieved by such a simple mechanism as a cell with only a single input and a single output.

10. *Cellular Networks and Manufacturing Processes*

Up to this point our discussion has centered upon the behavior of a single cell. But cells don't live in isolation, so for our (M,R)-framework to make contact with real cellular phenomena it's necessary for us to consider how to put many cells together into a cellular network and to consider the implications of the genetic components of the cells for the overall viability of the cellular cluster. As we shall see, the repair/replication features of our cells dramatically affect the way such a network can behave in the face of various types of outside influences.

Consider a collection of N "cells," each of which accepts a variety of inputs and produces a spectrum of outputs. Assume that at least one cell in the network accepts inputs from the external environment and that at least one cell produces outputs that are sent to this environment. Further, suppose that every cell accepts either an environmental input or has as its inputs outputs from at least one other cell in the collection; similarly, assume that each cell produces either an environmental output or has its output utilized as another cell's input. Such a network might look like Fig. 7.8, where we have taken N=5. Here we have the cells $M_1 - M_5$, together with the two environmental inputs ω_1 and ω_2, as well as the single environmental output γ_1.

It's reasonable to suppose that each cell in such a network will have a finite lifetime after which it will be removed from the system. When this happens, all cells whose input depends upon the output from the "dead" cell will also be affected, ultimately failing in their metabolic role as well. In Fig. 7.8, for instance, if the cell M_1 fails, then so will M_2, M_3, M_4 and M_5, all of whose inputs ultimately depend upon M_1's output. Any such cell whose failure results in the failure of the entire network is called a *central* component of the network.

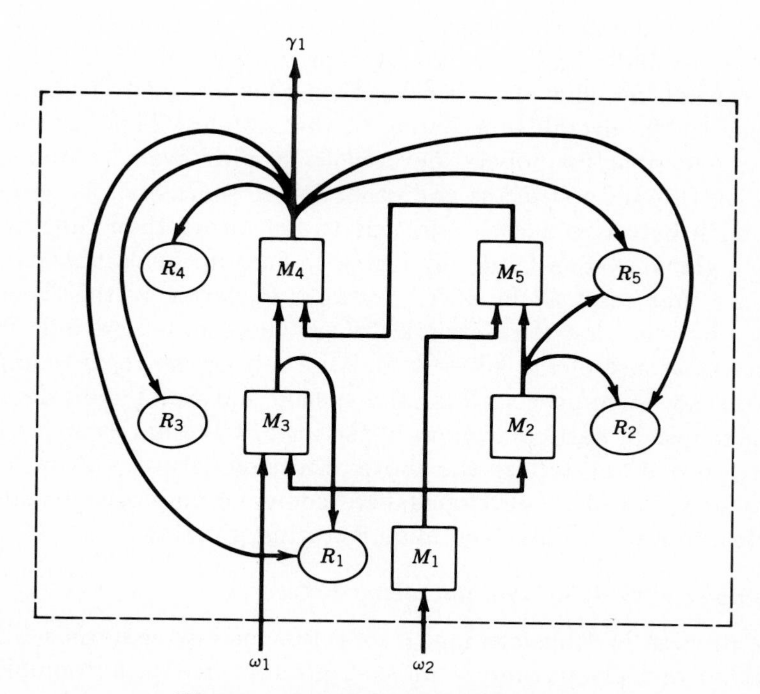

Figure 7.8 A Metabolic Network

Now let's suppose that we associate with each metabolic component M_i, a component $\mathcal{R}_i$ whose function is to repair M_i. In other words, when M_i fails, the repair component $\mathcal{R}_i$ acts to build a copy of M_i back into the network. The repair elements $\mathcal{R}_i$ are constituted so that each $\mathcal{R}_i$ must receive at least one environmental output from the network, and in order to function, $\mathcal{R}_i$ must receive all of its inputs. Thus, in Fig. 7.8 each $\mathcal{R}_i$ must receive the sole environmental output γ_1. Note also that by the second requirement, any cell M_i whose repair component $\mathcal{R}_i$ receives M_i's output as part of its input cannot be built back into the network. We call such a cell *nonreestablishable.* Thus the cell M_2 is nonreestablishable, but cell M_5 is *reestablishable.*

As already discussed, the repair components $\mathcal{R}_i$ are composed of the repair map P_f and the replication map β_f constructed as above from the cellular metabolism f. The elementary concepts introduced above already enable us to prove the following important results characterizing the interdependent nature of the components of an (M,R)-network:

NETWORK REPAIR THEOREM. *Every finite (M,R)-network contains at least one nonreestablishable component.*

COROLLARY. *If an (M,R)-network contains exactly one nonreestablishable component, then that component is central.*

This result shows that every (M,R)-network must contain some cells that cannot be built back into the system if they fail. Further, if there are only a small number of such cells, then they are likely to be of prime importance to the overall functioning of the system. This last result has obvious implications for policies devoted to keeping every component of a system alive (liberal politicians and other social reformers, please note!) It may be much better to allow some cells to fail rather than run the risk of incurring a global system failure by trying to prop up weak, noncompetitive components that can't all be saved anyway, as shown by the theorem. It should be observed that the Network Repair Theorem follows only from the connective structure of the network and the role of the repair components $\mathcal{R}_i$, making no assumptions about the specific nature of the cells or their metabolic maps. In particular, none of the linearity assumptions we invoked earlier are needed to establish the above results on networks. Now let's take a look at how the ideas developed here could be employed to study the overall structure of an industrial manufacturing process.

Example: Industrial Manufacturing Processes

The process of transforming a collection of raw materials like steel, glass, rubber and plastics into a finished product like an automobile offers us a perfect setting in which to consider the use of our (M,R)-metaphor outside the confines of cellular biology. Typically, such an enterprise involves a manufacturing process that we can think of as a map

$$f : \Omega \to \Gamma,$$

where Ω is a set of available inputs like raw materials, labor equipment, capital, etc., and Γ is a set of finished products. The "metabolism" f is a rule, or prescription, by which the input quantities are transformed into the desired finished goods by the manufacturing process. In this context we usually think of f as consisting of various types of processing elements like men and machines, together with the assembly plan by which the processing elements operate upon the input quantities from Ω. However this picture of manufacturing is incomplete in a variety of ways, not the least of which is that it contains no ready means to account for repair and replication which are essential ingredients of any truly automated "Factory of the Future." Let's look at how we might characterize some of the major features of such a manufacturing operation in terms of cellular quantities associated with a network of the sort described above.

Since any decent theory of manufacturing must be able to account for operational concerns of day-to-day production and planning—items like costs, capacity utilization and inventories—let's first examine the ways in which various aspects of cellular organization can act as counterparts of such features of manufacturing systems.

- *Direct Labor Costs*—Almost all cellular processes are energy driven. A natural measure of the cost of a process is the number of high-energy phosphate bonds, or molecules of ATP, required to carry out the process. If desired, this can be expressed in more physical units of kcal/mole. Such units, which can play the same role in the accounting of cellular processes that money plays in economics, seem a natural analogue of direct costs.

- *Indirect Labor Costs*—An important part of the cellular operation is devoted to actually producing ATP from ingested carbon sources (e.g., sugars or other foods). ATP is needed to drive the cycles that make ATP, and this might be the natural analogue of overhead or other types of indirect costs.

- *Material Costs*—Most of the raw materials required by a cell cannot, by themselves, get through the cellular membranes. They must be carried across by specific machinery ("permeases"), which must be manufactured by the cell. The cost to the cell of this manufacture and deployment of permease systems seems the most natural analogue of the material costs of an industrial manufacturing system.

- *Inventory Costs*—Cells do not generally maintain pools of unutilized materials or inventories. Rather, they seem organized around a "just-in-time" inventory strategy.

We can give similar cellular homologs for other manufacturing quantities of concern such as quality costs, reliability in finished product, capacity utilization and flexibility, but the foregoing list is sufficient to get the general idea. The interested reader is invited to consult the Notes and References for a more detailed discussion. Now let's look at how to tailor the (M,R) set up to the cause of abstractly representing such an industrial manufacturing operation.

We have seen above that the metabolic map $f: \Omega \to \Gamma$ can be thought of as the rule of operation of the plant. Consequently, by our earlier development we know that we can construct a repair map P_f and a replication map β_f directly from the firm's metabolism. However, there is a small technical point here that cannot be overlooked. Namely, we have considered the output of the manufacturing operation, the set Γ, to be composed of finished products like automobiles or TV sets. But the firm's repair map P_f must accept these items and somehow use them to reconstruct the metabolic map f. As it stands, of course, this idea is nonsense: TV sets or cars can't be directly transformed into hardware and software for manufacturing. To circumvent this difficulty, we simply assume that the output of the manufacturing operation is not the final finished product but rather its *monetary equivalent.* Thus, from now on we consider Γ to be just the set R_+ representing the gross revenues that the firm receives for production of their particular product mix, be it cars, TVs, wristwatches, or whatever.

Now recall from Chapter Three that the realization of the map f is abstractly equivalent to finding maps g and h, together with a space X, such that the diagram

$$\begin{array}{ccc} \Omega & \xrightarrow{f} & \Gamma \\ & g \searrow \quad \nearrow h & \\ & X & \end{array}$$

commutes, with g being onto and h one-to-one. In view of our assumptions about the space Γ, we can now attach direct physical meaning to these maps and to the set X. Since g is a map that transforms physical inputs into elements of X, it's natural to regard the X as a set of *physical products.* This interpretation of the elements of X is totally consistent with the role of the map h which is to transform elements of X into monetary revenues. So we can consider h as an abstract embodiment of the firm's *marketing and distribution* procedures. The technical conditions on the maps g and h mean that there is some input $\omega \in \Omega$ that can be processed by the firm's production operation g to achieve any desired product mix in X, and that no two distinct product mixes in X give rise to *exactly* the same gross revenues.

Using precisely the same line of reasoning, we can also consider realization of the repair and replication maps P_f and β_f in diagrammatic terms, and attempt to attach physical meaning to the resulting state-spaces and factor maps. Since this would take us too far afield from the major thrust of this book, we leave these considerations to the Discussion Questions and Notes and References.

Discussion Questions

1. Open-loop control is based upon the principle that we calculate the optimal control law $u^*(t)$ assuming that all aspects of the controlled system (dynamics, constraints, external forces, etc.) are known for the entire duration of the process. Feedback control is founded upon the notion that the control law should be calculated using the actual measured state of the system; i.e., the optimal law is based upon how the system is actually behaving. Discuss physical situations in which the assumptions underlying open-loop control might be satisfied. Consider also the disadvantages of using feedback control laws.

2. One of the growth areas in modern control theory is in the area of planning and regulating national economies. Consider the pros and cons of classical control theory in this context, in particular, feedback versus open-loop control. How would you go about setting up the dynamics, constraints and optimization criteria for such a process? How do such factors as political unrest, international balance of trade, currency exchange rates and tariffs enter into such a model?

3. In the Maximum Principle approach to optimal control it's necessary to solve a two-point boundary value problem, as well as to minimize the system Hamiltonian in order to determine the optimal control. In dynamic programming we need to determine the optimal value function and the optimal policy function for all possible initial states and all time horizons. Compare the computational work involved in each of these procedures, and discuss situations in which one is preferable to the other.

4. In the cosmology of A. N. Whitehead every step or process in the Universe involves both effects from past situations and anticipation of future possibilities. In particular, in Whitehead's philosophy it is admissible to postulate an influence of anticipation upon the statistical weights that play a part in quantum physics. This influence could change the probability distribution of the results of atomic reactions, and thus account for some of the differences between living and nonliving matter. Discuss this idea within the context of the anticipatory processes considered in this chapter.

5. One of the main reasons that anticipatory processes have not been studied to any great extent is the fact that any mode of system behavior,

regardless of how it is generated, can be *simulated* by a purely reactive system. In other words, the reactive paradigm of Newton is *universal.* Discuss whether or not this universality of the reactive paradigm justifies treating anticipatory processes by their simulation as a reactive system. In particular, consider this situation in connection with the Ptolemaic vision of planetary motion as a cascade of epicycles versus Kepler's model arising out of the heliocentric view.

6. The discussion in the text makes clear the point that a predictive model M of a system Σ is closed to interactions with the outside world to which the system is open. After a period of time, this fact will result in the model M being incapable of dealing with interactions that Σ is experiencing with its environment. This discrepancy will have the following effects: (1) the control will, in general, have other effects on Σ than those that were intended, and (2) the planned modes of interaction between Σ and its controller will be modified by these effects. These observations lead to what we usually term "side effects." Consider the role of side effects in planning operations in, for instance, corporate finance, social welfare or transportation networks. Discuss the degree to which side effects are an inherent aspect of *any* planning process. Examine how Gödel's Incompleteness Theorem, considered in Chapter Three, bears upon the issue of dealing with side effects.

7. In theoretical biology, the Principle of Function Change describes the phenomenon in which an organ originally evolved to perform one function changes over the millennia to perform a quite different function, one that was originally only an accidental feature of the original organ. For example, the evolution of the lung from the swim bladders of fish or the development of a seeing eye from an originally photosensitive patch of skin (a proto-eye) used for another purpose. Consider the Principle of Function Change from the perspective of designing systems that are insulated against side effects. Do you think it's possible to develop such a system? Why?

8. The development of the text emphasized the point that the metabolism, repair and replication subsystems operate on different time-scales. This fact introduces time-lags into the repair and replication operations relative to the metabolic time. Consider how these different time-scales interact to produce the global behavior of a network. In addition, consider time-lags that arise from the communication links in the network, i.e., lags associated with the flow of information from one cell to another or to the outside environment. Can you think of examples where such lags are critical to the operation of the network?

9. The Network Theorem showed that every (M,R)-network must contain at least one cell that cannot be repaired if it fails, and that if there is

only one such cell then its failure will destroy the entire network. In ecological circles, the concept of "resilience" is used to refer to the ability of an ecosystem to absorb unknown, and generally unknowable, disturbances and still maintain, or even increase, its viability. Discuss the possibilities for developing a *theory* of system resilience based upon the notion of reestablishable components in an (M,R)-network. In particular, consider the claim that the ecosystem is more resilient if it contains a large number of nonreestablishable components. Do you think this has anything to do with the idea of species extinction? Or species adaptation?

10. In Nature it's generally assumed that change of the cellular genotype (mutations) can only be directly brought about by environmental influences like radiation or by internal errors like DNA copying mistakes, but never by phenotypic traits, i.e., no Lamarckian inheritance. Consider the idea of phenotypic variations generating genotypic changes in the context of the manufacturing example discussed in the text. Does the idea of Lamarckian inheritance seem to make sense in this setting? What about in an economic context if we used an evolutionary paradigm? How would the system genotype and phenotype be defined in such a case? What about the possibility of *directed* mutations involving a direct feedback from the replication map β_f to the metabolism f? Can you think of any good examples where such mutations would play a role?

11. We have emphasized the formula

$$\text{adaptation} = \text{mutation} + \text{selection}$$

but have omitted any discussion of how to superimpose a selection criterion upon the (M,R)-framework developed in the text. Formally, this problem is a typical one faced in control theory; however, for (M,R)-networks we have a very different situation insofar as the interaction between the controls and system behavior is concerned. Furthermore, the primary goal of living systems is not so much optimality, but rather viability. Thus, the selection mechanism must serve two conflicting needs simultaneously: the need to *specialize* to exploit a particular eco-niche, and the need to *generalize* in order to remain viable under a variety of environmental disturbances and random mutations. Consider how you would go about developing such a selection criterion for different types of living systems, e.g., a cell, a firm, an economy, a society.

12. Consider a system composed of a collection of individuals. Discuss the contention that by optimizing the adaptability of each individual in the organization, we would kill the organization. Can you think of any examples illustrating this thesis? (*Hint:* Consider the human body.) What does this principle have to do with the controversy discussed in Chapter Six involving

the unit of selection (gene, individual, group) in evolutionary processes? Consider the old saying, "What's good for General Motors is good for the USA," in this regard.

13. Natural languages like English, Spanish and Arabic possess many of the features of an evolutionary system: a "metabolism" that transforms ideas (semantic content) into form (syntactic structure), a "repair" mechanism that acts to protect the language from random disturbances in its communicational role, and a "replication" mechanism that allows various types of "mutations" (new words, expressions, grammatical constructions, etc.) to enter the language. Consider how you would use the formalism of an (M,R)-system to model the evolutionary development of such a language. Do you think the same ideas would work for artificial languages like Esperanto or even a computer programming language like Pascal or Fortran?

14. In the Environmental Change Theorem we have seen that it's possible for the cellular metabolism f to be changed by various types of environmental disturbances $\omega^* \to \omega$. Discuss the situation in which such metabolic changes can be *reversed* by further changes in the environment. In particular, consider this reversibility problem in the context of carcinogenesis, as well as for problems of cellular differentiation and development. Relate this discussion to the notions of reachability considered in Chapters Two and Three.

15. In classical physics there are two complementary theories of optics: the *particle* theory introduced by Newton, which regards light as a collection of rays of individual photons, and the *wave* theory of Huygens, which thinks of light as the manifestation of the wavefronts of electromagnetic fields. These two theories are dual to each other in exactly the same manner that points and lines are dual in euclidean geometry. Consider the Maximum Principle and Dynamic Programming approaches to optimal control problems in this context, and determine which approach corresponds to a "point" theory and which to a "wave" approach. Can you identify the function describing the "wave front" for a control problem, as well as the function that describes the "particle" trajectory?

16. The Metabolism-Repair paradigm for cellular activity is a purely *functional* theory, i.e., it makes no reference to the material elements from which a cell is composed. This is a serious difficulty if one wants to make contact with the vast amount of experimental work being carried out in medical laboratories. Consider how you would go about linking-up these two vastly different views of a cell. Do you see any simple relationships between the functional notion of component and the structure that our methods of observation enable us to see in cells?

17. In the text we have discussed the complexity of the metabolic, repair and replication components of an (M,R)-system in terms of the number of entries that are needed to specify one coefficient in the expansion of the corresponding input/output map. Thus, for instance, the metabolic maps are specified in terms of the matrices $\{A_1, A_2, \cdots\}$, where each $A_i \in R^{p \times m}$ so the complexity of the map f is $O(pm)$, while the repair map P_f is given by the matrices $\{\mathcal{R}_1, \mathcal{R}_2, \cdots\}$, where each $\mathcal{R}_i \in R^{p \times pm}$ so that the complexity of P_f is $O(p^2m)$. Similarly, the complexity of the replicaion map β_f was seen to be $O(p^3m^2)$. These quantities depend only upon the number of metabolic inputs m and outputs p. Discuss and compare this concept of complexity with other possibilities such as the dimension of the corresponding state-spaces.

18. We have introduced the notion of a sequential machine in Chapter Two as consisting of an input set A, an output set B, a state-set X, and two maps: $\lambda\colon A \times X \to X$, the state-transition map and $\delta\colon A \times X \to B$, the output map. How would you reformulate the (M,R) set-up in terms of a sequential machine? In particular, is the state-set X finite? Or even finite-dimensional? Compare the formulation in terms of sequential machines with that developed via realization theory arguments as given in the text.

Problems and Exercises

1. Consider the scalar control process

$$\min_u J(u) \doteq \min_u \int_0^T (x^2 + u^2)\, dt,$$

where the state x and the control u are related through the linear differential equation

$$\dot{x} = u, \qquad x(0) = c.$$

a) Determine the Hamiltonian $H(x, u, \lambda, t)$ for this system.

b) Using the Maximum Principle, show that the optimal open-loop control law is given by

$$u^*(t) = c\frac{\cosh(t - T)}{\cosh T},$$

with the minimum value of the integral criterion being given by

$$J(u^*) = c^2 \tanh T.$$

c) Show that as $T \to \infty$, we have the asympototic control $u^*_\infty(t) = ce^{-t}$, with the corresponding minimal cost $J(u^*_\infty) = c^2$.

d) Solving the same problem using dynamic programming, show that the optimal value function satisfies

$$\frac{\partial I}{\partial T} = \min_v \left\{ c^2 + v^2 + v\frac{\partial I}{\partial c} \right\},$$
$$I(c,0) = 0.$$

e) Show that the minimizing v satisfies the equation

$$v^*(c,T) = -\frac{1}{2}\frac{\partial I}{\partial c},$$

and, as a result, we have

$$\frac{\partial I}{\partial T} = c^2 - \frac{1}{4}\left(\frac{\partial I}{\partial c}\right)^2.$$

f) Using the fact that I is a quadratic in c, i.e., $I(c,T)$ has the structure $I(c,T) = c^2 R(T)$, show that the function R satisfies the Riccati equation

$$\frac{dR}{dT} = 1 - R^2(T), \qquad R(0) = 0.$$

Hence, conclude that

$$I(c,T) = c^2 \tanh T,$$
$$v^*(c,T) = -c \tanh T.$$

g) Show that in the limit as $T \to \infty$, we have the simple optimal control law $v^*(c) = -c$, with the associated optimal cost $I(c) = c^2$.

h) How would you extend these results for the scalar case to the multi-dimensional setting of minimizing the integral

$$\tfrac{1}{2}\int_0^T [(x, Qx) + (u, Ru)]\,dt, \qquad Q \geq 0, \quad R > 0,$$

subject to the linear dynamics

$$\frac{dx}{dt} = Fx + Gu, \qquad x(0) = c?$$

Show that the solution is that given in the text, both by dynamic programming as well as by the Maximum Principle.

2. Consider the multidimensional control process

$$\min_u \int_0^T g(x,u)\,dt,$$

with the dynamics

$$\frac{dx}{dt} = h(x,u), \qquad x(0) = c.$$

a) Show that the Bellman-Hamilton-Jacobi equation for this system is

$$\frac{\partial I}{\partial T} = \min_v \left\{ g(c,v) + \left(h(c,v), \frac{\partial I}{\partial c} \right) \right\},$$
$$I(c,0) = 0.$$

b) Show that the Hamiltonian for the system is $H(c,v,\lambda,t) = g(c,v) + (\lambda, h(c,v))$. Consequently, derive the Maximum Principle from the B-H-J equation of part (a).

c) Show that

$$\frac{\partial I}{\partial T} = \min_v H,$$

and that

$$\lambda = \frac{\partial I}{\partial c},$$

i.e., the co-state vector λ can be interpreted as the marginal rate of change in the optimal cost when we change the initial state c. (*Remark*: This exercise shows that the dynamic programming optimal value function calculation and the minimization of the system Hamiltonian involve essentially the same computation, but via quite different directions.)

3. The following problem illustrates many of the features of the "dual control" problem in which at each moment the control can be used to either regulate the system state or to probe the system to learn more about its structure.

Assume we have a robot that moves in the grid on the next page. The robot is attempting to find its way through this environment to the "home" square located in the upper left-hand corner. There are two shaded squares which may or may not contain barriers that the robot is not allowed to pass through. Initially the robot doesn't know whether or not the barriers are actually present, so part of his control strategy is to learn about the presence or absence of these barriers, as well as to move toward the goal square. At the beginning the robot knows there is a barrier at ($x_1 = 1$, $x_2 = 2$) with probability 0.4, and a barrier at ($x_1 = 2$, $x_2 = 3$) with probability 0.5. The robot can always see one move ahead, i.e., if it is within one move of a barrier

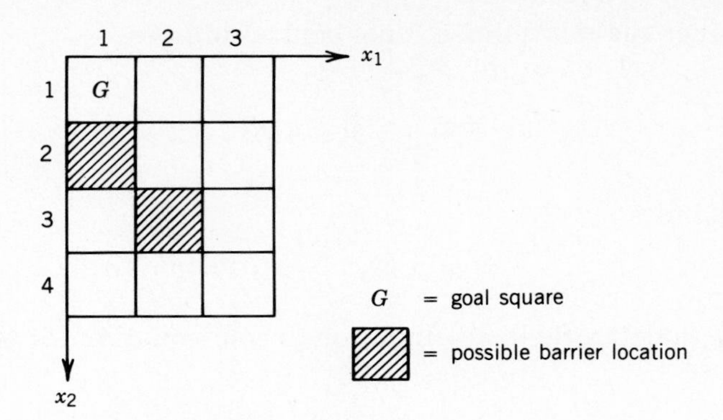

Robot Environment

Robot Environment

location it can always find out whether a barrier is there. For a price of 0.3 moves the robot can make an observation of all squares that are two moves away, where a move is defined to be either one horizontal or one vertical square away from the robot's current location. In other words, the robot can move or observe diagonally only in two moves. The robot's objective is to get to the home square while minimizing the sum of actual moves and penalties for observation.

a) Set up a dynamic programming formulation of this problem using the system state as

$$x = \begin{pmatrix} x_1 \\ x_2 \\ x_3 \\ x_4 \end{pmatrix},$$

where

x_1 = the robot's horizontal coordinate,

x_2 = the robot's vertical coordinate,

x_3 = the robot's state of knowledge about the barrier at $(x_1 = 1, x_2 = 2)$,

x_4 = the robot's state of knowledge about the barrier at $(x_1 = 2, x_2 = 3)$.

The components of the state can assume the values

$$\begin{aligned} x_1 &= 1, 2, 3, \\ x_2 &= 1, 2, 3, 4, \\ x_3 &= \text{P, A, Q}, \\ x_4 &= \text{P, A, Q}, \end{aligned}$$

where P, A, Q represent the robot knowing that the barrier is Present, Absent or Unknown, respectively.

For the admissible controls at each moment, take the finite set

$$U = \left\{ \begin{pmatrix} 1 \\ 0 \\ N \end{pmatrix}, \begin{pmatrix} -1 \\ 0 \\ N \end{pmatrix}, \begin{pmatrix} 0 \\ 1 \\ N \end{pmatrix}, \begin{pmatrix} 0 \\ -1 \\ N \end{pmatrix}, \begin{pmatrix} 0 \\ 0 \\ L \end{pmatrix} \right\},$$

where L denotes "make an observation," while N means "no observation is made."

b) Show that the system dynamics are given by

$$\begin{aligned} x_1(t+1) &= x_1(t) + u_3(t), \\ x_2(t+1) &= x_2(t) - u_2(t), \\ x_3(t+1) &= \int_3 [x_3(t), u_3(t), m_1(t), w_1(t)], \\ x_4(t+1) &= \int_4 [x_4(t), u_3(t), m_2(t), w_2(t)], \end{aligned}$$

where w_1 and w_2 are quantities taking on the values B or R, with B denoting the barrier is present and R meaning the barrier is absent, and m_1 and m_2 are quantities taking on the values 0 or 1 depending upon whether the presence or absence of the barrier will or will not be determined by the control being used. The quantity $\int_3$ is determined from the table below, while $\int_4$ is found from a similar table.

$x_3(t)$	$u_3(t)$	$m_1(t)$	$w_1(t)$	$x_3(t+1) = \int_3$
P	—	—	—	P
A	—	—	—	A
Q	N	0	—	Q
Q	N	1	B	P
Q	N	1	R	A
Q	L	0	—	Q
Q	L	1	B	P
Q	L	1	R	A

In the above table, the symbol "—" means that the value of $x_3(t+1)$ is the same for all values of the corresponding variable.

c) Work out the optimal contol policy, and show that the optimal decision for the robot involves making an observation rather than moving when he is in the states

$$x_1 = 1, \quad x_2 = 4, \quad x_3 = Q, \quad x_4 = P,$$
$$x_1 = 1, \quad x_2 = 4, \quad x_3 = Q, \quad x_4 = Q.$$

d) Solve the same problem but without allowing the possibility for the robot to make an observation. From this solution show that the "value of information" is 0.5 moves when in the first of the states in part (c), while it is 0.1 moves in the second. That is, the difference in the minimal expected number of moves needed to reach the goal from these two states is lower by these amounts when the robot is allowed to observe than when he cannot.

4. Assume we have the input/output relation

$$y_t = \sum_{i=0}^{t-1} A_{t-i} u_i, \qquad t = 1, 2, \ldots,$$

where the input sequence $\omega \in \Omega$ is *finite,* i.e.,

$$\omega = \{u_0, u_1, \ldots, u_N\}, \qquad u_i \in R^m, \qquad m - 1 \leq N < \infty.$$

a) Show that the coefficient matrices $\{A_i\}$, $A_i \in R^{p \times m}$ are uniquely determined if and only if the matrix

$$\begin{pmatrix} u_0 & u_1 & u_2 & \ldots & u_N & 0 & 0 & \ldots \\ u_1 & u_2 & u_3 & \ldots & 0 & 0 & 0 & \ldots \\ u_2 & u_3 & u_4 & \ldots & 0 & 0 & 0 & \ldots \\ \vdots & \vdots & \vdots & \ldots & 0 & 0 & \ldots & \\ u_N & 0 & 0 & \ldots & & & & \end{pmatrix}$$

has maximal rank.

b) Prove that the above condition is generic, i.e., is satisfied for an open, dense set of input sequences ω.

c) How would you extend this result to the case of an infinite number of inputs?

(*Remark:* This exercise shows that the metabolic map f of an (M,R)-system is generically uniquely determined by the cell's basal metabolism; hence, so are the cellular repair and replication maps P_f and β_f.)

5. In Chapter Three we considered the input/output relation $f(\omega) = \gamma$ given by

$$\omega = \{1, 0, 0, \cdots\},$$
$$\gamma = \{1, 1, 2, 3, 5, 8, \cdots\} = \text{ the Fibonacci numbers},$$

and saw that this "basal metabolism" admitted the canonical realization

$$F = \begin{pmatrix} 0 & 1 \\ 1 & 1 \end{pmatrix}, \quad G = \begin{pmatrix} 1 \\ 1 \end{pmatrix}, \quad H = (1 \quad 0).$$

a) Compute the repair and replication maps P_f and β_f that go along with this metabolism.

b) Suppose the environmental input ω shifts to $\hat{\omega} = \{1, 2, 1, 0, 0, \ldots\}$. Can the repair map neutralize this environmental change?

c) What ultimate effect does the above environment $\hat{\omega}$ have upon the system replication map β_f?

6. In Chapter Six we considered replicator dynamics that describe the change in genetic frequency distribution by assuming the selection principle that the relative change in frequency of a genotype is proportional to its advantage, i.e., if x_i represents the fraction of the total population having genotype i, and if $f_i(x)$ is the reward for genotype i when competing against the entire population, where Φ is the average reward for the entire population, then the replicator selection mechanism postulates that

$$\frac{\dot{x}_i}{x_i} \propto f_i(x) - \Phi.$$

How could you incorporate this selection principle into the (M,R)-framework of the text?

7. The Metabolic Repair Theorem of the text categorizes all those metabolic changes that can either be neutralized or stabilized by the repair map P_f. In general, if we have a change from the basal metabolism f^* to f, the system could go through a sequence like $f^* \to f \to \hat{f} \to \tilde{f} \to \cdots$. Show that this cycle is periodic of period k if and only if the metabolism f is a fixed point of the map $\Psi^k_{f^*,\omega^*}$, thereby generalizing the result given in the text. In particular, this result shows that in the space of metabolisms f, we have the possibility of three types of motion: a fixed point (equilibrium), a cycle of period $k < \infty$ (periodic motion), and aperiodic motion. Compare this result with the discussion of chaotic motion in Chapter Five or the classes of cellular automata dynamics presented in Chapter Two.

8. Show that for scalar inputs and outputs (ω^*, γ^*) the map Ψ_{f^*,ω^*} always equals the identity.

9. In Chapter Three we saw that if we have two *independent* linear systems Σ_1 and Σ_2 with transfer matrices W_1 and W_2, respectively, and we connect them in serial order, then the transfer matrix of the overall system is just the product $W_2 W_1$. In our (M,R) set-up if we consider the two subsystems f and P_f, then it's clear that the transfer matrix of P_f is not independent of the transfer matrix of f.

a) If the transfer matrix of f is W, what is the transfer matrix of P_f?

b) What is the transfer matrix of the overall system from f through P_f if the two subsystems are connected in series as indicated in the text?

c) Extend the results of parts (a) and (b) to include the system replication map β_f.

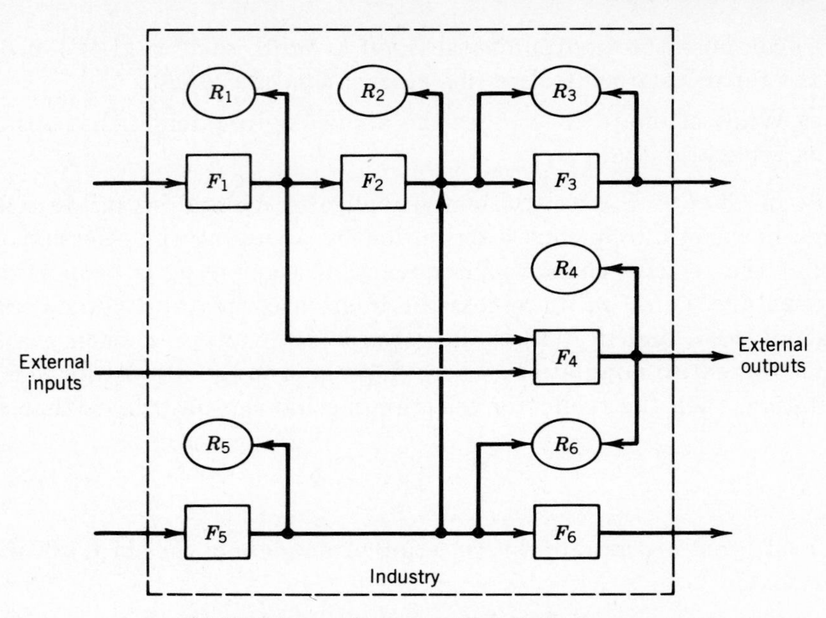

(M,R)-Network for a Global Industry

10. Consider the (M,R)-network for a global industry shown above.

a) Show that the firm F_6 is reestablishable but all other firms are non-reestablishable.

b) Are there any central components in this network?

c) In this network some of the firms devote their entire output to supporting other firms. Can you think of any real-life situations in which this type of behavior could naturally occur?

11. Suppose we have a cell whose design environment is given by

$$\omega^* = \{u_0^*, u_1^*, 0, 0, \ldots\}, \qquad u_i^* \in R^2, \qquad i = 1, 2,$$

where

$$u_0^* = \begin{pmatrix} 1 \\ 0 \end{pmatrix}, \qquad u_1^* = \begin{pmatrix} 0 \\ 1 \end{pmatrix}.$$

Further, assume the design output is

$$\gamma^* = \{y_1^*, y_2^*, y_3^*, \ldots\}, \qquad y_i^* \in R^2, \qquad i = 1, 2, \ldots,$$

with

$$y_1^* = \begin{pmatrix} 0 \\ 1 \end{pmatrix}, \qquad y_2^* = \begin{pmatrix} 2 \\ 0 \end{pmatrix} \qquad y_3^* = \begin{pmatrix} -2 \\ 2 \end{pmatrix}.$$

The remainder of the (infinite) output sequence is given by the relation

$$y_i^* = y_{i-2}^* - y_{i-1}^*, \qquad i = 4, 5, \ldots .$$

a) Show that the behavior sequence defining the basal metabolism is given by

$$f^* = \{A_1^*, A_2^*, \dots\}, \qquad A_i^* \in R^{2\times 2}, \qquad i = 1,2,\dots,$$

where

$$A_1^* = \begin{pmatrix} 0 & 1 \\ 1 & 0 \end{pmatrix}, \qquad A_2^* = \begin{pmatrix} 1 & -1 \\ 0 & 1 \end{pmatrix},$$

with the remaining elements of f^* being determined by the recurrence relation

$$A_i^* = A_{i-2}^* - A_{i-1}^*, \qquad i = 3,4,\dots .$$

b) Show that the canonical realization of the cellular metabolic subsystem is of dimension two and is given by

$$x_{t+1}^* = \begin{pmatrix} 0 & 1 \\ 1 & -1 \end{pmatrix} x_t^* + \begin{pmatrix} 1 & 0 \\ 0 & 1 \end{pmatrix} u_t^*, \qquad x_0^* = \begin{pmatrix} 0 \\ 0 \end{pmatrix},$$

$$y_t^* = \begin{pmatrix} 0 & 1 \\ 1 & 0 \end{pmatrix} x_t^*. \qquad (\Sigma^*)$$

c) Show that the repair system has $m_P^* = 4$ inputs and $p_P^* = 2$ outputs, with behavior sequence

$$P_{f^*} = \{\mathcal{R}_1^*, \mathcal{R}_2^*, \dots\}$$

given by the elements

$$\mathcal{R}_1^* = \left[\begin{array}{cc|cc} 0 & 0 & 0 & 1 \\ 0 & 0 & 1 & 0 \end{array}\right],$$

$$\mathcal{R}_2^* = \left[\begin{array}{cc|cc} 0 & 0 & 1 & -1 \\ 0 & 0 & 0 & 1 \end{array}\right].$$

The remaining elements are specified by the recurrence relation

$$\mathcal{R}_{i+2}^* = -(\mathcal{R}_i^* + \mathcal{R}_{i+1}^*), \qquad \mathcal{R}_i^* \in R^{2\times 4}, \qquad i = 1,2,\dots .$$

d) Construct the canonical model for the repair system as the two-dimensional realization given by

$$z_{\tau+1}^* = \begin{pmatrix} 0 & -1 \\ 1 & -1 \end{pmatrix} z_\tau^* + \begin{pmatrix} 0 & 1 \\ 0 & 0 \end{pmatrix} v_\tau^*, \qquad z_0^* = \begin{pmatrix} 0 \\ 0 \end{pmatrix},$$

$$w_\tau^* = \begin{pmatrix} 0 & 1 \\ 1 & 0 \\ 1 & -1 \\ 0 & 1 \end{pmatrix} z_\tau^*, \qquad \tau = 1,2,\dots . \qquad (\Sigma_P^*)$$

e) Show that the replication subsystem has the inputs and outputs

$$e_i^* = \mathcal{S}(A_{i+1}^*) \text{ and } c_{i+1}^* = \mathcal{S}(\mathcal{R}_{i+1}^*), \qquad i = 0, 1, 2, \ldots .$$

Thus, the replication system has $m_R^* = 4$ inputs and $p_R^* = 8$ outputs. Compute the the behavior sequence for the replication system as

$$\beta_{f^*} = \{\mathcal{U}_1^*, \mathcal{U}_2^*, \ldots\}, \qquad \mathcal{U}_i^* \in R^{8\times 4},$$

with

$$\mathcal{U}_1^* = \begin{bmatrix} 0 & 0 & 0 & 0 \\ 0 & 0 & 0 & 0 \\ 0 & 0 & 0 & 0 \\ 0 & 0 & 0 & 0 \\ 0 & 0 & 0 & 0 \\ 0 & 1 & 0 & 0 \\ 0 & 1 & 0 & 0 \\ 0 & 0 & 0 & 0 \end{bmatrix}, \qquad \mathcal{U}_2^* = \begin{bmatrix} 0 & 0 & 0 & 0 \\ 0 & 0 & 0 & 0 \\ 0 & 0 & 0 & 0 \\ 0 & 0 & 0 & 0 \\ 0 & 1 & 0 & 0 \\ 0 & 0 & 0 & 0 \\ 0 & -1 & 0 & 0 \\ 0 & -1 & 0 & 0 \end{bmatrix}.$$

Verify that the remaining elements of the behavior sequence are all linear combinations of the above two elements, leading to the conclusion that the canonical replication system has dim $\Sigma_R^* = 2$.

f) Construct the canonical model for the replication system as

$$q_{\sigma+1}^* = \begin{pmatrix} -1 & 1 \\ -2 & 0 \end{pmatrix} q_\sigma^* + \begin{pmatrix} 0 & 0 & 0 & 0 \\ 0 & 1 & 0 & 0 \end{pmatrix} e_\sigma^*, \qquad q_0^* = \begin{pmatrix} 0 \\ 0 \end{pmatrix},$$

$$c_\sigma^* = \begin{pmatrix} 0 & 0 \\ 0 & 0 \\ 0 & 0 \\ 0 & 0 \\ 1 & 0 \\ 0 & 1 \\ -1 & 1 \\ 1 & 0 \end{pmatrix} q_\sigma^*, \qquad \sigma = 0, 1, 2, \ldots . \qquad (\Sigma_R^*)$$

g) Suppose we want to characterize all those metabolic perturbations $f^* \to f$ that can be repaired by the genetic subsystem of the above cell. Show that this means that we must find all those metabolisms f such that $f - f^* \in \ker W$, where W is the triangular Toeplitz operator

$$W = \begin{pmatrix} w_1 & 0 & 0 & \cdots \\ w_2 & w_1 & 0 & \cdots \\ w_3 & w_2 & w_1 & \cdots \\ \vdots & \vdots & \vdots & \ddots \end{pmatrix},$$

with the action of the elements $w_i(\cdot)$ given by

$$w_i(Z) = \sum_{j=1}^{i} \mathcal{R}^*_{i-j+1}\, Z\, u^*_{j-1}, \qquad i = 1, 2, \ldots .$$

h) Using the elements $\omega^* = \{u_0^*, u_1^*, 0, 0, \ldots\}$ and $\{\mathcal{R}_i^*\}$ given above, prove that the kernel of the operator W for this cell is given by

$$\ker W = \left\{ Z_i \colon Z_i = \begin{pmatrix} x & x \\ 0 & 0 \end{pmatrix}, i = 1, 2, \ldots \right\}.$$

In the foregoing expression, the elements marked "x" are arbitrary numbers. Use this result to characterize all those metabolic perturbations $f = \{A_1, A_2, \ldots\}$ which the repair subsystem will be able to neutralize, i.e., perturbations that can actually be repaired by the cell's genetic component. In particular, explicitly display the repairable perturbations f. With the kernel of W given above, show that this means that the elements of a repairable metabolic perturbation must be of the form

$$A_i = \begin{pmatrix} x & x \\ a_i^{*\,21} & a_i^{*\,22} \end{pmatrix}, \qquad i = 1, 2, \ldots .$$

Here again "x" is arbitrary, while the element $a_i^{*\,kj}$ is just the (k, j)th component of the basal metabolism A_i^*. Thus, for this cell any metabolic perturbation that affects only the first row of the basal metabolism will be repaired by the cellular genetic machinery.

12. A *category* $\mathcal{A}$ is a family of objects $A, B, C, \ldots$, such that to each ordered pair of objects $(A, B) \in \mathcal{A}$, the set $H(A, B)$ of mappings from $A \to B$ is in $\mathcal{A}$. Further, it is assumed that $\mathcal{A}$ satisfies the axioms:

Axiom 1: If $f \in H(A, B)$ and $g \in H(B, C)$, then there is a unique map $gf \in H(A, C)$; gf is called the *composite* of f and g.

Axiom 2: If $f \in H(A, B)$, $g \in H(B, C)$, $h \in H(C, D)$, then we have $h(gf) = (hg)f$.

Axiom 3: To each object A in the category, we associate a unique mapping $i_A \in H(A, A)$ such that: i) for any object B and any $f \in H(A, B)$, $f i_A = f$; ii) for any object C and any $g \in H(C, A)$, $i_A g = g$.

A sequence $A, B, C, D, \ldots$ in a category will be called *normal* if no adjacent pair occurs infinitely often in the sequence. The category $\mathcal{A}$ will be called *normal* if, for any normal sequence of objects in $\mathcal{A}$, the associated sequence

$$H(A, B), H(B, C), H(C, D), \ldots$$

contains only a finite number of empty sets.

a) Show that if $\mathcal{A}$ is normal and possesses infinitely many objects, then to each object $A \in \mathcal{A}$ there is an object $Z \in \mathcal{A}$ such that $H(Z, A) \neq \emptyset$.

b) Prove that if $\mathcal{A}$ is normal, then any map $\alpha \in H(X,Y)$, $X, Y \in \mathcal{A}$, can be embedded as a metabolic map of an (M,R)-system, i.e., there exists a map $P_\alpha \in \mathcal{A}$ such that P_α can serve as the repair map in the (M,R)-system with α as the metabolic map.

(*Remark*: Different categories give rise to different classes of (M,R)-systems, i.e., to different abstract biologies. The foregoing results, and their many corollaries and extensions, strongly suggest category theory as a natural setting for comparing different types of (M,R)-systems. This point is pursued in detail in many of the papers cited in the Notes and References.)

13. In Discussion Question 18 we considered the connection between an (M,R)-system and a sequential machine. Let $M = \{A, B, X, \lambda, \delta\}$ and $M' = \{A', B', X', \lambda', \delta'\}$ be two such machines with the notation being as in the earlier discussion. We define a *homomorphism* between M and M' as a triple of mappings $\Psi = \{\psi_1, \psi_2, \psi_3\}$, where

$$\psi_1 : X \to X',$$
$$\psi_2 : A \to A',$$
$$\psi_3 : B \to B',$$

have the property that the diagrams

$$\begin{array}{ccc} X \times A & \xrightarrow{\lambda} & X \\ {\scriptstyle \psi_1 \times \psi_2}\downarrow & & \downarrow{\scriptstyle \psi_1} \\ X' \times A' & \xrightarrow[\lambda']{} & X' \end{array}$$

and

$$\begin{array}{ccc} X \times A & \xrightarrow{\delta} & B \\ {\scriptstyle \psi_1 \times \psi_2}\downarrow & & \downarrow{\scriptstyle \psi_3} \\ X' \times A' & \xrightarrow[\delta']{} & B' \end{array}$$

commute.

a) Use this definition of homomorphism between machines to prove the following theorem:

THEOREM. *Let $\mathcal{U}$ be a category that is closed under cartesian products, i.e., if $A, B \in \mathcal{U}$, then $A \times B \in \mathcal{U}$, and if f, g are maps in $\mathcal{U}$, then so is the map*

$f \times g$. Then the set of all sequential machines M whose sets $A, B, X \in \mathcal{U}$ and whose maps λ and δ are also in $\mathcal{U}$ forms a category with the maps $H(M, M')$ in this category being given by the homomorphism definition above.

b) Let $\mathcal{M}(\mathcal{U})$ denote the above category of machines over $\mathcal{U}$. Prove that if M is a machine induced by an (M,R)-system, with the map ψ_1 being onto and the maps ψ_2, ψ_3 one-to-one, then the machine M' is itself induced by an (M,R)-system.

c) Define a new category $\bar{M}(\mathcal{U})$ in the following manner: the objects of $\bar{M}(\mathcal{U})$ are the same as those of $M(\mathcal{U})$, whereas the set of maps $\bar{H}(M, M')$ consists only of those homomorphisms $\Psi = \{\psi_1, \psi_2, \psi_3\}$ such that ψ_1 is onto and ψ_2, ψ_3 are one-to-one. Show then that $\bar{M}(\mathcal{U})$ is a *subcategory* of $M(\mathcal{U})$. Thus, conclude that the totality of all (M,R)-systems over a category $\mathcal{U}$ is a subcategory of the associated category $\bar{M}(\mathcal{U})$.

14. Define a *simple* (M,R)-system on the category $\mathcal{U}$ to consist of the quadruple

$$\{\Omega, \Gamma, f, P_f\},$$

where Ω and Γ are objects of $\mathcal{U}$ such that $H(\Omega, \Gamma)$ is an object of $\mathcal{U}$, $f \in H(\Omega, \Gamma)$, and $P_f \in H(\Gamma, H(\Omega, \Gamma))$.

a) Prove that a sequential machine M on the category $\mathcal{U}$ represents a simple (M,R)-system if and only if M has an output-dependent state function, i.e., if and only if there exists a map $q \in H(B, X)$ such that

$$\lambda(x, a) = q(\lambda(x, a)).$$

b) Show that the sequential machine M on the category $\mathcal{U}$ represents an (M,R)-system if and only if the state-space X may be written as a subset of a cartesian product $X_1 \times X_2$ in such a way that there exist maps α, β such that

$$\begin{aligned}
&A.\ \pi_1(x) = \pi_2(x') \Rightarrow \delta(x, a) = \delta(x', a),\\
&B.\ \pi_1(\lambda(x, a)) = \beta(\delta(x, a)),\\
&C.\ \pi_2(\lambda(x, a)) = \alpha(\delta(x, a), \pi_2(x)),
\end{aligned}$$

where

$$\pi_1(x_1, x_2) = x_1, \qquad \pi_2(x_1, x_2) = x_2.$$

Note here that we have used $x \doteq (x_1, x_2) \in X_1 \times X_2$.

15. In his classic novel *Madame Bovary,* Flaubert tells the story of the gravedigger Lestiboudois who cultivates potatoes at the cemetery. For every body buried, Lestiboudois earns an amount M. Moreover, there are harvest revenues $f(S)$ depending upon the amount of land S under cultivation. The decision to bury a dead body diminishes the area S in an irreversible manner. Let $v(t)$ denote the rate of burials at the cemetery.

a) Show that Lestiboudois is faced with the following optimal control problem:

$$\max_v \int_0^\infty e^{rt}[f(S) + Mv(s)]\,dt,$$

subject to

$$\dot{S} = -av, \qquad S(0) = S_0, \qquad 0 \leq v \leq \bar{v}.$$

Here r is Lestiboudois' discount rate, a the area occupied by a grave, and $\bar{v}$ the maximal rate of funerals that can be carried out per unit time.

b) Assume that $f'(S) > 0$, $f''(S) < 0$. Then there exists a value $\hat{S}$ such that

$$f'(\hat{S}) = \frac{rM}{a}.$$

Using the Maximum Principle, show that the optimal gravedigging policy is to bury bodies with the maximal rate $\bar{v}$ as long as the cultivated area $S(t) > \hat{S}$, and to stop burying altogether as soon as the cultivated area exceeds $\hat{S}$ (assuming that $S_0 > \hat{S}$).

(*Remark*: A policy of this type is called a "bang-bang" policy for obvious reasons. The nature of the optimal burial policy is clear: the crossover point for Lestiboudois is reached when the discounted revenue of harvest equals the funeral revenue. But since Lestiboudois cannot control the rate of deaths, as long as the cultivated area is above the value $\hat{S}$, the gravedigger enjoys every funeral; when the cultivated area falls below $\hat{S}$, he mourns each additional death.)

Notes and References

§1. Classical optimal control theory dates back at least to the well-known "brachistochrone" problem involving the path that an object would follow in a minimal time descent between two points under gravitational attraction. It's also of interest to note that the theory of feedback control had its origins in the regulator developed by Maxwell to control the Watt steam engine. For an account of these and other elements of control theory in the words of their originators, see the volume

Bellman, R., and R. Kalaba, eds., *Mathematical Trends in Control Theory,* Dover, New York, 1964.

§2. The Maximum Principle is an outgrowth of techniques and ideas from the calculus of variations. For a discussion of the role of the classical variational calculus in giving rise to today's optimal control theory, see

Gelfand, I., and S. Fomin, *Calculus of Variations,* Prentice-Hall, Englewood Cliffs, NJ, 1963,

Dreyfus, S., *Dynamic Programming and the Calculus of Variations,* Academic Press, New York, 1965.

The classic work describing the Maximum Principle and its uses in optimization theory is

Pontryagin, L., V. Boltyanskii, R. Gamkrelidze, and E. Mischenko, *The Mathematical Theory of Optimal Processes,* Wiley, New York, 1962.

The determination of the optimal open-loop control law using the path of the Maximum Principle involves the solution of a two-point boundary-value problem. This can pose difficult numerical problems if the interval length of the process is large. One way of overcoming some of these difficulties is by use of embedding methods in which the original problem is replaced by a family of problems containing the original one. Relationships are then developed between adjacent members of the family, thereby leading to an *initial-value* procedure for moving from the trivial problem on a zero interval length to the length specified in the original problem. In many cases this approach stabilizes the original unstable situation. For a detailed account of how to carry out such an embedding in a variety of situations, see

Casti, J., and R. Kalaba, *Imbedding Methods in Applied Mathematics,* Addison-Wesley, Reading, MA, 1973.

Lee, E. Stanley, *Quasilinearlization and Invariant Imbedding,* Academic Press, New York, 1968,

Bellman, R. E., and R. E. Kalaba, *Quasilinearization and Nonlinear Boundary-Value Problems,* Elsevier, New York, 1965.

§3. The theory of dynamic programming was developed and popularized by Richard Bellman in an extensive series of books, papers and articles beginning in the mid-1950s. A representative sampling of this prodigious output is found in

Bellman, R., *Dynamic Programming,* Princeton U. Press, Princeton, 1957,

Bellman, R., and S. Dreyfus, *Applied Dynamic Programming,* Princeton U. Press, Princeton, 1962,

Bellman, R., and R. Kalaba, *Dynamic Programming and Modern Control Theory,* Academic Press, New York, 1965.

More recent works along the same lines are

Dreyfus, S., and A. Law, *The Art and Theory of Dynamic Programming,* Academic Press, New York, 1977,

Larson, R., and J. Casti, *Principles of Dynamic Programming, Parts I and II,* Dekker, New York, 1978, 1982.

Part II of the Larson and Casti works cited above is notable for its extensive treatment of the computational problems associated with dynamic programming and the many tricks and subterfuges that can be employed to overcome them.

§4. A thorough discussion of the ideas underlying anticipatory control processes is given in

Rosen, R., *Anticipatory Systems,* Pergamon, Oxford, 1985.

Of special note is the opening chapter of the above volume in which the historical genesis of the ideas developed here is traced.

The idea that what you do today is conditioned by where you are today, what you've done in the past, and what your expectations are for the future is not unknown outside of biology. In fact, the principle underlies what is termed "rational expectations" in economics, a topic that is treated in detail in

Sheffrin, S., *Rational Expectations,* Cambridge U. Press, Cambridge, 1983.

The idea of rational expectations seems to have first formally entered the economics literature in the path-breaking article

Muth, J., "Rational Expectations and the Theory of Price Movements," *Econometrica,* 29 (1961), 315–335.

For an easy introduction to the ideas underlying adaptive control processes as seen by the control engineering community, we recommend

Bellman, R., *Adaptive Control Processes: A Guided Tour,* Princeton U. Press, Princeton, 1961.

The chemical reaction example is found in the Rosen book cited earlier.

§5. For an extensive discussion of the way in which global failures can occur in the seeming absence of any local signs of difficulties, see the Rosen book above as well as the important article

Rosen, R. "Feedforward and Global System Failure: A General Mechanism for Senescence," *J. Theor. Biol.*, 74 (1978), 579–590.

§6. This taxonomy of adaptive systems is considered in greater detail in

Rosen, R., "Biological Systems as Paradigms for Adaptation," in *Adaptive Economic Models*, T. Groves and R. Day, eds., Academic Press, New York, 1974.

§7. The idea of a metabolism-repair system as an abstract metaphor for a living cell was first put forth by Rosen in the late-1950s. After a brief flurry of interest the idea seems to have gone into hibernation, a fate that we attribute to the absence of the proper system-theoretic framework within which to formulate properly the questions standing in the way of further development of the idea. Part of our exposition in this chapter has been directed toward substantiating this claim. Be that as it may, the original work in the area is summarized, together with references, in

Rosen, R., "Some Relational Cell Models: The Metabolism-Repair Systems," in *Foundations of Mathematical Biology*, Vol. 2, Academic Press, New York, 1972.

§8–9. Further elaboration of these results can be found in

Casti, J., "Linear Metabolism-Repair Systems," *Int'l. J. Gen. Sys.*, 14 (1988), 143–167,

Casti, J., "The Theory of Metabolism-Repair Systems," *Appl. Math. & Comp.*, to appear 1989.

§10. The idea of using (M,R)-networks as an abstract metaphor for a manufacturing enterprise arose in private discussions with R. Rosen. Some of the details are reported in

Casti, J., "Metaphors for Manufacturing: What Could it be Like to be a Manufacturing System?," *Tech. Forecasting & Soc. Change*, 29 (1986), 241–270.

DQ #4. We have seen that there need be no logical contradiction between the notions of causality and anticipation; i.e., an anticipatory decision procedure need not violate traditional ideas of causality. For further arguments along these lines, see

Burgers, J. M., "Causality and Anticipation," *Science*, 189 (1975), 194–198.

For introductory accounts of Whitehead's philosophy, we recommend

Whitehead, A., *Process and Reality: An Essay in Cosmology,* Cambridge U. Press, Cambridge, 1929.

Whitehead, A., *Science and the Modern World,* Macmillan, New York, 1925.

DQ #9. The notion of an ecosystem's resilience has been emphasized in a series of articles by C. S. Holling. A good summary of his views is found in

Holling, C. S., "Resilience and the Stability of Ecological Systems," *Ann. Rev. Ecol. Syst.,* 4 (1973), 1–23,

Holling, C. S., "Resilience in the Unforgiving Society," Report R–24, Insitute of Animal Resource Ecology, University of British Columbia, Vancouver, March 1981.

DQ #15. For an account of this duality which exactly mirrors the Maximum Principle vs. Dynamic Programming approach to the solution of optimal control problems, see the Dreyfus book cited under §2 above as well as

Hermann, R., *Differential Geometry and the Calculus of Variations,* 2d ed., Math Sci Press, Brookline, MA, 1977.

PE #3. The robot example is taken from the Larson and Casti book (Part II) cited under §3. This problem serves as the prototype for all adaptive control problems in which the control resource must be allocated in order to both control the system, as well as learn about it. This two-fold purpose of the control gives rise to the term *dual-control process* to describe this class of problems. For further details on this important kind of control process, see

Feldbaum, A. A., "Dual Control Theory-I," *Automation & Remote Cont.,* 21 (1961), 874–880.

PE #12. These category-theoretic results can be found in the Rosen work cited under §7 above.

PE #13–14. For further information on these issues, see

Arbib, M. A., "Categories of (M,R)-Systems," *Bull. Math. Biophys.,* 28 (1966), 511–517.

These are by no means the only efforts devoted to the use of category theory to formally characterize biological systems. See also

Baianu, I., and M. Marinescu, "Organismic Supercategories-I: Proposals for a General Unitary Theory of Systems," *Bull. Math. Biophysics,* 30 (1968), 625–635,

Ehresmann, A. C., and J. P. Vanbremeersch, "Hierarchical Evolutive Systems: A Mathematical Model for Complex Systems," *Bull. Math. Biology,* to appear, 1989.

PE #15. For many other examples of the use of control theory in off-beat settings, see the entertaining paper

Feichtinger, G., and A. Mehlmann, "Planning the Unusual: Applications of Control Theory to Nonstandard Problems," *Acta Applic. Math.,* 7 (1986), 79–102.

CHAPTER EIGHT

The Geometry of Human Affairs: Connective Structure in Art, Literature and Games of Chance

1. *Geometric Paradigms*

Without exaggeration, it's fair to say that virtually all the formal mathematical systems used to represent natural systems in the preceding chapters have been variants of the classical dynamical system-based Newtonian paradigm. We have discussed some of the reasons leading to the dominant position of the Newtonian world view, as well as some extensions and generalizations in the last chapter. Here we take an even bolder step away from Newton and discard the dynamical systems framework altogether, introducing a formal system based upon geometrical ideas of connectivity rather than dynamical ideas of change.

Rather than focus upon the twin concepts of particle and dynamical law that form the backbone of the Newtonian picture, here we shift attention to the notion of how the various pieces of a natural system fit together, and the manner in which this connective structure serves to characterize the global properties of the system. In short, we are concerned with the global *geometry* of the system instead of its *local* dynamics.

In order to capture the geometric essence of any natural system N, we must choose an appropriate formal geometric structure into which the observables of N can be encoded. It turns out to be useful to employ what is termed a *simplicial complex* as our formal mathematical framework. As will soon be made clear, a simplicial complex (or just plain "complex") is a natural generalization of the intuitive idea of a euclidean space, and is formed by interconnecting a number of pieces of varying dimension. The mathematical apparatus, which has its roots in algebraic topology, gives us a systematic procedure for keeping track of how the pieces fit together to generate the entire object, and how they each contribute to the geometrical representation of N.

Since the emphasis now is upon the *static* structure of N, we obtain information about the entire system N as opposed to the principally local knowledge inherent in the Newtonian dynamical view. The construction of the simplicial complex providing a "picture" of N can be carried out with virtually no assumptions about the analytic character of the observables and state-space representing N. This is exactly the sort of generality that we need to make headway in an effort to capture mathematically the major aspects of systems in the arts and humanities. Ample evidence of this claim will appear as we proceed.

2. *Sets and Relations*

The starting point for the construction of a geometrical formalism capturing the connective structure of N is at the most primitive level possible: finite sets and binary relations associating the elements of one set with those of another.

Let X and Y be two sets consisting of finite collections of elements,

$$X = \{x_1, x_2, \ldots, x_m\}, \qquad Y = \{y_1, y_2, \ldots, y_n\}.$$

We make no hypotheses about the nature of the objects in X and Y; they are just abstract elements. Further, let's suppose that we define (or are given) a rule λ that enables us to decide unambiguously whether an element $x_i \in X$ and an element $y_j \in Y$ are *related* according to the prescription specified by λ. Thus, λ is a kind of decision procedure: we feed a pair (x, y), $x \in X$, $y \in Y$ into λ, and out comes either a YES or a NO telling us whether or not x is λ-related to y. This procedure is graphically depicted in Fig. 8.1.

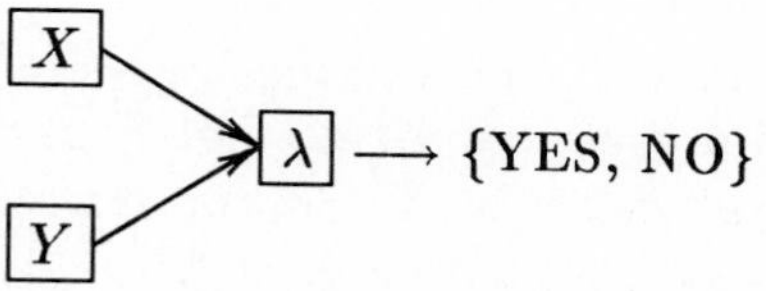

$$\begin{array}{l} \boxed{X} \searrow \\ \qquad\quad \boxed{\lambda} \longrightarrow \{\text{YES, NO}\} \\ \boxed{Y} \nearrow \end{array}$$

Figure 8.1 A Binary Relation λ between Sets X and Y

Since λ deals with pairs of elements, we call λ a *binary relation* on the sets X and Y.

Example

Let

$$\begin{aligned} X = \{\text{flowers}\} &= \{\text{daffodil, rose, carnation, tulip, pansy, orchid}\}, \\ &= \{x_1, x_2, \ldots, x_6\}, \end{aligned}$$

and let

$$\begin{aligned} Y = \{\text{colors}\} &= \{\text{red, yellow, green, blue, white}\}, \\ &= \{y_1, y_2, \ldots, y_5\}. \end{aligned}$$

A potentially interesting relation λ could be

flower type x_i is λ-related to color y_j if and only if there exists a strain of flower x_i having color y_j.

Since there normally exist only yellow daffodils, we would expect that $(x_1, y_2) \in \lambda$, i.e., by the rule λ daffodil (element x_1) is related only to the color yellow (element y_2). A similar argument for roses shows that $(x_2, y_1) \in \lambda, (x_2, y_2) \in \lambda$, $(x_2, y_5) \in \lambda$, and we can continue this process for all the other flowers in X.

As only certain pairs of elements from the cartesian product $X \times Y$ satisfy the condition of λ-relation, we see that, in general, $\lambda \subset X \times Y$. A compact way of expressing the relation λ is by means of an *incidence matrix* Λ. Let us label the rows of Λ by the elements of X and the columns by the elements of Y, and agree that if $(x_i, y_j) \in \lambda$, the (i,j) element of Λ equals 1, otherwise 0. So for the flower example above we would have the matrix Λ as

λ	red	yellow	green	blue	white
daffodil	0	1	0	0	0
rose	1	1	0	0	1
carnation	1	0	0	0	1
tulip	1	1	0	0	1
pansy	1	0	0	1	1
orchid	1	0	0	1	1

3. *Covers and Hierarchies*

Given a *single* set X, one of the most natural ways in which a binary relation emerges is by means of a *cover set* for X. Let Y be a set whose elements y_i are each *subsets* of the elements of X. So, for example, a typical element of Y might look like $y_1 = \{x_2, x_3, x_4\}$ or $y_2 = \{x_1, x_2\}$. We call Y a *cover* for X if

1) Each $x_i \in X$ is contained in at least one element of Y;

2) $\bigcup_j y_j = X$.

In the special case in which each x_i is contained in *exactly* one element of Y, we call the cover a *partition* of X.

Example

Consider the set of flowers X from the last section. Here we have

$$X = \{\text{daffodil, rose, carnation, tulip, pansy, orchid}\}.$$

The set

$$\begin{aligned} Y_1 &= \Big\{\{\text{daffodil, tulip, rose}\}, \{\text{carnation}\}, \{\text{pansy, rose, orchid}\}\Big\} \\ &= \{y_1^1, y_2^1, y_3^1\} \end{aligned}$$

constitutes a cover for X that is not a partition, since the element {rose} appears in two elements of Y. The set

$$Y_2 = \Big\{\{\text{daffodil, carnation}\}, \{\text{rose}\}, \{\text{tulip, orchid}\}, \{\text{pansy}\}\Big\}$$

is a cover that is also a partition, but the set

$$Y_3 = \Big\{\{\text{carnation, rose}\}, \{\text{orchid}\}\Big\}$$

is neither a cover nor a partition since $\bigcup_j y_j \neq X$.

A set cover naturally induces a binary relation λ in the following manner. Let X be a set and Y a cover set for X. Then we define the relation λ by the rule

$$(x_i, y_j) \in \lambda \text{ if and only if element } x_i \text{ belongs to the cover element } y_j.$$

So with the above set X and the cover set Y_1, we would have the incidence matrix Λ_1 given by

λ_1	y_1^1	y_2^1	y_3^1
x_1	1	0	0
x_2	1	0	1
x_3	0	1	0
x_4	1	0	0
x_5	0	0	1
x_6	0	0	1

Note that the incidence matrix representation gives us a simple test for determining whether a cover is a partition: it's a partition if each row of Λ contains a single nonzero entry.

The concept of a cover set enables us to introduce the notion of a hierarchy in a natural manner. Each element of the cover set Y is, in general, an aggregation of elements of X, and as a result, we can think of Y as a set existing at a level "higher" than that of X. If we arbitrarily label the level of X as level N, then Y is a level $N+1$ set. Since Y itself is a set, we can now consider a cover set Z for Y, and an associated binary relation $\mu \subset Y \times Z$. We would think of Z as a level $N+2$ set. This process can be carried out in the other direction as well, by regarding X as a cover set for some set W at level N-1. This idea of a hierarchy of sets and relations is depicted in Fig. 8.2.

Of special interest in this diagram are the diagonal relations like θ which enable us to go from one hierarchical level to another in a nontrivial way. It's clear that if the relation λ is given between X and Y at level N, and we induce the relation μ from the cover set A, then we can construct θ as $\theta \circ \lambda = \mu$.

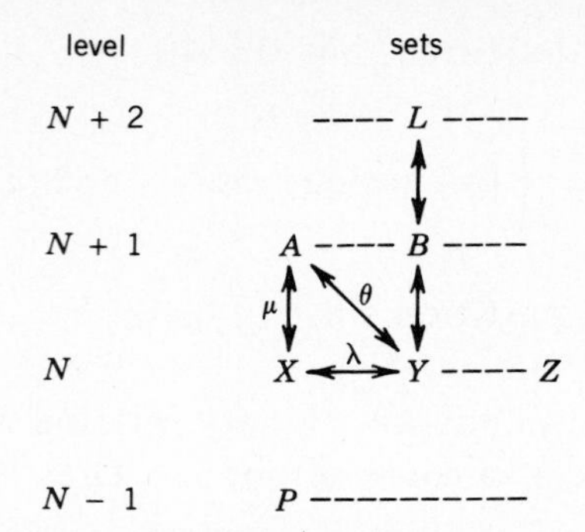

Figure 8.2 A Hierarchy of Sets and Relations

Primitive as the above ideas are, they already contain enough mathematical structure for us to say a few interesting things about matters in the arts and humanities usually thought to be outside the domain of mathematical investigation. Let's look at a couple of illustrations.

Example 1: Laughter and Tears

In his book *Multidimensional Man,* Atkin suggests that the process of evoking laughter or tears from a particular N-level situation corresponds to a movement either up to level $(N+1)$ for laughter or down to level $(N-1)$ for tears. His argument is that in order to be aware of witticisms present at level N, we must be able to contemplate new relationships on the N-level set by either rearranging existing elements or extending the elements to find new relationships between them—in short, by being aware of level $(N+1)$. Atkin's claim is that it is this sudden *jump* to the $(N+1)$-level set that generates a release of laughter.

In contrast to laughter, which is a widening of our horizons by a movement *up* the conceptual ladder, tears is a movement *downward* that shrinks those horizons. By moving up the hierarchy we see the possibility for new relationships, a potentially liberating situation; a movement downward contracts or eliminates the potential for new interactions, and tends to force us to think we are being imprisoned by the existing order with no way out.

As one of the many illustrations of the laughter/tears hypothesis cited in Atkin's book, let's look at a passage from Joseph Heller's classic work *Catch-22.* The N-level situation involves Yossarian's attempt to get Orr out of flying combat missions by having Doc Daneeka ground him on the basis that he's crazy. Doc states that this is only possible if Orr asks to be grounded but then adds that as soon as he's asked, he will not be able to ground Orr since the request itself would constitute evidence that he's not crazy!

In Heller's passage, the N-level consists of the *individuals* Yossarian and Orr. At the $(N+1)$-level, we have a set consisting of a number of descriptive words like Sane, Missions, Grounded, and Fit for Duty. Finally, at level $(N + 2)$ there is the set consisting of the single element Doc, since this is the agent who can decide whether or not a man at level N is a member of the $(N + 1)$-level element Fit for Duty.

The analysis of the scene and its humor comes from the fact that Yossarian thinks he is "covered" by the words Insane and Flying Missions at level $(N+1)$, and this would automatically mean that he cannot also be covered by the word Fit for Duty. But Doc reorganizes the cover set at level $(N+1)$ by saying that Yossarian's request is by itself sufficient to demonstrate that he's Sane, therefore covered by Fit for Duty. Here we see Yossarian's frustration at feeling trapped inside the N-level set and having his appeal to the $(N+2)$-level set rejected through a rearrangement of the $(N+1)$-level cover. Thus, if the reader identifies with Yossarian, he is brought to the verge of tears. But if the reader stands outside the book—at, say, level $(N+3)$—then he experiences the urge to laugh at this "Catch-22" situation.

Example 2: A Shakespearean Sonnet

Consider the following famous love poem by Shakespeare:

(1) Shall I compare thee to a summer's day?
Thou art more lovely and more temperate:
Rough winds do shake the darling buds of May,
And summer's lease hath all too short a date:
(5) Sometime too hot the eye of heaven shines,
And often is his gold complexion dimm'd;
And every fair from fair sometime declines,
By chance, or nature's changing course, untrimm'd:
But thy eternal summer shall not fade,
(10) Nor lose possession of that fair thou owest;
Nor shall Death brag thou wander'st in his shade,
When in eternal lines to time thou growest.
So long as men can breathe, or eyes can see,
(14) So long lives this, and this gives life to thee.

At level N, we take a set that consists of all the concepts referred to by the nouns in the poem. These are Thee (the beloved), May/summer, Sun, Fair (beauty), and Thy summer (the bloom time of the loved one). For a cover set Y^1 at level $(N+1)$, we must relate the N-level elements to general ideas, or properties, that they display. For these properties we could take

y_1^1 : 'being lovely' (lines 2 and 7),

y_2^1 : 'being temperate' (lines 2, 3 and 5),

y_3^1 : 'enduring time' (lines 4, 9 and 12),

y_4^1 : 'growing/diminishing' (lines 6, 7, 9, 11, 12 and 14).

If we let X be the N-level set, the incidence matrix for the natural binary relation λ relating X to its cover set Y^1 becomes

λ_1	y_1^1	y_2^1	y_3^1	y_4^1
Thee	0	1	1	1
May/summer	0	1	1	1
Sun	0	1	0	1
Fair	0	0	1	1
Thy/summer	1	0	1	1

It's important to observe that we can also generate other relations between these sets and the lines of the poem. Such relations would give insight into the structure of the poem at the corresponding hierarchical levels.

4. *Complexes and Connections*

The hierarchical analyses based upon the notion of cover sets, interesting as they appear, are really limited in what they can tell us about the overall manner in which the relation λ binds together the elements of the sets X and Y. For this type of information we need to obtain a multidimensional "picture" of λ. The way to do this is to represent λ geometrically by an object termed a *simplicial complex.*

Let us *abstractly* identify the elements of X with the *vertices* of the complex $K_Y(X;\lambda)$ representing λ. What this means is that we label the first vertex of K by the name "x_1," the second vertex has the name "x_2," and so on. Similarly, we name the *simplices* of K by the elements of Y. Thus, we call the first simplex "y_1," the second "y_2," and so forth. Note that there is no intrinsic ordering of these objects since the sets X and Y consist of unordered elements. Since the relation λ associates a *subset* of X with *each* element of Y, we can geometrically represent each $y \in Y$ by connecting its vertices (in X) into an abstract simplex. For example, if λ associates the vertices x_1, x_2, and x_4 with the element $y \in Y$, then we can geometrically represent y by the 2-simplex consisting of the filled-in triangle whose vertices are x_1, x_2 and x_4 (see Fig. 8.3).

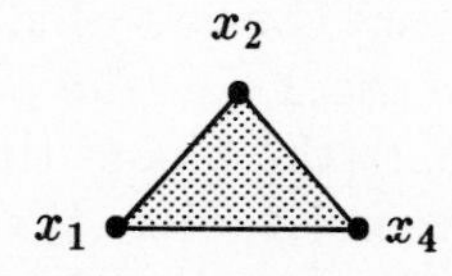

Figure 8.3 The 2-Simplex y

The same procedure can be used to associate an abstract geometrical object (point, line, triangle, tetrahedron, ...) with each $y \in Y$. The geometrical dimension of each such simplex equals the number of vertices comprising y minus 1.

Example

Let the relation λ be given by the incidence matrix

λ	x_1	x_2	x_3	x_4
y_1	1	0	1	1
y_2	0	1	1	0
y_3	1	0	1	0
y_4	1	1	1	1

Here y_1 is the 2-simplex consisting of the vertices x_1, x_3 and x_4; y_2 is the 1-simplex consisting of x_2 and x_3, and so on. Geometrically, Λ can be represented by the complex $K_Y(X;\lambda)$ shown in Fig. 8.4.

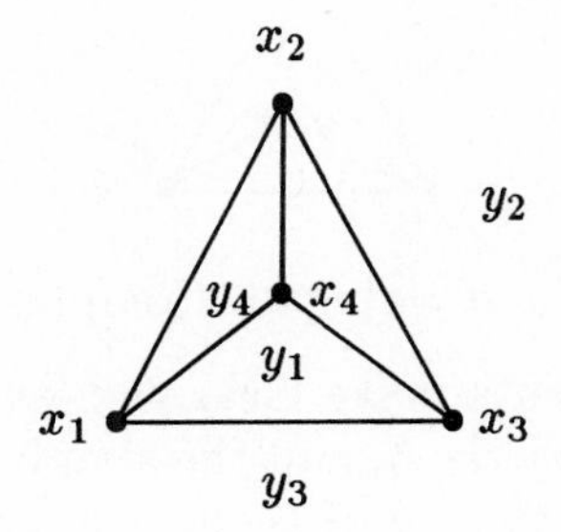

Figure 8.4 The Complex $K_Y(X;\lambda)$

Here y_4 is the solid tetrahedron composed of the faces, edges and vertices, whereas y_1, y_2 and y_3 are faces of y_4. Thus we see that the simplices of $K_Y(X;\lambda)$ are connected to each other by the sharing of vertices. For instance, y_3 is connected to y_1 by sharing the vertices x_1 and x_3 (i.e., the edge x_1–x_3), and y_1 and y_2 are connected by sharing the single vertex x_3 (i.e., the point x_3).

Just as there is no intrinsic ordering of the elements of X and Y and we could permute these objects at will and obtain the same abstract geometric structure, there is also no intrinsic reason to select X as the vertex set and Y as the simplex set; it's an arbitrary choice. If we interchange the roles of X and Y, we then obtain the *conjugate complex* $K_X(Y;\lambda^*)$. In terms of the incidence matrix Λ, interchanging X and Y corresponds to the operation of matrix transposition. Thus the incidence matrix for the conjugate relation λ^* is just Λ'. It should be noted, however, that the geometric structure of the conjugate relation λ^* is generally different from that of λ. We shall exploit this fact as we go along in order to obtain additional information about the connective structure inherent in λ.

Example

Consider the relation λ given in the preceding example. In this case the incidence matrix Λ' for the conjugate relation λ^* is

λ^*	y_1	y_2	y_3	y_4
x_1	1	0	1	1
x_2	0	1	0	1
x_3	1	1	1	1
x_4	1	0	0	1

and the complex $K_X(Y;\lambda^*)$ is geometrically shown in Fig. 8.5.

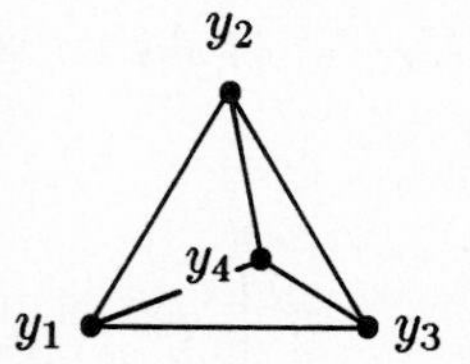

Figure 8.5 The Conjugate Complex $K_X(Y;\lambda^*)$

The preceding arguments make clear the fact that the binary relation $\lambda \subset X \times Y$, the incidence matrix Λ, and the simplicial complex $K_Y(X;\lambda)$ are equivalent objects; given any one of them, we can automatically construct the other two. Similarly, given λ we also automatically obtain the conjugate (or *dual*) relation λ^*. Which one of these objects we use depends upon considerations of the moment: Λ is convenient for computations; $K_Y(X;\lambda)$ is convenient for geometric insight; λ is useful for problem definition. Examples of all three approaches will be seen as we proceed, but for now let's look at how we can use the complex $K_Y(X;\lambda)$ to define additional objects telling us about the local and global geometric structure of λ.

5. *Structure Vectors and Eccentricity*

The simplices of $K_Y(X;\lambda)$ are connected to each other by sharing vertices; however, this does not mean that the simplices are connected *pairwise*. It's perfectly possible for two simplices y_i and y_j to have no vertices in common but yet be connected to each other by an intermediate chain of simplices that serves as a means for linking y_i and y_j. As a trivial example, consider the three 1-simplices y_1, y_2 and y_3 depicted in Fig. 8.6.

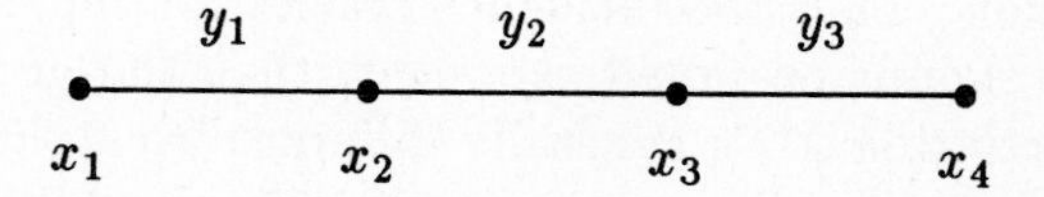

Figure 8.6 A Chain of 1-Connection

Here y_1 consists of the two vertices x_1 and x_2, with y_2 and y_3 defined similarly. The simplices y_1 and y_3 have no vertex in common; they are disconnected. Nevertheless, there is a chain of simplices consisting of y_1, y_2 and y_3 possessing the following properties: (1) y_1 shares a face with y_2; (2) y_2 shares a face with y_3. In other words, we can pass from y_1 to y_3 via the intermediate simplex y_2. The geometric dimension of the smallest face in this chain is 0 (the face consisting of either the vertex x_2 or the vertex x_3), so we call such a chain of connection a 0-chain. This idea of passing from one simplex to another via intermediate simplices forms the basis for the notion of *q-connection* in the complex $K_Y(X;\lambda)$.

Let the dimension of $K_Y(X;\lambda)$ be defined as $D =$ the dimension of the highest dimensional simplex in $K_Y(X;\lambda)$. For each integer q, $0 \leq q \leq D$, the relation of q-connection on the simplices of $K_Y(X;\lambda)$ is given by the following definition:

DEFINITION. *The simplices y_i and y_j are q-connected in $K_Y(X;\lambda)$ if there exists a sequence of simplices $\{y_{\alpha_i}\}$, $i = 1, 2, \ldots, r$ in $K_Y(X;\lambda)$ such that*

i) y_i shares a face of dimension m in y_{α_1}.

ii) y_j shares a face of dimension n in y_{α_r}.

iii) y_{α_k} and $y_{\alpha_{k+1}}$ share a face of dimension α_k.

iv) $q = \min\{m, \alpha_1, \alpha_2, \ldots, \alpha_{r-1}, n\}$.

In short, dimensionally speaking, q is the weakest link in the chain connecting y_i and y_j.

It's a minor exercise to verify that the relation of q-connection is an *equivalence relation* on $K_Y(X;\lambda)$ for each $q = 0, 1, 2, \ldots, D$. Consequently, q-connection partitions the complex into equivalence classes whose elements are all those simplices that are q-connected to each other. Such a relation gives us some insight into the global geometry of $K_Y(X;\lambda)$. If we call the equivalence relation λ_q, $q = 0, 1, \ldots, D$, we see that λ_q partitions the complex into Q_q disjoint classes, enabling us to define a structure vector Q whose components are the number of classes at each dimensional level q. Thus,

$$Q = (Q_D, Q_{D-1}, \ldots, Q_0).$$

In particular, note the case when each element in Q equals 1. In this case there is only a single equivalence class at each dimensional level, indicating that the complex is as tightly connected as it possibly can be with each simplex connected to every other through a chain of q-connection at every dimensional level. It's clear that if two simplices are q-connected for some value of q, then they are also p-connected for every $p \leq q$.

Another way of interpreting the structure vector Q is to imagine that you are equipped with a special pair of glasses that enable you to see only in dimensions q and higher. If you put on these special spectacles and look at a picture of $K_Y(X;\lambda)$ (assuming you could draw in all dimensions up to D), then you would see the complex split apart into Q_q disjoint pieces.

The simplest way to compute Q is to use the incidence matrix Λ. Forming the product $\Lambda\Lambda'$, the element in position (i,j) tells us how many vertices the simplices y_i and y_j have in common. The geometric dimension of this shared face is thus $(\Lambda\Lambda')_{ij} - 1$. With this information available for each pair (y_i, y_j), it's a simple matter to trace out the chain of connection between any two simplices and to determine the smallest dimensional face on this chain. As a measure of *obstruction* to a free flow of information from one part of the complex to another, we can use the *obstruction vector* $U = Q - (1)$, where (1) is the vector all of whose components equal 1.

Example

Let's return to the flower example of Section 2. The incidence matrix Λ is

λ	y_1	y_2	y_3	y_4	y_5
x_1	0	1	0	0	0
x_2	1	1	0	0	1
x_3	1	0	0	0	1
x_4	1	1	0	0	1
x_5	1	0	0	1	1
x_6	1	0	0	1	1

Forming the product $\Lambda\Lambda' - [1]$, where $[1]$ denotes the matrix all of whose entries are 1, we obtain

	x_1	x_2	x_3	x_4	x_5	x_6
x_1	0	0	–	0	–	–
x_2		2	1	2	1	1
x_3			1	1	1	1
x_4				2	1	1
x_5					2	2
x_6						2

where "–" indicates (-1). Since the matrix is symmetric, we show only the upper-triangular half. Note that in this example the simplices are the elements of X (the flowers) connected to each other by sharing the vertices in Y (the colors).

In the matrix $\Lambda\Lambda' - [1]$, the diagonal elements are the dimensions of the individual simplices x_i, and the off-diagonal elements are the dimensions of the faces shared by the simplices taken pairwise. An entry "–" means that the two simplices have no vertex in common. Performing the "q-analysis," we obtain

$$\begin{aligned} \text{at } q=2:&\quad Q_2=2, &&\{x_2,x_4\},\{x_5,x_6\},\\ q=1:&\quad Q_1=1, &&\{x_2,x_3,x_4,x_5,x_6\},\\ q=0:&\quad Q_0=1, &&\{\text{all}\}. \end{aligned}$$

Thus, at level $q = 2$ (sharing three or more colors) there are two disjoint components consisting of the flowers {rose, tulip} and {pansy, orchid}, each of which has three colors in common, but not the *same* three. Dropping to the level $q = 1$ (sharing two colors), all flowers except x_1 (daffodil) have at least two linking colors. At level $q = 0$, all flowers have at least one linking color. But note that this does **not** mean all flowers have at least one color *in common.* That this is not true can be easily seen from the incidence matrix Λ. What it *does* mean is that we can pass from one flower to another via a chain of flowers which *pairwise* have at least one common color. The structure vector for $K_X(Y;\lambda)$ in this case is

$$Q = (\overset{2}{2}\ 1\ \overset{0}{1}),$$

with obstruction vector $U = (\overset{2}{1}\ 0\ \overset{0}{0})$, indicating obstruction only at level $q = 2$.

We have already seen that the conjugate complex $K_X(Y;\lambda^*)$ comes along free of charge with the complex $K_Y(X;\lambda)$. Thus, we can perform a q-analysis on the conjugate complex, obtaining a second structure vector Q^*. The two objects Q and Q^* give us valuable insight into the manner in which the simplices in the two complexes connectively relate to each other by vertex sharing, albeit at a distance. This is information about the global geometric structure of the complexes. But what about the local structure? How do individual simplices fit into the overall geometry? Can we define a measure of the way specific simplices are integrated into the local geometry? This desire leads to the idea of *eccentricity.*

Consider a single simplex y. Intuitively, we would say that y is not well-integrated into the complex if it contains many vertices that are not shared with other simplices in the complex. So if dim $y = n$ (i.e., y is comprised of $n+1$ vertices), and if $m =$ the largest number of vertices that y shares with any other simplex in the complex, then the number $n - m$ is a measure of how "eccentric" y is as a member of $K_Y(X;\lambda)$. However, it's also reasonable to assume that this difference is more significant at lower-dimensional levels than at higher ones, so we normalize the difference $n - m$ by making it relative to m. Thus our final measure of eccentricity is

$$\text{ecc}(y) = \frac{n-m}{m+1},$$

where we have included the "1" in the denominator to avoid possible zero divisors for those y that share *no* vertices with any other simplex in $K_Y(X; \lambda)$.

Example: The Middle East Situation

To illustrate the above ideas, let's consider the Middle East crisis as a relation between the set of participants and the issues that divide them. For the set X we take

$$X = \{\text{issues}\} = \{x_1, x_2, \ldots, x_{10}\},$$

where

x_1 = autonomous Palestinian state on the West Bank and Gaza,
x_2 = return of the West Bank and Gaza to Arab rule,
x_3 = Israeli military outposts along the Jordan River,
x_4 = Israel retains East Jerusalem,
x_5 = free access to all religious centers,
x_6 = return of Sinai to Egypt,
x_7 = dismantle Israeli Sinai settlements,
x_8 = return of Golan Heights to Syria,
x_9 = Israeli military outposts on the Golan Heights,
x_{10} = Arab countries grant citizenship to Palestinians who choose to remain within their borders.

For the set of participants we take

$$Y = \{\text{participants}\} = \{y_1, y_2, \ldots, y_6\},$$

where

y_1 = Israel,
y_2 = Egypt,
y_3 = Palestinians,
y_4 = Jordan,
y_5 = Syria,
y_6 = Saudi Arabia.

The relation $\lambda \subset Y \times X$ specifying the way the participants interact with the issues will be

$(y_i, x_j) \in \lambda$ if and only if participant y_i is neutral or favorable toward issue (goal) x_j.

The incidence matrix for λ is

λ	x_1	x_2	x_3	x_4	x_5	x_6	x_7	x_8	x_9	x_{10}
y_1	0	1	1	1	1	1	0	0	1	1
y_2	1	1	1	0	1	1	1	1	1	0
y_3	1	1	0	0	1	1	1	1	1	1
y_4	1	1	0	0	1	1	1	1	1	0
y_5	1	1	0	0	1	1	1	1	0	0
y_6	1	1	1	0	1	1	1	1	1	1

Examination of the complex $K_Y(X;\lambda)$ shows that the most likely negotiating partner for Israel is Saudi Arabia, which is neutral or favorable on all issues except one (Israel retaining East Jerusalem). However, both Egypt and the Palestinians are nearly as likely candidates since they are simplices of dimension only one less than Saudi Arabia. As the Camp David talks demonstrated some years ago, Egypt is indeed a favored negotiating partner due also to psychological and other factors not incorporated into the above relation λ.

Focusing upon goals and issues, we find the high-dimensional objects in $K_X(Y;\lambda^*)$ being x_2 = return of the West Bank and Gaza to Arab rule, x_5 = free access to religious centers, and x_6 = return of the Sinai to Egypt. These goals are viewed as neutral or favorable by all six participants. Therefore they provide a good basis for a negotiated settlement of the conflict. This observation has also been borne out by the Camp David talks, as well as by subsequent developments.

The structure vector for the complex $K_Y(X;\lambda)$ is

$$Q = (\overset{8}{1}\ 1\ 2\ 1\ 1\ 1\ 1\ 1\ \overset{0}{1}),$$

with obstruction vector

$$U = (\overset{8}{0}\ 0\ 1\ 0\ 0\ 0\ 0\ 0\ \overset{0}{0}).$$

Thus the only obstruction to a free flow of information is at dimensional level $q = 6$, where the two components {Israel} and {Egypt, Palestinians, Jordan, Saudi Arabia} do not share any six-dimensional bridge. In fact, $q = 6$ is the first-dimensional level at which Israel enters the complex, and from level $q = 5$ downward all parties are fully connected to one another. Consequently, as long as the discussions are restricted to six issues or less, all parties are connected in the same component of the complex and the basis for a negotiated settlement exists. With more than six issues on the table, the parties become disconnected and an obstruction to a settlement arises.

Now let's look at the dispute from the viewpoint of the issues rather than the disputants, i.e., we examine the conjugate complex $K_X(Y;\lambda^*)$. In this complex the structure vector is

$$Q^* = (\overset{5}{1}\ 1\ 1\ 1\ 1\ \overset{0}{1}),$$

with obstruction vector

$$U^* = (\overset{5}{0}\ 0\ 0\ 0\ 0\ \overset{0}{0}).$$

Thus, there are no obstructions at any dimensional level, suggesting that all issues are tightly connected through sharing parties that are not unfavorably disposed to a given issue. This means that the highest-dimensional single issues in the complex would provide the likely starting point for a settlement which, in this case, are the issues involving the return of the West Bank and Gaza to Arab rule, free access to religious centers, and a return of the Sinai to Egypt, as noted earlier.

In terms of *individual* participants or issues, the only parties that show nonzero eccentricity are Israel (ecc$=\frac{1}{6}$) and Saudi Arabia (ecc$=\frac{1}{8}$), with all issues having eccentricity zero. These results are not surprising since the Middle East situation is basically Israel versus the Arabs, with Saudi Arabia being, for the most part, the least militant and most flexible Arab state.

While the foregoing example involves only a few states and a handful of issues, it should be noted that the method can be used for situations in which there are dozens or even hundreds of factors. Some indications along these lines will be seen later in our treatment of works of art, as well as in some of the case studies cited in the Notes and References.

6. *Complexity*

The matter of system complexity has been repeatedly addressed in earlier chapters, from both an objective measurement point of view and a subjective, observer-dependent perspective. Here we take up the issue again, this time focusing upon the static complexity of the complex $K_Y(X;\lambda)$. We develop an objective measure for the complexity of the complex that satisfies three basic axioms on our intuitive ideas about the complexity of an object composed of a number of subsystems. We shall leave to the Problems and Exercises the task of relating the notion of complexity developed here with those complexity concepts and measures presented in earlier chapters.

We adopt the following complexity axioms:

Axiom 1. A system consisting of a single simplex has complexity 1.

Axiom 2. A subcomplex (subsystem) has complexity no greater than that of the entire complex (system).

Axiom 3. The combination of two complexes to form a third has complexity no greater than the sum of the complexities of the component complexes.

Note that Axioms 1–3 implicitly assume that the complex in question is connected at level $q = 0$; i.e., the structure vector Q has $Q_0 = 1$. If not, we compute the complexity function for each of the disconnected components of the complex using the maximum of these numbers to represent the complexity of the entire complex. This is equivalent to regarding the complex as the parallel combination of its disconnected components.

A measure that satisfies the foregoing axioms and is readily computable from the structure vector Q is

$$\psi(K) = \frac{2}{(D+1)(D+2)} \sum_{i=0}^{D} (i+1) Q_i,$$

where $D = \dim K_Y(X; \lambda)$ and Q_i = the ith component of the structure vector Q. The factor $2/(D+1)(D+2)$ is introduced as a normalization to satisfy Axiom 1.

Example

Returning to the Middle East conflict, we can easily compute

$$\psi(K_Y(X; \lambda)) = \frac{52}{45},$$

and we find that the conjugate complex has complexity

$$\psi(K_X(Y; \lambda^*)) = 1.$$

Consequently, the complex $K_Y(X; \lambda)$, which focuses attention upon the participants, is somewhat "more complicated" than $K_X(Y; \lambda^*)$, which addresses the issues.

Having now developed some of the machinery needed to tackle questions of interest in the arts and humanities, let's proceed to show how the geometrical structures introduced above can be used in a variety of realistic circumstances.

7. *The Art of M. C. Escher*

Nowadays it's almost impossible to walk into the office of a scientist or mathematician without seeing an engraving or two by the well-known Dutch artist M. C. Escher (1898–1971). Escher is noted for the remarkable geometrical precision of his work as well as the deep connections with mathematical concepts, especially those in group theory. Here we examine one of his more famous works using q-analysis.

A good illustration of the use of simplicial complexes to capture abstract structure is provided by Escher's famous engraving *Sky and Water,* depicted in Fig. 8.7. Here we see a collection of what appear to be geese gradually being transformed into fish as the picture is scanned from top to bottom. At the same time, we also see a smooth transition from figure to ground as the shapes consituting the geese become background for the swimming fish. Our goal is to capture some of the structure of these transitions using the q-analysis apparatus discussed above.

The first step is to identify relevant sets X and Y and a relation λ that encapsulate some of the structure of *Sky and Water.* A bit of reflection soon leads to the view that the picture is really a statement about the relationship between various geometrical shapes (the birds, fish and their intermediate forms), and features that pertain to the identification of the shapes as being birdlike, fishlike or something in between. In Fig. 8.8 we have identified 39 different shapes that appear in the picture. So we let the set $Y = \{y_1, y_2, \ldots, y_{39}\}$ be composed of these shapes. For the set X we choose the following set of 12 features that play a prominent role in the picture.

Figure 8.7 M. C. Escher, *Sky and Water* (1938)

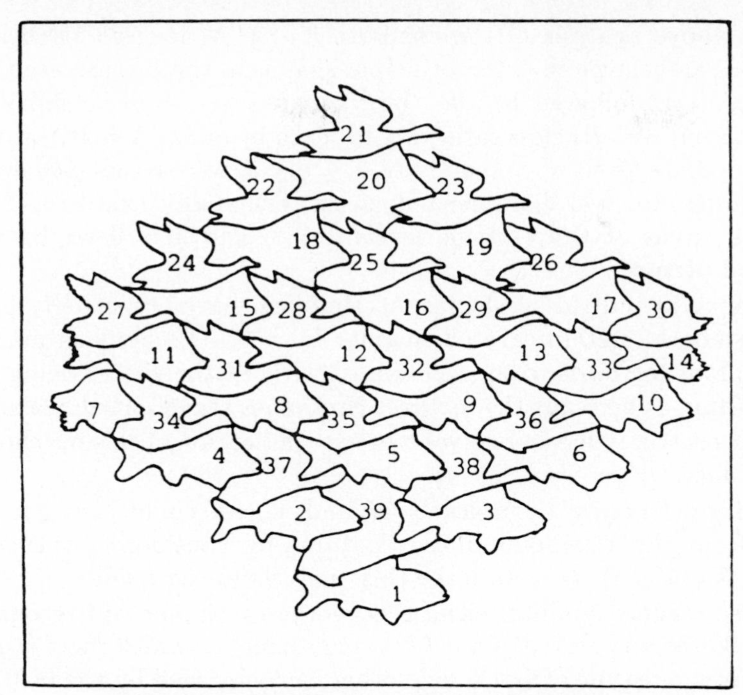

Figure 8.8 Shapes in *Sky and Water*

$$
\begin{aligned}
X &= \{x_1, x_2, \ldots, x_{12}\}, \\
&= \{\text{scales, mouth, gills, fish-tail, fins, fish shape, eye,} \\
&\qquad \text{duck shape, two wings, feathers, beak, legs}\}.
\end{aligned}
$$

For the relation $\lambda \subset Y \times X$, we make the obvious choice: $(y_i, x_j) \in \lambda$ if and only if shape y_i displays feature x_j. The incidence matrix for λ is given in the Appendix at the end of the chapter.

Using Λ, we form $\Lambda\Lambda' - U$ and obtain the following q-analysis:

$$
\begin{aligned}
q = 6: &\quad Q_6 = 1, \quad \{y_1 - y_6\}, \\
q = 5: &\quad Q_5 = 2, \quad \{y_1 - y_6, y_8 - y_{10}\}, \{y_{21} - y_{26}, y_{28}, y_{29}\}, \\
q = 4: &\quad Q_4 = 2, \quad \{y_1 - y_6, y_8 - y_{10}\}, \{y_{21} - y_{29}\}, \\
q = 3: &\quad Q_3 = 2, \quad \{y_1 - y_{13}\}, \{y_{21} - y_{29}\}, \\
q = 2: &\quad Q_2 = 2, \quad \{y_1 - y_{13}\}, \{y_{21} - y_{29}, y_{31} - y_{33}\}, \\
q = 1: &\quad Q_1 = 2, \quad \{y_1 - y_{13}\}, \{y_{21} - y_{33}\}, \\
q = 0: &\quad Q_0 = 1, \quad \{\text{all}\}.
\end{aligned}
$$

The above analysis of the complex $K_Y(X;\lambda)$ focuses attention upon the shapes and shows that the principal shapes in the picture are the "fish" shapes $y_1 - y_6$, followed by the "bird" shapes $y_{21} - y_{26}$, a fairly obvious conclusion but nevertheless satisfying to reach by our systematic procedures. At intermediate levels of connectivity, $1 \leq q \leq 5$, we see that *Sky and Water* breaks down into two disconnected pieces, essentially fishlike and birdlike shapes, whereas at the extreme levels $q = 6$ and $q = 0$ we have a fully integrated picture.

Looking at individual shapes, all simplices have eccentricity 0 indicating the smooth, well-integrated manner in which Escher has managed to progress from the birds to the fish almost without the viewer being aware of the transition as he scans the picture. No shapes stand out above any other, and our eccentricity measure gives a precise indication of this psychologically observed fact.

By interchanging the roles of X and Y, we could also analyze this picture from the viewpoint of the features by considering the conjugate complex $K_X(Y;\lambda^*)$. But we leave this analysis to the reader.

Those readers familiar with other works by Escher will recognize that *Sky and Water* is typical of many of his engravings in which there is a smooth passage from one type of figure to another, accompanied by a transition from figure to ground. The techniques introduced above, together with the deeper and more refined methods presented in the Notes and References, provide a basis for a systematic analysis of many aspects of Escher's style and form. We now turn to consideration of the structure present in another artistic style, that of another Dutch painter, Piet Mondrian.

8. *Connective Structure in the Work of Mondrian*

Piet Mondrian (1872–1944) was a Dutch artist who was a leading member of the de Stijl abstract art movement centered in The Netherlands. Mondrian's cubist period reached its peak with his works *New York City* and *Broadway Boogie Woogie,* first exhibited in 1943. Here we use our tools of q-analysis to study another one of his paintings of the same style. Figure 8.9 shows a black-and-white approximation to Mondrian's famous work *Checkerboard, Bright Colors,* which was briefly considered earlier in Discussion Question 5 of Chapter Two. This painting consists of a rectangular grid pattern of 256 squares each colored by one of the 8 colors indicated.

At first glance it appears that the colors are distributed more or less randomly on the canvas. Here we present an analysis by R. Atkin that aims to test this hypothesis in the sense that a random allocation of colors would display no discernible structure or pattern. The task is to see if the q-analysis techniques developed above can be used to tease out any hidden structure in the painting, thereby refuting the randomness hypothesis.

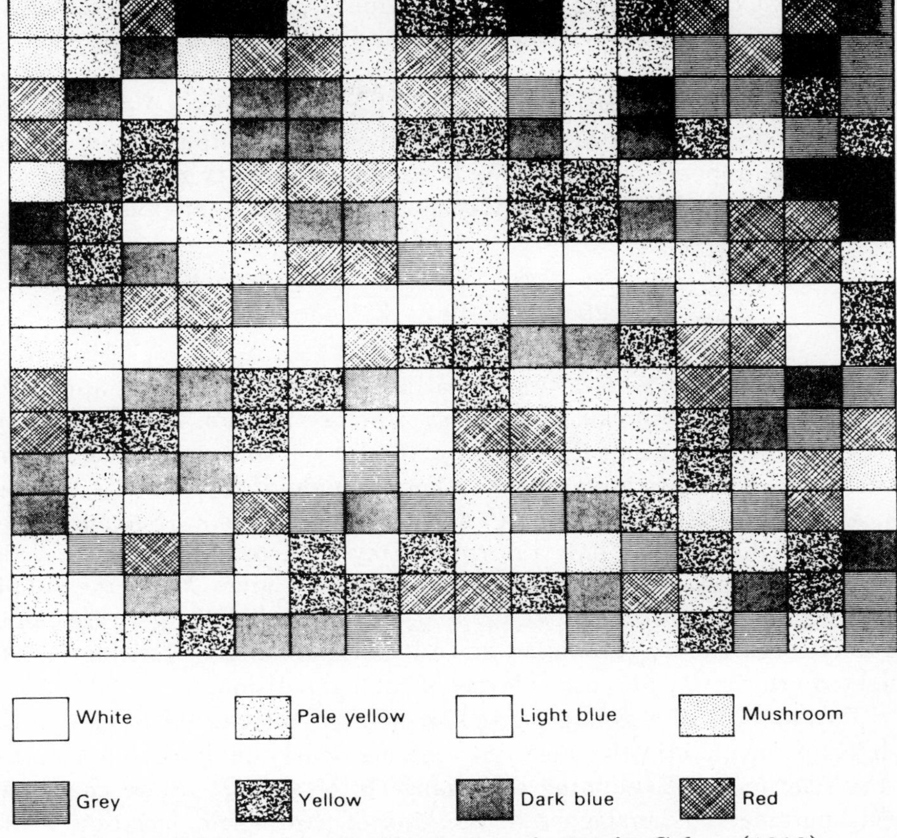

Figure 8.9 Mondrian's *Checkerboard, Bright Colors* (1919)

As usual, we begin by trying to identify relevant sets X and Y and a meaningful relation λ. In this case the choices are easy: the message of the painting (if it has one) is obviously a relationship between the squares on the canvas and the colors. Thus, let

$$\begin{aligned} X &= \{\text{squares on the canvas}\} \\ &= \{x_1, x_2, \ldots, x_{256}\}, \\ Y &= \{\text{colors}\} \\ &= \{y_1, y_2, \ldots, y_8\} \\ &= \{\text{white, grey, pale yellow, yellow, light blue,} \\ &\quad\ \text{dark blue, mushroom, red}\}. \end{aligned}$$

The relation $\lambda \subset Y \times X$ is that $(y_i, x_j) \in \lambda$ if and only if color y_i is used for square x_j. The 8×256 incidence matrix Λ can be easily (albeit tediously) filled in from Fig. 8.9, once we have agreed on an ordering of the squares.

Having Λ, we can proceed as before to compute the structure vectors Q, Q^*, as well as the eccentricities of the individual simplices. But let's take another approach.

Suppose we look at the canvas as one single large square X_1 and imagine that we define a relation $\lambda_1 \subset Y \times X_1$ that just counts the number of times that color y_i appears in X_1; i.e., λ_1 is no longer a binary relation but takes on integer values. So, for λ_1 the transposed incidence matrix is

λ_1^*	y_1	y_2	y_3	y_4	y_5	y_6	y_7	y_8
X_1	19	29	54	40	13	43	15	43

Here we have color y_1 (white) appearing 19 times, y_2 (grey) appears 29 times, and so on. We can now *induce* a binary relation λ by introducing a parameter θ and using the rule: $(y_i, X_1) \in \lambda$ if and only if color y_i appears at least θ times in square X_1. For example, if $\theta = 20$ only the pairs $(y_2, X_1), (y_3, X_1), (y_4, X_1), (y_6, X_1)$ and (y_8, X_1) are in λ. Setting $\theta = 20$ corresponds to "seeing" only the colors grey (y_2), pale yellow (y_3), yellow (y_4), dark blue (y_6) and red (y_8). This relation λ gives X_1 as the simplex consisting of the vertices $\{y_2, y_3, y_4, y_6, y_8\}$, making $\dim X_1 = 4$. We call the parameter θ a *slicing parameter,* since it has the effect of "slicing off" the weighted relation λ_1 at various levels of color repetition.

The concept of a slicing parameter is often very useful when we want to inject some quantitative ideas into the previously qualitative 0/1 pattern of the relation λ. Examining the change in Q and Q^* as we change the slicing parameter is analagous to the kind of parametric variation we saw earlier in Chapter Four. If we feel that looking at the painting as a single square gives too coarse a view, we could consider decomposing the square into subsquares and analyzing the connectivities introduced by various levels of θ. For instance, suppose we consider the painting decomposed into the 16 subsquares depicted in Fig. 8.10

X_{41}	X_{42}	X_{43}	X_{44}
$\vdots$			$\vdots$
$\vdots$			$\vdots$
X_{11}	X_{12}	X_{13}	X_{14}

Figure 8.10 Decomposition of the Canvas into 16 Subsquares

Defining the weighted relation λ_{16} as above by counting the number of appearances of each color in each subsquare, we obtain the incidence pattern

λ_{16}	X_{11}	X_{12}	X_{13}	X_{14}	X_{21}	X_{22}	X_{23}	X_{24}	$\cdots$	X_{44}
y_1	3	1	1	1	2	3	0	1	$\cdots$	1
y_2	2	3	3	4	0	1	0	3	$\cdots$	7
y_3	5	1	3	3	2	4	5	1	$\cdots$	1
y_4	1	4	2	4	2	4	3	3	$\cdots$	3
y_5	0	2	2	0	1	2	0	0	$\cdots$	0
y_6	2	3	2	2	5	1	2	2	$\cdots$	1
y_7	2	0	1	1	1	0	2	1	$\cdots$	0
y_8	1	2	2	1	3	3	4	5	$\cdots$	3

Slicing the above relation at $\theta = 1$ leads to a structure vector

$$Q = (\overset{15}{1}\ 1\ \cdots \overset{0}{1}),$$

for the complex $K_Y(X_{16}; \lambda_{16})$, where the colors of highest dimension are y_3 (pale yellow) and y_6 (dark blue), each of dimension 15. Thus, the color simplices require a space of at least 15 dimensions for their representation. Every other color is a polyhedron contained in the 16-vertex polyhedron representing y_3 (or y_6).

Turning attention to the conjugate complex $K_{X_{16}}(Y; \lambda_{16}^*)$, the structure vector is

$$Q^* = (\overset{7}{1}\ 1\ \cdots \overset{0}{1}),$$

with the square X_{13} being of highest dimension (7). Each of the remaining squares is a face of the X_{13} simplex.

If we slice the relation λ_{16} at $\theta = 2$, the structure vector for the complex $K_Y(X_{16}; \lambda_{16})$ becomes

$$Q = (\overset{12}{1}\ 4\ 3\ 2\ 1\ 2\ 2\ 1\ 2\ 1\ 1\ 1\ \overset{0}{1}).$$

The dominant colors are still y_3, y_4, y_6 and y_8, all of which are disconnected at dimension $q = 11$. The high number of components at the intermediate dimensional levels means that this view of the painting is much more rigid than at $\theta = 1$, indicating that the color distribution is artistically nonfree.

Finally, looking at the complex $K_{X_{16}}(Y; \lambda_{16}^*)$ when $\theta = 2$ yields the structure vector

$$Q^* = (\overset{5}{1}\ 5\ 3\ 1\ 1\ \overset{0}{1}),$$

also a significant change from the case $\theta = 1$. Further analysis shows that the squares with nonzero eccentricities tend to be concentrated along the bottom row of the painting.

For much greater detail on this type of analysis, we strongly urge the reader to consult the Atkin's book cited in the Notes and References.

9. *Chance and Surprise*

One of the great challenges to both science and philosophy is to provide a rational, systematic procedure to account for our perceived uncertainty regarding the occurrence of events in daily life. Classical probability theory provides one such approach, but is riddled with many well-known epistemological flaws and paradoxes; the theories of *fuzzy sets, satisficing* and *possibilities* represent recent attempts to provide a mathematically-based methodology to deal with some of the deficiencies in the traditional methods. Each of these approaches has at its heart the basic inequality

$$\text{uncertainty} \neq \text{randomness},$$

expressing the fairly evident fact that our experienced uncertainty about events and situations cannot usually be attributed to any kind of random mechanism at work, but rather seems to stem from a basic vagueness, or lack of information, either in the linguistic description or other circumstances surrounding the situation at hand.

Here we want only to indicate the manner in which q-analysis enables us to structure the notions of uncertainty, probability and surprise in a manner that provides a deeper insight into the basic difficulties involved, and to give some suggestions as to how a coherent theory of uncertainty might be developed.

Consider the situation of throwing a fair die four times in succession and observing the appearance or nonappearance of the face "6." Let's label the elementary events associated with each throw x_1, x_2, x_3 and x_4 and combine them into the set X of *elementary events.* To be more specific, x_i represents the event "6" on toss i, $i = 1, 2, 3, 4$. Now let's introduce the space Y of *compound* events associated with the entire experiment of four tosses. In this experiment there are 16 possible outcomes ranging from no occurrences of a "6" to all tosses resulting in "6." Thus we label the elements of Y as

$$Y = \{y_0, y_1, y_2, y_3, y_4, y_{12}, y_{13}, y_{14}, y_{23}, y_{24}, y_{34}, y_{123}, y_{124}, y_{134}, y_{234}, y_{1234}\},$$

where y_0 means no "6" occurred, y_2 means that a "6" occurred only on the second toss, and so forth.

If we take X as the vertex set for a complex and Y as the simplex set, it's easy to develop a relation λ by the rule that $(y, x) \in \lambda$ if and only if elementary event x is a component of compound event y. The incidence matrix for λ is is easily computed: there is a "1" in column x_i if and only if the integer i appears as one of the subscript numbers on simplex y, e.g., $(y_{13}, x_1) \in \lambda$ but $(y_{13}, x_2) \notin \lambda$.

Calculating the structure vector Q for the complex $K_Y(X; \lambda)$, we obtain

$$Q = (\overset{3}{1}\ 1\ 1\ \overset{0}{1}),$$

indicating that the complex has only a single component at each q-level. In fact, it's easy to see that we are really dealing here with the *single* simplex y_{1234} and all of its faces. This is exactly the kind of structure for which classical probability theory is the most appropriate tool for expressing our sense of likelihood as we now argue in more detail.

The complex $K_Y(X; \lambda)$ represents what the probabilist would term the *sample space* of the die-tossing experiment. However, in contrast to the usual view of events being dimensionless objects, the view presented above distinguishes strongly between the compound 0-events (y_1, y_2, y_3 and y_4), 1-events (y_{12}, y_{13}, etc.), on up to the single 3-event y_{1234}. *Before* the experiment of 4 tosses is performed, our sense of the likelihood of the outcome is measured by attaching numbers (probabilities?) to each simplex. *After* the experiment is complete, these numbers have rearranged themselves throughout the complex with all simplices now having value 0 except that single simplex corresponding to the actual outcome, whose value we conventionally set equal to 1. So the execution of the experiment corresponds to *traffic* on the complex. But such traffic can move freely about from one simplex to another only if the complex is sufficiently richly connected at all dimensional levels to support such a free flow of dimensionally significant numbers. Basically, this means we must have a structure vector each of whose components is 1, as above. The case when $K_Y(X; \lambda)$ consists of a single simplex and all of its faces is the simplest example of when this type of situation will occur.

In connection with the die-tossing experiment, the probabilist would attach the following *a priori* estimates to the elements of Y:

$$\mathcal{E}(y_0) = \frac{625}{1296},$$
$$\mathcal{E}(0 - \text{simplices}) = \frac{500}{1296},$$
$$\mathcal{E}(1 - \text{simplices}) = \frac{150}{1296},$$
$$\mathcal{E}(2 - \text{simplices}) = \frac{20}{1296},$$
$$\mathcal{E}(y_{1234}) = \frac{1}{1296},$$

which expresses his sense of likelihood based upon weighting the possible outcomes using the binomial distribution with its associated assumptions of independent trials and fixed probabilities for the elementary events x_i. After the experiment is complete, the above numbers have redistributed themselves over the complex coalescing on that one simplex corresponding to the actual outcome of the trial. But this rearrangement is possible only if the numbers associated with p-events can freely move about and "reaffiliate" themselves with events at all levels of connectivity p. This can happen only in structures with a single component at each level p.

The main point about connectivity is that if the numbers assigned to the events in Y are to represent our sense of the likelihood of the outcome of the experiment, then they must do so both before *and* after the experiment. But this requires the kind of free flow of traffic discussed above, a flow that is possible only if the complex of events is fully connected at all levels. Thus, we would conclude that classical probability theory will, in general, only reflect our sense of the likelihood of events for those structures possessing a single component at each dimensional level.

We note, in passing, that the redistribution of the numbers over the complex when we actually perform the experiment is the discrete analogue of the "collapse" of the quantum mechanical wave function, at least in the "Copenhagen Interpretation" of reality. In that theory, reality exists as the *potential* for the outcome of a measurement, with the various possible measurements weighted by the probability distribution specified by the Schrödinger wave equation whose solution is called the "wave function" for the system. Following the actual measurement, a definite value is obtained for the observed quantity, and the wave function "collapses" to the single value actually observed, all other possibilities now having probability zero.

Thus our complex of events, together with the associated likelihood numbers, is a discrete analogue of the wave function with the added feature that the possible events (simplices) have a dimensional character that must be respected when considering the redistribution of likelihoods following the experiment (observation). The quantum-mechanical implications of these dimensional factors have not been investigated as yet, with classical quantum mechanics confining itself to the same case as the probabilists, namely, complexes with $Q = (1\ 1\ \cdots 1)$.

One of the principal uses of conventional probability theory is to attach some numerical measure to our sense of how "surprising" the occurrence of a particular event would be. By the foregoing arguments, it's clear that the concept of surprise is intimately tied up with the level of connectivity between events in the sample space of possible outcomes. In particular, to develop a decent theory of surprises we need a measure for the "reachability" of a q-event σ_q from another base event σ_p^* (the Now!) in the complex. In what follows, we shall adopt the usual convention that an event (a simplex) is denoted by σ with its subscript indicating the dimension of the event represented by σ.

Intuitively, the "surprise value" surp σ_q of the event σ_q should possess the following properties:

1) The surprise value should reflect the level of connectivity between σ_q and σ_p^*; in particular, if there is no chain of q-connection between the two, then surp(σ_q)=0. That is, we cannot be *surprised* if the event σ_q cannot be experienced from σ_p^* if there is no appropriate dimensional path to move from the base event σ_p^* to σ_q.

2) The surprise value of σ_q should be greater if there are a large number of disjoint p-chains from σ_p^* to σ_q, since it is "more surprising" if a large number of p-chains go between the two events than if there are only a small number of such paths.

3) The surprise value should be smaller if $\dim \sigma_p^* = p$ is large, since it is "less surprising" that q-chains exist from σ_p^* to σ_q if σ_p^* has more q-faces.

If we let

$$n_q(\sigma_q, \sigma_p^*) = \text{the number of disjoint } q\text{-chains linking } \sigma_p^* \text{ to } \sigma_q,$$

then a measure of surprise of the event σ_q relative to σ_p^* is given by

$$\text{surp}(\sigma_q \mod \sigma_p^*) = \frac{n_q(\sigma_q, \sigma_p^*)}{p+1},$$

with the conditions

$$\text{surp}(\sigma_q \mod \sigma_p^*) = \begin{cases} 0, & \text{if } \sigma_q \text{ and } \sigma_p^* \text{ are in different components of the complex,} \\ 0, & \text{if } p < q, \\ 1/(p+1), & \text{if there are no loops.} \end{cases}$$

The last condition means that we do not distinguish loops beginning and ending at σ_p^* as different p-chains.

Example: Technological Disasters

An interesting and timely example of the use of surprise theory is given by Atkin, who considers the surprise value of a technological disaster like Three-Mile Island or Chernobyl. Let the vertices X of the complex $K_Y(X; \lambda)$ represent various technological features of the system under study. For a nuclear power plant, the elements of X might be things like the position of control rods, the level of coolants and the pressure in regulators. Let the simplex set Y consist of combinations of such features that we term a "property" or "behavior."

Initially all vertices are in the state OK, and we say that all is well. During the course of operation of the plant, some vertices turn into "anti-vertices"—i.e., their OK activity turns into "not-OK"—and the complex $K_Y(X;\lambda)$ turns into a new complex K^1. As the process of vertices shifting to/from OK$\longleftrightarrow$ not-OK unfolds, we have a progression

$$K \longrightarrow K^1 \longrightarrow K^2 \longrightarrow \cdots \longrightarrow K^D,$$

where the event σ = DISASTER belongs to K^D. We can now ask the question: Given the event (state) $\sigma_p^* \in K$, what is the value surp($\sigma \mod \sigma_p^*$) for $\sigma \in K^D$? Clearly, we would like to adjust our technology $K \cap_i K^i$ to make this number large.

10. *Simplicial Complexes and Dynamical Systems*

With the exception of this chapter, the dominant mathematical paradigm throughout this book has been Newton's legacy: ordinary differential equations. It's appropriate then to conclude with a brief indication of some linkages between the linear dynamical systems considered in Chapter Three and certain simplicial complexes of the type we have been examining in this chapter.

Consider the single-input linear system

$$x_{t+1} = Fx_t + gu_t, \qquad x_0 = 0, \qquad x_t \in R^n,$$

and its associated reachability matrix

$$\mathcal{C} = [g|\, Fg|\, F^2g| \cdots |\, F^{n-1}g].$$

Let us denote the elements of $\mathcal{C}$ as

$$\mathcal{C} = \{c_1, c_2, \ldots, c_n\}.$$

Now we form a new set Y from $\mathcal{C}$ by taking as elements all those combinations of subsets of elements from $\mathcal{C}$ that are linearly independent. So, for example, if c_1 and c_2 are linearly independent, we write this subset as $c_1 \wedge c_2$ and place this element in the set Y. In general, Y can contain at most $2^n - 1$ elements.

To define a relation λ on the product $Y \times \mathcal{C}$, we employ the rule $(y_i, c_j) \in \lambda$ if and only if the vector c_j belongs to the element y_i. As an example of this set-up, let $\Sigma = (F, g, -)$ be the system with

$$F = \begin{pmatrix} 0 & 1 & 0 \\ 0 & 0 & 1 \\ -\alpha_2 & -\alpha_1 & -\alpha_0 \end{pmatrix}, \qquad g = \begin{pmatrix} 0 \\ 0 \\ 1 \end{pmatrix}.$$

Then

$$\mathcal{C} = \begin{pmatrix} 0 & 0 & 1 \\ 0 & 1 & -\alpha_0 \\ 1 & -\alpha_0 & -\alpha_1 + \alpha_0^2 \end{pmatrix} = [c_1 | \, c_2 | \, c_3] \, .$$

The set Y is given by

$$Y = \{c_1, c_2, c_3, c_1 \wedge c_2, c_1 \wedge c_3, c_2 \wedge c_3, c_1 \wedge c_2 \wedge c_3\}.$$

The incidence matrix for the complex $K_Y(X; \lambda)$ is easily obtained as

λ	c_1	c_2	c_3
y_1	1	0	0
y_2	0	1	0
y_3	0	0	1
y_4	1	1	0
y_5	1	0	1
y_6	0	1	1
y_7	1	1	1

We can simply calculate the structure vector Q for the complex as

$$Q = (\overset{2}{1}\ 1\ \overset{0}{1}),$$

and for the conjugate complex $K_{\mathcal{C}}(Y; \lambda^*)$,

$$Q^* = (\overset{3}{3}\ 3\ 1\ \overset{0}{1}).$$

The above results concerning the structure vectors Q and Q^* are generalizable for *all* completely reachable single-input systems as the following result shows.

LINEAR STRUCTURE THEOREM. *The system* $\Sigma = (F, g, -)$ *is completely reachable if and only if the structure vectors for* $K_Y(\mathcal{C}; \lambda)$ *and* $K_{\mathcal{C}}(Y : \lambda^*)$ *have the forms*

$$Q = (\overset{n-1}{1}\ \ 1\ 1 \cdots \overset{0}{1}),$$
$$Q^* = (\overset{2^n-1}{2^n - 1}\ \ 2^n - 1\ \ 2^n - 1 \cdots 2^n - 1\ \overset{\beta}{1}\ 1\ 1 \cdots \overset{0}{1}),$$

where

$$\beta = \sum_{k=0}^{n-1} \binom{n-2}{k-1}, \qquad \binom{x}{-1} \doteq -1.$$

Example

Consider the system Σ defined by

$$F = \text{diag}\,(\lambda_1, \lambda_2, \lambda_3), \quad g = \begin{pmatrix} 0 \\ 1 \\ 1 \end{pmatrix},$$

which is not completely reachable. Computation of the incidence matrix shows that the element $c_1 \wedge c_2 \wedge c_3$ is not contained in Y since the three vectors

$$c_1 = \begin{pmatrix} 0 \\ 1 \\ 1 \end{pmatrix}, \quad c_2 = \begin{pmatrix} 0 \\ \lambda_2 \\ \lambda_3 \end{pmatrix}, \quad c_3 = \begin{pmatrix} 0 \\ \lambda_2^2 \\ \lambda_3^2 \end{pmatrix},$$

are not linearly independent.

The structure vector for $K_Y(C; \lambda)$ turns out to be

$$Q = (\overset{1}{3}\ \overset{0}{1}),$$

which is not of the form required for complete reachability by the Linear Structure Theorem. Thus Σ is not reachable.

These results give only the briefest hint at how ideas from algebraic topology can be used to capture the geometric structure of a dynamical system. For extensions to multi-input systems, as well as to nonlinear dynamics, the Notes and References should be consulted (see also Problems 6 and 9).

Discussion Questions

1. Suppose you wanted to characterize the difference between the spectrum of goods and services available at a major shopping center and those offered at a small country store. How could you capture this difference using sets and relations?

2. Discuss how you might employ hierarchies of sets and relations to structure a play like Hamlet as a relationship between the characters and the scenes.

3. We have seen that paintings can be given a simplicial structure that captures some of their abstract geometrical style. Do you think the same concepts can be employed to find a geometrical structure in works of music? How would you arrange the sets and relations in this case?

4. The measure of eccentricity discussed in the text suffers from the defect that a simplex will have small eccentricity if there is only a *single* other simplex with which it shares many vertices. This means that we could have two simplices that have many vertices in common but share very few vertices with any of the remaining simplices in the complex. Nevertheless, our measure of eccentricity would deem these simplices to be "well-integrated" into the complex. Develop and discuss alternate measures of eccentricity that would remove this deficiency.

5. The game of chess can be considered as a relation between the Black and White pieces and the squares of the board. Discuss various binary relations that can be used to define the essence of the game. Consider how the resulting complexes and structure vectors might be used in developing programs for computer chess playing. (*Hint:* There are at least two important relations here, the relation between the Black pieces and the squares, and the relation between the White pieces and the squares. In addition, consider the role of set covers and hierarchies in your analysis.)

6. The discussion in the text omits any consideration of *dynamics* on a simplicial complex. Consider the idea of a *pattern* π defined on K by attaching a number to each simplex $y \in K$, i.e., $\pi : Y \to R$. Discuss the pros and cons of using the change of pattern $\Delta\pi$ as a measure of dynamics on K. In particular, consider the constraints imposed on $\Delta\pi$ by the geometrical connectivity of the components of K. What is the connection between $\Delta\pi$ and the ordinary idea of a "force" as that term is used in physics?

7. Consider the problem of land usage in Manhattan. Define the sets

$$
\begin{aligned}
X &= \{\text{local geographic areas}\} \\
&= \{\text{Upper East Side, Upper West Side, Harlem, Midtown,} \\
&\qquad \text{Times Square, Garment District, Chelsea, Greenwich Village,} \\
&\qquad \text{Soho, Chinatown, Financial District}\} \\
&= \{x_1, x_2, \ldots, x_{11}\},
\end{aligned}
$$

$$
\begin{aligned}
Y &= \{\text{activities}\} \\
&= \{\text{retail trade, cultural amenities, residential, entertainment,} \\
&\qquad \text{light manufacturing, heavy industry, financial/business}\} \\
&= \{y_1, y_2, \ldots, y_7\}.
\end{aligned}
$$

Let the relation $\lambda \subset Y \times X$ be given by $(y_i, x_j) \in \lambda$ if and only if activity y_i takes place in area x_j.

a) Determine the incidence matrix for λ using your knowledge of Manhattan.

b) Calculate the structure vectors Q and Q^* for the complexes $K_Y(X;\lambda)$ and $K_X(Y;\lambda^*)$.

c) Compute the eccentricities of the simplices for these complexes.

d) Discuss the *interpretations* you can give to these calculations; i.e., what do these numbers tell you about the way goods and services are distributed throughout Manhattan?

8. Investigate the uses of q-analysis as a pattern recognition algorithm in the sense of template matching. Imagine that a pattern is displayed on a rectangular grid, and establish a binary relation between the X and Y coordinates of the cells that constitute the displayed pattern. For example, the square below might be represented by the incidence matrix

λ	0	1	2	3	4	5	6	7	8	9
0	0	0	0	0	0	0	0	0	0	0
1	0	0	0	0	0	0	0	0	0	0
2	0	0	0	0	0	0	0	0	0	0
3	0	0	1	1	1	1	1	0	0	0
4	0	0	1	0	0	0	1	0	0	0
5	0	0	1	0	0	0	1	0	0	0
6	0	0	1	0	0	0	1	0	0	0
7	0	0	1	1	1	1	1	0	0	0
8	0	0	0	0	0	0	0	0	0	0
9	0	0	0	0	0	0	0	0	0	0

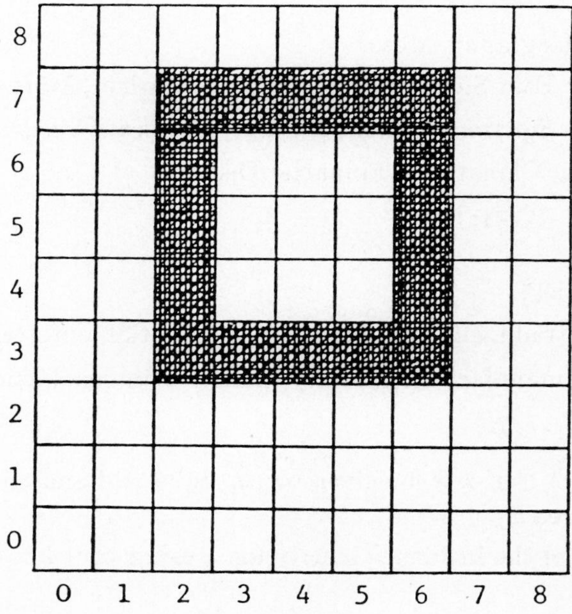

a) Examine the possibility of distinguishing between a square, a triangle and a circle by means of the structure vectors Q and Q^* of their respective simplicial complexes.

b) The geometric nature of a pattern must be invariant under translations, rotations and uniform contractions/dilations. Discuss these conditions in terms of properties of the incidence matrix Λ for the foregoing types of geometrical patterns.

c) Discuss how the above procedure might be developed into a template-matching program.

9. Two simplices σ_p and σ_r in a complex K are said to be *q-near* if they share a face of dimension q, i.e., if they have $(q+1)$ vertices in common. Reformulate the concept of q-connection in terms of q-nearness.

10. The classical Newtonian view of time can be represented by the diagram

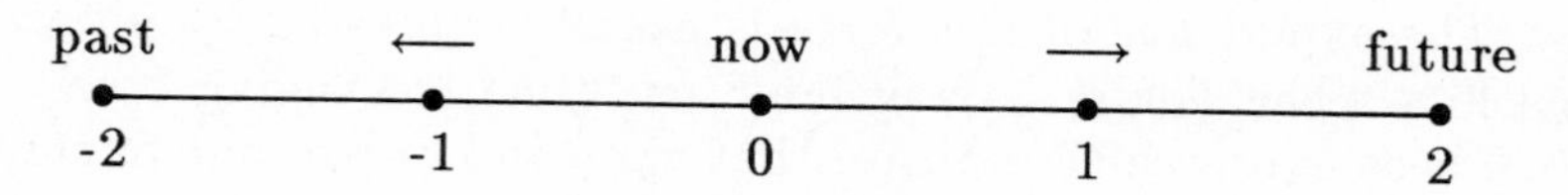

where the numbers represent the measurement of specific moments of time (τ_0) and time intervals between the moments (τ_1). This picture represents an elementary sort of simplicial complex having an infinite set of vertices with the numbers attributed to each vertex forming a pattern τ_0 on the vertices, i.e., the 0-simplices. The numbers assigned to the edges joining the vertices form another pattern τ_1, referred to as the time interval between successive measurements. Thus, in the Newtonian view the time pattern is the *graded* pattern $\tau = \tau_0 \oplus \tau_1$.

The above representation makes it clear why we refer to Newtonian time as a *linear* concept associated with a complex K consisting of a set of 1-simplices that are 0-connected. When the Newtonian time-axis is used to represent a set of observed events, the idea behind it is to somehow produce a "clock" whose time moments (the vertices) can be put into a 1–1 correspondence with the set of events. The pattern τ_0 describes the "Now" events, while τ_1 describes the interval pattern.

a) Discuss why the Einsteinian view of relativistic time has the structure $\tau = \tau_3 \oplus \tau_4$, so that time moments are now represented by 3-simplices with intervals corresponding to 4-simplices. What possible interpretation can you attach to the fact that such relativistic time has also 1- and 2-simplex intervals between appropriate "Now" moments?

b) Consider the following view of "experienced" time: let K be a simplicial complex of events. A pattern τ_p represents the "Now" moments of p-events and can be thought of as a simple 0/1 function on the p-simplices of K, with the single nonzero value being attached to the particular p-event which is the "Now" moment. Then the "next" p-event in K is experienced by a change in this pattern, so that the "1" moves to a different p-simplex. This movement involves the connectivity of K, since one p-event cannot follow another unless there is available a $(p+1)$-interval connecting the two. Thus, the change $\delta\tau_p$ is a *p-force* in the structure of events representing our sense of moving time.

Normally, when we speak of time we refer to the Newtonian pattern $\tau = \tau_0 \oplus \tau_1$. Thus, if an individual experiences time traffic of dimension p, he finds it culturally necessary to replace his pattern by the Newtonian τ as

$$\underset{\text{(experienced time)}}{\tau_p \oplus \tau_{p+1}} \quad \longrightarrow \quad \underset{\text{(Newtonian time)}}{\tau_0 \oplus \tau_1}$$

This picture means that the individual has to force his $(p+1)$-perception of the time interval down into the one-dimensional interval of the Newtonian pattern. The experience of this force is usually expressed by phrases such as "time flies when you're having fun," or "time is hanging heavy on my hands." These expressions indicate that τ_{p+1} and τ_1 are out of step. Discuss the relevance of the above view of multidimensional time in your own experiences.

c) If we assume that the experience of a p-event corresponds to the recognition of a p-simplex in K, and there is a 1–1 correspondence between the 1-simplices of K and the Newtonian reference frame, then the time interval for the gap between one p-event and another will be proportional to the quantity $(p+1)(p+2)/2$, the number of edges in the least connection between two p-simplices. For example, if the Newtonian unit interval τ_1 is 1 day, then a 6-event would require $(6+1)(6+2)/2 = 28$ days to occur ("arrive").

Discuss the appropriateness of assuming that a p-event occurs by way of the edges that make up the $(p+1)$-interval, bridging the gap between the current and previous p-events. Consider other possibilities and their implications for the interval between p-events.

11. In the Middle East conflict example in the text, calculate the complexity measure for the complex given there, and discuss the thesis that a resolution of the dispute will require actions by all parties to reduce the complexity of both the complex and its conjugate.

How would you suggest extending this example to develop a more general conflict resolution scheme?

12. Consider how you might organize a sets/relations framework to measure the relative strengths of competing teams in sports such as baseball, football or basketball.

13. Imagine you are given a time-series of data $(t_i, y(t_i))$, $i = 1, 2, \ldots, n$. Determine how you might structure this data as a binary relation between two sets. Compute the structure vectors associated with this complex and compare these numbers with the usual correlation coefficients obtained from a least-squares regression analysis of the data.

14. Consider the primary color triangle whose vertices are Red, Blue and Green (the 0-simplices). The 1-simplices (Violet, Turquoise and Yellow), as well as the 2-simplex (White), are formed by combining the primary colors. So each simplex in this complex is a face of the single simplex $W =< R, B, G >$, and a person whose color vision is capable of seeing three primary colors at once (and their combinations) will be able to see the entire spectrum of visible colors.

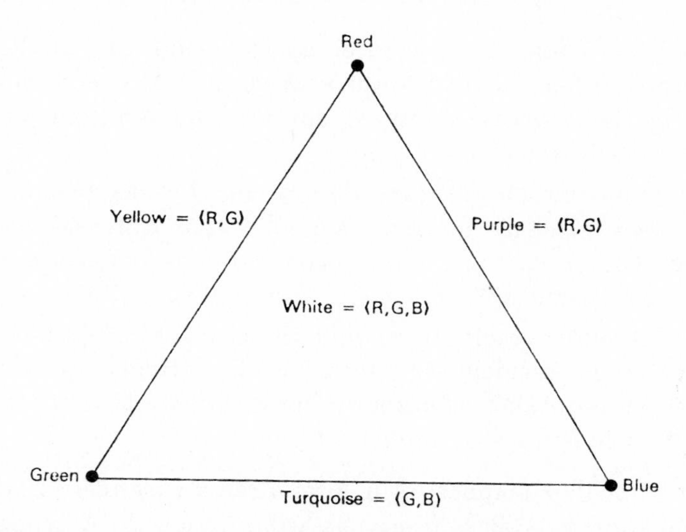

Now imagine an experiment with a subject whose color vision is only one-dimensional; i.e., he can see only two colors at a time (or any combination of these two primary colors). For the sake of definiteness, assume the subject cannot distinguish the color red. Suppose the experiment consists of showing the subject flashcards composed of the seven colors indicated, each appearing randomly but with equal frequency. When a card is displayed, the subject tells us what color he sees, and we pass on to the next card. With a subject possessing normal two-dimensional color vision, we would expect him to identify accurately each color, and our records would show each color as being named one-seventh of the time. What result would you expect from the subject who cannot see red? Explain your answer in terms of connectivity patterns in the color simplex.

15. The North American power-generating network consists of around 6000 individual generating stations. If the set of generating stations is con-

sidered as the N–level in a hierarchy of sets describing the interconnections in the network, try to develop a range of sets and connections from levels N–4 to N+4 describing this system. How many elements do you think there are in the sets at each level?

Imagine that there is a major power failure and that it's necessary to try to restore service immediately. Interpret what it means for a failure to occur in terms of changes in the geometry of the backcloth associated with the complexes at each level of the network.

Discuss how the foregoing considerations relate to the theory of surprises developed in the text.

Problems and Exercises

1. Consider the following conjecture: up to a relabeling of the rows and columns of the incidence matrix Λ, the associated binary relation is uniquely determined by the structure vectors Q and Q^* obtained from the simplicial complex generated from Λ.

a) Prove that this conjecture is false by constructing a counterexample; i.e., display two incidence matrices Λ and $\widehat{\Lambda}$ that differ by more than a permutation of their rows and columns but nevertheless generate the same structure vectors Q and Q^*.

b) The preceding result shows that the elements of Q and Q^* do not form a complete, independent set of invariants for incidence matrices Λ under the group of permutations. Consider other invariants that can be adjoined to Q and Q^* to constitute a complete set.

2. The set of all p-simplices in K form a *chain* C_p under formal addition and multiplication by scalars from an abelian group J. A typical chain in C_p is

$$c_p = m_1\sigma_p^1 + m_2\sigma_p^2 + \cdots + m_r\sigma_p^r, \qquad m_i \in J.$$

We can make C_p into an abelian group by defining $c_p + c_p'$ and αc_p, $\alpha \in J$ in the obvious way. Combining every such group C_p, $p = 0, 1, \ldots, n$, we obtain the *chain group*

$$C_\bullet = C_0 \oplus C_1 \oplus \cdots \oplus C_n.$$

With every p-chain c_p we can associate a $(p-1)$-chain, the *boundary* of c_p, denoted ∂c_p. The boundary operator ∂ is defined on p-simplices by

$$\partial\sigma_p = \partial < x_1 x_2 \cdots x_{p+1} >= \sum_i (-1)^{i+1} < x_1 x_2 \cdots \hat{x}_i x_{i+1} \cdots x_n >,$$

where "$\hat{x}_i$" means that the vertex x_i is to be omitted from the term.

a) Show that $\partial\colon C_p \to C_{p-1}, \quad p = 1, 2, \ldots, n$.

b) Prove that ∂ is nilpotent; i.e., $\partial^2 = 0$.

c) Show that ∂ is a homomorphism, i.e.,

$$\partial(c_p + c_p') = \partial(c_p) + \partial(c_p'),$$
$$\partial(\alpha c_p) = \alpha\partial(c_p), \qquad \alpha \in J.$$

d) Prove that $\partial(C_p) = B_{p-1}$ is a subgroup of C_{p-1}.

e) Show that $\partial(B_{p-1}) = 0$ in C_{p-2}.

3. Those p-chains c_p whose boundaries vanish ($\partial c_p = 0$) are called *p-cycles,* denoted Z_p.

a) Show that the p-cycles form a subgroup of C_p.

b) Show that B_p, the *bounding cycles,* form a subgroup of Z_p.

c) Define the *factor group* $Z_p/B_p = H_p$ using the following equivalence relation in Z_p: $z_p \sim z_p'$ if and only if $z_p - z_p' \in B_p$. Show that $H_p =$ ker ∂/im ∂. The groups H_p are called the *Betti groups* of the complex K, or the *homology groups.*

d) If $H_p \neq 0$,i.e., there are cycles z_p that are not bounding cycles, then H_p as an additive group is isomorphic to a certain number of copies of J, the number being equal to the number of generators (cycles z_p that are linearly independent over J). These numbers, denoted β_p, are called the *Betti numbers* of the complex K. Show that if the complex is arcwise connected, then $\beta_0 = 1$ and that if K possesses k connected components, $\beta_0 = k$.

e) Show that the first component of the structure vector Q is such that $Q_0 = \beta_0$, though, in general, $Q_i \neq \beta_i$, $i > 0$.

4. Consider the elementary complex K, consisting of the three 1-simplices σ_1^1, σ_1^2, and σ_1^3, with vertex set $\{x_1, x_2, x_3\}$.

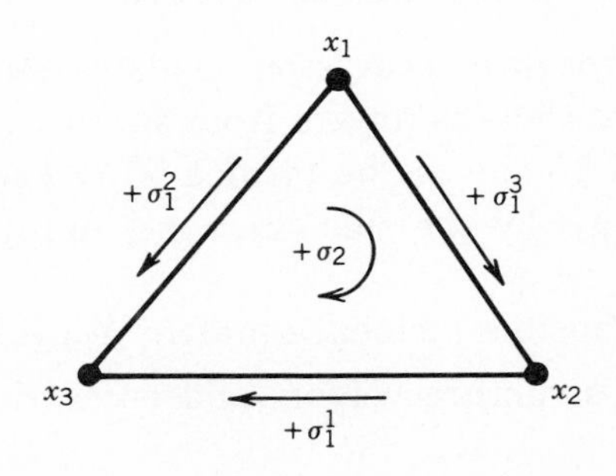

a) Show that K has a nontrivial 1-cycle.

b) Show that $H_0 \cong 1$, $H_1 \cong J$ and that $\beta_0 = 1$, $\beta_1 = 1$.

c) Assume that we modify K to include the 2-simplex $\sigma_2 = < x_1 x_2 x_3 >$, i.e., we "fill in" the triangle. Show now that for this new complex $H_1 \cong 0$.

d) Interpret the above result in terms of the number and types of "holes" present in the complex. Show that the Betti number β_i measures the number of i-dimensional holes in K.

5. Consider the Middle East example of the text. Compute the homology groups and the Betti numbers for this complex. What interpretation can you give to the results of your computation?

6. Consider the single-input linear system

$$x_{t+1} = Fx_t + gu_t.$$

a) Show that complete reachability is equivalent to the associated complex having trivial homology, i.e.,

$$H_0 \cong 1, \quad H_i \cong 0, \quad i > 0.$$

Thus the system is reachable if and only if it has no "holes."

b) The *Euler characteristic* of a complex is defined in terms of the Betti numbers as

$$\chi(K) = \sum_{i=0}^{n} (-1)^i \beta_i, \qquad n = \dim K.$$

Show that the single-input system is completely reachable if and only if $\chi(K) = 1$.

7. In our consideration of binary relations $\lambda \subset Y \times X$, we have adopted an "all-or-none" position on whether or not an element $(y_i, x_j) \in \lambda$. Suppose now that we assign a number Λ_{ij} between 0 and 1 to measure the degree to which we *believe* that (y_i, x_j) belongs to the relation λ. Consider how to modify the idea of a chain of q-connection in this "fuzzy sets" situation and discuss what it could mean for two simplices to be "q-connected."

8. A model predator–prey ecosystem is displayed in the directed graph on the next page. Here there is an arc from species i to species j if i feeds upon j. Let the set of 15 species be termed $X = \{x_1, x_2, \ldots, x_{15}\}$. Define a predator relation λ_{PRD} by the rule $(x_i, x_j) \in \lambda_{\text{PRD}}$ if and only if x_i is a predator of x_j.

a) Determine a plausible incidence matrix Λ_{PRD} for this ecosystem.

b) Determine the structure vectors and eccentricities for the complex $K_X(X; \lambda_{\text{PRD}})$.

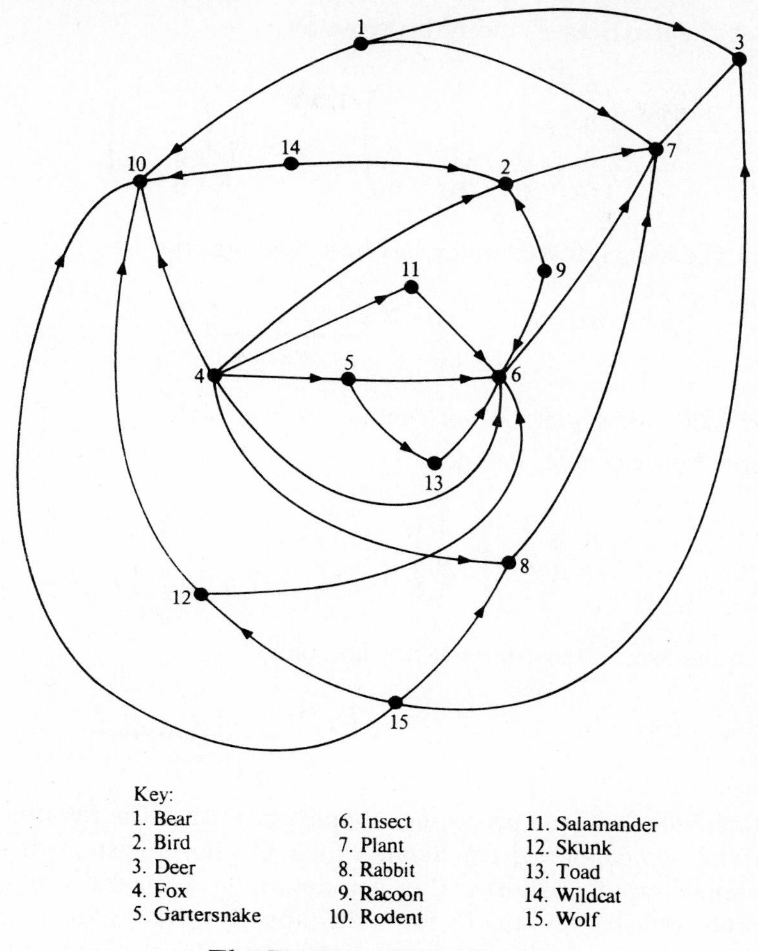

The Predator–Prey Graph

c) Define an analogous relation λ_{PRY} for the prey and calculate the same quantities.

d) Compute the complexities of the pair of complexes $K_X(X; \lambda_{\text{PRD}})$, $K_X(X; \lambda_{\text{PRY}})$ and their conjugates. Interpret the numbers obtained in terms of the food web structure.

e) Show that λ_{PRD} has the nontrivial homology $H_0 \cong 1$, $H_1 \cong J$, while λ_{PRY} has trivial homology. How would you interpret this result?

9. Now consider the *multi-input* system

$$x_{t+1} = Fx_t + Gu_t, \qquad F \in R^{n \times n}, \qquad G \in R^{n \times m}, \qquad m > 1.$$

a) Let the matrices F and G be given by

$$F = \begin{pmatrix} 2 & 4 & 1 & 1 \\ 0 & -1 & 0 & 1 \\ 0 & 0 & -3 & -2 \\ 0 & 0 & 0 & 1 \end{pmatrix}, \qquad G = \begin{pmatrix} 0 & 1 \\ 1 & 1 \\ 0 & 0 \\ 0 & 0 \end{pmatrix}.$$

Show that the associated complex has homology groups

$$H_0 \cong J, \qquad H_1 \cong \underbrace{J \oplus J \oplus \cdots \oplus J}_{\text{21 copies}}.$$

Show also that this system is *not* completely reachable.

b) For the reachable system

$$F = \begin{pmatrix} 3 & 0 & 1 \\ 2 & 1 & 0 \\ 1 & 1 & 1 \end{pmatrix}, \qquad G = \begin{pmatrix} 1 & 0 \\ 0 & 1 \\ 0 & 0 \end{pmatrix},$$

show that the associated complex has homology

$$H_0 \cong J, \qquad H_1 \cong 0, \qquad H_2 \cong \underbrace{J \oplus J \oplus \cdots \oplus J}_{\text{10 copies}}.$$

c) Conclude from the preceding computations that the relationship between trivial homology and reachability for multi-input systems is different than for single-input processes. Can you determine a general result for calculating the homology groups H_i for multi-input systems? (*Note:* This exercise is quite difficult, requiring advanced techniques from algebraic topology, e.g., exact sequences and the Mayer-Vietoris sequence.)

10. Write a computer program to calculate the structure vectors Q and Q^* as well as the eccentricities of individual simplices, given the incidence matrix Λ.

11. Consider a complex $K_Y(X;\lambda)$ and its conjugate. Show that the homological structures of these two complexes are isomorphic, i.e., $H_i \cong H_i^*$, $i = 0, 1, \ldots$.

12. Define the *complementary complex* K^C to K by the following rule: extend the vertex set to include all the vertices $\bar{x}_i$ that are the *negations* of those x_i in K. Thus, the incidence matrix $\bar{\Lambda}$ of K^C is obtained from Λ by changing the "1s" to "0s," and vice-versa. Show that K and K^C have exactly the same connective structure; i.e., the structure vectors are such that $Q = Q^C$, $Q^* = Q^{*C}$.

13. a) Using the measure of surprise developed in the text, show that if σ_q^1 and σ_q^2 are disjoint events

$$\text{surp}\ (\sigma_q^1 \cup \sigma_q^2 \mod L) = \text{surp}\ (\sigma_q^1 \mod L) + \text{surp}\ (\sigma_q^2 \mod L),$$

where L is some base subcomplex of K.

b) Consider the special case when $K = \bigcup_i \sigma_0^i$ is a space of 0-events only. Let E be a subset of K. Show that

$$\begin{aligned} \text{surp}\ (E \mod K) &= \sum_{\sigma_0 \in E} \text{surp}\ (\sigma_0^i \mod K) \\ &= \sum_{\sigma_0 \in E} \text{surp}\ (\sigma_0^i \mod \sigma_0^i) \\ &= \text{card}\ E, \end{aligned}$$

so that surp $(K \mod K) = \text{card}\ K$. Thus, in the space of 0-events

$$\text{prob}\ (E \mid K) = \frac{\text{surp}\ (E \mod K)}{\text{surp}\ (K \mod K)},$$

i.e., we recover the standard "relative frequency" rule from elementary probability theory using our definition of surprise in this very special setting.

c) How would you use surprise theory to recapture the usual notion of "independent" events? That is, if H and K are disjoint (independent) complexes with A and B being subcomplexes of H and K, respectively, how would you *define* the quantity surp $(\sigma_q^i, \sigma_q^j) \mod (A \times B)$, where $\sigma_q^i \in H$, $\sigma_q^j \in K$, in order to recapture the usual rule $P(A \cap B) = P(A)P(B)$ used to express independence of the 0-events A and B in elementary probability theory?

Notes and References

§1–3. The idea of representing binary relations between abstract sets as simplicial complexes, and then using the abstract geometry of the complex to study interesting relationships in the arts and humanities, is due primarily to R. H. Atkin. For an introduction to his ideas, including the examples of the text, see

Atkin, R. H., *Multidimensional Man,* Penguin, London, 1981.

§4. The study of the geometrical properties of a space by looking at the algebraic properties of a suitable representation forms the core of the subject matter of algebraic topology. Good introductions are

Giblin, P., *Graphs, Surfaces and Homology,* Chapman and Hall, London, 1977,

Singer, I. M., and J. A. Thorpe, *Lecture Notes on Elementary Topology and Geometry,* Scott Foresman, Glenview, Illinois, 1967.

§5. The notion of q-connection as a means for studying the global geometry of a complex is due to R. Atkin. For an extensive account of how this concept is used in a variety of applied settings, see his pioneering work

Atkin, R. H., *Mathematical Structure in Human Affairs,* Heinemann, London, 1974.

A summary of the basic ideas of q-analysis, along with many examples, is given in

Casti, J., *Connectivity, Complexity and Catastrophe in Large-Scale Systems,* Wiley, Chichester and New York, 1979.

The example on the Middle East crisis is based on discussions with M. Shakun and is considered in more detail in

Casti, J., "Topological Methods for Social and Behavioral Systems," *Int. J. Gen. Syst.,* 8 (1982), 187–210.

§6. The matter of system complexity for simplicial complex descriptions is considered further in the Casti volume cited for §5. The axiomatic view of complexity in terms of subsystems is developed in great detail from a highly algebraic point of view in

Rhodes, J., "Applications of Automata Theory and Algebra," Lecture Notes, Dept. of Mathematics, U. of California, Berkeley, 1971.

For a more readily accessible account of Rhodes' ideas, together with many examples, see

Gottinger, H., *Coping with Complexity,* Reidel, Dordrecht, 1983.

§7. The analysis of *Sky and Water* given in the text follows the treatment given in

Johnson, J., *Combinatorial Structure in Digital Pictures,* NERC Remote Sensing Project, Discussion Paper No. 1, Center for Configurational Studies, The Open University, Milton Keynes, UK, January 1985.

Escher's work has by now acquired a certain cult following, at least in academic scientific circles. A good account of the many-sidedness of Escher's work for both science and art is provided in the symposium volume

Coxeter, H. S. M., M. Emmer, R. Penrose, and M. L. Teuber, eds., *M. C. Escher: Art and Science,* North-Holland, Amsterdam, 1986.

In this same connection, see also

Ernst, B., *The Magic Mirror of M. C. Escher,* Ballantine, New York, 1976,

Gardner, M., "The Art of M. C. Escher," in *The Mathematical Carnival,* Knopf, New York, 1975, pp. 89–103,

J. L. Locher, ed., *The World of M. C. Escher,* Abrams, New York, 1971.

§8. Mondrian's work is examined in far greater detail in the Atkin book cited under §5 above from which our treatment is but a brief excerpt.

§9. The degree to which the *mathematical* theory of probability accurately reflects our sense of the likelihood of events has been a hotly debated issue in the philosophy of science for many years, with the number of incompatible "meanings" of probability rivaling that of the number of interpretations of quantum mechanics. For an introductory account of the various positions, see

Black, M., "Probability," in *Encyclopedia of Philosophy,* P. Edwards, ed., Macmillan, New York, 1967.

For a stimulating view of the whole issue of probability and its role in measuring uncertainty as developed over a scientific lifetime's worth of pondering the matter, see

Good, I. J., *Good Thinking: The Foundations of Probability and Its Applications,* U. of Minnesota Press, Minneapolis, 1983.

In recent years a number of alternatives to classical probability theory have been put forward to account for uncertainty. Prominent among these alternate theories is the notion based on a *fuzzy set,* in which the usual 0/1 set membership function is replaced by a membership function in which an element's degree of membership in the set can be any number between 0 and 1. This seemingly small change leads to large consequences and dramatic implications for our view of the uncertainty vs. randomness question. For a detailed description of fuzzy sets and the likelihood "calculus" that the theory engenders, see the compendium

Yager, R., ed., *Fuzzy Sets and Applications: Selected Papers by L. A. Zadeh,* Wiley, New York, 1987.

In this same regard, see

Shafer, G., *A Mathematical Theory of Evidence,* Princeton U. Press, Princeton, 1976,

Dubois, D., and H. Prade, *Possibility Theory,* Plenum Press, New York, 1988,

Kruse, R., and K. D. Meyer, *Statistics With Vague Data,* Reidel, Dordrecht, 1987,

Klir, G., and T. Folger, *Fuzzy Sets, Uncertainty, and Information,* Prentice-Hall, Englewood Cliffs, NJ, 1988,

Kickert, W., *Fuzzy Theories on Decision-Making,* Nijhoff, Leiden, 1978,

Negoita, C., and D. Ralescu, *Applications of Fuzzy Sets to Systems Analysis,* Birkhäuser, Basel, 1975.

The idea of characterizing classical ideas of probability by means of special types of simplicial complexes is developed further in the Atkin book cited under §1–3. The theory of surprises, together with the technological disaster example outlined in the text, follows the development in

Atkin, R. H., "A Theory of Surprises," *Env. & Planning-B,* 8 (1981), 359–365.

§10. A more extensive discussion of these results that enables us to link up the ideas of Chapters Three and Eight can be found in

Casti, J., "Polyhedral Dynamics and the Controllability of Dynamical Systems," *J. Math. Anal. Applic.,* 68 (1979), 334–346.

Extension of the basic idea of the above paper to the much more difficult case of multi-input systems has been made by G. Burstein in an unpublished work.

DQ #2. For an example of how to use hierarchies of relations to say something meaningful about Shakespeare's immortal play *A Midsummer Night's Dream,* see the Atkin book cited under §1–3.

DQ #5. The game of chess is considered from a q-analysis point of view in the Atkin book noted in §5, as well as in

Atkin, R. H., *et al,* "Fred CHAMP, Positional Chess Analyst," *Int. J. Man-Mach. Stud.,* 8 (1976), 517–529,

Atkin, R. H., "Positional Play in Chess by Computer," in *Advances in Computer Chess I,* M. Clarke, ed., Edinburgh U. Press, Edinburgh, 1976, pp. 60–73.

DQ #6. The idea of dynamics on a complex as being a change of pattern is considered in great detail in

Atkin, R., *Combinatorial Connectivities in Social Systems,* Birkhäuser, Basel, 1977.

DQ #10. The role of simplicial complexes in providing an alternate formulation of much of classical and modern physics, including the issue of time, is explored in the works

Atkin, R., "A Homological Foundation for Scale Problems in Physics," *Int. J. Theor. Phys.,* 3 (1970), 449–466,

Atkin, R. H., "Time as a Pattern on a Multi-Dimensional Structure," *J. Soc. & Biol. Struct.,* 1 (1978), 281–295.

DQ #15. This discussion of the failure of electrical power networks is carried on to much greater lengths in the article

Gould, P., "Electrical Power Failure: Reflections on Compatible Descriptions of Human and Physical Systems," *Env. & Planning-B,* 8 (1981), 405–417.

PE #1. For an explicit counterexample showing the conjecture is false, see

Earl, C., "A Complex Is Not Uniquely Determined by Its Structure Vectors," *Env. & Planning-B,* 8 (1981), 349–350.

PE #2–4. For more information on these bread-and-butter items from elementary algebraic topology, see the volumes cited earlier under §4.

PE #6 and PE #9. These results appear in the Casti paper already referenced under §10.

PE #8. This food web example has been used in many places as a starting point for a variety of graph-theoretic and q-analysis investigations. The food web itself appears to date back to

Burnett, R., *et al., Zoology: An Introduction to the Animal Kingdom,* Golden Press, New York, 1958.

For more details on the way in which q-analysis sheds light upon the hidden structures in the web, see the Casti volume cited under §5 as well as the paper

Doreian, P., "Analyzing Overlaps in Food Webs," *J. Soc. & Biol. Structures,* 9 (1986), 115–139.

PE #11. The conclusion of this problem is known as Dowker's Theorem and first appeared in

Dowker, C. H., "Homology Groups of Relations," *Annals of Math.,* 56 (1951), 84–95.

PE #12. For a discussion of the complementary complex, see Appendix D of the Atkin book cited in §5, as well as the volume by Atkin noted in DQ #6.

PE #13. The surprise paper by Atkin referenced under §9 contains much more information on the relationship between the notion of surprise discussed here and the familiar concepts from elementary probability theory.

Appendix

The incidence matrix for the Escher engraving *Sky and Water* is as follows:

λ	x_1	x_2	x_3	x_4	x_5	x_6	x_7	x_8	x_9	x_{10}	x_{11}	x_{12}
y_1	1	1	1	1	1	1	1	0	0	0	0	0
y_2	1	1	1	1	1	1	1	0	0	0	0	0
y_3	1	1	1	1	1	1	1	0	0	0	0	0
y_4	1	1	1	1	1	1	1	0	0	0	0	0
y_5	1	1	1	1	1	1	1	0	0	0	0	0
y_6	1	1	1	1	1	1	1	0	0	0	0	0
y_8	0	1	1	1	1	1	1	0	0	0	0	0
y_9	0	1	1	1	1	1	1	0	0	0	0	0
y_{10}	0	1	1	1	1	1	1	0	0	0	0	0
y_{11}	0	0	0	1	1	1	1	0	0	0	0	0
y_{12}	0	0	0	1	1	1	1	0	0	0	0	0
y_{13}	0	0	0	1	1	1	1	0	0	0	0	0
y_7	0	1	1	0	1	0	1	0	0	0	0	0
y_{21}	0	0	0	0	0	0	1	1	1	1	1	1
y_{22}	0	0	0	0	0	0	1	1	1	1	1	1
y_{23}	0	0	0	0	0	0	1	1	1	1	1	1
y_{24}	0	0	0	0	0	0	1	1	1	1	1	1
y_{25}	0	0	0	0	0	0	1	1	1	1	1	1
y_{26}	0	0	0	0	0	0	1	1	1	1	1	1
y_{28}	0	0	0	0	0	0	1	1	1	1	1	1
y_{29}	0	0	0	0	0	0	1	1	1	1	1	1
y_{31}	0	0	0	0	0	0	1	1	1	0	0	0
y_{32}	0	0	0	0	0	0	1	1	1	0	0	0
y_{33}	0	0	0	0	0	0	1	1	1	0	0	0
y_{27}	0	0	0	0	0	0	1	1	1	1	1	0
y_{30}	0	0	0	0	0	0	0	0	1	1	0	0
y_{34}	0	0	0	0	0	0	0	1	0	0	0	0
y_{35}	0	0	0	0	0	0	0	1	0	0	0	0
y_{36}	0	0	0	0	0	0	0	1	0	0	0	0
y_{37}	0	0	0	0	0	0	0	1	0	0	0	0
y_{38}	0	0	0	0	0	0	0	1	0	0	0	0
y_{39}	0	0	0	0	0	0	0	0	0	0	0	0
y_{14}	0	0	0	1	0	0	0	0	0	0	0	0
y_{15}	0	0	0	0	0	1	0	0	0	0	0	0
$\vdots$						$\vdots$						
y_{20}	0	0	0	0	0	0	0	0	0	0	0	0

CHAPTER NINE

How Do We Know?: Myths, Models and Paradigms in the Creation of Beliefs

1. *Ordering the Cosmos*

In his *Metaphysics,* Aristotle remarked that "all men by nature desire to know." This desire springs from a need to somehow organize our observations of Nature, as well as the events and activities of daily life, into some kind of coherent pattern that we call "reality." Each domain of knowledge has its own observables, and we generally can't predict the observables in one domain from knowledge of those in another; hence, our need to create many realities in order to make a lawful ordering of the cosmos. In this volume our reality-generating mechanism has been that handmaiden of theoretical science, the mathematical model. However, it would be remiss in such a work to leave the reader with the impression that science as we know it today is the only, or even the best, tool for this task. Historically, Western science is a johnny-come-lately on the reality generation scene, having arisen in the Middle Ages as a response to the inability of the competition to offer a satisfactory explanation of the Black Death. Consequently, in this chapter we shall delve a bit more deeply into some of the philosophical issues underlying the relationship between realities and models, and at the same time take a critical look at the degree to which the concepts, methods and techniques of science are any better at the Reality Game than competing brands of medicine.

At this juncture it's worthwhile to reflect for a moment on the idea of an external reality as we've been using that term. In everyday language, the notion of an external reality comes down to a belief that the world around us enjoys an independent existence. In other words, lamps, TVs, cars and atoms are all 'out there' whether we observe them or not. According to this belief, the universe is composed of the collection of such independently existing objects. Loosely speaking, this common sense belief constitutes what is usually thought of as "objective" reality and forms one of the pillars upon which modern science is precariously perched. Recent results in the philosophy of quantum theory question this view. According to a famous result due to Bell, either one may retain the concept of an objective reality, or one may keep the concept of no faster-than-light physical effects—but not both. Recent experiments by Aspect in France have validated the theoretical predictions of Bell's result, leaving us in the uncomfortable position of either having to throw away objective reality or having to discard the key element in Einstein's theory of relativity, thereby leaving us open to a host of time travel and causality paradoxes. Most physicists and practicing scientists are loathe to dispense with the idea of an objective reality; nevertheless, this seems to be the only recourse Nature leaves open if we want to avoid the difficulties of acausal effects. Fortunately for most practical day-to-day matters, there is no harm done in clinging to the illusion of an objective reality and the practice of science at the meso- and macro-level goes on much as before. However, when moving into the microworld, it's well to bear in mind that it is not the *fact* that there may be no reality "out there" that is important, but rather that there may be as many "realities" as you want. This point will surface periodically in our subsequent discussion.

In virtually every culture the traditional means for structuring experiences is the *myth.* In my dictionary, the term myth is traced back to the Greek *mythos* meaning "word," in the sense that it is the decisive or definitive statement on the subject. A myth presents itself as an authoritative account of the facts which is not to be questioned, however strange it may seem. According to the mythologist Joseph Campbell, myths serve several quite distinct purposes:

- *Metaphysical*—Myths awaken and maintain an "experience of awe, humility and respect" in recognition of the ultimate mysteries of life and the universe.
- *Cosmological*—Myths provide an image of the universe and explanations for how it works.
- *Social*—Myths validate and help maintain an established social order.
- *Psychological*—Myths support the "centering and harmonization of the individual."

It will become clear as we proceed that the idea of "objective" science is one of Western man's most widely accepted myths. Myths need be neither true nor false, but only useful fictions; however, we should not confuse myths with the kind of ordinary fiction that has entertainment value alone and does not pretend to be true. The opposite side of the coin of *mythos* is *logos,* the Greek term for an account whose truth can be demonstrated and debated. This duality should be kept in mind as we compare the role of science as myth with various competitors like religion, astrology, extraterrestrial interventions and Zen.

Throughout this book we have sought answers from science, not mythology. Thus we claim there is indeed something more to science as it is practiced than just a story taken to manifest some aspect of cosmic order. Basically our position is that science involves myths of a special type, namely, myths that are *predictive, empirically testable and cumulative.* In the following pages we shall amplify upon these conditions and try to explain a few of the reasons for the success of modern science, as well as point out its glaring deficiencies as a reality-structuring mechanism. But for now it's sufficient to think of the connection between science and myth as

$$\text{science} = \text{myth} + \text{discipline}.$$

2. *Models and Theories*

In works on the philosophy of science, many different meanings are attached to the term "model." There are *physical* models such as the well-known billiard ball model for the behavior of an enclosed gas; there are *logical* models involving a collection of symbolic entities satisfying a particular set of axioms and theorems; there are *mathematical* models which are symbolic representations of quantities appearing in physical, social, and behavioral systems. Our focus has been solely upon the latter class of models, and we have been distinctly pragmatic in that we have considered a model of a natural system N to be the representation of the observables and equations of state of N in the elements of a formal mathematical system F. To a certain degree this view edges dangerously close to that held by the now-defunct school of *logical positivists,* who claimed that the only meaningful statements were those that could be translated into observables, and that the semantic content of a statement was identical with the prescription for verifying it. Although we are sympathetic to some aspects of the positivist program, our leanings are more toward the school of thought usually termed *instrumentalism,* which holds that models should not be judged by their truth or falsity but rather by their usefulness as algorithms for correlating observations and making predictions. Furthermore, we hold no brief for the extreme view that theoretical terms can (and must) be translated into equivalent observational terms. Thus, for us models are tools for reality organization, i.e., a tool for ordering experiences rather than a *description* of reality.

If models are only a means for ordering our experiences, then it follows that there can be many different models of the same experiences—hence, many alternate realities, at least to the degree that we see reality in our models. This position is made strikingly clear in the linguistic work of Sapir and Whorf, who considered the role of natural—rather than mathematical—language as a tool for ordering and describing experiences. The Sapir-Whorf hypothesis asserts that the way one sees the world depends upon the structure of one's language. If true, such a contention would imply that there is no correct way of seeing the world, or alternately, different languages give rise to different concepts.

As support for their claims, Whorf spent considerable time studying the languages of several American Indian tribes and found, for instance, that the Hopi language has one noun that covers everything that flies, with the exception of birds which as a class are covered by another noun. Thus, in Hopi the only distinction that can be made between flying objects is between birds and nonbirds. Similarly, Eskimo languages have different words for expressing various gradations of snow—wet, dry, powdery, icy, and so on, distinctions that can be made in everyday English only with difficulty or by using complicated constructions. Consequently, it can be argued that the snow "seen" by an Eskimo is literally not the same snow that I see now on the street as I look out my study window.

The foregoing examples are manifold, and they form the basis for the claim that language is the prime force in shaping our conceptions of the world. For science it means that no individual is free to describe Nature with absolute impartiality but is constrained to certain modes of interpretation even when he thinks himself most free. Of course, we can argue that the "language" of mathematics that we have adopted provides a universal tongue that enables us to describe nature in as impartial and universal way as can be imagined, surely far better than any natural language. There is some merit to this argument. But the same principle that underlies the Sapir-Whorf hypothesis for natural languages also applies to mathematical descriptions, since every formal system contains its own symbols, axioms and provable truths, though we know from Gödel's Theorem that there are other truths that cannot be expressed within that system. So in the final analysis we must recognize that just as it is with objective reality, so it is too with objective science, and there is no possibility of attaining the nirvana of objectivity untainted by the "prejudices" built into our method of describing nature, be it linguistic or mathematical. The best we can hope for is to illuminate various facets of natural systems using different, and inequivalent, formal descriptions.

The traditional philosophy of science literature distinguishes carefully between a scientific *theory* and a *model*. Roughly speaking, a theory is an interpreted mathematical proposition (in the terms of Chapter One, a *decoding*), whereas a model is generally thought of as something like the billiard ball model of a gas mentioned above. Throughout the book we have blurred this distinction, primarily because we have steadfastly held to the position that the only types of models we're interested in are mathematical models. With this view of our model class, there is very little remaining to separate a theory from a model, at least at the practical level that we are considering in this book. For us a model is the mathematical manifestation of a particular theory, and we are happy to leave the debate about exactly what constitutes that theory to the philosophical literature. Nevertheless, it's worthwhile to dip into that literature briefly and examine some of the issues surrounding the criteria for a good theory and the standards that a theory must meet before it can be adopted by the scientific community.

In what follows we shall take the semantic view and regard a theory as being synonomous with a *family* of related models, where model has our usual interpretation of a formal mathematical system. Theories like relativity, quantum mechanics, evolution and continental drift that have come to have a high level of universal acceptance, and by common agreement constitute what are usually termed "good" theories, generally have a number of identifying fingerprints:

• *Consistency*—Good theories contain no self-contradictory statements. So, for example, a good theory of gravitation would have no statements of the type that a ball rolls both up and down an inclined plane under seemingly identical experimental circumstances. Or a good theory of genetics does not involve situations in which identical genotypes give rise to radically different phenotypical forms.

• *Noncircular*—The claims of a good theory are not subtley buried in its premises. In other words, there are no circular arguments contained in good theories.

• *Cumulative*—A good theory is somehow "better" than previous theories in the sense that it contains all the predictions of the earlier theories but still adds something new. It is in this way that the Special Theory of Relativity is better than Newtonian gravitation, since all the claims of the older theory are contained as special cases of Einstein's extension. Thus a good theory always encompasses all the observational evidence of its predecessor and still manages to add something new.

• *Testable*—Good theories always make predictions or claims that are testable by experiment. This criterion of *falsifiability* has been emphasized by Popper as the line of demarcation between scientific and nonscientific theories, and has been used to dismiss the claims of astrologers, Marxists, and creationists that their brand of medicine qualifies to be labeled "scientific."

Despite the foregoing rather strict requirements, the scientific graveyard is littered with the corpses of "good" theories that never managed to gain acceptance. This leads us to wonder whether there are any criteria or standards beyond those imposed above for the acceptance of a scientific theory. In his pioneering work on the ways of science and scientists, Thomas Kuhn identified at least two additional conditions that a theory must meet before it can be accepted by the scientific community.

One of these conditions is that the new theory must answer a significant number of questions answered by the old theory and yet provide insight into issues that the old theory cannot or does not address. Note that this doesn't mean that the new theory must answer *all* the questions answered by the old theory; as we shall see in a moment, it's perfectly possible for the new theory and the old theory to have only a substantial intersection rather than completely overlap each other. For example, in the theory of ideal gases Boyle originally thought of the gas molecules as little coiled springs that, when compressed, exerted a strong repulsive force. Newton improved upon this picture by imagining the gas particles as little perfect spheres that were mutually repelled by an inverse square law. Both of these theories are inherently *static,* for they regard the gas as a sort of sponge that shows elasticity when compression is removed. A radically different theory was put forth by Bernoulli, who thought of the now familiar billiard ball model of molecular motion in which the gas molecules careen about in the container, bouncing off one another much as a collection of randomly moving balls on a billiard table. This is a *dynamic* theory of gas behavior. The Bernoulli theory clearly answers many of the same questions about the pressure-temperature-volume relationship of gas as the older theories, but it also contains the possibility for addressing substantial new issues by virtue of its dynamic view of the origin of the observed behavior. However, there are questions that the dynamic theory cannot address, questions that are within the purview of the earlier theories such as the elasticity constants of the "springs" in the Boyle view or the "gravitiational" constant of the spheres in Newton's picture. But since such questions aren't important for addressing the major issues surrounding the behavior of gases, they don't work to the disadvantage of the new theory insofar as its acceptance by the scientific community is concerned. What's important are the questions that are answered and the new questions that are generated, not those that are igonored. This observation leads to Kuhn's second criterion for theory acceptance.

New theories, like their predecessors, always generate new questions as well as answer old ones. An important component of the success of a new theory, or at least its acceptance, is that it provide enough new puzzles to keep scientific research alive. In other words, a new theory can't be too successful in answering the old questions or else it won't leave enough material for the next generation of scientists to keep themselves busy. In essence, the lifeblood of science is unsolved problems, and if the current world view doesn't leave enough problems it won't be accepted by the scientific community.

This is a situation radically different from the way science is usually pictured in the media and cinema, where the popular vision is of scientists as seekers after all truths, large and small; nevertheless, science is done by scientists, and scientists are no more immune to the psychological and sociological influences of their culture than the politician, mechanic or businessman. So it's perhaps not surprising that we see such subjective factors enter into the acceptability of a scientific theory.

These social and psychological considerations led Kuhn to postulate his now famous idea of a "paradigm" to explain the workings of science and, most important, the manner in which the scientists' view of the world undergoes discontinuous periodic shifts. Let's now examine the core of the "paradigm" concept in a little more detail.

3. *Paradigms, Revolutions and Normal Science*

A scientific paradigm is basically a theory plus a set of presuppositions involving physical or mental models, a symbolism and the other social and technical trappings of the time. To use a somewhat fanciful metaphor, imagine science as the *terra incognita* of the ancient geographers and mapmakers. In this context a paradigm may be thought of as a crude sort of map in which territories are outlined, but not too accurately, with only major landmarks like large rivers and prominent mountains placed on the map. From time to time explorers venture into this ill-defined territory and come back with accounts of native villages, desert regions, minor rivers, and so forth, which are then dutifully entered onto the map. Often such new information is inconsistent with what was reported from earlier expeditions, so it's periodically necessary to redraw the map in accordance with the current best estimate of how things stand in the unknown region. Furthermore, there is not just one mapmaker but many, each with his own set of sources and data on the lie of the land.

As a consequence, there are a number of competing maps of the same region, and the adventurous explorer has to make a choice of which map he will believe before embarking upon an expedition to the "New World." Generally, the explorer will choose the old, reliable firm of mapmakers, at least until gossip and reports from the exploration community show too many discrepancies between the standard maps and what has actually been observed. As these discrepancies accumulate, eventually the explorers shift their allegiance to a new firm of mapmakers whose picture of the territory seems more in line with the observational evidence.

This exploration fable gives a fair picture of the birth and death of a scientific paradigm. Kuhn realized that revolutionary changes overturning old theories in science are not in fact the normal process of science. Nor do theories start small and grow more and more general as claimed by Bacon. Nor can they ever be axiomatized as asserted by Newton and Descartes. Rather, for most scientists major paradigms are like a pair of spectacles that they put on in order to solve puzzles. Occasionally a paradigm shift occurs when the spectacles get smashed, and they put on a new pair that turns everything around into new shapes, sizes and colors. Once this shift takes place, a new generation of scientists is brought up wearing the new glasses and accepting the new vision as "true." Through these new glasses scientists then see a whole new set of puzzles.

The paradigms have great practical value for the scientist just as maps have value for the explorer: without them he wouldn't know where to look or how to plan an experiment (expedition) and collect data. This observation brings out the crucial point that there is no such thing as an "empirical" observation or fact; we always see by interpretation, and the interpretation we use is given by the prevailing paradigm of the moment. In other words, the observations and experiments of science are made on the basis of theories and hypotheses contained within the prevailing paradigm. As Einstein put it, "the theory tells you what you can observe." According to Kuhn's paradigmatic view of scientific activity, the job of "normal" science is to fill in the gaps in the map given by the current paradigm, and it's only seldom, and with great difficulty, that the map gets redrawn when the normal scientists (explorers) turn up so much data that doesn't fit into the old map and the map begins to collapse of its own weight. But what happens during these times of paradigm crisis?

Imagine we are at the initial stages of such a crisis, where the old paradigm can't account for certain anomalies, strange observations, and the like. Two new theories emerge that offer different explanations for these aberrations. These theories represent different maps, or sets of spectacles, i.e., different realities. After a period of competition, one of these theories begins to gain the acceptance of the scientific community. The reasons may not be objective but may revolve about matters like simplicity, elegance, social position of its adherents or government science policies. This support leads to experiments that then "corroborate" the theory; the more evidence that accumulates, the more supporters the theory gathers, especially among the "young Turks" in the scientific community. Soon "reality" begins to take on the vision of the new theory, and scientists begin to universally see and test for certain features of this reality and ignore others.

But what if the community had given its initial support to the other, competing theory? According to Kuhn, in that event "reality" would have taken a quite different turn, and the scientific view of the world would have been seen through that pair of spectacles rather than the first. This means that there is no such thing as scientific "progress," at least not in the sense that one paradigm builds upon its predecessor. Rather, the new paradigm turns in an entirely different direction and as much knowledge is lost from the old paradigm as is gained from the new. Instead, we "know" a *different* universe.

If Kuhn's thesis is true, then it also destroys one of the main pillars of the scientific method, since the whole idea of a scientific experiment rests upon the assumption that the observer can be essentially separate from the experimental apparatus that tests the theory. Kuhn contends that the observer, his theory and his equipment are all essentially an expression of a point of view, and the results of the experimental test must be an expression of that point of view as well. This position effectively shows that science is not *objective,* but at the same time we know that science is not totally subjective either since paradigms are eventually overthrown. So what is the relationship of the scientist to the universe he observes? This question leads us to a consideration of what it means to say a model constitutes an explanation of Nature rather than just a description.

4. *Explanations vs. Descriptions*

Many would argue that the basic business of science is to be able to say "because." Planets travel around the Sun in ellipical orbits *because* they obey the inverse square law of gravitation; diamonds are the hardest substance known *because* their crystalline structure resists compression more strongly than any other; it's raining today *because* at the current temperature and pressure the outside air cannot hold all the water vapor present. It's in the nature of an *explanation* to answer the question "why?" On the other hand, a *description* just gives an account of "what is," without entering into any underlying reasons for why things are as they are. For the most part we have been concerned with descriptions, and our modeling philosophy has been directed primarily at providing formal structures that describe what is without being too concerned with whether they are even true in the explanatory sense of the term "true." However, it's patently clear that an explanation is far preferable to a description in just the same way that a formal proof is much better than a heuristic plausibility argument. Consequently, it's worth taking a little time to consider some of the differences between an explanation and a description and, at the same time, shed a little light on how we can distinguish a model of a natural system from a simulation.

In common parlance, an explanation involves giving an account of an already *known* fact on the basis of logical conclusions drawn from well-established general theories. Often there is some sort of hierarchical structure of explanations, with the fact to be "explained" residing at a higher level than the components of the explanation. Such a hierarchy, of course, lies at the heart of the reductionist program for scientific explanation in which *all* phenomena are ultimately explainable by some kind of elementary or rock-bottom level of facts or entities. In any case, the essence of an explanation is to answer the question "why?" at one level by a "because" constructed from entities and logical operations at a lower level. This is the pattern we see, for example, in the explanation of the temperature of an enclosed gas in terms of the kinetic behavior of its constituent molecules, the explanation of gross national product in terms of the contributions of individual industries, and in innumerable other situations.

From the above considerations it's evident that there's an intimate connection between giving an explanation for a process and providing a causal description of it; in fact, the two are virtually indistinguishable. For this reason it's of interest to note that the idea of a causal description is at least as old as the ancient Greeks, in whose philosophies we find a number of attempts to provide a causal ordering to events. The dominant causal structure for over two millennia was put forth by Aristotle, who identified four inequivalent "causes" in his treatment of physical phenomena. According to Aristotle, the "because" of things is some combination of the following elementary, inequivalent, and comprehensive causal categories:

- *Material Cause*—Things are as they are because of the material elements from which the system is constructed.
- *Efficient Cause*—Things are as they are because of the physical work or energy that went into making them as they are.
- *Formal Cause*—Things are as they are because of the plan or design according to which they were constructed.
- *Final Cause*—Things are as they are because of the desire of an external agent to have them take their observed form.

In Aristotle's epistemology, the highest place was reserved for the last category, final causation which, amusingly enough, was banished from polite scientific discourse by the paradigm that finally overthrew Aristotle's, Newtonian mechanics. Since it's of some interest to see how this elimination of final cause was orchestrated by Newton, let's look at the situation in a bit more detail.

In Newton's world of particle mechanics, every possible "why" question is answerable in terms of two aspects of the particle system: (1) the positions and velocities of all the particles as measured from some appropriate reference frame, and (2) the external forces imposed upon the particles. In mathematical terms, the state $x(t)$ representing the positions and velocities of all particles of the system at time t is given by

$$x(t) = x_0 + \int_0^t g_\alpha(x(s), f(s))\, ds,$$

where x_0 is the initial state, $f(\cdot)$ represents the external forces acting upon the system, and $g_\alpha(\cdot, \cdot)$ is the law of motion of the system, including the constitutive parameters α representing the force of gravity, masses of particles, and so on. In Aristotelian terms, we can use the above framework to answer the question "why is the system in the state $x(t)$?" in the following manner:

- *Material Cause*—The system is in the state $x(t)$ because it was in the state x_0 at time $t = 0$.
- *Formal Cause*—The particles are in the state $x(t)$ because the system's constitutive parameters assumed the value α.

- *Efficient Cause*—The state is now $x(t)$ because the integral operator $\int_0^t g_\alpha(\cdot)\, ds$ acted upon previous states to bring the system to the state $x(t)$.

Note that there is no need, and no room either, for any sort of final cause in Newton's epistemology. It's for this reason that the idea of final cause, or teleology, has been turned into an unfashionable, even crankish notion in modern scientific thought. Here we see a perfect example of the earlier claimed "different worlds" of competing paradigms in the Kuhnian sense.

Classical science regards the casting out of teleology as a significant step forward and a selling point for the Newtonian paradigm. Our view is distinctly bleaker. Although holding no brief for teleology in its strict form of the future generating the present, we do believe there is a type of quasi-teleological principle at work in all systems involving living organisms, and that this principle lies at the core of the difference between the physical sciences, on the one hand, and the life, social and behavioral sciences, on the other. By banishing teleology from scientific discussion, Newton in effect threw out the baby with the bathwater, eliminating all notions of anticipation and self-reference from scientific consideration. We have given extended treatment in Chapter Seven as to how we think it's possible to reintroduce this most important of Aristotelian causes, final cause, back into a major extension of the Newtonian *Weltanschauung,* or "world view." Now let's turn to the matter of predictions and the use of models as tools for data correlation.

5. *Predictions, Simulations and Models*

In naïve modeling exercises, a typical sequence of events involves construction of some sort of formal mathematical model, then using the model to "postdict" the results of past experiments as well as to "predict" the outcomes of future observations. In the event these post- and predictions are in close enough agreement with what is actually observed, the formal system is then declared a *model* of the process under investigation. Is it really? Is this all it takes to be deemed a model of a natural phenomenon? Shouldn't there be something more than just agreement with the data? After all, Ptolemy and his epicycles provided at least as accurate agreement with observed planetary positions as the ellipses of Copernicus and Kepler, yet no one today seriously contends that the edifice of epicycles piled upon epicycles is a *good* model of the solar system. At best such a structure provides a *simulation* of planetary motion, but surely not a model. So what's the difference between a simulation and a model? And if there is a difference, can we produce any procedures for separating one from the other?

The perceptive reader will recognize this problem as the same one we faced in Chapter Three, where we considered the input/output pattern of a black box, and considered how to construct an *internal* model that matched this behavior. This model was nothing more than a simulation of the actual system as we shall endeavor to show below. This simulation vs. model problem is so ubiquitous in the applied modeling literature that it's well worth our taking some time to examine the differences between the two, as well as to put forward a few ideas upon which answers to the above questions can be based. Perhaps the simplest way to outline the difference between a model and simulation is to follow an example originally put forth by Rosen.

Consider a formal neural network described by a collection of $nm + r$ formal neurons. By arranging the neurons into an $n \times m$ rectangular array below a row of r output neurons, it's well-known that we can represent the neural net by a finite-state mathematical machine $\mathcal{M} = (S, A, B, \lambda, \delta)$, where S is the state-set, A is the set of inputs, B is the set of outputs and $\lambda\colon S \times A \to S$, $\delta\colon S \times A \to B$ are the machines state transition and output maps, respectively. Here we have S, A and B being sets of cardinalities n, m and r, respectively. By suitably *restricting* the inputs to the neural net, the behavior of the net can mirror the behavior of $\mathcal{M}$. Or looked at from the perspective of the machine, we say the machine $\mathcal{M}$ constitutes a *realization,* or a *model,* of some particular behavior of the net.

Note the following key aspects of the above situation:

1) There is a clear and direct correspondence between the states of the machine and the states of the net, as well as direct correspondences between the inputs (outputs) of $\mathcal{M}$ and the inputs (outputs) of the net.

2) The machine $\mathcal{M}$ is *simpler* than the net it models, in that the net is capable of many behaviors that the machine is not. Thus, given a machine $\mathcal{M}$, there exists a neural net that mirrors the behavior of $\mathcal{M}$ if we impose restrictions on the inputs and outputs of the net.

These properties are characteristic of one system being a genuine *model* of another instead of being just a *simulation,* and are intimately tied up with the notion of system similarity discussed in Chapter One. In essence, the modeling relation is transitive in that if X is a model for Y and Y is a model for Z, then X is a model for Z.

Now consider the case in which we want to use $\mathcal{M}$ to *simulate* another machine $\mathcal{A} = (\mathcal{S}, \mathcal{U}, \mathcal{Y}, \mu, \nu)$. For simplicity, assume that $\mathcal{A}$ is just a state-transition machine so that the behavior of $\mathcal{A}$ is characterized solely by the map $\mu\colon \mathcal{S} \to \mathcal{S}$, i.e., $\mathcal{U}$ = singleton, $\mathcal{Y} = \mathcal{S}$, ν = identity. In order to simulate $\mathcal{A}$ by $\mathcal{M}$, we must *encode* the states of $\mathcal{A}$ into the input strings A^* of $\mathcal{M}$, and we must *decode* the output strings of $\mathcal{M}$ into the states of $\mathcal{A}$. This means we must find maps

$$g\colon \mathcal{S} \to A^*, \qquad h\colon B^* \to \mathcal{S},$$

such that the diagram

$$\begin{array}{ccc} \mathcal{S} & \xrightarrow{\mu} & \mathcal{S} \\ g\downarrow & & \uparrow h \\ A^* & \xrightarrow[\psi]{} & B^* \end{array}$$

commutes, where ψ is determined by the initial state of $\mathcal{M}$ as well as by the maps λ and δ.

Let's now take a look at the relevant aspects of this set-up for our simulation vs. model comparison.

1) The relation between $\mathcal{A}$ and its simulator $\mathcal{M}$ does not preserve structure. Roughly speaking, the "hardware" of $\mathcal{A}$ is mapped into the "software" of $\mathcal{M}$, and the hardware of $\mathcal{M}$ has no relation whatsoever to that of $\mathcal{A}$. In fact, if we try to describe the hardware of $\mathcal{M}$ as *hardware*, $\mathcal{A}$ will disappear completely.

2) In general, $\mathcal{M}$ is *bigger* than $\mathcal{A}$, usually much bigger; the state-set S of $\mathcal{M}$ must usually be much larger than $\mathcal{S}$ if $\mathcal{M}$ is to simulate $\mathcal{A}$. This is in stark contrast to our earlier situation when $\mathcal{M}$ served as a model of the neural network.

It's possible to summarize this modeling vs. simulation distinction by saying that the modeling relation is transitive, whereas the simulation relation is not. So if A simulates B and B simulates C, then it does not follow that A simulates C; in fact, A is only a simulation of a simulation of C. (The reader is invited to reexamine Problem 1 of Chapter One in this connection.) In category-theoretic terms, the modeling relation is *functorial,* but the simulation relation is not.

Now let's return to the question of whether it matters that a simulation is not structure-preserving. Assume that in the natural system of interest N there is some property P of concern and we have a simulation $\mathcal{M}$ of N. Then if P is preserved by $\mathcal{M}$, fine; if not, then we're in the kind of trouble that would never arise if $\mathcal{M}$ were a model of N rather than just a simulation. For example, if we're interested in the motion of an arm, then a computer simulation of arm movement will certainly not possess the property P of arm motion. But a robot constructed to mimic human arm movement will be a legitimate *model* of arm motion and, hence, will possess property P. Thus the robot represents a model of arm motion, whereas the computer program is only a simulation. The degree to which this very obvious distinction matters in practice remains open and, to some extent, dependent on the particular situation. Nevertheless, the distinction is clear, mathematically testable, and bears heavily upon the question of model credibility. In our opinion, these factors argue for the issue to receive somewhat more attention than it has in the applied modeling literature.

6. *"Good" Models and "Bad"*

Earlier we examined some criteria for a "good" theory and then noted that from a semantic point of view we could consider a theory to be a collection of related models, where model here denotes the kind of formal system that we have centered attention on in this book. Now we want to investigate the related issue of what constitutes a "good" model, together with the delicate issue of how to validate or confirm a proposed model of a specific natural process. To serve as a focal point for our deliberations, let's return to that most ancient of modeling problems, planetary motion, and consider the two competing models put forth by Ptolemy and Copernicus.

Ptolemy's model of the solar system had the planets moving in orbits that were described by a collection of superimposed epicycles, with the Earth at the center of the system. From a predictive standpoint, this structure gave remarkably good accounts of where the planets would be at any time, and was successfully used by many generations to predict eclipses, comets and, most importantly, lunar positions influencing the flooding of the Nile. The discovery of new solar bodies and their influence on the motion of others could always be accommodated in this model by piling on another epicycle or two to make the behavior of the model agree with the observed data. In this sense, the model was somewhat akin to the process of fitting a polynomial through a set of data points; we know that the fit can be made exact, provided that we're willing to take a polynomial of high enough order.

The major change introduced by Copernicus was to assume in his model that the Sun rather than the Earth lies at the center of the solar system. This assumption, later translated into mathematical form by Kepler and Newton, led to the now familiar picture of the planets moving about the Sun in elliptical orbits, and has generated a highly accurate set of predictions about future planetary positions. However, it's of the greatest importance to note that from a predictive standpoint the Copernican model is no better than the Ptolemaic. In fact, if anything the epicycles of Ptolemy provided a mathematical framework that could make *more accurate* predictions of orbits than the ellipses of Kepler and Newton. Again this is analogous to the situation with curve-fitting by polynomials: you can't do better than to match the data *exactly,* and the Ptolemaic model was capable of doing this to a higher degree than even the Copernican. So why was the Ptolemaic view discarded in favor of Copernicus? The obvious answer is that the Copernican view was *simpler*; it contained fewer *ad hoc* assumptions and led to a less cumbersome, more straightforward mathematical formalization. Fundamentally, Ptolemy's model of planetary motion was cast aside on aesthetic grounds, not on the grounds that the new model was in better agreement with the observations or gave better predictions about future positions of the planets. This kind of subjective criterion is one that we shall see more of in a later section when we consider competition between models.

The above scenario for choosing one model over another has been formalized in the principle now termed Occam's Razor which asserts, roughly speaking, that one should not multiply hypotheses beyond what is required to account for the data. This principle leads to the corollary that everything else being equal (which it never is), take the simplest model that agrees with the observations. Of course, this leads immediately to another issue that we have come back to a number of times in this volume—the problem of complexity. Given two models, by what criterion do we say one is simpler (or more complex) than the other? Nevertheless, *simplicity,* however you choose to define it, is certainly one of the primary components of a "good" model. But there are others.

In the Copernicus vs. Ptolemy competition, we saw another important criterion for a "good" model: it must agree to a reasonable degree of accuracy with *most* of the observed data. Here we emphasize the term "most," since it's important in this connection to bear in mind the earlier remark that the theory tells you what you can observe, and for us theory is a collection of models. Consequently, if the aesthetic reasons for choosing a model are strong enough, then it's only necessary for the model to agree with the majority of the observations.

As an example of the foregoing precept, consider Newton's model of gravitation as embodied in the inverse square law and his Third Law, $F = ma$. This led to predictions about the orbit of Mercury that were not in agreement with observation. Nevertheless, this model of gravitation is probably the most successful model in the history of science, and it needed only some minor fine-tuning by Einstein to account for this anomaly. In fact, one might say it's the hallmark of a "good" model when the data that it **doesn't** agree with lead to the next scientific revolution. Thus, *good, but not necessarily perfect, data agreement* is a second bench-mark test for a "good" model.

The third element in our Holy Trinity of good modeling is *explanatory power.* A "good" model should be in a modeling relation to the natural system it represents and not be just a simulation. This means that there should be some interpretable connections between the entities of the formal system comprising the model and the physical entities characterizing the natural system under study. Since we have already spent some time elaborating our view of the differences between an explanation and a description, it's not necessary to repeat those arguments here other than to note that the notion of explanatory power forces us to confront head-on the problem of how we would go about validating a proposed model. To address this issue, we again return to matters central to twentieth-century philosophy of science.

7. *Validation and Falsification*

The problem of the validation of a scientific theory or a formal model was brought into sharp focus by the extreme position taken by the logical positivist movement in the 1930s. According to the positivists, a scientific theory is not a representation of the world but a shorthand calculational device for summarizing sense-data. The positivists asserted that the scientist should only use concepts for which he can give "operational definitions" in terms of observations. Thus, according to this position, length and time are not attributes of things in the world but relationships defined by specifying experimental procedures. The culmination of the positivist position was the notorious *verification principle* which demanded that only empirical statements verifiable by sense-experience have meaning (formal definitions and tautologies are also meaningful but convey no factual information). Consequently, by this thesis most statements in philosophy, and all those in metaphysics, ethics and theology, were said to be neither true nor false, but meaningless; i.e., they state nothing and merely express the speaker's emotions or feelings. The difficulties with this position are by now well chronicled and we won't enter into them here. In our quest for insight into the model validation question, the importance of the positivist position is the implied claim that a model is valid only if it can be tested (verified) by experimental procedures; otherwise, it's not only invalid, it's nonscientific.

A major outgrowth of the verification thesis was put forward by Popper in his view that scientists use models not just as explanations of natural phenomena, but as probes to provoke nature into behavior which the scientist can recognize as an observation. This observation can then be used to: (1) provide further inductive evidence in favor of the model, or (2) provide a counter-example to the claims of the model. It is the second case that Popper termed *falsifying* the model.

Popper's basic position is that science is not at all in the business of validating models, but rather of falsifying them. His claim is that real scientific models are set forth in a way that spells out observations and predictions that can be tested experimentally. If the prediction fails, then the model is falsified and must be abandoned or completely rethought; if a model passes its crucial test, it's not validated but only "corroborated," and the process of testing must go on.

An interesting and important consequence of Popper's theory is that controversies about the merits of competing models shouldn't really exist. When faced with two competing models to represent a given situation, scientists would simply choose the "better" one—the one that could survive the toughest tests. As we have already noted, Kuhn realized that in practice this is impossible. The parties in paradigm debates speak such different languages and wear such different glasses that even if they look at the same clock, they won't agree on which way it is running.

Kuhn concluded that Popper's falsification standard for scientific validation was a myth and that, in fact, most scientists hardly ever see an anomaly in the observations as a challenge to the paradigm that lies behind it. Just as astrologers claim that when their predictions fail, it is due to factors like too many variables in the system, too much uncertainty, etc., and view this failure as in no way disproving their cosmic model, so it is also with meteorologists who make the same excuses when their predictions of sunny weekends turn into torrential downpours.

Thus, although the theory of falsifiability of models looks logically attractive, it seems to have little to do with how science actually operates and certainly doesn't appear to form the basis for any objective procedure for model validation. When all is said and done, we are left in much the same position we found ourselves in earlier when considering standards for theory acceptance. There are no purely objective criteria by which we can select one model over another, and in practice the choice as often as not hinges upon psychological and sociological factors having little or nothing to do with the empirical content of the model. So let's take a longer look at the way real-life science functions in separating the wheat from the chaff in picking models to back.

8. *Competition among Models*

Scientific theories and their models prove their mettle in a forum of debate. Most of us are familiar with the kind of school-days debate in which one party takes the affirmative of an issue while another supports the opposite view. But what is the nature of the debate by which models are accepted or rejected? There are at least three basic ground rules.

First of all, each of the competing models must be an available alternative. By this we mean that the model must be known to the scientific community, and it must be developed to the extent that its advantages are explicit. A model or theory must be a serious contender in its own time, which means that it must be able to explain the facts in question, as well as either fit into the current scientific framework or else have something going for it that makes it worth considering even though it violates some basic assumptions. It should also be noted here that competing models, unlike school-days debates, are usually not contradictory in that one is not the denial of the other. Thus, in contrast to the familiar debating situation where all it takes to show one side right is to show the other side wrong, in the modeling derby it could very well be that all sides are wrong.

The second basic ground rule of scientific debate is that the arguments and evidence offered in support of a model must be submitted to evaluation according to the standards of the scientific community. What sorts of reasons do the scientific community accept? It depends to a large extent upon the particular subject matter. It's somewhat easier to say what kinds of reasons are *not* acceptable, a detailed treatment of which we shall defer to the next section.

Our final ground rule is that the winning model must prove itself better than its competitors. It's not enough that there are reasons that seem to favor it, and it's not enough that there is solid evidence that seems to confirm it. It has to be shown that the model has more going for it than the other candidate models. What are the identifying features of such a model? Just those features we outlined earlier in our discussion of model credibility: simplicity, agreement with the known facts, explanatory power and predictive capability.

Although the primary filtering mechanism of modern science is the debate outlined above, we shouldn't discount the social and psychological factors that also enter into a model or theory's acceptance or rejection. To begin with, the deck is always stacked in favor of the existing orthodoxy since this is the paradigm by which we see theoretical constructs. In addition, the existing power structure in the scientific community generally has a large professional, psychological and financial stake in protecting the *status quo*. After all, who wants to see a lifetime's worth of work on developing a chemical analysis technique or a delicate asymptotic approximation wiped out overnight by by some young hotshot's computer-based substitute? This ego factor is only exacerbated by the system of research grants administered by federal agencies which, more often than not, reward Kuhn's "normal" science and strongly penalize adventuresome revolutionaries. This process is further reinforced, of course, by the promotion and tenure policies in the universities in which advancement is contingent upon publication of lots of "potboiler" articles and the hustling of grants as much as by painstaking scholarly activity. All these factors taken together introduce an enormous bias into the model acceptance process, a bias strongly leaning toward preservation of the existing orthodoxy, if at all possible. This situation calls to mind Max Born's famous aphorism that "new theories are never accepted, their opponents just die off." This has never been more true than in today's world of science.

Nevertheless, there is no shortage of theories about how the world is ordered, many of them at least clothed in the trappings of science. A glance at the shelves of your favorite bookshop will turn up an almost endless array of volumes on UFOs, hollow-earth theories, creationism, astral projection, extraterrestrials, telekinesis, numerology, astrology, spoon-bending, pyramidology and the like, many of which lay claim to being legitmate alternatives to mainstream science. This is not to mention the various types of religions such as Scientology, Christian Science, or even more exotic forms given to beliefs involving speaking in tongues, handling of snakes, infant damnation and other practices. We shall deal with religion later, but for now by way of contrast, and also to underscore the difference between science and the "pseudosciences," let's take a harder look at some of the features of these alternative belief systems.

9. *Science and Pseudoscience*

Most working scientists and every editor of a science journal has had the experience of receiving a bulky package in the mail plastered with stamps and addressed in a semilegible scrawl containing the results of a lifetime's worth of work claiming to square the circle, create perpetual motion, refute the theory of relativity or in some way definitively resolve another outstanding issue in science. Generally such efforts come accompanied with rambling accounts of why the scientific community has misunderstood the basic issues involved, and how the author has seen the way clear to set the record straight using what he claims is the scientific method of gathering data, generating hypotheses and doing experiments. After a time, the experienced eye learns to recognize the telltale signs of such cranks and pseudoscientists, and learns to never, ever, under any circumstances enter into any sort of rational dialogue with such people over the merits of their positions. Our aims in this section are not to lay out the complexities of the scientific process but to point out that there are ways of operating that are not found in science but are found in pseudoscience. These ways have to do with the reasons for proposing alternative hypotheses, with what are accepted as facts to be explained, with what counts as supporting evidence and with what counts as a theory.

In what follows we detail a checklist of criteria for identifying pseudoscience when you see it. Anyone who meets even one of these conditions is practicing in pseudoscience.

• *Anachronistic Thinking*—Cranks and pseudoscientists often revert to outmoded theories that were discarded by the scientific community years, or even centuries, earlier as being inadequate. This is in contrast to the usual notion of crackpot theories as being novel, original, offbeat, daring and imaginative. Many of them actually represent a return to a world view that was dismissed by the scientific world as being too simplistic or just plain wrong years earlier. Good examples of this kind of crankishness are the creationists who link their objections to evolution to catastrophism, claiming that geological evidence supports the catastrophic rather than uniformitarian view of geological activity they associate with evolution. The argument is anachronistic insofar as it presents the uniformitarianism-catastrophism dichotomy as if it were still a live debate.

• *Seeking Mysteries*—Scientists do not set out in their work looking for anomalies. Max Planck wasn't looking for trouble when he carried out his radiation emission experiments, and Michelson and Morley certainly were not expecting problems when they devised their experiment to test for the luminiferous ether. Furthermore, scientists do not reject one theory in favor of another solely because the new theory explains the anomalous event. On the other hand, there is an entire school of pseudoscience devoted to enigmas and mysteries be they the Bermuda Triangle, UFOs, Yetis, spontaneous combustion or other even more offbeat phenomena. The basic principle underlying such searches seems to be that "there are more things on heaven and earth than are dreamed of in your philosophies," coupled with the methodological principle that anything that can be seen as a mystery ought to be seen as one.

• *Appeals to Myths*—Cranks often use the following pattern of reasoning: start with a myth from ancient times and take it as an account of actual occurrences; devise a hypothesis that explains the events by postulating conditions that obtained at that time but which no longer hold; consider the myth as providing evidence for support of the hypothesis; argue that the hypothesis is *confirmed* by the myth as well as by geological, paleontological or archaelogical evidence. This is a pattern of reasoning that is absent from the procedures of science.

• *Casual Approach to Evidence*—Pseudoscientists often have the attitude that sheer quantity of evidence makes up for any deficiency in the quality of the individual pieces. Further, pseudoscientists are loathe to ever weed out their evidence, and even when an experiment or study has been shown to be questionable, it is never dropped from the list of confirming evidence.

• *Irrefutable Hypotheses*—Given any hypothesis, we can always ask what would it take to produce evidence against it. If nothing conceivable could speak against the hypothesis, then it has no claim to be scientific. Pseudo-science is riddled with hypotheses of this sort. The prime example of such a hypothesis is creationism; it is just plain not possible to falsify the creationist model of the world.

• *Spurious Similarities*—Cranks often argue that the principles that underpin their theories are already part of legitimate science, and see themselves not so much as revolutionaries but more as poor cousins of science. For example, the study of biorhythms tries to piggyback upon legitimate studies carried out on circadian rhythms and other chemical and electrical oscillators known to be present in the human body. The basic pseudoscience claim in this area is that there is a similarity between the claim of the biorhythm theorists and the claims of the biological researchers, therefore biorhythms are consistent with current biological thought.

• *Explanation by Scenario*—It's commonplace in science to offer scenarios for explanation of certain phenomena, such as the origin of life or the extinction of the dinosaurs, when we don't have a complete set of data to construct the exact circumstances of the process. However, in science such scenarios must be consistent with known laws and principles, at least implicitly. Pseudo-science engages in explanation by scenario *alone,* i.e., by mere scenario without proper backing from known laws. A prime offender in this regard is the work of Velikovsky which states that Venus' near collision with the Earth caused the Earth to flip over and reverse its magnetic poles. Velikovsky offers no mechanism by which this cosmic event could have taken place, and the basic principle of deducing consequences from general principles is totally ignored in his "explanation" of such phenomena.

- *Research by Literary Interpretation*—Pseudoscientists frequently reveal themselves by their handling of the scientific literature. They regard any statement by any scientist as being open to interpretation, just as in literature and the arts, and such statements can then be used against other scientists. They focus upon the words, not on the underlying facts and reasons for the statements that appear in the scientific literature. In this regard, the pseudoscientists act more like lawyers gathering precedents and using these as arguments rather than attending to what has actually been communicated.

- *Refusal to Revise*—Cranks and crackpots pride themselves on never having been shown to be wrong. It's for this reason that the experienced scientific hand never, under any circumstances, enters into dialogue with a pseudoscientist. But immunity to criticism is no prescription for success in science for there are many ways to avoid criticism: write only vacuuous material replete with tautologies; make sure your statements are so vague that criticism can never get a foothold; simply refuse to acknowledge whatever criticism you do receive. A variant of this last ploy is a favorite technique of the pseudoscientist: he always replies to criticism but never revises his position in light of it. They see scientific debate not as a mechanism for scientific progress but as a rhetorical contest. Again the creationists serve as sterling examples of the power of this principle.

The major defense of pseudoscience is summed up in the statement: "anything is possible." Earlier in this chapter we considered the question of competition between models and theories and drew up a few ground rules by which the competition is generally carried out in legitimate scientific circles. Let's look at how the pseudoscientist with his "anything is possible" shield enters into such competition.

In the competition among theories, the pseudoscientist makes the following claim: "Our theories ought to be allowed into the competition because they may become available alternatives in the future. Scientists have been known to change their minds on the matter of what is and is not impossible, and they are likely to do so again. So who is to say what tomorrow's available alternatives may be?" In other words, anything is possible! This argument clearly violates our first rule of scientific debate. The fact that a theory may become an available alternative in the future does not constitute a reason for entering it into the competition today. Every competitor now must be an available alternative now. The pseudoscientist suggests that we may as well throw away the current scientific framework since it will eventually have to be replaced anyhow.

By referring to a future but as yet unknown state of science, the cranks are in effect refusing to participate in the competition. This would be all right if they didn't at the same time insist on entering the race. It's as if one entered the Monaco Grand Prix with a jet-propelled car and insisted on being allowed to compete because, after all, someday the rules may be changed to a jet car race!

The pseudoscientist also worms his way into the competition by putting the burden of proof on the other side. He declares to the scientific community that it is up to them to prove his theory wrong, and if they cannot do so he then states that his theory must be taken seriously as a possibility. The obvious logical flaw is the assumption that not proving a theory impossible is the same thing as proving it is possible. Although the principle of innocent till proved guilty may be defensible in a court of law, scientific debate is not such a court. The reason why the pseudoscientist thinks he can put the burden of proof on the scientists can be traced to his mistaken notion of what constitutes a legitimate entry in the debate. The pseudoscientists think that the scientific method places a duty on the scientific community to consider *all* proposed ideas that are not logically self-contradictory. So, to ignore any idea is to be prejudiced.

Finally, we note that the pseudoscientists often act as if the arguments supporting their theory are peripheral to the theory. Thus, they fail to see that what makes a theory a serious contender is not just the theory, but the theory plus the arguments that support it. Cranks think that somehow the theory stands on its own and that the only measure of its merits for entering the competition is its degree of outlandishness. Hence, they think the scientific community has only two choices: admit their theory into the competition or else prove it to be impossible. So we conclude that when it comes to defending a theory or model in scientific debate, there is just no room for the "anything is possible" school of pseudoscience. No room, that is, unless the theories are supported by scientifically-based arguments.

Belief systems outside science come in many forms, some of them covered by the general umbrella of pseudoscience. However, by far the most interesting and important alternate to a scientific ordering of the world is that provided by the principles and tenets of organized religion. From the beginnings of Western science in the Middle Ages, there has been a sort of undeclared guerilla war being waged between the church and the scientific community on the matter of who is the keeper of true knowledge about the nature of the cosmos. In the next section we examine this conflict as our final statement about the alternate realities by which we shape our existence.

10. *Science, Religion and the Nature of Belief Systems*

In the Reality Game, religion has always been science's toughest opponent, perhaps because there are so many surface similarities between the actual practice of science and the practice of most major religions. Let's take mathematics as an example. Here we have a field that emphasizes detachment from worldly objects, a secret language comprehensible only to the initiates, a lengthy period of preparation for the "priesthood," holy missions (famous unsolved problems) to which members of the faith devote their entire lives, a rigid and somewhat arbitrary code to which all practitioners swear their allegiance, and so on. These features are present in most other sciences as well, and bear a striking similarity to the surface appearances of many religions. In terms of similarities, both scientific and religious models of the world direct attention to particular patterns in events and restructure how one sees the world. But at a deeper level there are substantial differences between the religious view and that of science.

Let's consider some of the major areas in which science and religion differ:

- *Language*—The language of science is primarily directed toward prediction and control; religion, on the other hand, is an expression of commitment, ethical dedication and existential life orientation. So even though we have superficial surface similarities at the syntactic level, the semantic content of scientific and religious languages are poles apart.

- *Reality*—In religion, beliefs concerning the nature of reality are presupposed. This is just the opposite of the traditional view of science, which is directed toward *discovering* reality. Thus, religion must give up any claims to truth, at least with respect to any facts external to one's own commitment. In this regard, the reality content of most religious beliefs is much the same as in the myths considered earlier in the chapter. Fundamentally, what we have in science is a basic belief in the comprehensibility of the universe, a belief that is not necessarily shared by many religions

- *Models*—Although both scientific and religious models are analogical and used as organizing images for interpreting life experiences, religious models also serve to express and evoke distinctive attitudes, and thus encourage allegiance to a way of life and adherence to policies of action. The imagery of religious models elicits self-commitment and a measure of ethical dedication. These are features completely anathema to the role of models in science. In religion, the motto is "live by these rules, think our way and you'll see that it works." The contrast with science is clear.

• *Paradigms*—In the discussion of paradigms we saw that scientific paradigms were subject to a variety of constraints like simplicity, falsification and influence of theory on observation. **All** of these features are absent in the selection of a religious paradigm.

• *Methods*—In science there is a method to get at the scheme of things: observation, hypothesis and experiment. In religion there is a method, too: divine enlightenment. However, the religious method is not repeatable, nor is it necessarily available to every interested investigator.

So both pseudoscience and religion provide alternate reality structuring procedures that are radically different in character from those employed in science. It's of interest to ponder the point as to why there is such a diversity of brands of nonscientific knowledge, especially in view of the claims of virtually every sect that its own brand of medicine is uniformly most powerful.

Our view on this matter is quite simply that science, religion or pseudoscience cannot give a product that is satisfactory to all customers. The wares are just not attractive enough. In some cases the beliefs are not useful in the way that people want to use them. For example, many people have a deep-seated psychological need for security and turn to conventional religion for myths of all-powerful and beneficent Beings who will attend to these needs to be protected. Science with its mysterious and potentially threatening pronouncements about black holes, the "heat death" of the universe, evolution from lower beings, nuclear holocausts, and so on, offers anything but comfort to such primal needs and, as a result, loses customers to the competition. Basically, beliefs thrive because they are useful, and the plain fact is that there is more than one kind of usefulness.

To the practicing scientist, the foregoing observations may come as a sobering, if not threatening, conclusion because they seem to put into jeopardy the conventional wisdom that the road to real truth lies in the "objective" tools of science, not the subjective, romantic notions of believers and crusaders. But if we accept the fact of alternative and equally valid belief systems, we are inexorably led full circle back to the position that there are many alternate realities, not just within science itself but outside as well, and the particular brand of reality we select is dictated as much by our psychological needs of the moment as by any sort of rational choice.

In the final analysis, there are no complete answers but only more questions, with science providing procedures for addressing one important class of such questions. We conclude with the admonition that the only rule in the Reality Game is to avoid falling into that most common of all human delusions, the delusion of one reality—our own!

Discussion Questions

1. If one accepts the Kuhnian picture of science as a succession of paradigms, it's difficult to imagine such a thing as scientific "progress" since successive paradigms don't usually encompass all the observations accounted for by their predecessors. How can you reconcile this notion of no progress with the obvious advances in medicine, communication, transportation, and so on, over the past decades?

2. In connection with the distinction made in the text between a "model" of a natural system N and a "simulation," do you think the following representations are models or simulations:

a) A linear system $\Sigma = (F, G, H)$ describing an electrical circuit?

b) A scaled-down version of an airplane wing in a wind tunnel?

c) A robot designed to make certain welds on an automobile assembly line?

d) The Navier-Stokes equation describing laminar flow of a fluid?

e) A regression curve describing the behavior of the Dow-Jones stock index?

3. Consider the claims of some religions such as Scientology or the Christian Scientists that their brand of faith is based on "scientific" principles. Also consider the same claim made by political ideologies such as Marxism. Is there any interpretation, other than *pro forma,* by which you could term these belief systems "scientific" as we have used that term?

4. In quantum theory, at least in the classical Copenhagen interpretation, there are two contradictory models for describing the behavior of an object: a *particle* model by which the object displays behavior characteristic of point particles, and a *wave* model by which the object displays properties like interference that are associated with waves. Is there any way you could see either of these models as providing an "explanation" for the behavior of an object? Do either of these visions of the object provide the basis for describing an external reality or are they only tools for predicting the results of measurements?

5. Give an account of why each of the following beliefs would be considered pseudoscience: astrology, graphology, numerology, dowsing, orgone therapy, ESP, the prophecies of Nostradamus, Adam and Eve, theosophy, psychoanalysis, psychokinesis, Noah's Ark and creationism.

6. We have given great weight to the Law of Parsimony, or Occam's Razor, as a guiding principle in the selection of models for natural phenomena.

How can you reconcile this principle with the evermore complex paradigms of modern particle physics, evolutionary biology, paleontology and other branches of science? For example, in particle physics it was once thought that the only basic particles were protons, electrons and neutrons. Today the prevailing orthodoxy involves a host of quarks with different "colors," "flavors" and "orientations." This view is clearly not as simple as the earlier model. Give reasons why, despite this added complexity, the quark model has replaced the earlier picture.

7. In science we often see intense competition to obtain credit for a discovery. Prominent examples are Darwin's rushing to print to achieve priority over Wallace, Watson and Crick's race with Pauling to unravel the genetic code and the current competition between Montagnier and Gallo for priority in discovery of the AIDS virus. To what degree do you feel these psycho-sociological factors enter into the acceptance of one paradigm and the rejection of another?

8. Modern Western culture has traditionally based its view of reality upon two pillars: materialism and cartesian duality. There is an external reality composed of "things," and there is a separation between these things of the world and the things of the observing mind, i.e., reality is divided into the *res extensa* and the *res cogitans* in cartesian terms. To what degree have the discoveries of modern physics (Bell's Theorem, for example), neurophysiology, computer science and linguistics altered this vision of reality? Is there any way we can interpret or identify the altered states of consciousness of the mystic with the alternative realities of the scientist?

9. The position that we have advocated on the question of how to mirror Nature in a formal system is rather close to the philosophical view termed "instrumentalism." Discuss the following objections that have been raised against this approach to modeling Nature and man:

a) There is too much prominence given to *formal* analogies, and not enough attention to *substantive* ones, i.e., analogies between observables in different natural systems N.

b) The approach is unable to provide for the extension of theories; i.e., it can't generate new encoding/decoding operations or extend observables.

c) The approach neglects the importance of physical models because it's not concerned with the process of discovery.

d) Instumentalism asserts that explanation is tantamount to prediction, i.e., they are equivalent concepts.

Problems and Exercises

Dear Reader:

If you've worked even a substantial fraction of the Problems and Exercises from the preceding chapters, you're entitled to a well-deserved rest; if you haven't, then please go back and do so now! The Discussion Questions and Problems and Exercises form an integral part of the message of this book, and if you want to hear it, you'll have to make an active effort by trying your hand at doing, not just listening and reading. Fortune favors the prepared mind, and the only preparation for the art and science of modeling is practice, so please don't imagine that the skills of good modeling practice can be achieved without making the effort of thinking. It will be an exercise in developing your own alternate realities as well as creating the basis for understanding those seen by others.

Notes and References

§1. According to Bertrand Russell, "It is not what the man of science believes that distinguishes him, but *how* and *why* he believes it." This is the how and why that sets science apart from other types of belief systems like religion, dogmatism and pseudoscience. For introductory discussions of these and other matters pertaining to the role of science as a belief system, see

Stableford, B., *The Mysteries of Modern Science,* Routledge and Kegan Paul, London, 1977,

McCain, G., and E. Segal, *The Game of Science,* 4th ed., Brooks and Cole, Monterey, CA, 1982.

Spradlin, W., and P. Porterfield, *The Search for Certainty,* Springer, New York, 1984.

An authoritative account of myths and their place in Western civilization can be found in

Campbell, J., *The Masks of God: Occidental Mythology,* Viking, New York, 1964.

§2. Good accounts of the types of models used by science and the way in which theories rise and fall are given in

Oldroyd, D., *The Arch of Knowledge,* Methuen, London, 1986,

Barbour, I., *Myths, Models and Paradigms,* Harper & Row, New York, 1974.

Benjamin Whorf was a chemical engineer whose views on the relationship between human thought and language can be summed up in the postulates: (1) all higher levels of thought are dependent on language, and (2) the structure of the language that one habitually uses conditions the manner in which one understands his environment. An account of the origin and evolution of his views are found in

Whorf, B., *Language, Thought and Reality,* MIT Press, Cambridge, MA, 1956.

§3. The classic account of the Kuhn's views on the origin of scientific paradigms, as well as a response to some of the criticisms of his ideas, is

Kuhn, T., *The Structure of Scientific Revolutions,* 2d ed., U. of Chicago Press, Chicago, 1970.

§4–5. The reconsideration of Aristotelian causes as a basis for questioning the Newtonian paradigm, especially in the context of biological systems, is taken up in the work

Rosen, R., "On Information and Complexity," in *Complexity, Language and Life: Mathematical Approaches,* J. Casti and A. Karlqvist, eds., Springer, Heidelberg, 1986.

The modeling vs. simulation arguments using neural nets and mathematical machines is examined further in

Rosen, R., "Causal Structures in Brains and Machines," *Int. J. Gen. Syst.,* 12 (1986), 107–126.

§6. The rise of the heliocentric view of Copernicus and its supplanting of the geocentric model of Ptolemy has been recounted in many places. One of the best is

Kuhn, T., *The Copernican Revolution,* Harvard U. Press, Cambridge, MA, 1957.

Copernicus' life and times are presented in graphic detail in the novel

Banville, J., *Dr. Copernicus,* Martin Secker & Warburg, Ltd., London, 1976.

§7. The position of the positivists and the many objections raised against their view of scientific knowledge and how to get it are chronicled in the Oldroyd volume referred to under §2.

Popper has been a key figure in the twentieth-century philosophy of science, and his views on the aims of science and the means for attaining those aims have been put forth in many forums. Two excellent sources for the ideas, as well as for information about the man himself, are the following:

Popper, K., *Unended Quest,* Fontana, London, 1976,

Magee, B., *Popper,* 2d ed., Fontana, London, 1982.

For a more detailed account of Popper's views in his own words, see

Miller, D., ed., *A Pocket Popper,* Fontana, London, 1983.

§8. The way in which competing models of a phenomenon are judged in a forum of debate is most properly a topic within the sociology, rather than philosophy, of science. Good accounts of this process are found in

Merton, R. K., *The Sociology of Science: Theoretical and Empirical Investigations,* U. of Chicago Press, Chicago, 1973.

Case studies of the process are provided by the works

Fisher, C. S., "The Death of a Mathematical Theory," *Arch. Hist. Exact Sci.,* 3 (1966), 137–159,

Fisher, C. S., "The Last Invariant Theorists," *Arch. Euro. Soc.,* 8 (1967), 216–244.

Raup, D., *The Nemesis Affair,* Norton, New York, 1986,

Sheldrake, R., *A New Science of Life,* 2d ed., Anthony Blond, London, 1985.

Other accounts of the way in which some models survive and others die are found in the Kuhn volume listed under §3, as well as in the works noted under §2.

§9. Pseudoscience is now flourishing as never before, with the volume of material available seemingly only limited by the capacity of the printing presses to supply the public's insatiable appetite for easy explanations and quick "scientific" fixes. Our treatment of this topic follows that in

Radner, D., and M. Radner, *Science and Unreason,* Wadsworth, Belmont, CA 1982.

Martin Gardner, former editor of the Mathematical Games column of *Scientific American,* has been a tireless crusader against the debasement of science by pseudoscience. Some of his work in this direction is reported in

Gardner, M., *Science: Good, Bad and Bogus,* Prometheus, Buffalo, NY, 1981,

Gardner, M., *Order and Surprise,* Prometheus, Buffalo, NY, 1983,

Gardner, M., *The New Age: Notes of a Fringe Watcher,* Prometheus, Buffalo, NY, 1988.

For an account of the history of the Velikovsky controversy, see

Bauer, H., *Beyond Velikovsky: The History of a Public Controversy,* U. of Illinois Press, Urbana, IL, 1984.

§10. The contrast between religion and science as competing belief systems is brought out in sharp detail in the Stableford and Barbour books cited under §1 and §2 above. Another work along the same lines is

Peacocke, A., "A Christian Materialism?", in *How We Know,* Michael Shafto, ed., Harper & Row, San Francisco, 1985.

DQ # 6. For a blow-by-blow account of the development of the current state of elementary particle physics, see

Crease, R., and C. Mann, *The Second Creation,* Macmillan, New York, 1986,

Sutton, C., ed., *Building the Universe,* Basil Blackwell, Oxford, 1985.

DQ #8. The relationship between the altered states of consciousness of the mystic and the realities of the modern physicist is explored in the popular volumes

Capra, F., *The Tao of Physics,* 2d ed., Shambhala, Boulder, CO, 1983,

Zukav, G., *The Dancing Wu Li Masters,* Morrow, New York, 1979.

Some of the same territory has also been scouted in the more recent works

Krishnamurti, J., and D. Bohm, *The Ending of Time,* Gollancz, London, 1985,

Weber, R., *Dialogues with Scientists and Sages,* Routledge and Kegan Paul, London, 1986.

INDEX

A

B

C

D

E

F

G

H

S

T

U

V

W